Sec

12/93

GENETICS AND
ANIMAL BREEDING

IVAR JOHANSSON

JAN RENDEL

Genetics and Animal Breeding

Translated from the Swedish by
MICHAEL TAYLOR, B.SC. (READING), MS (AMES, IOWA)

OLIVER & BOYD
EDINBURGH

ISBN 0 05 001665 2

OLIVER AND BOYD
Tweeddale Court Edinburgh EH1 1YL
A Division of Longman Group Ltd

First English edition 1968
Reprinted 1972
This is a translation of the Swedish book
Ärftlighet och husdjursförädling published
in 1963 by L. T's Förlag, Stockholm
© 1968, English translation, Oliver and Boyd
All other language rights reserved
by the Authors

Printed in Hong Kong by
Dai Nippon Printing Co., (International) Ltd.

Preface

The Swedish edition of *Genetics and Animal Breeding* was published in 1963 by L.T's Förlag, Stockholm, under the title *Ärftlighet och husdjursförädling*. It was not written as a textbook for particular college courses but was intended to be a general and fairly popular survey of the present knowledge about the genetics of farm animals and the methods of their improvement by breeding. It was hoped, however, that the book would prove useful as an introduction to animal genetics for students at agricultural and veterinary colleges, for vocational teachers and advisers in animal husbandry, for veterinarians in the field as well as practical breeders. For the benefit of readers without previous training in general, or population, genetics three chapters were devoted to an elementary presentation of some of the basic principles in this field of science.

Before being translated, the Swedish edition of the book was thoroughly revised. We have brought the text up to date as far as possible and have replaced local Swedish considerations with points of more interest to English-speaking readers. The English edition follows quite closely the plan and subject treatment of the original Swedish version of the book. In making our revisions, we have had excellent help from several colleagues. Dr Joseph Edwards of the Milk Marketing Board, Thames Ditton, Surrey has read Chapter 1, 'Historical'; his colleague, Dr L. K. O'Connor, has read Chapter 10 on 'Udder development; milking rate, yield and composition of milk'; Dr H. P. Donald of the Animal Breeding Research Organisation, Edinburgh, has read Chapter 5 on 'Multiple births and twin research'; and Dr Mikael Braend of the Norwegian Veterinary College, Oslo, has read Chapter 7 on 'Inheritance of blood characteristics, Basic results and practical applications' and Chapter 8, 'Inherited defects and disease resistance'. Since neither of us had much experience of poultry genetics, Dr John C. Bowman, Professor of Animal Production, University of Reading (formerly Chief Geneticist, Thornber Ltd, Research Division), kindly agreed to rewrite Chapter 13 on 'Production characters in poultry', and he has also read other sections dealing with poultry breeding. Furthermore, Dr Alan Robertson of the Institute of Animal Genetics, Edinburgh, has read the complete English manuscript and given us many valuable comments and suggestions. We have carefully considered the comments of our referees and are exceedingly thankful for the help so generously given to us. May we point out, however, that the authors, and not the referees, are responsible for any shortcomings and faults which may be revealed by the reviewer.

Since the book is intended as a general survey of the field rather than as a textbook for advanced study, extensive lists of literature references have been omitted. Instead a few selected references are given after each chapter, mainly to books and reviews for further study and also to some original reports which we thought were of special interest. Many more authors are referred to in the text, and from the name of the author and the year of publication it is a simple matter to find the complete reference in *Animal Breeding Abstracts*.

Finally, our sincere thanks to Mr Michael Taylor who has translated our book into English, his native language. Thanks are due not only for conscientious work but also for his cheerful response to suggestions from ourselves and the referees.

Uppsala, March 1966

IVAR JOHANSSON
JAN RENDEL

Contents

PREFACE	v
1. HISTORICAL. *Ivar Johansson*	1
2. REPRODUCTION AND PHYSICAL BASIS OF INHERITANCE. *Ivar Johansson*	14
Reproduction	16
The oestrus cycle	20
Pre-natal development and period of gestation	21
Hormonal regulation of sexual functions	25
The physical basis of inheritance	30
Cell division and reduction division	32
The chromosome number of farm animals	36
3. MENDELIAN INHERITANCE. *Ivar Johansson*	41
Mendel's laws of inheritance	41
Simple Mendelian inheritance	42
Dihybrid and polyhybrid segregation	45
Modified Mendelian ratios	49
Quantitative characters	53
Linkage and crossing over	55
Sex determination and sex linked inheritance	58
Gene interaction and phenocopy	64
Maternal influence	65
Mutations	67
Mutation frequency; artificially induced mutations	72
The viability of mutants	75
Genes and gene function	77
Biochemical genetics	84
4. THE PRINCIPLES OF POPULATION GENETICS. *Jan Rendel*	89
Some statistical concepts	89
Probability	89
Variation	91
Covariation	94
Qualitative characters	96
Gene frequency and genetic equilibrium	96
Changes in gene frequencies	101
Changes in gene and zygote frequencies due to population size and inbreeding	104
Quantitative traits	110
The additive effect of genes and deviations due to dominance and gene interaction	112
The variation of quantitative characters	116
The concept of heritability	118
Selection	133

CONTENTS

5. MULTIPLE BIRTHS. TWIN RESEARCH. *Ivar Johansson* — 136
- Multiple births in normally monotocous animals — 136
  - The heritability and frequency of multiple births — 138
  - Disadvantages of multiple births — 141
- Variation in litter size in polytocous animals — 143
- The use of monozygous twins in research — 145

6. THE INHERITANCE OF EXTERNAL TRAITS. *Jan Rendel* — 158
- Colour inheritance — 158
  - Pigment formation — 159
  - The genetics of colour in rodents. Gene symbols — 160
  - The colour of farm animals — 164
- Other external traits — 180
  - Horns — 180
  - Hair characteristics — 182
  - Plumage in fowls — 183

7. INHERITANCE OF BLOOD CHARACTERISTICS. BASIC RESULTS AND PRACTICAL APPLICATIONS. *Jan Rendel* — 185
- Blood antigens — 186
  - Definitions and technique — 186
  - Erythrocyte antigens of cattle — 188
  - Blood antigens in cattle twins — 191
  - Blood antigens in other farm animals — 194
- Biochemical blood characteristics — 196
- Genetic variation of enzymes — 199
- Blood groups as genetic markers in breed studies — 201
- Practical applications — 202
  - Determination of parentage — 202
  - Diagnosis of monozygosity in cattle — 204
  - Diagnosis of fertility in freemartin heifers — 205
  - Blood groups and disease — 205
  - Blood groups and production performance — 206
- Concluding remarks — 212

8. HEREDITARY DEFECTS AND DISEASE RESISTANCE. *Jan Rendel* — 214
- Defects and anatomical malformations — 214
  - Experimental induction of congenital defects — 215
  - Severity of action of mutant genes — 216
  - Defects of definite genetic origin — 217
  - Defects with hereditary disposition but mode of inheritance unknown — 227
  - Homologous defects in different species — 228
  - 'Accidents' in development — 228
- Disease resistance — 229
  - Resistance to nutritional deficiencies — 229
  - Climatic sensitivity — 230
  - Resistance to infectious diseases — 231
  - The biological basis of resistance — 235
- Concluding comments — 236

CONTENTS

9. STERILITY AND LOW FERTILITY. *Ivar Johansson* - 238
Malformations of the genitalia - 239
   Gonad hypoplasia - 239
   Other genital malformations in males - 241
   Underdevelopment of the female genitalia - 241
   Intersexuality - 242
Gametic sterility - 245
Hereditary disposition for reduced fertility - 248
   Variations in the fertility of male animals - 249
   Variations in female fertility - 253
Inbreeding and fertility - 259
'Stress' as a cause of reproductive disturbances - 260
Length of gestation - 262

10. UDDER DEVELOPMENT, MILKING RATE, YIELD AND COMPOSITION OF MILK. *Ivar Johansson* - 264
Udder and teats - 264
The milking rate - 272
The milk and fat yield of cows - 279
   Some non-genetic causes of variation in yield - 281
   Persistency of yield - 283
   The heritability of the lactation yield - 284
   Herd differences and trends within herds - 284
   The heritability within herds - 287
Milk composition - 290
   Temporary variations in butterfat content - 290
   Causes of variation which operate in the long term - 291
   Genetic variation - 292
   Genetic differences in protein composition - 294
   Correlation between the various milk constituents and between the quantity and composition of the milk - 297

11. BODY SIZE AND CARCASS TRAITS. *Jan Rendel* - 300
Birth weight - 300
Pre-weaning growth - 301
Post-weaning growth rate - 302
   Growth rate and feed conversion in pigs - 302
   The growth rate of cattle and sheep - 304
Changes in body proportions during growth - 306
   The effect of plane of nutrition on body proportions - 307
The influence of heredity on body size - 310
Meat quality - 314
   Quality demand and methods of assessment - 314
   Carcass quality in pigs - 319
   Carcass quality in cattle - 322
   The effect of sex on carcass quality - 322
Concluding remarks - 324

12. WOOL PRODUCTION AND FUR QUALITY. *Jan Rendel* - 328
Types of wool hair - 328
Wool quality - 330
Factors which influence the yield and quality of wool - 332
   The yield of wool - 332
   Wool quality traits - 335
   Genetic relationship between wool traits - 337
Pelt characteristics - 337

13. PRODUCTION CHARACTERS IN POULTRY.
John C. Bowman — 340

Egg production characters — 340
  Fertility and hatchability — 341
  Egg production — 342
  Egg-size — 346
  Viability — 347
  Body-weight and feed consumption — 349
  Egg quality characters — 350
    Egg-shape — 350
    Shell quality — 351
    Shell colour — 352
    Albumen quality — 353
    Blood and meat spots — 354
    Yolk characters — 355
    Miscellaneous characters — 355
Meat production traits — 356
  Growth rate — 356
  Feed efficiency — 358
  Viability — 358
  Carcass quality — 358
  Feather and skin characters — 358
Other poultry — 359

14. BREEDING METHODS.
Ivar Johansson — 361

Species, breeds, strains, lines and families — 362
A review of different mating systems — 363
Inbreeding and crossing in the light of experiments — 368
  Poultry — 369
  Pigs — 373
  Sheep and goats — 377
  Cattle — 378
Theoretical explanations of inbreeding depression and hybrid vigour — 382

15. ESTIMATION OF BREEDING VALUES.
Ivar Johansson — 387

Testing suspected carriers for unfavourable genes — 387
Breeding value for quantitative traits — 391
  Definition of breeding value concept — 392
  Various possibilities of assessing the breeding value of an individual — 393
    Individual performance — 398
    The pedigree (performance of ancestors) — 399
    Collateral relatives — 403
    Progeny test — 406
  Selection indices for several quantitative traits — 419

16. SELECTION RESPONSE AND SELECTION METHODS.
Jan Rendel — 424

Changes in the thorax length of fruit flies (*Drosophila melanogaster*) — 424
Selection experiments with mice — 426
Selection for single traits in the fowl — 426
Conclusions from the experiments with *Drosophila*, mice and fowls — 431
Selection response on different planes of nutrition — 432
Selection for production traits in farm animals — 436
  The selection intensity and generation interval — 436
  Selection methods — 438
  Individual and family selection — 442
  Selection methods for exploiting non-additive genetic variation — 446
The genetic change in populations of farm animals — 451

CONTENTS

17. SOME SPECIAL PROBLEMS IN BREEDING AND
    SELECTION. - - - - - - - 454
    Ivar Johansson (pp. 454 to 465) and *Jan Rendel* (pp. 466 to 474).
Horses - - - - - - - - 454
Cattle - - - - - - - - - 455
   Progeny-testing and selection among tested bulls - - - 456
   Selection of dams of bulls - - - - - - 459
   The selection of young bulls for testing - - - - 461
   Selection of cows within herds - - - - - 462
   The organisation of breed improvement within A.I. units - 462
   Breeding for beef traits - - - - - - 464
   Crossbreeding - - - - - - - - 465
Pigs - - - - - - - - - 466
   Selection based on sib and own performance - - - 467
   Organised crossbreeding - - - - - - 470
Poultry - - - - - - - - - 470
   Selection for egg production - - - - - - 470
   Selection for meat - - - - - - - 472
   Random sample tests - - - - - - - 472

18. RETROSPECTS AND PROSPECTS. *Ivar Johansson* - 475
Some results of selection combined with improved environment - 475
Prospects for further progress - - - - - - 478

INDEX - - - - - - - - - 481

1 Historical

From time immemorial, the breeding of domesticated animals has involved selection of one kind or another. Even until comparatively modern times the basis of the selection was an accumulation of experience, passed down from one generation of breeders to the next. Unfortunately, many erroneous ideas and superstitions were also perpetuated, such as the belief in maternal impression and telegony.

Maternal impression implies that events, or things, which the mother happens to see during mating or pregnancy are transmitted to the foetus which will then be born with a corresponding change of characteristic. The appearance of red calves in a breed of black cattle would be explained, for example, by the fact that the mothers, at a critical moment, happened to see a red roof or a red sunset. This belief must be very old, since reference is made to it in the First Book of Moses. It is said that Jacob was shepherding Laban's flocks and made the following proposal to his master: that he, as shepherd, should have all the spotted and speckled lambs and kids which were produced when no spotted or speckled animals were used for breeding. Jacob is said to have induced the desired colouring in the offspring by cutting white stripes on sticks of poplar, almond and plane tree which he then laid in the stream where the animals came for water: 'And the flocks conceived before the rods and brought forth flocks ring-straked, speckled and spotted' (Gen. xxx.39). Even today there are, without doubt, many people who believe that a child can be born with a birth-mark if the mother happens to be frightened by a fire during pregnancy.

Telegony is the supposed influence of a female's first mate on the offspring of her later matings with other males. It has long been the superstition in dog breeding that if a pure-bred bitch happens to be mated to a mongrel, then its value for breeding would be reduced in all future matings, even with pure-bred dogs. Even the eminent CHARLES DARWIN refers in his book, *Animals and Plants under Domestication* (London, 1868), to several supposed cases of telegony in horses and pigs. In one case referred to by Darwin, a black and white Essex sow was mated to a wild boar and gave birth to wild type pigs, as was to be expected. However, when the wild boar was dead and could not be mistaken for the father, the sow was mated to a boar of her own breed. This pure-bred boar, which had not previously thrown offspring with the colour of wild pig, produced, together with this sow, pigs which showed the wild pig colouring. That matings within a breed can give rise to pigs with the characteristics of wild pigs is not entirely unusual; and there is good reason to believe that the Essex breed, at the middle of the nineteenth century, was not so consolidated that segregation could not give rise to offspring with wild pig characteristics, even if the sow had never been mated to a wild boar. Darwin also mentions the case of Lord Morton's famous hybrid, from a chestnut mare and a male quagga. Not only the hybrid, but also the offspring subsequently produced by the mare when mated to a black Arabian sire, were more plainly barred across the legs than even the pure quagga. Darwin concluded, 'There can be no doubt that the quagga affected the character of the offspring subsequently begot by the black Arabian horse.'

Many breeding experiments have been conducted with the animals under strict control, but no one has succeeded in producing a single case of telegony. Earlier, attempts were made to explain the phenomenon in different ways. It was suggested, for example, that sperms could influence the womb and bring about permanent changes which would affect the foetus of subsequent pregnancies, or that sperms could survive in the womb throughout pregnancy and fertilise ova released after the termination of pregnancy. Both these hypotheses are in opposition to all known facts. The most likely explanation of the phenomena mentioned would be that there has been a segregation of recessive factors in both the sire and dam.

Another erroneous idea, which is more difficult to disprove, is that of the inheritance of acquired characters. Reference will be made to this later.

BREED FORMATION, PRODUCTION RECORDING, AND BREEDING FOR TYPE

ROBERT BAKEWELL (1725–1795) is generally acknowledged as the first great pioneer in animal breeding. At the age of thirty-five he took over his father's estate at Dishley in Leicestershire. The estate was about 360 acres, of which 90 were arable and the remainder pasture; and it was here that he carried out his breeding of horses, cattle and sheep. In each of these species he laid the foundation of a new breed: the Shire horse, the Longhorn cattle and a new Leicester breed of sheep. Today the Longhorn breed is practically extinct, the Shire horses are few in number and the Leicester sheep are not numerous either. Both the Longhorn cattle and Leicester sheep were typical meat breeds.

Bakewell's great contribution to animal breeding was that he tried out new methods of breeding and succeeded so well that he gained many followers, not only in his own country but in many parts of the Western world. He purchased animals from many different places and with these he began his breeding, selecting only those individuals which he thought most suitable for his aim. He then inbred intensively for several generations in order to consolidate the type, rejecting all animals with undesirable characters. Bakewell appears to have been the first to conduct systematic progeny testing of rams and bulls. This he achieved by hiring out selected male animals to other breeders for a given fee, Bakewell retaining the right to inspect all the progeny. The terms of the contract even included how the progeny were to be reared, for example, that only home-grown fodder was to be used. Those males which gave the best progeny were used by Bakewell later in his own herds. Since his neighbours eventually began to have better flocks of sheep by using Bakewell's rams, he formed the Dishley

Society—an association of sheep breeders. The members undertook not to hire out rams for service for less than a fee of 50 guineas per season.

In 1783 Bakewell had a young farm pupil called CHARLES COLLING, who later, together with his brother Robert, founded the Shorthorn breed which during the nineteenth century became the most famous of the cattle breeds. It was used in the formation of a number of other breeds, among which was the Swedish Red and White cattle. Like Bakewell, the Colling brothers conducted intensive inbreeding, as is shown by the ancestry of the bull Comet in Fig. 1:2. The degree of inbreeding of Comet corresponds to almost three generations of full-sib mating. The bull, which was born in 1804, was subsequently sold by Charles Colling for the record price, at that time, of 1000 guineas.

Fig. 1:1. Robert Bakewell (1725–1795), pioneer in the field of practical animal breeding.

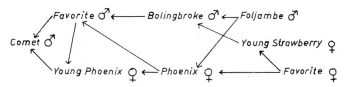

Fig. 1:2. Pedigree of the Shorthorn bull, Comet, born in 1804 in Charles Colling's herd. The inbreeding coefficient of 0·47 corresponds to nearly 3 generations of successive full-sib mating (cf. p. 107).

The formation of new breeds began with Bakewell and continued during the whole of the nineteenth century. Even today new breeds are being formed, but not at anything like the same rate as earlier. Inbreeding has almost always been a factor, to a greater or less degree, in the foundation of breeds. Results have varied, but usually it has been found that this method of breeding carried with it rather great risks of degeneration. It is known, for instance, that the Charles Colling's herd suffered from reduced fertility; and complete sterility was not at all unusual.

The first of the official herd books for the new breeds appeared about the end of the eighteenth century. A herd book for the English thoroughbred horses was started in 1791 and the first volume (*The General Stud Book*) was published in 1808. The first herd book for cattle—for the Shorthorn breed—was *Coates Herd Book* which first appeared in 1822. Somewhat later, herd books were published for the different breeds of horses in France and Germany, and herd books for cattle appeared about the middle of the century. The first official herd book for Dutch Friesian cattle was published in Holland in 1875.

The early herd books contained only information about the pedigree of the animals. During the latter half of the nineteenth century methods were developed for measuring and recording the conformation and production of the animals; and milk recording of cows began in a number of herds.

About 1890, three methods for the rapid determination of the fat content of milk were developed, quite independently, by LINDSTRÖM, Sweden (1889), BABCOCK, U.S.A. (1890) and GERBER, Germany (1893). The methods were, in principle, quite similar. In 1895 the first association of farmers for milk recording on farms was formed in Denmark. This type of recording spread quite quickly to all milk producing countries. The recording of growth rate, food consumption and carcass quality of pigs began in Denmark in 1907 and in Sweden in 1923. Pig testing in Sweden also included the recording of the number of pigs per litter and the litter weight at three weeks of age. Production tests were also introduced into breeding flocks of sheep and poultry. Trap nesting of laying hens appears to have been introduced first in Austria in 1879.

The recording of production was the first really notable event in animal breeding after the time of Bakewell. During the nineteenth century the only criteria for selection were the external characteristics of the animals, with the exception of the English thoroughbred where selection was based on the results of racing. It follows therefore that conformation received a great deal of attention. Charles Colling is reputed to have said that there were one hundred people suitable for the post of Prime Minister for every one person who was competent to judge Shorthorn cattle.

It is easier to assess the worth of draught horses, beef cattle, sheep and pigs on the basis of conformation than to judge by inspection the production capacity of dairy cattle. Milk production must be measured, as far as possible, by exact methods in order to provide a reasonably accurate basis for selection. Methods have been developed for measuring the growth rate, food conversion and carcass quality of beef cattle, sheep and pigs, the wool quality of sheep and so on. Such methods are the object of continual investigation to achieve improvements and discover new methods for increasing the accuracy of selection. Subjective methods of judging animals are perhaps still necessary as a complement to the objective methods of measuring performance, but their importance has greatly decreased. Besides their direct use in selection, the data collected have been used in investigations into the influence of inheritance and various environmental factors on the variation of production traits, and a great deal of knowledge of practical importance has been gained.

After the introduction of herd books, it was the practice to decide upon a standard type for each breed, and individuals were required to conform to this standard type as closely as possible. An animal which deviated widely from the standard type could not be registered in the herd book. To breed for a type associated with a particular function, e.g. riding or draught horses, beef or dairy cattle, wool or fleece type of sheep, is naturally desirable; but the enthusiasts of type breeding went considerably further. Clydesdale horses were required to have long sloping pasterns and long fetlocks, even though this latter point was associated with a number of obvious disadvantages. Bulls and cows of the Swedish Friesian breed had to have a certain shape of head and curvature of the horns; and not only was a general colour pattern required, but it was also considered to be of great importance that the black colour did not extend below the front knee or the hock. Shropshire sheep, to be typical of the breed, had to have wool growing almost completely over their eyes. This formalism in animal breeding undoubtedly resulted in a great deal of damage, and the consequence of considering so many unessential details decreased the effectiveness of selection for the economically important traits. As will be shown later, it is quite possible for one-sided selection for a particular type of conformation to result in a weakening of the animals' vitality due to a disturbance of the physiological functions. The background to the extreme of type breeding was, most likely, the ease with which a homogeneous herd could be produced with respect to outward appearance—if not in any other way then 'by the butcher's knife', as one of the most energetic opponents of type breeding, the Norwegian, CHRISTIAN WRIEDT, expressed his view. The type breeders were of the belief that by decreasing the variation of external appearance, they would also decrease the variation in such characteristics as growth rate, milk production and butter fat; which is of course by no means the case. Formalism has gradually decreased in importance, perhaps primarily because of the economic pressure on animal production after World War II, but it has not completely disappeared.

BREEDING THEORIES BEFORE 1900

There was very little theoretical background to the developments just described, and it was mainly the animal breeders themselves who probed their way forward by trial and error. The method of Bakewell and the Colling brothers, of conducting close consanguineous breeding for several generations, did not infallibly lead to success for each and all. On the contrary, many did not succeed but no one could say why they failed. During the whole of the nineteenth century the idea was prevalent that acquired characters were inherited, and the breeding theories of that day were dominated by this idea.

The French zoologist LAMARCK (1744–1829) put forward a theory of evolution in 1809, according to which the species underwent progressive adaptation to prevailing environmental conditions. The development of different organs, e.g. the extremities of reptiles, the neck of the giraffe and the disappearance of the tail in humans were believed to be due to 'use or disuse'. If the organ in question was used regularly, it developed strongly; whereas if it was used little or not at all, then it receded; these alterations were registered, according to the theory, in the germ cells and thereby transmitted to the progeny. If this were the case, then there would presumably be no limit to the possibilities of increasing the speed of race-horses or the milk production of cows, and the advances ought to be uniform for a given environmental pressure.

CHARLES DARWIN (1809–1882), who in 1859 published his book *The Origin of Species* did not draw any definite line between inherited and non-inherited variation but simply assumed that part of the variation was, or tended to be, inherited. The development of species, according to Darwin, was the result of natural selection; the better adapted individuals coming out better in 'the struggle for existence' and leaving more progeny than those less well adapted. With great accuracy and persistency Darwin collected impressive evidence for the idea that an evolution takes place from the lower to the higher organisms and he showed how natural selection probably directs the course of this evolution. To show how new genetic variation arises was reserved for his followers. However, this does not lessen the great contribution which Darwin made to the theory of evolution.

The first to make a determined stand against the idea of 'inheritance of acquired characters' was the German zoologist, AUGUST WEISMANN, who in 1892 presented his theory of 'the continuity of the germ plasm'. Weismann distinguished between the soma (body) and the germ plasm. According to his theory, there is an early separation in the animal embryo between the germ plasm and the somatoplasm; the former was supposed to retain its character throughout the life of the individual. The germ plasm is not affected by influences from the soma, the implication being that it is biologically immortal, whereas the soma is a passing mortal structure. The development of different organs of the body may be modified in one direction or another, but these modifications are not registered in the germ plasm. Weismann was also among the first to associate the hereditary determiners with the chromosomes of the cells. However, his views were based more on abstract reasoning than on his own experiments, even though he did carry out a considerable number. For example, he amputated the tails of rats for many successive generations in order to find out whether it would have any effect on the development of tails in the progeny. During the course of the experiment all rats were born with normally developed tails.

In the first half of the nineteenth century there developed in Germany the so-called 'principle of constancy', according to which a breed of animals could, by several generations of selection be brought to a stage where it would breed true for a large number of characteristics. The homogeneity of the breed depended on the number of generations it had been subjected to pure breeding and selection, and it was supposed gradually to approach complete constancy. This is, of course, a great exaggeration of what really happens.

Another theory, which was based upon the Lamarckian idea of inheritance of acquired characters and the 'principle of constancy', claimed that the breed as well as the individual is a product of the soil and climate in the area where it has been developed. When a breed has lived under the same environmental conditions for many generations, it achieves a high degree of adaptation and uniformity, and thrives better under these conditions than breeds which may be introduced from other regions. The concept of the breed as a 'product of the soil' was vigorously taught by Professor PROSCH (Copenhagen, 1861) as a warning against the importation of foreign breeds into the Northern countries. Although the concept contains a nucleus of truth, it was confused by false interpretations and lack of foresight. The environmental conditions for farm animals have changed very markedly with the general development of agriculture in Western Europe. The indigenous breeds in comparatively small areas—breeds which in spite of the process of natural selection were rather heterogeneous—had therefore to be changed by artificial selection or crossbreeding, or replaced by imported stocks, in order to fit the new conditions. One outstanding example is the rapid spread and great success of Friesian cattle in Great Britain.

FRANCIS GALTON (1822–1911) introduced statistical methods into the study of heredity, and he is generally considered as the founder of *Biometry*. In his book *Natural Inheritance* (1889) he presented his view on heredity, based on experiments with sweet peas and analysis of data on various characters in man, mainly stature and eye colour. He studied each character separately, but he was interested more in laws which applied to whole populations than in segregation ratios. Galton distinguished between 'natural' (or inborn) and 'acquired peculiarities' and between 'heritages that blend and those that are mutually exclusive'; but he pointed out that there is no marked dividing line between the latter two groups. He emphasised that the acquired characters are not hereditary. As examples of blending and of mutually exclusive characters respectively he mentions skin colour and eye colour in man.

In analysing the inheritance of stature in man, Galton converted the height of adult women to that of men by using the correction factor 1·08 on the female data, and then he calculated the 'regression' of offspring on the corresponding parental averages (mid-parents). He found that the deviation of the offspring from the population average (P) was 2/3 of the mid-parent's deviation from the same average, half of that deviation was supposed to be due to each one of the parents, and he formulated 'the law of regression' as follows: 'The Deviation of the Sons from P are, on the average equal to one-third of the deviation of the Parent from P, and in the same direction.' It may be noted that Galton's term *regression* originally meant the 'step back' from the mid-parent towards the population average, 1/3 in the case referred to, rather than the increase in stature of the offspring for each unit increase in the average stature of the parents, which was 2/3. However, the phenomenon of regression cannot be considered as a 'law' of heredity. This was shown very clearly by WILHELM JOHANNSEN about fifteen years later; he found that within 'pure lines' of beans, where there is no genetic variation, the regression of offspring on parents is zero. When regression is found, its magnitude depends on the relative amounts of genetic and environmental variation within the populations.

In a later publication, GALTON (1897) announced 'a statistical law of heredity that appears to be universally applicable to bisexual descent'. He formulated this 'law' as follows: 'The two parents contribute between them, on the average, one-half (or 0·5), of the total heritage of the offspring, the four grandparents one-quarter or $(0·5)^2$, the eight great-grandparents one-eighth or $(0·5)^3$, and so on. Thus the sum of the ancestral contributions is expressed by the series $(0·5) + (0·5)^2 + (0·5)^3$ etc., which, being equal to 1, accounts for the total heritage.' Basically, this is the same as the 'law of regression'. The results describe the statistical relation between offspring and parents or more distant ancestral generations in a random breeding population, with regard to characters in which additive gene effects account for the whole variation. The correlation between an offspring and each one of his parents would then be 0·5, that between an offspring and each one of the grandparents would be 0·25, etc. Squaring the correlation coefficients and adding the squares (or coefficients of determination) for each generation we obtain the ancestral contributions stated by Galton. In his data on stature, the standard deviations of the dependent and the independent variable were approximately equal, and therefore the coefficients of correlation and regression would also be equal. When part of the variation of a character is due to environmental influences, Galton's calculations would be applicable only to the genetic variation. However, Galton's 'law of ancestral inheritance' is still erroneous. An individual obtains his total heritage from his parents; nothing can be passed over from grandparents or more distant ancestors except through the parents. If the heredity of each one of the parents were completely known, nothing would be gained by going farther back in the ancestry; the partial correlation (cf. p. 94) between the offspring and any one of the ancestors beyond the parents, when these are held constant, would be zero.

Galton's 'law of ancestral inheritance' fitted well to the old conception of the 'percentage of blood' as a measure of the relationship between the animals in a pedigree. The amount of truth, and of misunderstanding, is about the same in both these concepts.

Nevertheless, although Galton gave a wrong interpretation to his results, not yet knowing Mendel's laws of segregation and recombination of hereditary units, his work is of great interest, and he must be considered one of the early pioneers in the study of quantitative inheritance.

THE BEGINNING OF ANIMAL GENETICS

In 1900 the laws of inheritance, which had been formulated thirty-five years earlier by the Austrian abbot and naturalist, GREGOR MENDEL, were rediscovered and intensive work began in many countries to test them on different species of animals and plants. The English geneticist, WILLIAM BATESON (1861–1926), appears to have been the first to demonstrate the Mendelian inheritance of qualitative characters in farm animals. In 1902 he published his work on the inheritance of comb shape in hens and, in collaboration with Saunders, a paper on the inheritance of hornedness and polledness in cattle. Incidentally, it was Bateson who in 1906 suggested the name *genetics* for that section of biological research which was then developing. The terms *gene, genotype* and *phenotype* (the observable type) were first used by WILHELM JOHANNSEN in 1909 in his book *Elemente der exakten Erblichkeitslehre*.

At first it was uncertain how Mendel's laws could be applied to characters that showed continuous or quantitative variation. The English talked about 'blending inheritance' in connection with such characters, and it was to them that Galton applied his law of regression. NILSSON–EHLE presented, in a short communication in 1908 and more fully in his doctorate thesis the following year, the results of breeding experiments with wheat and oats, in which he studied, among other things, the inheritance of colour of the kernels. He showed that in wheat the red colour of the kernels depended on three genes which were inherited independently of each other, different combinations of these genes giving rise to different grades of red colour. Shortly after, EAST published similar results from his study of the inheritance of ear length in maize. Both the kernel colour of wheat and ear length of maize showed themselves to be typical examples of quantitative inheritance. In crossing two extreme types, progeny are obtained which are midway between the values of the two parents. The first generation of crossbreeds (F_1) are thus intermediate. When the F_1 individuals are self-fertilised, or mated with each other, the next generation (F_2) shows an increased variation. This is exactly what would be expected from Mendelian inheritance when the character depends on several genes. 'Blending inheritance' can thus be explained on the basis of Mendel's laws of inheritance.

The results of these early experiments created great optimism among research workers studying the inheritance of characters in the higher animals. In quick succession came hypotheses concerning the number of genes which influenced such quantitative traits as the milk yield of cows per recording year, or lactation, and the egg production of hens. It soon became evident, however, that these hypotheses did not at all correspond to reality. The number of genes which must be reckoned with is very large and their interrelationship complicated, and in addition there is considerable environmental variation with regard to many of these characters.

That animal breeding up to the middle of the 1930s was only slightly affected by the results of genetic research was not due to the lack of interest from the breeders but rather to the fact that the animal geneticists had not been able to arrive at methods suitable for the breeders to apply in practice. There was a marked change, however, when research was intensified and important results of practical value were obtained in several fields, e.g. population genetics combined with inbreeding, crossbreeding and selection experiments, in addition to research on monozygous twins, blood grouping and disease resistance.

There was a sharp difference of opinion in England, in the first two decades of this century, between biometry, represented first by Galton and later his successor, the eminent statistician KARL PEARSON (1857–1936), and the experimental geneticists, the foremost of whom was Bateson. Eventually it became clear that there was no real contradiction between the Mendelian and the statistical approach to the study of genetics. The groundwork for the integration of these two sciences was laid by the Englishman, RONALD A. FISHER and the

Fig. 1:3. Dr Jay L. Lush, born 1896, Professor of Animal Breeding at Iowa State University, USA. Pioneer in the application of population genetics to animal breeding.

American, SEWALL WRIGHT. In 1918 Fisher published a paper on the correlation between relatives with the assumptions of Mendelian inheritance; and a few years later Sewall Wright published a series of papers on the genetic effect of different mating systems and on correlation and causation. In the later papers Wright developed his method of path coefficients, which is closely related to Fisher's analysis of variance. Wright's path coefficients played an important part in population genetics research with farm animals, but they are used now mainly to demonstrate causal relationships. Analysis of variance, on the other hand, has increased in importance in all fields of biological research. As early as 1908 the English mathematician, HARDY and the German physician and geneticist, WEINBERG, had formulated, quite independently of each other, what is now called the *Hardy–Weinberg law*, which together with the work of Fisher and Wright is the foundation of population genetics.

The pioneer in the application of population genetics to animal breeding is the American, JAY L. LUSH; his contribution to animal breeding has been recognised by honorary degrees of Universities in the United States and in several European countries. A general theory has been developed which can be applied to the analysis of the inheritance of quantitative characters as well as to estimating the breeding value of animals and the effect of selection. With the aid of population genetics it is possible to determine which breeding and selection methods can be expected to give the best result for a given set of conditions. Animal breeding is no longer an art but an applied science in the same way as plant breeding, and can now be conducted along strictly scientific lines.

THE ERA OF ARTIFICIAL INSEMINATION

Reproductive physiology is another branch of science which has developed rapidly since the turn of the century. Some of the more important findings in the endocrinology of reproduction will be briefly referred to in the next chapter. The achievements with the techniques of artificial insemination are of special interest from a genetic point of view. During the last two decades cattle breeding has been revolutionised in Europe and North America by the rapid spread of A.I. and it has gained considerable importance in the breeding of other farm animals also.

The first successful artificial insemination on record was made by the Italian biologist LAZZARO SPALLANZANI with dogs in 1780. The pioneer in the practical application of A.I. to farm animals is the Russian veterinary physiologist ELIA IVANOV, who started his work in this field in 1899. Ten years later Ivanov became Director of the State Laboratory for investigations into sperm physiology and artificial insemination. There was a standstill in the work during World War I and the following years of political turmoil, but in 1923 the work was started again on a large scale. Artificial vaginas were constructed for the collection of semen and methods were developed for the dilution of semen and preservation outside the body. About 1930 large numbers of cows and ewes were artificially inseminated in the U.S.S.R. The method was also successfully applied to mares and sows, although to a limited extent because of the greater difficulties of maintaining the fertility of semen *in vitro*. In the early 1930s investigations with A.I. were begun by WALTON and HAMMOND at Cambridge University; and in 1942 the first Cattle A.I. centre in Great Britain was founded as a Farmer Cooperative at Cambridge, with DR JOSEPH EDWARDS as chairman. In Denmark the first A.I. association had already been established in 1936.

After 1939 the commonly used diluent for bull semen was buffered egg yolk and by storing the semen at a temperature of 5°C satisfactory fertility could be maintained for four or five days. A discovery of tremendous importance for the application of A.I. in animal breeding was reported in 1949 by POLGE, SMITH and PARKES working at the National Institute for Medical Research, London. They found that bull semen could be deep-frozen to $-79°C$ and later thawed without any serious effect on fertility, provided a certain amount of glycerol was added to the diluent before freezing. They used solid CO_2 and alcohol for freezing and storage, but later the use of liquid nitrogen has made it possible to store semen at about $-196°C$. Calves have been produced by the use of semen which has been kept deep-frozen for ten years or more and it seems probable that the storage period can be increased indefinitely without any severe effect on semen fertility. The latest development is to deep-freeze diluted semen as droplets (pellets) on solid CO_2, each droplet containing the number of spermatozoa needed for one insemination. The pellets can be stored at -79 or $-196°C$. Immediately before insemination a pellet is transferred to the insemination syringe, a small amount of dilution fluid is then added and the solution is deposited at the proper site in the female genitalia.

Bull semen can be diluted to 10–100 times the original volume, depending on the sperm concentration of the ejaculate. When semen is collected once or twice weekly from bulls of good fertility it would be possible, at least theoretically, to produce more than 100 000 calves per bull annually. According to PERRY (1961) one A.I. bull in the U.S.A. had been bred to more than 140 000 cows during his lifetime and two bulls had each over 1200 milk recorded daughters. In Great Britain one bull has produced 2218 daughters with milk records.

From the standpoint of breed improvement the greatest advantage of A.I. lies in the possibility of forming large breeding units with centralised direction of progeny testing and mating combinations. A large number of bulls can be progeny tested with satisfactory accuracy at an early age, thereby making effective selection possible. After completion of the tests the best bulls can be utilised to the full extent. When crossbreeding is desirable, the matings are much easier to organise with artificial insemination than with natural mating. For example, importation of semen from one district, or country, to another costs much less than the importation of bulls.

The A.I. technique has spread rapidly in dairy cattle breeding. In the countries mentioned below the percentages of the total number of dairy cows artificially inseminated in 1964 were as follows:

Denmark	98	U.S.A.	45
Holland	70	Canada	20
England and Wales	65	U.S.S.R.	65
France	57	Czechoslovakia	85
West Germany	45	Hungary	80
Finland	60	East Germany	65
Norway	50	Japan	93
Sweden	65	Israel	95

Artificial insemination of sheep and pigs is slowly gaining ground, but progress is retarded by the relatively high cost as compared with the value of the animals and the cost of natural service. In horse breeding, A.I. is resorted to mainly in order to increase the number of progeny from valuable stallions and in cases where natural matings are ineffective due to certain abnormalities of the mares. Artificial insemination is used quite extensively in turkey breeding and also for breeding hens when they are kept in laying cages.

The question has been raised whether it would be possible to increase the rate of reproduction not only of the mammalian male but also of the female. Polyovulation can be induced by parenteral administration of gonadotropic hormones, but the increase in the number of viable young born is usually very slight because of the pronounced increase in intra-uterine mortality when the number of fertilised eggs rises above the level which is normal for the species. Comprehensive research has been carried out in order to find feasible methods of polyovulating ewes and cows and of transplanting the fertilised eggs into recipient females for development into full-term foetuses. By such procedures it would be possible to increase the number of progeny from genetically valuable donors and to utilise less valuable recipients as acting mothers.

In vitro fertilisation of the eggs has not been possible so far, and it is therefore necessary for fertilisation to take place in the donors, with transplantation within a few hours into recipients which are synchronised with the donors with regard to the stage of the sexual cycle. There are two main difficulties in this procedure: one is to recover the fertilised eggs, and the other is to transfer these eggs to the proper site for development in the recipient. Many successful transplantations have been made in sheep and a few in cattle. Until recently the recovery of the fertilised eggs involved major surgery (laparotomy), but there are now on record cases where the eggs have been flushed out from the intact genitalia of cows a certain number of days after fertilisation and then transferred to the distal end of the uterine horn of the recipients in which oestrus had occurred at the same time as, or one day later than, in the donor. SUGIE (1965) has reported two successful transplantations by the use of special instruments for non-surgical technique. Although these findings will probably open up new possibilities for practical application, the method of egg transplantation cannot be expected to reach the same importance as that of artificial insemination. Theoretically, it is possible to harvest thousands of eggs from an individual cow but their recovery and transplantation is laborious and expensive, even when it can be done without surgery. Methods for deep-freezing and storing the eggs without harmful effects must be developed before the egg transplantation technique can be applied as a routine procedure in cattle breeding.

SELECTED REFERENCES

BERGE, S. 1959. Historische Übersicht über Zuchttheorien and Zuchtmethoden bis zur Jahrhundertwende. In *Handbuch d. Tierzüchtung*, 2: 1–23. Hamburg.
GALTON, F. 1897. The average contribution of each several ancestor to the total heritage of the offspring. *Proc. R. Soc.*, 61: 401–413.
LUSH, J. L. 1945. *Animal Breeding Plans*, 3rd ed. (443 pp.). Iowa State Univ. Press, Ames, Iowa.
PAWSON, H. C. 1957. *Robert Bakewell* (200 pp.). London.
PERRY, E. J. 1960. *The Artificial Insemination of Farm Animals* (430 pp.). New Brunswick, N.J.
SUGIE, T. 1965. Successful transfer of fertilised bovine egg by non-surgical technique. *J. Reprod. Fert.*, 10: 197–201.

2 Reproduction and Physical Basis of Inheritance

The life cycle of an individual begins when the egg cell (ovum) of the female is fertilised by a sperm (spermatozoon) from the male; the starting point is the fertilised ovum. A rapid sequence of cell divisions, and an involved specialisation of the daughter cells leads to the formation of the different germ layers and organs of the body, and at birth the new individual is ready to meet the rigours of the outer world. In mammals the development of the embryo takes place inside the body of the mother, whereas with birds it takes place mainly outside the mother by incubation of the egg. After birth in the case of mammals, or hatching in the case of birds, the development of the body continues, but at a slower rate, until the individual is fully grown. Some time before this stage is reached comes the onset of puberty with the awakening of sexual drive and the beginning of the production of active gametes. Real sexual maturity, when the animal is sufficiently developed for use in breeding, comes somewhat later. In mammals the interval between puberty and sexual maturity is greater in females than in males.

There is a certain stage of the life cycle when the individual makes its greatest contribution in animal production. This age varies not only with the species of animal but also with the breed, nature of the product, the influence of the environment and the marketing conditions. The general trend during the past twenty years has been towards a more rapid turnover of animals than previously, especially with respect to poultry and pigs, and also with cattle intended for meat consumption. The dairy cow does not reach her highest level of production until she is 5–7 years old.

After this maximum is reached, senescence starts and the production capacity decreases. In assessing the production traits of animals, consideration must always be taken of their age so that a valid comparison can be made between individuals

In modern animal production, no animals are kept until they die of old age—on the contrary, they are slaughtered much earlier. There are, however, in the literature many reports of horses and cattle which have lived to a great age and retained their reproductive capacity for a surprisingly long time. What the maximum age for different species of animals can be is uncertain, but there is evidence that horses can live to be 30–40 years old and cattle 20–30 years. A cow of the Swedish Red and White breed is recorded to have calved when she was 22 years old and showed signs of heat even up to the 24th year of age.

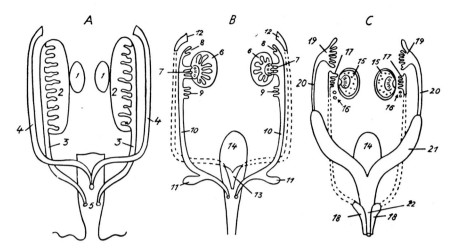

Fig. 2:1. Sex differentiation in mammals. Diagram of (*A*) Indifferent stage, (*B*) Male and (*C*) Female differentiation.
1. Genital ridge. 2. Pronephros. 3. Wolffian ducts. 4. Mullerian ducts. 5. Common urinogenital vestibule. 6. Testicle. 7. Epididymis. 8 and 9. Remains of pronephric ducts. 10. Vas deferens. 11. Scrotum. 12 and 13. Remains of Mullerian ducts. 14. Bladder. 15. Ovary. 16, 17 and 18. Remains of Wolffian ducts. 19. Fallopian tube. 20. Oviduct. 21. Uterus. 22. Vagina.

(BANE and BONADONNA, 1958. *Handbuch der Tierzüchtung*, I. Verlag Paul Parey, Hamburg)

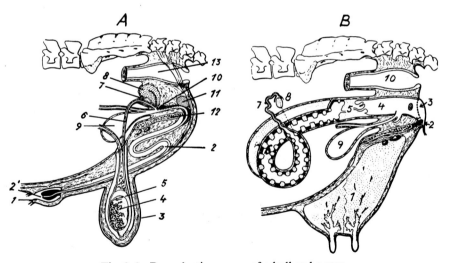

Fig. 2:2. Reproductive organs of a bull and a cow.
(*A*) Bull. 1. Prepuce. 2. Penis. 3. Scrotum. 4. Testicle. 5. Epididymis. 6. Vas deferens. 7. Ampulla of vas deferens. 8. Vesicula seminalis. 9. Bladder. 10. Prostate gland. 11. Cowper glands. 12. Pubic bone. 13. Rectum.
(*B*) Cow. 1. Udder. 2. Clitoris. 3. Vulva. 4. Vagina. 5. Cervix. 6. Uterine horn. 7. Fallopian tube. 8. Ovary. 9. Bladder. 10. Rectum.

(BANE and BONADONNA, 1958. *Handbuch den Tierzüchtung*, I. Verlag Paul Parey, Hamburg)

REPRODUCTION

The reproductive organs

In the development of the embryo the genital organs remain relatively undeveloped even when many of the organs and systems of the body are differentiated for their future functions (Fig. 2:1A). The gonads appear as ridges (genital folds) in the anterior-dorsal region of the abdominal cavity, one on each side of the median line. Alongside the gonads are the first rudiments of the kidneys (pronephros) together with their outlets, the Wolffian ducts, and close to these the Mullerian ducts. In cattle the differentiation of the gonads begins during the

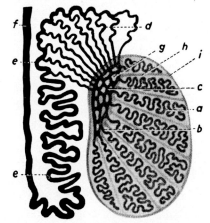

Fig. 2:3. Diagram of the testicle and epididymis.
(*a*) Convoluted seminiferous tubules and (*b*) straight part, (*c*) and (*d*) efferent ducts, (*e*) epididymis, (*f*) vas deferens, (*g*) and (*h*) interstitial cells, (*i*) outer membrane of testicle.
(TRAUTMANN and FIEBIGER 1958; *Handbuch der Tierzüchtung*, I. Verlag Paul Parey, Hamburg)

third month of foetal life. In the male embryo the genital ridges develop into the testicles, part of the pronephros forms the epididymis, and each Wolffian duct develops into a vas deferens. The Mullerian ducts become vestigial (Fig. 2:1B). In the larger mammals the testicles have already descended into the scrotum at birth. The penis is developed from a bud-shaped tubercle in the posterior part of the genital region; the corresponding structure in the female is the clitoris, situated in the lower part of the vagina just inside the vulva. The genital ridges in the female embryo develop into the ovaries, and the Mullerian ducts develop into the Fallopian tubes, uterus and vagina, whilst the Wolffian ducts become vestigial (Fig. 2:1c).

If sexual differentiation does not proceed normally but is checked or otherwise altered at an early or a later stage, the result is an intersex (a form between male and female). In the literature, pronounced intersexuality is often called *hermaphroditism*. 'True hermaphroditism' means that the animal has both testicles and ovaries more or less well developed. However, such animals are invariably sterile. The occurrence of intersexes is relatively frequent in pigs or goats but is unusual in horses or cattle. If one or both of the testicles remain in the abdominal cavity the result is a *cryptorchid*. The function of the scrotum in mammals is as a heat regulator for the testicles; the temperature inside the body cavity is too high for the production of living sperm. A knowledge of sexual sex differentiation

Fig. 2:4. Schematic drawing of a mammalian sperm. × approx. 1500.
(BANE and BONADONNA, 1958. *Handbuch der Tierzüchtung*, I. Verlag Paul Parey, Hamburg)

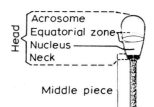

facilitates the understanding of the different degrees of intersexuality which can appear in animals (cf. p. 242).

Fig. 2:2 schematically shows the normally developed sexual organs of bulls and cows. It will be seen from the figure that the penis of bulls is bent in an S shape, as it is also in other ruminants and in boars, but not in stallions. When the penis is erected, prior to mating, the S-bend is straightened out and the length of the penis is increased. In cows the horns of the uterus are longer than the body of the uterus; this is still more pronounced with sows and other polytocous animals. In the mare the horns and body of the uterus are of about equal length. The mucous membrane of the uterus of the cow is covered by a number of mushroom-like projections, caruncles, which are enlarged during pregnancy and connect with the saucer-shaped projections from the foetal membrane, the cotyledons. A caruncle with the attached cotyledon is called a *placentom*. The form of the contact area between the outer foetal membrane (chorion) and the uterine mucosa, together forming the placenta, varies from one species to another. The placenta serves for the exchange of substances, both nutrients and waste products, between the mother and the foetus. From each horn of the uterus a Fallopian tube leads to the funnel-shaped infundibulum (Fig. 2:1c) which receives the eggs released at each heat period.

In poultry, only the left ovary and Fallopian tube (oviduct) develop; the right ovary and oviduct remain only as functionless rudiments. The Fallopian tube, urinary duct and end of the intestine open directly into the cloaca. Both testicles are developed in the male bird but are retained in the body cavity. The penis is absent in the cockerel but in ducks and geese both the sperm ducts end in a penis-like structure which is especially well developed in the goose.

The primary function of the gonads is the formation of gametes but they also discharge sex hormones into the blood which are of importance for the regulation of the sexual functions.

The testicles are comparatively large and egg-shaped, the glandular portion being composed of convoluted seminiferous tubules which are lined on the inside with an epithelium from which the sperms are produced (Figs. 2:3 and 2:17). These tubules join to form larger ducts which discharge into the single coiled tube, the ductus epididymis (Fig. 2:3e), a storage place for the sperms, which go through a ripening process there. The epididymis passes into the vas deferens which opens into the urethra.

Interspersed between the seminiferous tubules of the testes are Leydig cells which are responsible for the production of the male sex hormone, androgen (Fig. 2:17).

A spermatozoon consists of a relatively small head and a long tail (Fig. 2:4) by means of which the sperm is able to propel itself in a suitable medium, e.g. the spermatic fluid or the secretion from the female sex organs. Semen is the total fluid which the male ejects during the act of mating (Fig. 2:5). The semen of stallions and boars is very diluted with secretions from the accessory glands (the seminal vesicles, the prostate gland, and the cowper glands), whereas the semen of bulls and rams is much less diluted. In the epididymis the sperms lie tightly packed together and practically immobile, but after mixing with the secretion from the accessory glands they exhibit considerable motility. The average

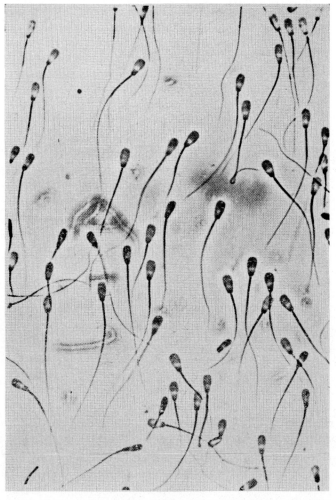

Fig. 2:5. Microphotograph of bull sperm. Magnified about 500 times. (BANE 1962)

volume of semen and concentration of sperm in a normal ejaculate from rams, bulls, stallions, boars and cockerels are as follows:

	Volume (cc)	Sperm concentration (Million per cubic mm)
Rams	1	2·5
Bulls	5	1·0
Stallions	70	0·10
Boars	200	0·15
Cockerels	0·2	4·0

The ovaries of the female are rounded bodies attached by the broad ligament to the dorsal wall in the sublumbar region of the body cavity. The outer layer of the ovary is made up of the germinal epithelium with a very large number of

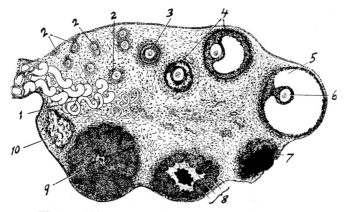

Fig. 2:6. Diagrammatic transverse section of an ovary.
1. Blood vessel. 2. Primary follicle. 3. Growing follicle. 4. Graafian follicle. 5. Mature follicle where the egg (ovum) (6) is about to be discharged. 7. Recently ruptured blood filled follicle. 8. Developing yellow body (corpus luteum). 9. Fully developed corpus luteum. 10. Corpus luteum undergoing absorption. (PATTEN 1927)

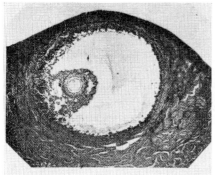 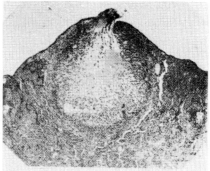

Fig. 2:7. Section of mammalian follicles.
Left: shortly before ovulation; *right*: during ovulation when the follicle has ruptured, releasing the ovum and *liquor folliculi*. (ENDERS 1952)

primary follicles, each of which contains a potential ovum. After the onset of puberty a number of these primary follicles develop into 'egg sacs' or Graafian follicles, as illustrated in Fig. 2:6. At the right of the drawing can be seen a mature follicle enclosing the ovum which is attached to the inside wall by a stalk of cells. When the follicle approaches maturity, the wall becomes thinner and finally ruptures, releasing the egg together with the follicular fluid. The egg is taken up by the infundibulum and passes into the Fallopian tube. Fig. 2:7 shows micro-photographs of a follicle before and during rupture. Ovulation takes place in the mare at the end of the heat period, whereas sows and ewes ovulate during the latter half of the heat and cows 12–14 hours after the end of the heat period. In all these species of animals ovulation takes place irrespective of whether the female has been mated or not. Rabbits and mink on the other hand ovulate following an orgasm which is a consequence of the mating. The female rabbit ovulates about 12 hours and the female mink 36–48 hours after mating. Monotocous animals, as a rule, release only one egg at each ovulation, whereas polytocous species usually release 50–100 per cent more eggs than the number of embryos which complete the intra-uterine development. Following ovulation the empty follicle is filled with coloured cells, which form the yellow body called the *corpus luteum*.

The female sex hormone, oestrogen, is produced by the cells of the Graafian follicle and is stored in the follicular fluid. During its active period the yellow body produces another hormone, progesterone, which has importance for the implantation of the fertilised egg and the nourishment of the developing foetus.

The oestrous cycle

The sexual activity of many species of wild mammals is strongly associated with the time of the year; the animals have a particular mating or breeding season. This is also the case with a number of domesticated animals, e.g. sheep and horses. In the Northern latitudes sheep have their breeding season during late autumn and horses during spring and early summer, whereas cattle and pigs breed throughout the year, domestication having eliminated the association between breeding and season of the year. The heat period appears at regular intervals except when the animal is pregnant. This applies to species with a distinct breeding season as well as to species which have regular heat periods throughout the year. The sexual activity of females follows a clearly defined cyclic pattern (Fig. 2:8). Each species has its own characteristic oestrous cycle, defin ed as the time interval from the beginning of a heat to the next following heat. In females in good health the cycle occurs with great regularity. The oestrous cycle is divided into phases beginning with pro-oestrus, when the follicles in the ovaries are enlarging and the uterine mucosa thicken and become congested with blood. This is followed by oestrus, when the female will mate and ovulation takes place; and finally by metoestrus, when the yellow body develops in the ruptured follicle and the uterine mucosa are prepared to receive the fertilised egg. If fertilisation takes place, the metoestrus is followed by the period of pregnancy; but if fertilisation has not taken place, the cycle proceeds into a fourth stage called dioestrum, when the yellow bodies cease to function, in order that a new heat period can begin. The mucous membrane of the uterus then goes into a period of quiescence. Those animals which have a definite breeding season end the final oestrous cycle for the season with a period of sexual

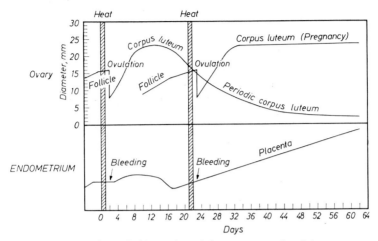

Fig. 2:8. Schematic illustration of the oestrous cycle of the cow. The diagram shows how the developments in the ovaries are followed by changes in the endometrium. (BANE and BONADONNA, 1958. *Handbuch der Tierzüchtung*, I. Verlag Paul Parey, Hamburg)

quiescence called 'anoestrus'. Table 2:1 shows the length and normal variation of the oestrous cycle for the larger mammals; with hormonal disturbances the cycle can be appreciably shorter or longer than shown.

Table 2:1 *Age at puberty, length of oestrous cycle and heat period*

	Age at puberty months	Oestrous cycle days		Heat period days	
		Mean	Range	Mean	Range
Horses	10–12	22	20–30	6	3–10
Cattle	7–11	21	16–25	1	1/2–1
Sheep and Goats	6–8	17	14–21	1	1–2
Pigs	5–7	21	18–24	2	1–3

The age of reaching puberty shows considerable variation, primarily due to nutritional causes and other environmental influences, but also as a consequence of breed and individual differences. The length of the oestrous cycle and the length and intensity of the heat period are subject to variation due to similar factors, as well as the time of year and age of the animals.

Pre-natal development and period of gestation

After the female has mated with a fertile male, fertilisation takes place in the upper part of the Fallopian tube, where the ovum is surrounded by a large number of sperm. The transport of the sperm from the place where they are deposited in the female genital organ to the Fallopian tube takes only a short time (2–15 minutes), aided by rhythmic contractions of the vagina, cervix and uterus. In mares the sperm are placed mainly in the uterus, in cows and ewes in the lower part of the vagina, and in sows in the cervical canal. Entry into the ovum is probably achieved by the sperm's own power of movement. The viability of the sperm in the female genital tract is quite limited: in cows, ewes and sows about 24 hours and in mares 2–3 days. In the case of the hen, however, the sperm can remain viable in the oviduct for 2–3 weeks. The viability of the unfertilised ovum in the above species is believed to be not more than 12–24 hours. The synchronisation of the heat and ovulation restricts mating to a period which is most favourable for fertilisation. On the average, mating during the latter part of the heat results in the highest percentage of fertilisation. This applies particularly to mares, which have relatively long heat. In cows, a satisfactory percentage of fertilisation can be achieved by inseminating during the first 6 hours after the end of the heat, since ovulation does not take place for a further 6–8 hours.

Several sperm may enter the outer layer of the ovum but only one will accomplish the fertilisation. The pronuclei of the sperm and egg come together and eventually fuse with each other (Fig. 2:9(2)). Shortly after this the fertilised egg, called a zygote, goes into the first division, which is completed within 24 hours after fertilisation (Fig. 2:9(3)). By repeated cell divisions a mass of cells, called the *morula*, is formed (Fig. 2:9(4)). A cavity develops inside the morula which then becomes a blastocyst (Fig. 2:10). The duration of the different stages of embryo development varies according to the total length of the embryonic period. In cows and ewes the zygote moves down into the uterus about 4 days after fertilisation when it is in the 8–16 cell stage. Following further cell divisions, the next stage is a specialisation of cells to form the embryonic membranes, in the manner shown in Fig. 2:11. The embryo is surrounded by three membranes: the outer chorion, inside which is the *allantois*, the outer layer of which is joined with the chorion to form the *allantochorion*; and finally the inner *amnion*, which together with the inner layer of the allantois forms the *allantoamnion*. In this way the embryo is enclosed in two fluid filled sacs. When development of the embryo has reached this stage, the allantochorion eventually makes contact with the developed uterine mucosa, and implantation gradually takes place. In the cow it takes place towards the end of the first month after fertilisation and is completed in the third month; in sows implantation takes place in the 12–24th day of pregnancy.

The organ which provides the means of exchange of substances between the mother and the foetus is called the placenta. This consists of the foetal part (allantochorion) and the maternal part which is formed from the mucous membrane of the uterus. Horses and pigs have the diffuse type of placenta whereby the chorionic villi make close contact with the uterine mucosa. Ruminants have the cotyledonary type of placenta. The quantitative rate of development of the foetus proceeds as illustrated by the following figures. At 45 days of age the foetus of cattle weighs only about 3 g. By 3 months of age it has reached a weight of about 1 kg (Fig. 2:12), at 200 days 10 kg, at 230 days 18 kg, and at 260

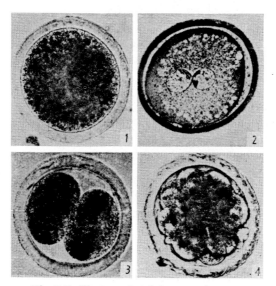

Fig. 2:9. Photographs of the ovum in cows: *1* before and *2* at fertilisation, *3* the two-cell stage and *4* the morula stage. During all these stages the ovum, and later the daughter cells, are surrounded by a thin membrane (zona pellucida). (WINTERS *et al.* 1942)

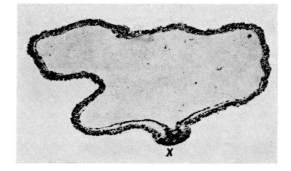

Fig. 2:10. Blastocyst of a cow, 12 days after fertilisation; *x* marks the embryonic knob. (WINTERS *et al.* 1942)

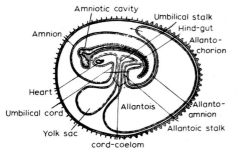

Fig. 2:11. Schematic illustration of the foetus and foetal membranes in the cow, 27 days after fertilisation. (ZEITSCHMANN and KRÖLLING 1945)

days 31 kg. The development of hair begins at 150 days and is complete at around 230 days of age. The average length and variation of the gestation period for different species of mammals, obtained from various sources, is presented in Table 2:2.

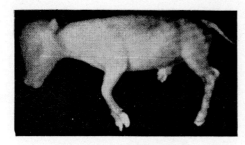

Fig. 2:12. Foetus of a calf 100 days old.

Table 2:2 Gestation period of some species and breeds of mammals: mean and standard deviation in days

	Mean	Standard deviation
Horses		
Arabian and English thoroughbred	337	9·4
Trakehnen	331	—
Belgian	333	9·4
Cattle		
Dutch Friesian	279	5·4
Jersey	278	—
Guernsey	283	—
Ayrshire (Sweden)	284	5·8
Ayrshire (U.S.A.)	278	—
Shorthorn	282	—
Aberdeen-Angus	282	—
Hereford	285	—
Charollais	288	—
Brown Swiss	290	—
Simmental	289	—
Sheep		
Karakul	151	—
Shropshire	146	2·2
Southdown	144	—
Merino	149	2·2
Lincoln	149	—
Pigs		
Danish and Swedish Landrace	115	1·5
Large White	114	2·4
Middle White	113	—
Berkshire	115	—
Goats		
Saanen	154	—
Anglo-Nubian	151	3·3
Toggenburg	150	—
Rabbits	31	0·83
Foxes: silver fox and blue fox	52	0·97
Mink	50	4·68

It is apparent from the table that within each species the early-developing and early-maturing breeds have a shorter period of gestation than later developing breeds. The normal variation can be expected to lie within the limits of $\pm 3s$ from the given mean (cf. p. 92).

In poultry the development of the embryo begins immediately after fertilisation of the egg and at laying has reached the blastula stage. When the egg is thereafter stored at 20°C or lower temperatures the development of the embryo ceases but begins again at the commencement of incubation. The incubation time for the eggs of hens is 19–21 days, turkeys 26–30, geese 26–33 and ducks 26–32 days.

The hormonal regulation of sexual functions

The various functions of the body are regulated by the nervous system and by hormones. Hormones are produced in ductless, or endocrine, glands which secrete directly into the blood stream for transmission to the various tissues. Particular tissues are sensitive to a particular hormone, which is therefore specific in exerting its effect on the function of a particular organ.

As was mentioned earlier, the sex glands have a double role, partly the production of sex cells, or gametes, and partly the production of sex hormones.

It has been assumed that the control of the body's endocrine system is delegated to the pituitary, or hypophysis, a small gland located at the base of the cerebrum. The gland lies in a cavity in the sphenoid bone and consists of three lobes: the anterior lobe, the intermediary lobe and the posterior lobe. The anterior lobe produces the gonadotrophic hormones, which include the follicle-stimulating hormone (FSH) and the luteinising hormone (LH), together with a number of other hormones which serve in the regulation of the body functions, such as the luteotrophic hormone (LTH), which stimulates the yellow body to produce progesteron, growth hormone (GH), thyrotrophic hormone (TSH) and adrenocorticotrophic hormone (ACTH). The posterior lobe also produces a number of hormones among which can be mentioned oxytocin, which exerts its effect on the smooth muscles of the milk-producing glands and the uterus.

It is now considered that the master control of the body's endocrine system is not the pituitary itself but a part of the brain called the hypothalamus, situated above the pituitary and connected to it by a stalk (Fig. 2:13). The function of the anterior lobe is regulated by hormonal control from the hypothalamus, which receives nervous impulses from the different parts of the body and thus from the outer world. The influence of light on reproductive functions, for example, is communicated via the optic nerve to the hypothalamus; this in turn activates the anterior lobe of the pituitary to produce gonadotrophic hormones. The hormones of the posterior lobe are presumably produced in the hypothalamus and transported to the posterior lobe for storage. In some cases the secretion of a hormone is instantaneous as a result of a nervous reflex; examples of this will be given later.

Not only light but also nutrition—both the quantity and the composition of it—and the physical treatment of the animal can influence the sexual functions in a neuro-hormonal way. Shock or fright occasioned by violent handling or unusual environmental conditions can reduce sexual drive or cause other disturbances in the reproductive performance. In the following a review is given of those hormones which are directly important for the sexual organs and mammary gland functions.

I. *Pituitary hormones.* The pituitary hormones have the chemical character of proteins. Only oxytocin is fully known with regard to its chemical composition and it can be produced synthetically. The gonadotrophic hormones (FSH and

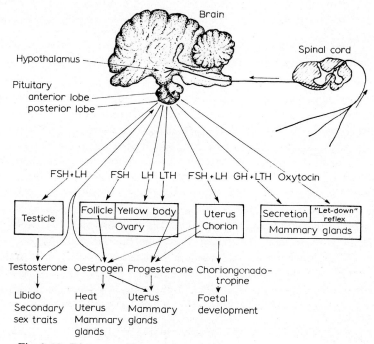

Fig. 2:13. Diagram to illustrate the origin and effect of the pituitary and sex hormones.

LH), however, can be produced in more or less pure form and it is known that they both are glyco-proteins, i.e. they contain a simple sugar in addition to protein.

A. *Gonadotrophic hormones*

1. *Follicle-stimulating hormone (FSH).* The most important function of FSH in the female is to activate the ovaries to follicular growth and maturation of eggs. A small amount of LH must however be present in order that FSH may achieve its full effect. In the male, FSH causes growth of the testis and induces spermatogenesis.

2. *The luteinising hormone (LH).* In the female it is responsible for ovulation by rupture of the follicle and for the production of a yellow body (luteinising). In the male, LH stimulates the Leydig cells in the testicles to produce the male sex hormone.

3. *The luteotrophic hormone (LTH)*. This governs the production of progesterone by the yellow body as long as is necessary during the normal course of pregnancy.

Both FSH and LH are secreted simultaneously from the pituitary, but vary in amount according to circumstances. During pro-oestrus and early oestrus the secretion is mainly FSH which causes the follicle to grow. Thereafter the amount of FSH decreases and the amount of LH increases, thereby effecting ovulation. It is important that the relative amounts at each stage correspond to the requirements needed to achieve the effect. FSH and LH are not specific to species, with regard to their effect, even though certain differences in chemical composition exist. The relative amounts in which these two gonadotrophic hormones are secreted during corresponding stages of the oestrous cycle varies between the species.

Gonadotrophic hormones are produced also by the chorion or by the uterus during certain stages of pregnancy. The urine of pregnant women contains, especially 30–80 days after the beginning of pregnancy, large amounts of choriongonadotrophin, which has mainly a LH-effect. On the other hand the blood serum from pregnant mares (40–140 days pregnant) is rich in gonadotrophin which has both FSH and LH effect, the balance being more towards FSH. Gonadotrophic preparations from the urine of pregnant women (PU) and from pregnant mares blood serum (PMS) have found wide application in medicine.

B. *Hormones which influence milk production*

It is clearly established that the growth of the mammary glands at the onset of puberty is due to an oestrogenic hormone, that the building of functional alveoli takes place under the influence of the yellow body hormone, progesterone, and that certain pituitary hormones are responsible for the secretion of milk at the beginning and during the course of the lactation period. There are, however, differences of opinion as to which of the pituitary hormones is most important in this respect.

1. *Prolactin*, which is supposed to be the real milk secretion hormone, is apparently identical with luteotrophin (LTH). At least it has been shown in a number of species that this latter hormone can stimulate milk secretion when the mammary glands have reached a certain stage of development.

2. *Growth hormone (GH)* has a pronounced stimulating effect on milk secretion. This has been shown both in laboratory animals and larger domesticated animals.

3. *Thyrotrophic hormone (TSH)* influences the thyroid gland production of the hormone thyroxin which has importance in regulating the rate of metabolism in the body and thereby milk production.

4. *Posterior lobe hormone oxytocin*. The neuro-hormonal reflex which comes into play in the evacuation of the mammary glands can be described in the following manner. When the nipples of the lactating female are subject to a light massage due to the sucking of the offspring, or by massage immediately before machine or hand milking, a nervous impulse passes from the teats to the hypothalmus which then communicates with the posterior lobe of the pituitary.

This reacts immediately by discharging a dose of oxytocin into the blood which in less than half a minute reaches the mammary glands where it stimulates the smooth muscle fibres which surround the alveoli where the milk is formed. The muscle cells contract and the milk is pressed down into the cavity of the nipples from where it can be obtained by suction, as in sucking or machine milking, or by the squeezing of the nipples as in hand milking. This neuro-hormonal reflex has been called the mammary or 'let-down' reflex. During mating a similar reflex occurs; nervous excitation during orgasm is led via the hypothalamus to the posterior lobe of the pituitary which immediately gives off oxytocin into the blood. As soon as the hormone reaches the muscle layer of the uterus, rhythmic contractions begin, by which the sperms are transported up through the uterus and into the Fallopian tubes. In rabbits and mink the same excitation induces the anterior lobe of the pituitary to secrete gonadotrophic hormones into the blood, which when they reach the ovaries stimulate first the growth of the follicle and later the ovulation.

The potency of preparations which contain pituitary hormones are given in international units (I.E.), which correspond to a certain weight of a standard preparation. The effect is measured by biological tests on experimental animals.

II. *The sex hormones.* Those hormones which are produced in the ovaries of the female and in the testicles of the male are chemically closely related to one another. They contain only carbon, hydrogen and oxygen and are similar in their chemical structure to cholesterol, a substance found in animal fat and in the bile. The term *oestrogen* for the female sex hormones and *androgen* for the male hormones are collective terms for a number of substances which differ more or less from each other in their chemical composition and in their biological activity. Several such substances with high hormonal activity are now produced synthetically. In general, oestrogenic and androgenic substances have a low solubility in water and are therefore dissolved in oil for injection (intramuscular). There are, however, preparations which have a good effect with oral administration, mixed in the food, or implanted under the skin in the form of tablets.

1. *Female sex hormone (oestrogen)* is produced in the wall of the follicles in the ovary, especially when the follicles are under the influence of FSH and are in a state of rapid growth. Small quantities are produced also in the adrenal glands and in the testicles of the male. The hormone has an effect on the vagina and mucous membrane of the uterus in preparation for mating (Fig. 2:13). The secretion of mucous increases and a certain amount of mucous discharge usually takes place from the vagina during heat. It is also responsible for the psychological adjustment which finds expression during heat and the most important result of which is the willingness to mate. As was mentioned previously, the same hormone influences the mammary glands where, during the period of sexual maturity, it causes a vigorous growth and ramification of glandular tissue. Oestrogenic hormone is also produced by the foetal membrane during pregnancy; the pregnant mare excretes large amounts in the urine. Several different oestrogenic substances are produced synthetically. The most widely used are *diethylstilbestrol* and *hexoestrol*. In the U.S.A. they have found wide application

for mixing in the food of young stock reared for slaughter, because they increase the rate of growth and reduce food consumption per kg growth. Large amounts of oestrogen have the effect of sterilisation since among other things the activity of the pituitary is impeded.

2. *Progesterone* is produced by the yellow bodies in the ovaries when these are fully active. After the uterus has first been subject to the influence of the oestrogenic hormone, the progesterone further prepares the uterus for reception of the fertilised egg and ensures the normal course of pregnancy. The hormone has a corresponding effect on the mammary glands, where it completes the development of the glandular system begun under the influence of oestrogenic hormone so that at the end of pregnancy the mammary glands are ready for the secretion of milk. Progesterone is produced also by the foetal membranes; and this appears to be the reason why in a number of species the yellow bodies recede or can cease functioning during certain stages of pregnancy without adversely affecting the development of the foetus.

3. *Male sex hormone (androgen)*. The active substance testosterone can be produced synthetically. The hormone is responsible for the development of the male secondary sex characters (masculine body type and temperament) as well as sexual drive. In fact, the earlier a male animal is castrated the less masculinity at maturity. In the female, small quantities of male sex hormone are produced by the adrenals and ovaries.

There is no absolute connection between the capacity of the testicles to produce normal sex cells (gametes) and their production of male sex hormone.

If testosterone is given to a male animal in large quantities for a long period, it has the effect of checking the pituitary production of gonadotrophic hormones, and so can have the effect of preventing the production of sperms.

The potency of sex hormones and progesterone preparations is indicated by the weight in milligrammes of the active substance contained in the preparation.

The hormonal control of the oestrous cycle in the larger domesticated animals and the regularity of the occurrence of heat can be explained in the following manner. During pro-oestrus the ovaries are influenced by the follicle-stimulating hormone (FSH), which leads to a rapid growth of the follicles and, as a consequence, increased production of oestrogen. When the oestrogen in the blood has risen to a certain level, the heat period begins. The high level of oestrogen checks the production of FSH and, instead, more LH hormone is produced; this stimulates ovulation and the development of the yellow bodies in the ovaries. Under the influence of luteotrophic hormone (LTH) the yellow bodies produce progesterone which prevents new follicles from maturing and prepares the mucous membrane of the uterus for the fertilised egg. If fertilisation has taken place, the yellow body remains and continues to function; otherwise it recedes (Fig. 2:8), and this means that production of FSH can again increase, growth of new follicles takes place, and a new heat follows. An oestrous cycle is thereby completed.

For several reasons hormone treatment of reproductive disorders is seldom used by veterinarians. In the first place, it is relatively expensive. Furthermore, it is necessary to know the animal's own hormonal status in order to decide which hormone or hormones to use and the dosage. It is not easy to determine this status. A wrong dose can do more harm than good. Finally, it should be considered that hormonal disturbances, e.g. cystic ovaries are, to some extent, hereditary and therefore the causative genes should not be preserved in the population.

THE PHYSICAL BASIS OF INHERITANCE

All tissues of the body consist of cells and more or less of an intercellular matrix. The cells have different size and shape according to the function they fulfil in the body. They contain a living substance, *protoplasm*, and their characteristic properties are that they can resorb and metabolise nutrients, grow, react to external stimuli, and multiply themselves. A cell is surrounded by a cell membrane, inside which the *cytoplasm* and the *nucleus* can be distinguished. At one

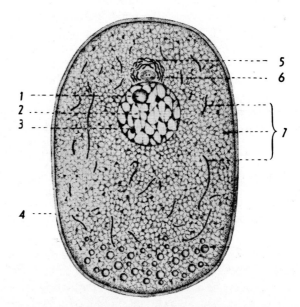

Fig. 2:14. Schematic illustration of an animal cell.
1 nucleolus, *2* nuclear sap, *3* chromatin, *4* cell membrane, *5* Golgi bodies, *6* centrosome, *7* mitochondria (thread-like structures in cytoplasm).

pole of the cell nucleus lies the so-called *centrosome* which is of importance in cell division. In the nucleus of the resting cell one or more *nucleoli* are found, together with a large number of small particles which because of their affinity to certain colouring agents are called *chromatin granules* (Fig. 2:14). At particular stages of cell division these appear as thread-like structures, *chromosomes*, in which it may be possible to distinguish between a central coiled string and thickened regions called *chromomeres*. The size and shape of the chromosomes vary from one species to another and during the different stages of cell division. There are also differences between particular chromosome pairs within the same species. With few exceptions, however, the number of chromosomes is constant within the species. Like the cells themselves, the individual chromosomes have the capacity to multiply by division in step with the cells, whilst retaining their characteristic properties.

The study of cells, their construction and function, is called *cytology*. Research workers in this field had already, before the rediscovery of Mendel's laws, come to the conclusion that the vehicles of inheritance were associated with the chromosomes. Shortly after the turn of the century American cytologists indicated that the segregation of the chromosomes in the production of gametes was undoubtedly the mechanism for Mendelian segregation. Cytogenetics is today an important field of research, in which by study of the structure of the cells, first and foremost the chromosomes, explanations are sought for observed hereditary phenomena. The real break-through contribution in this field was

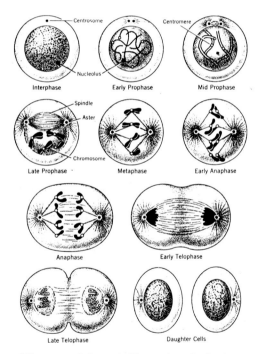

Fig. 2:15. Schematic illustration of mitosis.
1 Cell in the resting or *interphase* condition, *2* three different stages of the *prophase*, the chromosomes become apparent and the cell prepares for division, *3 metaphase*, the chromosomes arrange themselves in the equatorial plane of the cell, *4 anaphase*, the chromatid pairs have separated and move to opposite poles of the spindle, *5 telophase*, the chromatids have moved to their respective poles and the cell divides into two daughter cells each of which has the same chromosome constitution as the mother-cell, *6* cell division is completed.
(Reproduced by permission from E. J. GARDNER: *Principles of Genetics*, 1960. John Willey & Sons, Inc., New York)

made by the 'Morgan School' (Morgan, Bridges, Sturtevant and Muller) at Columbia University, New York, during the period 1910–1928. Impressive evidence was presented for the linear order of genes on the chromosomes; and deviations from the ratios expected according to Mendel's laws could be explained by cytologically observable changes in the distribution of the chromosomes or alterations in the structure of individual chromosomes. Today newer methods of investigation are available to the cytogeneticist, e.g. phase-contrast and electron-microscopes, and tissue cultures which make it possible to study in more detail the alterations referred to above, even in domestic animals.

In order to facilitate the understanding of Mendelian inheritance, a short outline of the mechanism of cell division and the production of gametes in the higher animals is given below.

Cell division and reduction division

New somatic cells are produced by mitotic division by which each daughter cell receives the same chromosome complement as the mother cell. In the production of gametes, on the other hand, the number of chromosomes is reduced to half the normal number for the species. This process is called reduction division or *meiosis*. Somatic or body cells are diploid, that is, they have $2n$ chromosomes, whereas the germ cells or gametes are haploid and have n chromosomes.

Cell division (mitosis). Cell division (Fig. 2:15) commences with the so-called *prophase* when the chromatin appears arranged in a certain number of chromosomes which when stained can be fairly easily observed under the microscope. At the beginning of the prophase each chromosome is divided along its length into two halves, called *chromatids*. These are attached to each other only at one point, the *centromere*, which may be located at the end of the chromosome (telocentric) in the middle (metacentric), or somewhere between. The centromere appears to be the centre for the movement of the chromosomes during cell division. As the process proceeds, a shortening and thickening of the chromosomes can be observed. Towards the end of the prophase the nuclear membrane disappears and the chromosomes arrange themselves on the equatorial plane of the cell, which marks the beginning of the next phase, the *metaphase*.

Metaphase. During the prophase the centrosome (Fig. 2:15 (2 and 3)) has divided into two daughter centrosomes which move to opposite poles of the cell. Radiating from them appear rays, or 'threads', to form a spindle between the respective poles and the chromosomes in the equatorial plane; in each chromosome a 'thread' goes from the centrosome to the centromere. Towards the end of the metaphase the centromere divides, but the two chromatids are still closely bound to each other.

During the next stage, the *anaphase*, the chromatids separate, each chromatid moving to an opposite pole as an independent chromosome. Thereafter the spindle disappears, the chromosomes lengthen and in the end can no longer be identified. This, the last phase of the mitotic division, is called the *telophase*. Finally, the whole cell divides into two halves each with its own nucleus. The daughter cells then go into a longer or shorter period of rest, called the *interphase*.

PHYSICAL BASIS OF INHERITANCE 33

In mitosis each daughter cell receives exactly the same chromosome complement as the mother cell both in quality and quantity.

Reduction division (meiosis). In the production of germ cells (gametes) a halving of the number of chromosomes takes place, which enables the chromosome number to be kept constant from generation to generation. In the zygote

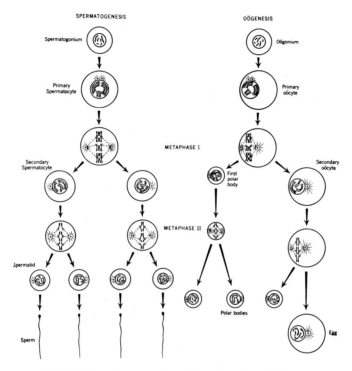

Fig. 2:16. Schematic illustration of the réduction division (maturation) in gametogenesis; spermatogenesis is the formation of sperms and oögenesis the formation of ova.
(Reproduced by permission from E. J. GARDNER: *Principles of Genetics*, 1960. John Willey & Sons, Inc., New York)

each chromosome from the father corresponds to a chromosome from the mother. The paternal and maternal chromosomes in a pair are said to be *homologues* of each other. Meiosis consists of two nuclear divisions and two cell divisions, but only one division of the chromosomes. The mother cells are called *primary gametocytes*. The process, illustrated by Fig. 2:16, is as follows.

During the first nuclear division of meiosis one can for the most part differentiate between the same stages as with mitotic division, but with certain essential differences. The prophase is considerably longer and is usually divided into different stages. During the first stage (*leptotene*) the chromosomes resume their shape but they are not separated along their length as in mitosis. In the next stage (*zygotene*) homologous chromosomes conjugate together so that an exact pairing of corresponding regions (genes) is achieved. It appears as if they have fused into one chromosome and are said to be bivalent. This pairing of the homologous chromosomes is called *synapsis*. During the third stage (*pachytene*) a lengthwise division of the chromosomes takes place and under the microscope they appear as four threads in a 'bundle'.

During the next stage (*diplotene*) the homologous chromosomes partly separate from each other but are held together at certain points, *chiasmata*, where the chromatids twist around each other. The number of chiasmata may vary from one to four or five or even more. It has been shown that an exchange of segments between homologous chromosomes takes place almost anywhere along the chromosomes where chiasmata can be observed. The chiasma is a visible expression of 'crossing-over' with exchange of segments between two homologues of the four chromatids (see p. 55). During the diplotene stage the homologous chromosomes are held together by chiasmata, whereas the two chromatids of each divided chromosome are held together by the still undivided centromere.

During the final stage of the prophase (*diakinesis*) the chromosomes shorten, prior to the disappearance of the nuclear membrane. Then follows phases corresponding to those in mitotic division. A spindle is formed and the tetrad chromosome 'bundles' arrange themselves in the cell equatorial plane. The chromatids are still united by the centromeres and therefore move together to the same pole during the anaphase. The other two chromatids move to the opposite pole. As a consequence of the exchange of chromatin between the homologues, which now separate they are not quite the same chromosomes which recently paired with each other during synapsis, even though the alteration may be very slight. During the final phase (telophase) the division of the cell nucleus and the cell is completed whereby two secondary gametocytes are formed. These have only half the normal number of chromosomes for the species.

During the metaphase of the second meiotic division, the chromosomes arrange themselves as usual in the equatorial plane of the cell, and the centromeres divide so that each chromatid has its own centromere and thereby becomes a complete chromosome. The two chromosomes, previously chromatids, move to opposite poles. The daughter cells are haploid like the mother cells since the real reduction division took place at the first meiotic division when the secondary spermatocytes were formed. The second division in the production of spermatids is a mitotic division of the previously reduced chromosome complement.

The paternal and maternal chromosomes in the primary gametocytes can be combined in different ways in the gamete; and it is apparently purely a matter of chance to which pole they go during the nuclear division. The number of combinations possible is 2^n where n is the haploid number of chromosomes. Since cattle have 30 chromosome pairs the number of possible combinations of paternal and maternal chromosomes is over a thousand million. On account of the exchange of segments, and thereby genes, between homologous chromosomes during meiosis, the number of possible gene combinations reaches still higher figures.

The formation of germ cells—gametogenesis. Fig. 2:16 illustrates the different stages of gametogenesis: in the male, when sperm are produced (spermatogenesis), and in the female, when eggs (ova) are produced (oogenesis).

Spermatogenesis is further illustrated in Fig. 2:17 which shows a cross section through one of the tubules of the testicle during cell division. On the inside wall of the tubule can be seen the mother cells, the diploid spermatogonia, which by mitotic division give rise to primary spermatocytes (cf. Fig. 2:16). These in turn

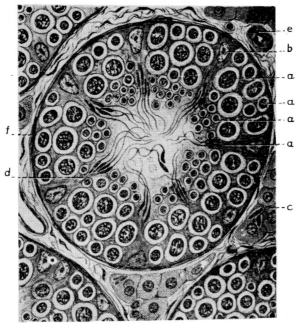

Fig. 2:17. Section through a seminiferous tubule.
a Cells in various stages of spermatogenesis: the smaller cells, nearest the lining of the tubule are spermatogonia, the larger ones are primary spermatocytes; the smaller ones nearer the centre are secondary spermatocytes and spermatids. *b* and *d* support cells, which aid in the development of spermatids into spermatozoa. *c* Leydig cells. *f* wall of the seminiferous tubule. *e* blood vessel. (TRAUTMANN and FIEBIGER 1952)

go through meiotic division to form secondary spermatocytes, which then divide again and produce spermatids which eventually become sperm or spermatozoa.

From Fig. 2:16 metaphase I, it can be seen how the homologous chromosomes pair with each other in the equatorial plane of the cell, one chromosome from the father and one from the mother in each pair. Each chromosome divides into two chromatids which are held together by the centromere, the two homologues being held together by the twisting of the chromatids so that a bundle with four chromosome threads is formed. At the next cell division (metaphase II in Fig. 2:16) the centromeres divide so that the chromatids can separate and move to their respective poles as complete chromosomes. When the second division is completed, therefore, two spermatids have been formed from each secondary spermatocyte; and after a maturation process and without further division they are formed into sperm. Four haploid spermatozoa are thus formed from each diploid primary spermatocyte.

In investigations with bulls and rams, by the use of modern isotope technique it has been found that the complete course of spermatogenesis from primary spermatocyte to completely developed sperm takes 48 days and that the average age of the sperm in an ejaculate is approximately 70 days. Van Demark and his co-workers have calculated that a two-year-old bull of the Friesian breed can produce approximately three million sperm per day per gm of testicular tissue. Sperm production is thus a highly effective process.

Oogenesis proceeds, in principle, in the same manner as spermatogenesis, with the difference, however, that only one egg is formed from each primary oocyte. During meiosis two small abortive cells, called polar bodies, are rejected; they have no known function to fill after they have been formed. The first polar body is produced when the primary oocyte divides, that is about the time of ovulation. The rejection of the second polar body in the second meiotic division does not take place until sperm has entered the egg and the fertilisation process thereby begun.

During both the cell divisions of oogenesis the chromosome complement is quantitatively the same for the polar body and the ovum, but the cytoplasm goes almost completely with the ovum. Mammalian ova must have a certain amount of nutrients which can be used during the first stages of development of the fertilised egg, i.e. before nourishment can be obtained from the uterus. The ovum is quite small, with a diameter of about 140 μ ($= 0.14$ mm); but it is very much larger than the sperm head, which measures only $3-5$ μ in length and $2-3$ in breadth. The head of the sperm corresponds most nearly to the nucleus of the ovum and contains the same amount of 'inheritance'. Thus the father and mother contribute in the same degree to the inheritance of the individual.

It has been calculated that a newly-hatched hen has in her ovaries 50–100 thousand oogonia. Only a very small number of these go through complete oogenesis during the reproductive period of the individual; many degenerate and disappear. Apparently, after hatching, or birth, no new oogonia are produced.

After ovulation and a fertile mating, the ovum is surrounded in the Fallopian tube by a large number of sperm. Many of these usually penetrate into the outer layer of the ovum (*Zona pellucida*), but as a rule only one sperm penetrates through the membrane of the ovum. The tail of this sperm disappears, the nucleus lies side by side with the nucleus of the ovum and they eventually fuse with each other. Fertilisation has thus taken place and the fertilised egg with $2n$ chromosomes is ready to begin the first division.

The chromosome number of farm animals

Determination of the chromosome number of animals has been a difficult task for cytologists, mainly due to the fact that the techniques evolved for plant material were not satisfactory when applied to mammals and birds. However, during the last ten years or so much progress has been made both in cytological technique and in optical equipment. Cells at different stages of division can be obtained from tissue cultures, by using material from the blood (leucocytes), skin, bone marrow or internal organs. By adding various substances, the cell division can be accelerated or retarded and the chromosomes can be dispersed

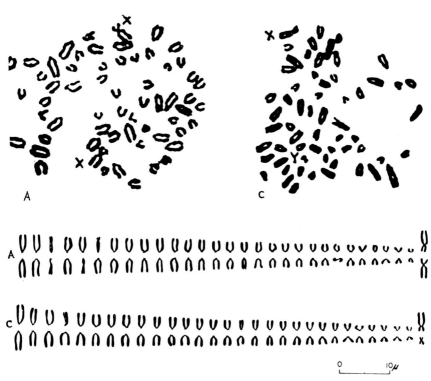

Fig. 2:18. The diploid chromosome set of cattle during the metaphase of mitotic division.
Above: squash preparation of cells from a culture of kidney tissue, on left (A) from a cow, and on right (C) from a bull. *Below*: corresponding karyogram with the autosomes arranged according to size and the sex chromosomes, marked X, on the right. (After SASAKI and MAKINO 1962)

within the cell by addition of hypotonic solutions. A squash preparation is used instead of the earlier method of imbedding and sectioning of the tissues. The cells are squashed between the object glass and the cover slide, whereby the chromosomes appear distinct and well distributed in the cells; the risk of mixing chromosomes from different cells is also eliminated.

It is usual to distinguish between sex chromosomes, which are wholly or partly responsible for the development of primary sex characteristics in either a male or female direction, and the other chromosomes, usually called autosomes. In mammals the female has two sex chromosomes similar in appearance, indicated by XX, whereas the male has one X and one Y chromosome. If the haploid complement of autosomes is indicated by A, then the diploid chromosome complement of females can be written $XX + 2A$ and males $XY + 2A$. The female transfers to all her ova one X chromosome, and she is said to be homogametic. Only half of the sperm produced by the male, on the other hand, contain the X chromosome, the other half containing the Y chromosome. The male is said to be heterogametic. In birds the position is reversed, the males having the chromosome constitution $XX + 2A$ and are the homogametic sex, whereas the female is heterogametic, i.e. only half the ova contain an X chromosome.

Fig. 2:18 shows a squash preparation of cells cultures of kidney tissue from cows and bulls together with the chromosomes from the picture arranged in pairs, according to the assumed homology, into a karyogram with the sex chromosomes on the right. The chromosomes are represented in the metaphase of mitotic cell division (cf. Fig. 2:15, p. 31); the two chromatids are still held together by the centromere, which in all the autosomes is terminally located though in the X chromosomes it is a short distance from the end. By comparing a number of karyograms from tissue cultures from the same species it is possible to obtain an idea as to which chromosomes are homologues. As far as cattle are concerned, this has not been entirely successful, since the morphological differences between chromosomes are very small. The sex chromosomes are, however, quite distinct from the autosomes.

The chromosome number of birds is much larger than that of mammals, and the chromosome set shows the peculiarity that a relatively small number of chromosomes are rather large, macrochromosomes, whereas the remainder are small, microchromosomes, and extremely difficult to count. The result of earlier counts was that the diploid chromosome number of fowls was 12, but YAMASHINA (1944) increased the figure to 78 for cocks and 77 for hens. NEWCOMER (1957) suggested that only the 12 large chromosomes can be regarded as 'genuine chromosomes', and that the others, called 'chromosomoids' by Newcomer, varied in number and with staining behaved differently from the real chromosomes. VAN BRINK (1959) is of the opinion that Newcomer's interpretation is wrong due to the technique he used in preparing the material. According to van Brink, the total number of chromosomes for cocks is between 62 and 84, with 78 being the most likely number; he found it impossible to make an exact count of the microchromosomes. Six or seven pairs are regarded as macrochromosomes and, if these are arranged according to size (Fig. 2:19), the fifth pair should be the sex chromosomes XX. None of the research workers mentioned has been able to demonstrate the presence of a Y chromosome in the heterogametic sex in the fowl or other species of poultry, but van Brink does not reject the possibility that the Y chromosome may be among the unidentified microchromosomes. Later SCHMID (1962) SCERBAKOV (1964) and OWEN (1965) reached the conclusion that not only the macrochromosomes but also the smaller elements are chromosomal in nature, and that the X-chromosome of the female has a Y-partner. We shall assume in what follows that this interpretation is correct. The chromosome set of the male will therefore be denoted $XX + 2A$ and that of the female by $XY + 2A$. In order to avoid confusion the sex chromosomes of poultry are often denoted by ZZ and ZW but the introduction of this notation can hardly be regarded as having any particular advantages.

The diploid number of chromosomes in human beings and the common species of domestic animals are given below. In the case of poultry the earlier

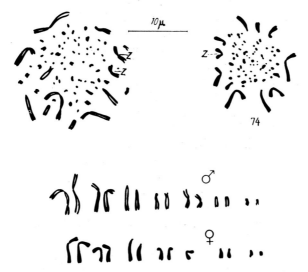

Fig. 2:19. The diploid chromosome set of the fowl during the metaphase of mitotic division.
Above: drawing after the microscopic picture of a squash preparation: *left* from a male, *right* from a female (74). The sex chromosomes have been indicated with Z.
Below: karyogram of the seven largest chromosomes from the same preparation, the fifth pair are the sex chromosomes. (After J. M. van Brink 1959)

practice of giving the chromosome number inclusive of the microchromosomes has been followed, though their numbers may be somewhat uncertain (Ford and Woollam, 1964).

Man	46	Dog	78
Horse	64	Fox	38
Ass	62	Cat	38
Cattle	60	Mink	30
Goat	60	Fowl	78
Sheep	54	Duck	80
Pig	40	Turkey	82
Rabbit	44	Dove	80

The chromosome constitution of a species is a well balanced system, in which practically no deviations are possible without endangering the normal development and viability of the zygote. Irregularities can occur in isolated mitotic divisions in the somatic tissues, but as a rule the chromosome-defective daughter cells are eliminated, since they lose the capacity to reproduce themselves. In certain cases, however, colonies of cells with widely deviating chromosome numbers can establish themselves in initially normal tissue; this eventually may result in a malignant tumour, where the body completely loses control over the growth of the cells.

Disorders in meiosis usually have the result that no fertile gametes are formed or that the zygote dies at an early stage. In man it has been shown that individuals with certain development anomalies have an excess number of chromosomes, e.g. monogloid idiots, and that in other anomalies a chromosome has been lost.

The haploid chromosome set of a germ cell with its complement of genes is called a *genom*. The diploid fertilised egg receives one genom from the mother and one genom from the father. In a number of invertebrates the haploid zygote is a common phenomenon, e.g. in the honey bee the males (drones) are haploid but the females (the queen and workers) are diploid. Among farm animals the turkey is the only example in which unfertilised ova may undergo partial and even complete embryonic development by parthenogenesis. In the small Beltsville turkey it was discovered that foetal development took place in a number of eggs from turkey hens which had never been in contact with turkey cocks (OLSEN and MARSDEN, 1956); though many of the embryos died at an early stage of incubation. By selection it was possible to increase the frequency of parthenogenesis quite considerably and also to increase the viability of the embryos. In 1965 Olsen reported that the frequency of embryos attaining a size equivalent to normal 21–28–day embryos had increased from 9·1 per cent in 1954 to 32·1 in 1963, when 106 parthenogenetic poults hatched from 10,060 incubated non-fertilised eggs. All the poults were typical males and about 25 per cent survived to maturity. Twenty-five parthenogenetic males have sired normal poults of both sexes. Cytological investigations showed that the chromosome number of the parthenogenetic offspring corresponded approximately with the diploid number for the species. This is regarded as being due to a doubling of the haploid chromosome number of the egg, possibly because the second polar body did not separate.

SELECTED REFERENCES

BANE, A. and BONADONNA, T. 1958. Fortpflanzung und Fortpflanzungsstörungen der Haussäugetiere. *Handbuch der Tierzüchtung*, Vol. 1, 59–168. Hamburg.

COLE, H. H. and CUPP, P. T. 1958, 1959. *Reproduction in Domestic Animals*. Vols. 1 and 2 (651 and 451 pp. respectively). New York.

HENRICSON, B. and BÄCKSTRÖM, L. 1964. A systematic study of the meiotic divisions in normal and subfertile or sterile boars and bulls. *J. Reprod. Fert.*, 9: 53–64.

OWEN, J. J. T. 1965. Karyotype studies on Gallus domesticus. *Chromosoma*, 16: 601–608.

SALISBURY, G. W. and VAN DEMARK, N. L. 1961. *Physiology of Reproduction and Artificial Insemination of Cattle* (639 pp.). San Francisco.

VAN BRINK, J. M. 1959. L'expression morphologique de la digamétie chez les Sauropsides et les Monotrèmes. *Chromosoma*, 10: 1–72.

WHITE, M. J. D. 1961. *The Chromosomes*, 5th ed. (100 pp.). London.

3   Mendelian Inheritance

Although there is no clearly defined boundary, it is usual to distinguish between the *qualitative* and *quantitative characteristics* of animals. Examples of *qualitative characters* are the presence or absence of horns in cattle, black and red hair colour, blood types and a number of congenital malformations such as bulldog calves or crooked or bent legs in pigs. These characters are more or less clearly alternative so that on the basis of their manifestation individuals can be alloted to well-defined classes which do not overlap with each other. In general the inheritance of these characters is relatively simple and environment plays only a minor role in their variation. Characteristics from this group are usually selected for demonstrating how Mendelian inheritance functions.

Quantitative characteristics show continuous variation between the extreme types, and the mean type is usually the most frequent (cf. Fig. 4:1). Live weight, height of withers, milk yield and fat percentage of milk are all examples of quantitative characters, and as a rule they are determined by a large number of genes with relatively small individual effect (see p. 118). Furthermore, these characters are usually influenced by a number of external factors, and the continuous variation is therefore a result of the combined effect of genetic and environmental causes.

The term *environment* includes all non-genetic factors which can influence the individual from the time the egg (ovum) is fertilised, e.g. nutrition, management, climate, hygiene, etc. Differences in the manifestation of a character in different individuals, brought about by differences in environment, are called *modifications*. Quantitative characters are, in general, considerably more modifiable than qualitative.

Regardless of the group of characters to be dealt with, a distinction is made between the *phenotype*, which is the external, observable type or characteristic, and the *genotype*, which represents the genetic constitution of the type or characteristic. It is the genotype that is all-important in assessing the value of an individual for breeding. In one way or another the genotype must, however, be assessed on the basis of the phenotype of the individual itself or its relatives, since the genotype cannot be the object of direct measurement.

MENDEL'S LAWS OF INHERITANCE

When GREGOR MENDEL began his epoch-making crossbreeding experiments he chose to work with a self-fertilising plant, the garden pea. He decided to use genetically constant lines differing from one another in a few, well-pronounced qualitative characteristics, such as the form and colour of the seeds, colour of the flowers, etc. He crossed these lines with each other and established that the product of the first cross (F_1) was not intermediate between the parents but in all cases the progeny resembled one or the other of the parents with respect to the characteristic studied. When the F_1 individuals were reproduced by self-fertilisation the next generation (F_2) showed a segregation of those characters which had 'disappeared' in the F_1 and in addition new combinations of the parental (P) characters appeared. F is an abbreviation of *filial* and P is an abbreviation of *parents*. Mendel found that segregation and recombination took place according to certain numerical proportions and from this he drew the conclusion that each character was determined by something which he called a germinal unit or factor which remained unchanged from one generation to the other. One of the contrasting characters could, however, be *dominant* over the other; the other was then said to be *recessive*. This did not mean that the germinal unit for the recessive trait was in any way altered.

In the spring of 1865 Mendel presented the results of his experiments in two lectures to the Natural History Society in Brünn (now Brno) in Czechoslovakia. The following year the lectures were published in the annals of the Society, but they attracted very little attention at the time. In 1900 three botanists, DE VRIES in Holland, CORRENS in Germany and TSCHERMAK in Austria, independently of each other, came to the same conclusions as Mendel, and it was then that Mendel's work was really brought to the attention of the scientific world. Mendel's discovery, which laid the ground for the whole of the modern science of genetics, is usually summarised in the following two laws or rules:

1. *The law of segregation.* The characteristics of an individual are determined by pairs of genes, but the gametes contain only one gene from each pair.

2. *The law of independent assortment.* The genes combine at random with each other both at the formation of gametes and at fertilisation. Limitations of the validity of this law will be discussed later.

Mendel's laws will be first illustrated with some examples from animal breeding and later supplemented with the results of newer investigations. According to the normal practice, the genes will be represented by letters. When the genes of a pair are different, one will be represented by a capital letter and the other by a small letter. It is usual to use a capital letter for the dominant and a small letter for the recessive gene (cf. p. 44). When an individual has identical genes in a particular pair, represented for example by RR, it is said to be *homozygous*; when the genes are different, e.g. Rr, it is said to be *heterozygous* with regard to this pair. The genes R and r are *alleles* of each other and occupy a corresponding place, called a *locus*, on two homologous chromosomes.

Simple mendelian inheritance (Monohybrid segregation)

The colour of the wild mink is brown, with somewhat varying intensity of pigmentation. Ranch-bred animals with this colour-phase are usually called standard mink. Several new colour phases have appeared through sudden changes (mutations) of colour genes. The majority of the mutant genes are recessive to their standard alleles but dominant mutants are also known. One of the recessive colour-phases is called platinum or silverblue. Homozygosity for the platinum gene (p) changes the brown colour of the standard mink to bluish grey. When homozygous standard mink (PP) is mated to platinum (pp) all the progeny in the F_1 ($= Pp$) are standard colour, and when a large number of F_1 animals are mated *inter se*, approximately 75 per cent of the offspring will show the standard colour and 25 per cent the platinum colour. The gene for standard colour may be considered as completely dominant over the recessive platinum gene. As shown in Fig. 3:1, the phenotypic segregation ratio in F_2 is 3 standard: 1 platinum; but the genotypic ratio is 1 PP:2 Pp:1 pp. Whether a certain standard-coloured mink is a carrier of the platinum gene or not can be tested by mating it to a few platinum females. If it begets at least seven standard type offspring from such matings, the chance is only 1:100 that the standard male is heterozygous (Pp); normally this number of offspring is obtained in two litters. The birth of one or several platinum kits would prove that the male is a carrier of the platinum gene. If a large number of offspring of a carrier is obtained when mated to the recessive type we would expect approximately 50 per cent standard and 50 per cent platinum coloured kits. Mating between F_1 animals and either of the parental types is called *back-crossing*.

The English Shorthorn breed consists of red, white and roan animals. When

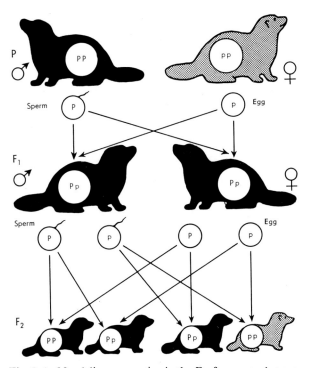

Fig. 3:1. Mendelian segregation in the F_2 after a cross between dominant standard and recessive platinum mink. (By courtesy of R. M. SHACKLEFORD, 1957. *The Genetics of the Ranch Mink*. New York)

red animals are mated together the progeny are all red, and similarly white animals mated together give rise to only white progeny. Both self-colour types are therefore homozygous for hair colour genes. However, when red are mated with white the progeny are all roan, and in matings between roan animals segregation takes place according to Mendel's law to give 1 red; 2 roan; 1 white

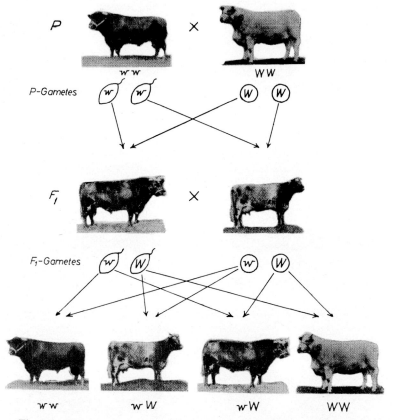

Fig. 3:2. The inheritance of red, white or roan coat colour in the Shorthorn breed.

(Fig. 3:2). The inheritance of roan colour can be best explained by assuming that the white Shorthorn is homozygous for a gene W, which in a single dose Ww results in a mixing of pigmented and white hairs. When the progeny of a cross lies between the two parents, it is usually referred to as *intermediate inheritance*. The degree of roanness in the F_1 individuals from a crossing between red and white Shorthorns can vary within very wide limits and in this special case it is more a question of *partial dominance* than of intermediate inheritance. In partial dominance it may be difficult to know which gene should be regarded as dominant, and which as recessive. The most important thing, however, is to decide whether simple Mendelian inheritance occurs or not.

Variation in the degree of dominance is a common occurrence. The dominance may appear to be complete, but with careful investigation it is quite common to find that even the recessive gene makes itself apparent to some extent. Polledness in cattle, for example, is not completely dominant over hornedness but the heterozygote, as a rule, has protuberances on the bones of the skull and, especially in the male animals, these protuberances usually correspond to loose scurs. In general the variation in the degree of dominance for a given gene depends upon the residual genotype of the individual, i.e. the genetic environment in which the gene operates. An illustration of this type of inheritance is a type of feathering in hens which goes under the name of 'frizzle'. The whole of the

Fig. 3:3. *Left*: Frizzle fowl. *Right*: normal feathering.

feathering is very loose and the shafts of the contour feathers are recurved, the outer surface being concave (Fig. 3:3). The 'pure' frizzle hen is homozygous for a dominant gene F, but the dominance is only partial. It has been shown that a certain gene m modifies the dominance of the F gene and in the homozygous condition mm almost completely overrides the F gene in the heterozygotes (Ff).

Even certain environmental factors can influence the degree of dominance. In cattle, black is dominant to red but quite often it is possible to see a difference between the homo- and heterozygote with respect to the intensity of the black colour. When the animals are in a poor condition, due to nutritional deficiencies or disease, the intensity of pigmentation decreases so that the gene for red colour 'breaks through' more strongly in the heterozygotes.

A single individual cannot have more than two allelic genes, e.g. Aa or Bb, but within the population there can be several so-called 'multiple alleles', which can be combined in pairs in different ways. Multiple alleles and their importance will be the subject of a later discussion (pp. 69 and 76).

Simultaneous segregation of two or more gene pairs
(Dihybrid and polyhybrid segregation)

Reference has been made to the inheritance of the brown colour of the standard mink and the recessive platinum colour of the mutant type (*pp*). Another recessive mutant is called aleutian (*aa*), which is much darker than the platinum mink. When platinum and gunmetal-coloured aleutian minks are crossed with each other, all the progeny in the F_1 are standard-coloured *AaPp* dihybrids. If the F_1

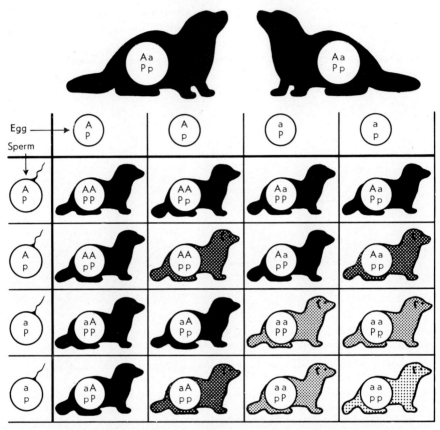

Fig. 3:4. Illustration of dihybrid segregation in the F_2 after crosses between silver blue and aleutian mink. The dihybrids segregate after mating with each other in the ratio 9 standard:3 silver blue:3 aleutian:1 sapphire. (Courtesy of R. M. SHACKLEFORD 1957. *The Genetics of the Ranch Mink*, New York)

animals are then mated with each other, segregation in the F_2 is obtained according to the ratio 9 standard *A–P–*:3 silver blue *A–pp*:3 aleutian *aaP–*: 1 sapphire *aapp*. Fig. 3:4 illustrates this segregation. The F_1 males and females form four different gametes *AP*, *aP*, *Ap* and *ap* in equal numbers and these gametes can be combined in $4 \times 4 = 16$ possible ways. All animals which receive at least one *A* gene and one *P* gene are standard-coloured due to the dominance of these genes. Animals that are homozygous for the *a* and *p* genes are decidedly lighter than the parent types. The double recessive, so-called 'sapphire' mink, has for many years been one of the most coveted colour-types on the fur market. The example shows how it is possible by crossing genetically different types to obtain in the F_2 new combinations of the genes which entered the cross. Mink farmers have, by planned breeding, carried out such crosses to produce new colour types.

One more example of dihybrid segregation will be presented. The single comb characteristic of Leghorn fowls is recessive to both pea comb and rose comb. If

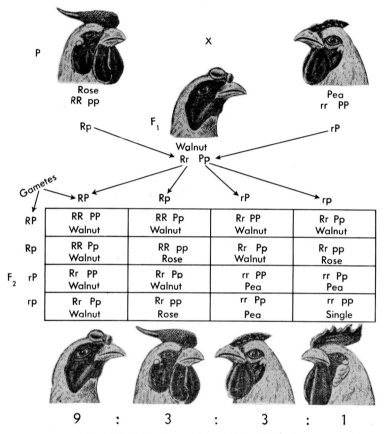

Fig. 3:5. Inheritance of comb shape in fowls.

a breed with rose comb (RR), e.g. Wyandotte, is crossed with the Leghorn (rr) then all F_1 individuals develop rose comb (Rr) on account of the dominance of the R gene. When these F_1 birds are mated with each other the proportions in the F_2 are 3 rose comb:1 single comb. The same conditions apply when the Leghorn is crossed with a breed with the pea comb (PP), e.g. Brahma; P dominates over p and in the F_2 the proportions obtained are 3 pea comb:1 single comb. If, however, the Brahma is crossed with the Wyandotte, all the individuals in the F_1 have a completely new comb type called the walnut comb ($PpRr$). It is the interaction of the genes P and R which produce this effect. When these F_1 individuals are mated with each other, four different phenotypes are produced in the F_2 in the following proportions: 9 walnut P-R-:3 pea P-rr:3 rose ppR-:1 single $pprr$. The single comb is the double recessive type. The number of different genotypes is however $3 \times 3 = 9$. Fig. 3:5 shows, in diagrammatic form, the segregation in the F_2 from which the number of genotypes and phenotypes may be calculated.

If parent types which differ from each other in 3 gene pairs are crossed, trihybrid segregation is obtained in the F_2. Assume that the sapphire mink is crossed with the royal pastel, a relatively light brown colour type which is completely recessive to the standard type. The genotype of the royal pastel is denoted by bb. The parental types are thus $aappBB$ and $AAPPbb$, and consequently all progeny in the F_1 will be standard coloured trihybrids, $AaPpBb$. Eight different gametes are then formed by the F_1 males and females in equal numbers APB, APb, ApB, aPB, Apb, aPb, apB and apb. The number of possibilities of combination at fertilisation is $8 \times 8 = 64$, but the number of different genotypes is only $3 \times 3 \times 3 = 27$. The number of different phenotypes is the same as the number of different gametes, viz. 8, in the following proportions: $27A$-P-B-:$9A$-P-bb:$9A$-ppB-:$9aaP$-B-:$3A$-$ppbb$:$3aaP$-bb:$3aappB$-:$1aappbb$. The last colour type, the triple recessive is called winterblue. In order to study the different combinations more closely a two-way block diagram for the F_2 generation can be drawn which will have 64 squares. This is rather laborious and the same result can be achieved in a simpler way.

In calculating the proportions between the different phenotypes in the F_2 one begins with the proportions obtained for each gene pair. In dominant inheritance these proportions are 3:1. The dominant type is indicated with a capital and the recessive with a small letter, thus in a cross between platinum × aleutian mink $(3P + 1p)(3A + 1a) = 9AP + 3Ap + 3aP + 1ap$, or 9 standard:3 platinum:3 aleutian:1 sapphire. In the case of the triple hybrids, e.g. $AaPpBb$, the proportion in the F_2 are calculated according to $(3A + 1a)(3P + 1p)(3B + 1b) = 27APB + 9APb + 9ApB + 9aPB + 3Apb + 3aPb + 3apB + 1apb$, the last-mentioned being the triple recessive. If four gene pairs are involved the number of brackets must be increased by a further $(3 + 1)$. Observe that the letters here do not indicate genes but phenotypes.

Table 3:1 The combination possibilities in the F_2 when the F_1 individuals are heterozygous for the indicated number of gene pairs

Pairs of genes	No. of gametes	Combination possibilities	No. of genotypes	Homozygous combinations
1	2	4	3	2
2	4	16	9	4
3	8	64	27	8
4	16	256	81	16
10	1 024	1 048 576	59 049	1 024
n	2^n	4^n	3^n	2^n

Table 3:1 shows the number of possible combinations and different genotypes in the F_2 after crossing, when the parents differ from each other for the indicated number of gene pairs (n).

The number of possible combinations increases rapidly with the number of gene pairs. In domestic animals the number of heterozygous gene pairs is very large, no doubt several thousand. It is no wonder therefore that no two individuals with the exception of one-egg twins, are genotypically or even phenotypically alike.

Modified Mendelian ratios

If the F_2 generation is small, the result of segregation can deviate, due to chance, quite considerably from the expected ratios of genotypes and phenotypes according to Mendel's laws. This situation will be dealt with in a later section (p. 90). Deviations from the segregation ratios hitherto discussed can also have biological reasons and it remains then to find the explanation. Fig. 3:6 shows some examples of modified ratios in dihybrid segregation which apply only to the phenotypes since the genotypes segregate entirely in accordance with the 'laws'. The deviations are thus only apparent.

Scheme 1 shows the usual dihybrid segregation 9:3:3:1, which applies in complete dominance. According to scheme 2 the inheritance is intermediate and the phenotype is determined by the number of genes with a positive effect on the trait, the effect of the genes being additive (cf. p. 113). In the 16 possible combinations in the F_2 the number of 'plus genes' can vary from 0 to 4. If the number of similar squares in the scheme are added, it will be found that 1/16 of the individuals have 4 plus genes, 4/16 have 3 plus genes, 6/16 have 2 plus genes and 4/16 have one plus gene, whereas 1/16 do not have any plus genes. The numerical ratio between the phenotypic classes is thus 1:4:6:4:1.

Scheme 3 applies to duplicate recessive genes where *aa* has the same effect as *bb*. Several examples are known in mink breeding of the same colour type depending upon different gene constitutions. The two brown types, royal pastel and imperial pastel, cannot be phenotypically distinguished from each other. Crossing between them results only in dark standard mink in the F_1, but segregation takes place in the F_2 according to the ratio 9 standard:6 pastel:1 light brown. The pastel group, however, contains 3 royal pastel and 3 imperial pastel.

Scheme 4 shows the segregation in the F_2 after crossing between White Leghorn and White Wyandotte fowls. The White Leghorn is homozygous for a dominant gene *I* which prevents the formation of pigment in the skin and feathers, whereas the Wyandotte is homozygous for a recessive gene *c* which in the homozygous condition results in a white plumage. All the F_1 individuals are white but in the F_2 segregation takes place according to the ratio 9*I–C–*:3*I–cc*: 3*iiC–*:1*iicc*. Only the genotypes *iiC–* show the pigmentation and the rat o between the phenotypes becomes therefore, 13 white:3 pigmented. When the gene *I* is present, *C* cannot exert any effect. *I* is said to be *epistatic* to *C* and *C hypostatic* to *I*. In population genetics the term *epistasis* is extended to cover all non-linear interaction effects between genes at different loci (cf. p. 115).

Finally, scheme 5 shows how a double recessive can be epistatic to dominant genes at other loci. If rabbits with wild colour (*agouti*) are crossed with albinos

cc, which are not carriers of the wild colour gene A, then the progeny in the F_1 will all be wild coloured and in the F_2 segregation takes place according to the ratio 9 wild coloured, $A–C–:3$ black, $aaC–:4$ white, $(3A–cc + 1aacc)$. In the genotype $A–cc$ the wild colour gene cannot manifest itself because the formation of pigment requires the presence of the gene C.

In all these examples it is only the phenotypic segregation which deviates from the normal dominant and recessive ratios. However, deviating proportions can occur even in respect of the genotypes. The most usual cause is that a certain genotype is not viable but dies in the embryonic stage and is therefore not observed. Several such cases are known.

An example of this is the once famous platinum fox which did not breed true but always gave rise to one-third silver foxes. When platinum foxes were mated with each other the litters at birth were 25 per cent smaller than after matings between silver foxes or platinum × silver. The explanation is that the gene for platinum colour is lethal in a double dose and the homozygous platinum embryos die at an early stage and are resorbed in the uterus. Fig. 3:7 shows this situation. The gene for platinum colour w^p is dominant over the 'normal gene' for silver w and all viable platinum foxes have the genotype ww^p; the homozygote $w^p w^p$ is not viable. In mating $ww^p \times ww^p$ the proportions between the viable phenotypes are consequently 1 silver $ww:2$ platinum ww^p.

The deviations from the usual proportions in Mendelian segregation can be of a more complicated nature depending upon, for example, linkage between genes which are on the same chromosome, complicated interaction situations between genes, disturbances in the chromosome mechanism, and so on. These conditions will be dealt with in more detail later.

As was mentioned previously, no sharp dividing line can be drawn between qualitative and quantitative characters. A number of characteristics are determined by major genes (*oligogenes*); but one or more other genes (modifiers) can bring about a certain amount of variation in the manifestation of the character. Some examples are given below.

The general pattern of white markings in the Friesian breed of cattle depends on a recessive gene which shows clear Mendelian inheritance, the dominant allele producing self colour. But the extent of the white depends on a large number of minor genes (*polygenes*), which individually have so little effect that no segregation ratios can be demonstrated. The area of pigmentation in the Friesian breed varies from predominantly black to predominantly white animals (Fig. 8:5). The gene for white is said to have *variable expressivity*. The extent of the white pattern shows quantitative variation and can be measured.

It can also happen that an oligogene, which can be dominant or recessive, does not produce its effect in combination with certain other genes or under certain environmental conditions. The gene is then said to have *incomplete penetrance*. Penetrance is expressed as a percentage which shows how large a proportion of individuals of a particular genotype manifest the trait. If in a mating between the recessive homozygotes $aa \times aa$ only 3/5 of the aa progeny phenotypically express the aa genotype, the penetrance is said to be 60 per cent. The other 40 per cent

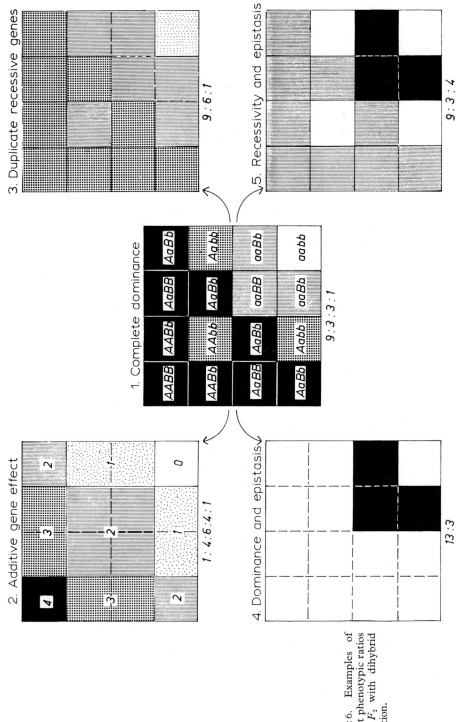

Fig. 3:6. Examples of different phenotypic ratios in the F_2 with dihybrid segregation.

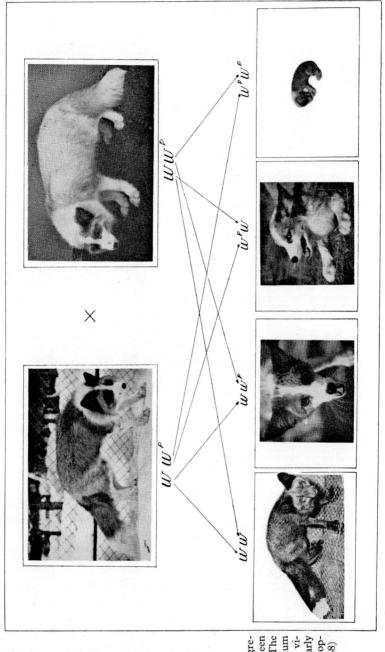

Fig. 3:7. Mendelian segregation in matings between platinum foxes. The homozygous platinum gene foetuses are not viable and die at an early stage of foetal development. (JOHANSSON 1948)

of the aa individuals are so like Aa and AA that they cannot be distinguished from them. The main reason for the variation in penetrance seems to be the general genetic constitution, i.e. the genetic environment for aa. Quite often, however, incomplete penetrance has been used as an explanation for aberrant ratios between phenotypes in a sample of field data, or a small scale experiment, without any testing of the assumptions which have been made. Especially when the assumed penetrance is low, there may be good reasons to doubt the validity of the hypothesis of monofactorial inheritance. GARDNER (1960), points out that the terms penetrance and expressivity 'are sometimes used to make ignorance respectable in cases where the reasons for the failure of expression or variation in phenotype are not known'.

The terms *oligogene* and *polygene* have been introduced and will be used henceforth for purely practical reasons; they say nothing about either the dimensions or the physiological importance of the genes. The term oligogene refers to a gene, the effect of which on a given trait is so clearly expressed that Mendelian segregation can be established. Polygenes do not show such marked effects, presumably because they are numerous and their individual effects are usually quite small, and also because the character in question is modifiable within wide limits. Thus oligogenes are associated with qualitative, and polygenes with quantitative characters. Just as no sharp dividing line can be drawn between qualitative and quantitative characters, it is impossible to draw a definite line between oligogenes and polygenes. It is no doubt common for a gene which acts as an oligogene on a qualitative character to act also as a polygene on one or more quantitative characters. This does not, however, detract from the utility of the terms.

QUANTITATIVE CHARACTERS

The extent of white fur in Dutch rabbits shows a very wide variation; they vary from almost completely black to almost completely white, as can be seen from Fig. 3:8A. We will assume that this variation is determined by four pairs of genes $(A_1a_1A_2a_2A_3a_3A_4a_4)$. Then, if relatively light coloured animals $(a_1a_1a_2a_2a_3a_3A_4A_4)$ are mated with relatively dark $(A_1A_1A_2A_2A_3A_3a_4a_4)$, represented by Nos. 5–6 and Nos. 13–14 in Fig. 3:8A, it can be expected that the intermediate F_1 generation will show an average amount of white as represented by Nos. 8–9. When these F_1 individuals are mated with each other, segregation takes place in the F_2 so that the whole of the colour scale shown in Fig. 3:8A will be represented. The segregation is illustrated in Fig. 3:8B. Each column in the diagram corresponds to a pigmentation class and the classes to which the parental types belong have been marked by the thick black lines. The majority of the F_2 individuals will fall into the groups between the parental types but some will be lighter than the lightest parental type and some darker than the darkest parents. The situation where a number of F_2 individuals show more extreme manifestation of a given trait than the parental generation is referred to as *transgressive segregation*. The segregation of extreme types is relatively common when many genes influence a character, i.e. polygenic inheritance. The probability of transgressions is of course greater when the parental types are relatively close to the mean of the population than when they are nearer to the extremes.

Quantitative characters, such as body size, milk yield, fat content of milk, and egg production of hens, are inherited similarly, but two important differences come into the picture: (1) these characters are strongly modified by environmental variation, not least by nutrition; (2) the characters mentioned are, without doubt, determined by a much larger number of genes which probably have different degrees of effect and exert a complicated interaction with each other. Any attempt to show by the use of genetic symbols how such characters are inherited is a waste of time; statistical methods must be applied in order to estimate the importance of the different causes of variation (cf. p. 118).

Fig. 3:8A. Variation in the extent of white fur in Dutch rabbits. The standard type of the breed is represented by Nos. 8 and 9.

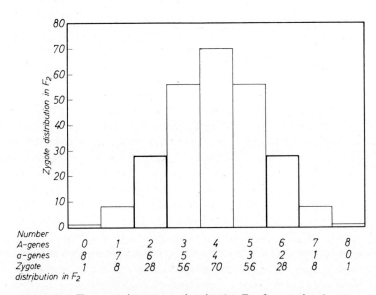

Fig. 3:8B. Transgressive segregation in the F_2 after mating between relatively dark and relatively light-coloured Dutch rabbits with the assumption that segregation depends on four gene pairs. The frequencies drawn with heavier lines represent the parental types.

LINKAGE AND CROSSING OVER

Cytological research has shown that there is complete agreement between the behaviour of homologous chromosomes during meiosis and Mendel's first law, according to which each gene pair divides so that the gametes contain only one gene from each pair. Similarly Mendel's second law corresponds to the free combination of chromosomes, one from each pair, in the formation of gametes and a similar free combination of gametes at fertilisation.

Genes located on the same chromosome do not show independent assortment but tend to go together in inheritance; they are more or less strongly linked to

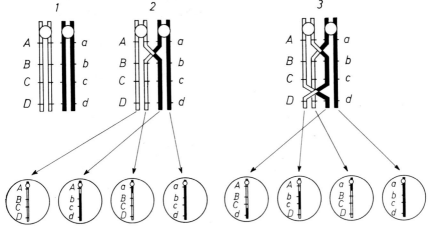

Fig. 3:9. Schematic illustration of crossing over and exchange of genes between the chromatids of homologous chromosomes. (See text and cf. p. 34.)

each other. If there were no exchange of material between homologous chromosomes, linkage would be complete and each chromosome would then appear to be a single gene. However, this is not the case. BATESON and PUNNET discovered that in crossing different varieties of sweet pea certain genes tended to be inherited together, thus deviating from Mendel's second law. The explanation was given some years later by MORGAN and his co-workers after they had found that the number of chromosome pairs and the number of linkage groups in the common fruit fly (*Drosophila melanogaster*) were the same. The formation of chiasmata during the first meiotic division was assumed to be the cytological basis for the phenomenon: an exchange of chromatin substance takes place between the chromatids when they are twisted round each other. A break occurs at the place where the chromatids cross each other; the one chromatid end fastens to the corresponding end of the other chromatid. This interchange of corresponding segments (genes) between homologous chromosomes is called *crossing over*. The strength of the linkage between different genes depends on the distance between them, genes lying close to each other being more strongly linked than those which are far apart. At least one chiasma must be formed for the homologous chromosomes to be held together during the 'four strand stage'; more than four or five chiasmata are believed to be exceptional.

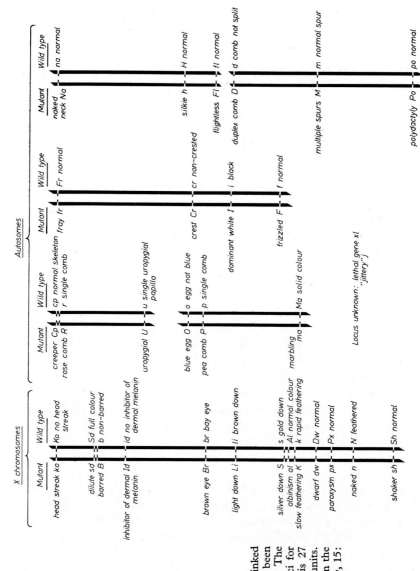

Fig. 3:10. Chromosome map of linked genes in the fowl. The same scale has been used for all six pairs of chromosomes. The crossing-over percentage between loci for *Id* and *Br* on the X chromosomes is 27 units and between *al* and *K*, 1·6 units. (Mainly after F. B. HUTT: New loci in the sex chromosome of the fowl, *Heredity*, 15: 97–110, and other publications.)

The process of crossing over is represented schematically in Fig. 3:9. Part 1 in the diagram shows two homologous chromosomes which have both divided lengthwise to form two pairs of chromatids; no crossing over has yet taken place. In part 2 a single cross-over has taken place, in this case in the middle of the chromosome; and in part 3 a double crossing over has taken place between two chromatids and a single crossing over with a third chromatid. In the diagram, only one allele is shown in the upper part of the chromosome in the single crossing over; in fact it is whole blocks of the chromosome that are exchanged, each block containing many genes. The conjugating chromosomes have 'mated' with each other so that each gene locus on the one chromosome 'mates' with the corresponding gene locus on the other chromosome. Crossing over does not always occur at the same place; and when crossing over does take place at a given

Fig. 3:11. Schematic illustration of a segment of a salivary gland chromosome of the fruit fly. The numbers indicate identified 'gene regions'.

point, the immediate vicinity of the chromosome is prevented from crossing over, presumably because the chromosome has a certain degree of inflexibility, which minimises the possible distance between two chiasmata.

The crossing over percentage can be calculated by selecting suitable genotypes for crossing, and this percentage is usually constant for a population unless influenced by special factors. The crossing over percentage is used as a measure, called *centimorgan*, of the distance between gene loci. The relative order of gene loci on the chromosome can be deduced from the crossing over percentages. If genes A and B are linked and A shows linkage to a third gene C, then it follows that C and B are also linked. If the crossing over between A and B is 25, between B and C 15 and between A and C 40 per cent, then the genes must lie in the following order $A \leftarrow (25\%) \rightarrow B \leftarrow (15\%) \rightarrow C$. On the other hand, if the crossing over between A and C is only 10 per cent then the order is $A \leftarrow (10\%) \rightarrow C \leftarrow (15\%) \rightarrow B$. On the basis of this type of experiment chromosome maps have been made for a number of species of plants and animals. A chromosome map for fowls is shown in Fig. 3:10; it includes 6 linkage groups which corresponds to the haploid number of 'large chromosomes' (cf. p. 39). The largest linkage group includes 14 different loci.

Of all the mammals, the best analysed is the mouse, in which 19 linkage groups have been established; the haploid chromosome number is 20. Isolated cases of linkage between the colour genes in mink have been shown, and in humans and pigs linkage is known to exist between certain blood group loci, as well as between a number of genes for inherited disorders.

The nuclei of cells in the salivary glands of *Drosophila* contain unusually large chromosomes, and this has made it possible to localise many genes to definite regions of a given chromosome (Fig. 3:11). Thereby cytological evidence for the linear order of the genes on these chromosomes is also obtained. There is no similar evidence for the order of genes in the larger animals.

Fig. 3:10 shows that the percentage of crossing over between different loci on the same chromosome can vary between very low values and 50 per cent, the latter corresponding to independent assortment. Crossing over of 25 per cent between two genes, e.g. A and B, means that, being together in the parental generation, they are distributed between different gametes only half as often as they would be if they were on different chromosomes.

Concerning the importance of linkage between genes it can be said that linkage apparently reduces the independent assortment of genes at each reduction division, but from a long term point of view it is a source of new gene combinations that can be tested in natural or artificial selection. When the linkage is not absolute, the opposite combination is formed at an increasing rate the higher the crossing over percentage. If, for example, A is linked to B on the one homologue and a is linked to b in the other, a crossing over between them of 10 per cent means that an increasing number Ab and aB genes become linked to each other. There is, of course, crossing over between Ab and aB but this operates on a smaller number of combinations until a stage is reached when A is coupled to b in the population as often as to B. When this stage is reached, the result of segregation is the same as if the genes were on different chromosomes. If it could be shown, for example, that the gene for a given blood group locus was coupled to a certain defect, e.g. hairlessness in calves, this would be of practical importance for selection only if the linkage was very strong. If the coupling was weak, then far too many mistakes would be made if one attempted to identify carriers of the defect gene by blood group investigations.

SEX DETERMINATION AND SEX LINKED INHERITANCE

The proportion of males to females at birth, the sex ratio, is usually given as the percentage of all births that are male. The sex ratio of all species of farm animals lies very close to 50 per cent. Significant deviations occur, but these are small when the calculation is made on a large number of individuals. For example, the sex ratio for cattle has been calculated to be 51·5, for horses 49·6, for pigs 52·3 and for poultry at hatching 49·2 per cent. In small or medium sized herds the deviations, over a limited period of time, can be rather large, due to chance.

As was mentioned earlier, sex is determined by the chromosome complement of the zygote. In mammals the female is homogametic, the male heterogametic. With birds the situation is reversed. On this point there is complete agreement between the results of cytological and genetic experiments. The inheritance of sex follows the same simple scheme as with back crossing of a heterozygote (Aa) to the recessive homozygote (aa). The theoretically expected sex proportions are 1 male:1 female.

Significant deviations from the ratio 1:1 can be explained in different ways. On the basis of the sex ratio of aborted and stillborn foetuses, it is generally assumed that the sex ratio at fertilisation in mammals is considerably higher than 50 per cent, but that the ratio is reduced during the course of pregnancy by a higher rate of foetal deaths for the male sex. The reason why the sex ratio at fertilisation is relatively high may be that sperms carrying the Y chromosome are quicker and more resistant to the uterine environment than the sperms carrying the X chromosome and therefore able to fertilise a larger number of ova. However, from a practical point of view the deviations can be disregarded.

Various attempts have been made to separate the male- and female-determining sperm from each other, e.g. by centrifuging the semen and then using the different fractions of the semen in artificial insemination. It has been asserted that in some cases the sex ratio had been shifted in a given direction, but it is uncertain whether there is any clear evidence for this. In working with a comparatively small number of individuals the deviations from the normal sex ratio must be quite large if they are to justify definite conclusions.

Several investigations show that the sex is determined by a balance between factors which operate in the male or female direction. In *Drosophila* the female-determining genes are located on the X chromosomes, whereas the autosomes (A chromosomes) operate in the male direction. Normal females and males have the chromosome complement $2X+2A$ and $XY+2A$ respectively. As a result of disturbances in the reduction division flies may appear with other proportions between the X and A chromosomes. Individuals with the chromosome constitution $2X+3A$ are pronounced 'intersexes' because the male-determining factors take the lead at a certain stage of development but too late to develop normal males. The situation is the same even when a Y chromosome is present at the same time as the two X chromosomes. Zygotes with the chromosome constitution $XO+2A$ develop outwardly into normal males, but are sterile. It has also been shown that homozygosity for a particular autosomal gene can transform females of the type $2X+2A$ to phenotypic males, but they are sterile. In this connection it can be mentioned that in goats homozygosity for the gene for polledness usually makes a genetic female into an intersex; with regard to polledness this gene is dominant but in sex differentiation it operates as a recessive.

As was mentioned previously, in birds the male is homogametic (XX) and the female heterogametic (XY). This is also the situation in butterflies. With the gypsy moth (*Lymantria dispar*) it is the X chromosomes which are male-determining, whereas the cytoplasm of the eggs is female-determining, possibly due to the influence of the Y chromosome. Individuals with the chromosome constitution $XX+2A$ and $XY+2A$ normally develop into males and females respectively. The male- and female-determining factors can, however, have different strength in different species. In crosses between species, therefore, the balance between male- and female-determining factors may be disturbed, so that intersexes occur even with apparently normal chromosome constitution. In the silk moth it is the Y chromosome which is the active female-determiner.

The above example shows that the male- and female-determining factors may be located on different sex chromosomes (X or Y), or on the autosomes, in different species, even when the species are closely related, e.g. the gypsy moth and the silk moth. The species appear to have one thing in common; that normal sex differentiation depends on a balance between the male- and female-determining factors and that disturbance of this balance leads to intersexuality. Investigations show that this is also the case with humans and domestic animals.

Pronounced intersexuality is often called *hermaphroditism*. In humans and domestic animals it is usual to distinguish between 'genuine hermaphroditism', when testicular and ovarian tissues are developed in the same individual, and 'pseudohermaphroditism' which does not affect the gonads but only the mating organs and those parts of the genitalia developed from the Mullerian and Wolffian ducts (cf. Fig. 2:1, p. 15). All intermediate stages between these types occur. There is no known case of the same animal being able to produce simultaneously both fertile ova and sperm. All genuine hermaphrodites are sterile and, in general, so are pseudohermaphrodites.

It has been found that in human beings certain types of intersexes have deviating sex chromosome constitutions. In 1942 KLINEFELTER described a syndrome complex in men which was manifested after the age of puberty. The testicles were small and spermatogenesis could not be demonstrated, whereas the mammary glands developed and relatively large amounts of gonadotrophins were excreted with the urine ('Klinefelter's syndrome'). Later, other research workers have shown that men with this syndrome, as a rule, have two or more X chromosomes; the most common is XXY but types with $XXXY$, $XXXXY$ and $XXYY$ have also been found. Apparently the presence of two Y chromosomes does not alter the picture of this defect. On the other hand, if an individual has only one X chromosome but no Y chromosome, i.e. XO, development apparently goes in the female direction but as a rule the individual lacks secondary female sex characteristics ('Turner's syndrome'). In mice, however, individuals with XO constitution show normal female development.

The Klinefelter syndrome appears to be fairly common. In an investigation which included 10 725 new-born males it was found that approximately 0·2 per cent conformed to the cytological picture of the syndrome. An XXY individual can arise as a consequence of non-disjunction during meiosis (failure in the reduction division with regard to the sex chromosomes). If, for example, at the meiotic division of a spermatocyte, both the X and Y chromosomes go to the same sperm and this later fertilises a normal ovum, the resulting zygote has the chromosome constitution $XXY + 2A$. The other sperm from the same secondary spermatocyte will not have received any sex chromosome (O); and when this sperm fertilises a normal ovum, the resulting zygote will be $XO + 2A$. In humans the Y chromosome is thought to be the active male-determiner. Zygotes of the type $XXX + 2A$ develop in the female direction but display certain disorders, the so-called 'metafemales'.

A sex chromatin in the somatic cells from various tissues has been demonstrated in several species of mammals, including humans, cattle, goats and pigs. The sex chromatin appears as a rounded, strongly staining body which lies on the inside of the nuclear membrane; it normally occurs in females but not in males. In studies of individuals with excess X chromosomes it has been found that the number of sex chromatin bodies is always one fewer than the number of X chromosomes, and it is assumed that the active substance in the excess X chromosomes is transformed into sex chromatin. Normally the female, like the male, should have only one fully active X chromosome in any cell, though different cells may have different X's active. In the study of intersexes an examination of the sex chromatin can give information as to the extent to which intersexuality is due to alterations in the distribution of chromosomes.

In the literature several cases have been reported of sex reversal in fowls, where the female has changed over to being a male. Initially the hens were fully normal both in appearance and behaviour, including egg laying. Gradually they altered in the male direction, finally producing sperm, mating with females and producing progeny. However, the fertility was very low and the incubation losses high. The progeny consisted entirely of cockerels, including those chicks which died before hatching but were sufficiently developed to be sexed. According to NEWCOMER and co-workers (1960) the spermatogonia from cocks which were previously hens have only one X chromosome, as have normal hens; but they have varying degrees of excess autosomes. In spermatogenesis, some sperm are produced which have an X chromosome and a varying number of autosomes, and other sperm which only have a varying number of autosomes but no X chromosome. It is presumed that only sperm carrying the X chromosome are viable; which accounts for the low fertility. The explanation usually given for a fully grown hen changing sex is that the left ovary is destroyed by a tumour or in some other way, and that the rudimentary right gonad then develops into a functional testicle.

The sex chromosomes are carriers of other genes than those which influence sex differentiation. This is primarily the case with the X chromosomes but in a number of animals and in humans it has been found that some such genes are located also on the Y chromosome. Genes which occur exclusively on the X chromosome are said to show sex-linked inheritance. This phenomenon has long been recognised in poultry but very few cases are known in domestic mammals.

According to the chromosome map for fowls which is given in Fig. 3:10, there are more known gene linkages on the X chromosomes than in any other chromosome pair. This has been of practical importance in sex-linked crosses and in the formation of autosexing breeds. The traits and genes which are important in this respect are barring (B) versus self-colour (b), silver (S) versus gold colour (s) and late feathering (K) versus early feathering (k) in chicks.

Since the cock is homogametic (XX), it can be either homozygous or heterozygous for the X-linked genes, whereas the hen, which is heterogametic, has X-linked genes only in a single dose (hemizygous).

Fig. 3:12 illustrates a cross between a barred hen and a homozygous black cock. In the F_1 barred cocks and black hens are obtained. This is referred to as criss-cross inheritance. After hatching, the difference between the sexes can be seen immediately, since the male chicks that will later exhibit the barred plumage have a light patch on the back of the head and are relatively light coloured on the abdomen and feet, whereas the female chicks have down of an even dark colour. If an F_2 generation is produced then 50 per cent of the birds are barred and 50 per cent are black in both sexes. In the reciprocal cross between the homozygous barred males and black females, all the F_1 birds are barred, and in the F_2 segregation takes place in the ratio 3 barred:1 black, but in this case all the males and 50 per cent of the hens will be barred.

The gene for silver plumage can be used for sexing in crosses with a gold breed (ss), e.g. when the Light Sussex hen (S–) is mated with a cock of the Rhode Island Red or New Hampshire (ss). The male chicks (Ss) are much lighter and have a different pattern from that of the female chicks (s–).

The White Leghorn laying breed have a sex linked gene (k) for early feathering in the chicks, whereas the medium heavy English and American breeds have a corresponding dominant allele for late feathering (K). Early feathering has many advantages since it is correlated with earlier maturity and the chicks are less liable to chilling than those with late feathering. By crossing hens from a breed with late and cocks from a breed with early feathering, pullet chicks are obtained in the F_1 with the early feathering trait and cock chicks with late feathering. The newly hatched chicks can be sexed with about 95 per cent accuracy. Those

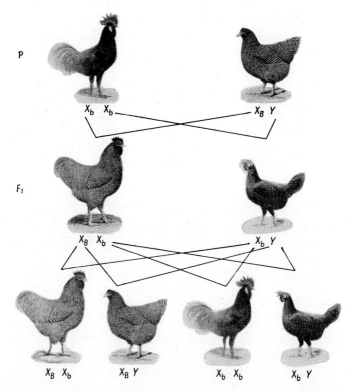

Fig. 3:12. Sex-linked inheritance of barring in the fowl. Crosswise inheritance in the F_1 when a black cock is mated with a barred hen. (*Handbuch der Tierzüchtung*, II. 1959).

cases which are uncertain can be examined again when the chicks are 10 days old. In order to utilise sex-linkage as a method of sex determination the dominant gene must always be introduced with the hemizygous female bird, whereas the male must be homozygous for the recessive gene.

Autosexing breeds have been produced by crossing the barred Plymouth Rock with a 'Gold Breed'. Repeated back crossing is then made between the barred and the non-barred progeny until males are obtained which are homozygous and females which are hemizygous for the barred pattern as well as gold colour. The chicks show a clear sex difference: the males have two B genes and are appreciably lighter than the females. In this way the *Legbar* has been produced by introducing the B gene into the Brown Leghorn and the *Cambar* by introducing it into the Campine (Fig. 3:13). These autosexing breeds, however, have not been particularly popular since they have been somewhat inferior to the highly specialised laying breeds, e.g. White Leghorn.

Sex-linked inheritance has also been demonstrated in turkeys and geese. If the white Italian gander is crossed with a goose from the Scania or Toulouse breeds, the sex-linked colour inheritance can be used for sexing the goslings.

Fig. 3:13. Day-old Cambar chicks with the autosexing barred pattern. The male chick which is homozygous for the barred gene is much lighter coloured than the hemizygous female chick.

Up to now the dog is the only domestic mammal in which sex-linked inheritance has been clearly demonstrated but there are indications that it may occur in horses also. *Haemophilia* in dogs, as in humans, is due to a recessive sex-linked gene.

Sex-linked inheritance ought not to be confused with *sex-limited manifestation* of certain traits, e.g. the milk production of cows and the laying capacity of hens. In so far as these traits are genetically determined, it can be assumed that the responsible genes are as likely to be located on the autosomes as on the X chromosome. It is quite common for characters to be manifested in different degrees in the males and females, e.g. body size, intensity of pigmentation, and temperament. In the red and brown cattle breeds, the Jersey and Ayrshire breeds for example, the bulls are on the average more strongly pigmented than the cows. This sex dimorphism is hormone-dependent; the traits are *sex-influenced*. Even the degree of dominance of certain genes is influenced by the sex. Polledness for example, shows stronger dominance in cows than in bulls (cf. p. 180).

GENE INTERACTION AND PHENOCOPY

Gene interaction can mean either interaction between alleles or interaction between genes at different loci. The first case is concerned with dominance, complete or incomplete, and overdominance when the heterozygotes are superior, in one way or another, to both the homozygotes. The occurrence of overdominance has been clearly demonstrated with respect to a number of oligogenes, and it is assumed to play an important role in such quantitative traits as viability and fertility. It is possible that it is due either to a dosage effect, one gene having a more advantageous effect than two genes of the same type, or to both the alleles complementing each other in some way. When interaction between genes at different loci (epistasis) is present the effect of a single gene depends upon which other genes it co-operates with. For example, homozygosity for the gene A, or its allele a, may have a decidedly more advantageous (or disadvantageous) effect in some combinations than in others. Both dominance and epistasis alike can bring about deviations from the gene's additive effect in a positive or negative direction. It has seldom been possible to pin-point that in domestic animals certain definite gene loci are especially important for non-additive inheritance; but by statistical analysis of suitable material it is possible to get an idea of their contribution to the total variation of certain traits. This, together with the question of the interaction between inheritance and environment will be discussed in more detail in Chapter 4.

It has been shown that, for a number of traits, the same effect can be obtained by certain changes of environment as with changes in the genetic constitution; this is called *phenocopy*. There are many examples of phenocopy in domestic animals, especially congenital malformations. LANDAUER has studied the congenital defect rumplessness in the fowl. In some cases it is genetically determined, in others it can be produced by external influences during incubation of the egg. Genetically determined rumplessness may be due to either a dominant gene or homozygosity for a recessive gene. A similar defect can be obtained by shaking the egg at a critical period of the foetal development, or by injecting certain substances, such as insulin. Using rodents (mice, rats, guinea-pigs, rabbits) as laboratory animals, it has been possible to produce phenocopies of a number of genetically determined deformities by injecting the mother with such substances as insulin, cortisone, nicotine and large doses of sulphonamide during the first half of pregnancy. Deficiencies of certain essential substances, e.g. vitamins, in the nutrition of the pregnant female can bring about deformities in the foetus. A further discussion of this subject will be given in Chapter 8.

Maternal influence

During the intra-uterine development and the suckling period the mother has a great influence on the development of the young, as is manifested in their weight at birth and weaning. A few examples will be given. KOCH and CLARK (1955) made an interesting analysis of data from the U.S. Range Livestock Experiment Station, Montana, comprising 4533 birth and weaning weights of Hereford calves during the period 1926–1951. The data were adjusted—for sex to heifer basis, and for weaning age to a standard of 182 days. Corrections were made for the age of the dam and differences between years and breeding lines were eliminated. The following half sib correlations were then obtained:

	Birth weight	Gain from birth to weaning	Gain from weaning to yearling age
Maternal half sibs	0·26	0·34	0·09
Paternal half sibs	0·09	0·05	0·10

The figures indicate that the maternal environment has a considerable influence on the birth weight of the calves and on their gain from birth to weaning, but little on the gain from weaning to one year of age.

In pigs, the birth weight decreases with the number of young born, and in cattle and sheep the birth weight of twins is 25–30 per cent lower than that of single-born individuals, but the difference tends to decrease with rising level of nutrition of the mothers. With good feeding conditions after birth (and weaning), twins and singles reach the same body size at maturity.

Data are available from reciprocal crosses between breeds of large and small body size. JOUBERT and HAMMOND (1958) made reciprocal crosses between South Devon and Dexter cattle; the live weight of mature bulls and cows of the former breed is about 2·5 times that of the Dexter. The difference in birth weight of the reciprocal crosses was about 10 per cent of the average birth weight of the two parental breeds (mid-parent weight). When the calves had reached 9 months of age the percentage difference in weight was gradually diminishing. DONALD et al. (1962) analysed data on 1015 viable calves in an experiment in which reciprocal crosses were made between the British Friesian (F), Ayrshire (A), and Jersey (J) breeds. The average birth weight of contemporary controls were: F, 39·2 kg; A, 32·9 kg; and J, 22·3 kg, calculated as the weight of heifer calves from primiparous cows. The mean birth weight of the crossbred offspring deviated from the mid-parent in the direction of the maternal birth weight; the deviation was, on an average, 10·6 per cent of the difference between the parental and maternal breeds. This figure indicates the magnitude of the maternal influence on birth weight in the crosses mentioned. Dams of a small breed produce offspring smaller than expected (mid-parent), whilst dams of the larger breed produce offspring heavier than expected. DICKINSON (1960), working on data from the same experiment, studied the development of the crossbred calves until two years of age and found that the difference between the reciprocal crosses disappeared before one year of age. Twelve different measurements and the live weight of the calves were taken at certain intervals and it was found that the more mature characters (at the age of one month) were the least affected by the maternal environment before birth.

Investigations have been made with sheep as to the maternal effect in reciprocal crosses and also when fertilised eggs have been transplanted between large and small breeds. HUNTER (1957) crossed the large Border Leicester and the small Welsh Mountain breeds; the average weight of the ewes in the latter breed is slightly more than 50 per cent of the former. The cannon bone of crossbred lambs from Border Leicester ewes was 0·49 kg heavier and 0·37 cm longer at birth than the corresponding figures for crossbred lambs from Welsh mothers. Comparable figures for the maternal influence when fertilised eggs were transferred from one breed to another were 0·88 kg and 0·33 cm. The maternal influence on the early maturing cannon bone length disappeared before 8 months of age but at this stage 17 per cent of the variation in live weight was still due to maternal influence. DICKINSON et al. (1962) carried out two series of transplantation experiments in sheep. In the first series fertilised eggs were transplanted reciprocally between the Lincoln and Welsh Mountain breeds, and in the second 37 pure-bred Lincoln and 28 Welsh lambs were born as a result of transferring eggs to nulliparous Scottish Blackface ewes. Components of variance were computed and it was found that the lamb's genotype accounted for 72 per cent of the variation in birth weight and 97 per cent of the variation in cannon bone length; maternal influence was responsible for 20 and 1 per cent respectively.

There is no indication in these experiments that the maternal influence during the gestation period has any effect on the body size of the full grown animals.

However, in more extreme crosses the situation may be different. WALTON and HAMMOND (1938) found that in reciprocal crosses between the world's largest and smallest horse breeds, the Shire horse and the Shetland pony, the weight of the foal at birth was three times greater when carried by a Shire mare than when the mother was a Shetland pony, and that this difference persisted when the foals had attained maturity. Similar results have been obtained in reciprocal crosses between the Flemish Giant and the small Polish rabbit. The mature weight of the former is about three times that of the latter breed. When the crossbreds were carried by a doe of the large breed the F_1 animals weighed 533 g more at 10 months of age than when they were carried by a Polish doe

(JOHANSSON and VENGE, 1952). The difference in body weight and skeletal measurements between the reciprocal F_1 groups was more pronounced towards maturity than at birth and weaning.

It has been suggested (WALTON and HAMMOND, 1938) that not only the intrauterine environment but also cytoplasmic influences may be responsible for the rate of foetal growth. In order to elucidate this problem VENGE (1950 and 1953) carried out experiments with the transplantation of fertilised rabbit eggs from donors of a large breed to females of a small breed, and the reverse. In addition, eggs from reciprocal crosses between genetically different colour types were transplanted to females of a medium-sized breed. In the latter case no clear difference could be established between the young of F_1 reciprocals in respect of growth rate or body size at maturity. These experiments, therefore, did not provide any evidence in support of the idea that the offspring inherit more from the dam than from the sire. The most likely explanation of the size difference between the young from reciprocal crosses between large and small breeds appears to be the differences in nourishment during the foetal and suckling periods. The genotype of the offspring is determined at fertilisation and the maternal influence during gestation is not, in principle, different from the environmental influences at later stages.

The occurrence of a number of congenital defects has been found to be correlated with the physiological status of the mother. WRIGHT (1924) found that in a strain of polydactylous guinea-pigs the penetrance and expressivity of the responsible genes were partly influenced by the age of the mother. The frequency of the defect was highest among foetuses born of mothers less than one year old. A similar tendency has been found for various skeletal abnormalities in mice. The frequency of some anomalies of the nervous system and the skeleton in humans tends to increase with increasing age of the mother. To what extent maternal age influences the frequency of congenital defects in domestic animals is not known, but quite certainly the health and nutritional state of the mother are important for the development and viability of the young.

MUTATIONS

There are many known examples of the sudden appearance of variants in the different species of animals, variants which are clearly heritable and which in certain cases have given rise to new breeds. Examples of this are the Angora rabbit, the Rex rabbit, the one-hoofed pig and the Ancon sheep. A number of extreme types of pigeon and dog have no doubt originally arisen in the same way, though they have been further altered by selection during many generations. As far as is known, Ancon sheep first appeared in the United States in 1791 where they gave rise to a new breed. It is no longer in existence. The sheep had normal bodies but unusually short legs (Fig. 3:14) the advantage of which, it was claimed, was that they could not jump over even comparatively low fences. Later the same type appeared also in Norway, where it was described by WRIEDT.

One of the rediscoverers of the Mendelian laws, the Dutchman HUGO DE VRIES, published early in this century a large work in which he pointed out that the changes within species and breeds were sudden and drastic and that such changes appeared with different frequency at different periods of time. DARWIN too had noticed such sudden changes, or 'sports', but he attached greater importance to the small heritable changes, on which natural selection works, for the formation of species and breeds, especially in plants and animals under natural conditions. De Vries based his idea primarily on the study of *Oenothera Lamarckiana* (evening primrose), which he observed both under natural conditions and in his

Fig. 3:14. Ram and ewes of the Ancon breed: the ewe on the left has normal length legs. (*Handbuch der Tierzüchtung*, II. 1959)

own experimental garden. With low but regular frequency there appeared new types, so strikingly different from the mother type that de Vries assumed that they would give rise to new species. He called these changes *mutations* and the new types *mutants*. It has been shown later, however, that these extreme variants were due not to alterations of genes, but in some cases to changes in the chromosome number and in other cases to crossing over between different chromosome complexes. Mutation research had begun, however, and has been progressing ever since.

Mutation in its wide sense means every change in the heritable substance which is not due to segregation or recombination of previously existing genes. It is usual to differentiate between *gene mutations* (point mutations) and *chromosomal changes*, involving either the number of chromosomes or their structure. There is no sharp dividing line between gene mutation and chromosome structure changes, since a certain amount of overlapping occurs between these two

MUTATIONS

categories. The greatest interest in animal genetics has been, at least until now, centred on gene mutations.

A gene mutation occurs at a certain gene locus or a part thereof. A gene can be inactivated, as is the case when gene C, which is responsible for the animal's capacity to form pigment (melanin) in the skin, hair and feathers, mutates to c. Those individuals which are homozygous for the latter gene (cc) become albinos. The gene C can, however, be altered in many different ways so that a series of multiple alleles is built up. In the rabbit at least five different alleles are known which influence the intensity of pigmentation: C full pigmentation, c^{ch} chinchilla, c^m sable, c^h himalayan and c albino (see Fig. 6:3 p. 163). These alleles can be combined in pairs in the individuals and the number of different combinations is

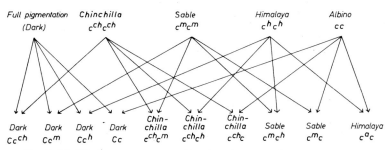

Fig. 3:15. The heterozygous combinations in the allelic series for degree of pigmentation in rabbits. Full pigmentation = CC

$0.5n(n+1)$, where n denotes the number of alleles in the series. The heterozygous combinations are shown in Fig. 3:15; to them can be added the homozygous combinations, which are as many as the number of alleles. In the example shown the alleles have been arranged according to the degree of reduction in pigmentation, each gene being more or less dominant over all the following genes.

In some cases one or more genes may be lost when a piece of the chromosome breaks off during cell division. If this involves only a very small chromosome segment, it can be called a *point mutation*; but if the segment is large, it is more a case of structural alteration of the chromosome (*deletion*). The loss of a piece of the chromosome usually has lethal consequences in the homozygote, and if the piece is relatively large the effect may be lethal already in the heterozygotes. In many cases a point mutation is a reversible process, i.e. if a gene A mutates to the allele a then it can mutate back to A, although as a rule mutations in this direction take place much less frequently. In the majority of cases a new mutant gene is recessive to the earlier alleles, but dominant mutants do occur. Concerning the nature of gene mutations reference may be made to page 84.

Structural chromosomal changes are a consequence of a break occurring at one or more places on the chromosome. In some cases rejoining takes place and if none of the gene loci has been damaged the chromosome behaves normally after the rejoining. In other cases, however, the parts of the chromosome separate from each other (*fragmentation*); in which cases several things can happen:

1. The chromosome may break into two parts, one of which contains the

centromere. The other part lacks a centromere and is therefore said to be *acentric*. The part containing the centromere behaves like a normal chromosome at cell division, whereas the acentric part most likely passes into one or other of the daughter cells and is absorbed there. In this way a greater or lesser part of the chromosome is lost and the result is a so-called *deficiency*. If several breaks occur a middle portion of the chromosome may be lost and, if the outer portions then rejoin, a *deletion* has taken place. In both cases a portion of the chromosome, with all the associated genes, has been lost.

2. The opposite of deletion is *duplication*. This can occur during the crossing over process when a segment lost from one chromosome is added to another chromosome. If a gamete with a chromosome duplication unites with a normal gamete, the zygote has those genes in triplicate which were on the duplicated chromosome segment. A duplicated recessive gene may then show dominance over a single dose of a normally dominant gene.

3. Another possibility is that the separated chromosome segment rotates through 180° and then rejoins in its original place. This is called *inversion*. No genes are lost in this case, but the genes have an altered position in relation to each other, and this can alter their effect. This is called the *position effect* of the genes.

4. In some cases, parts of non-homologous chromosomes can change places with each other, a situation known as *translocation*. Apparently only segments with newly broken ends can join with non-homologous chromosomes. This does not however exclude the possibility of a large segment of a chromosome joining another chromosome which has lost only a small segment.

An inversion or translocation chromosome will not be homologous with the corresponding normal chromosome along the whole of its length; which leads to difficulties at the reduction division. In order that the homologous chromosome parts may be able to conjugate, the inversion chromosome must form a loop, around which the other chromosome must arrange itself as shown in Fig. 3:16. If the loop is large, crossing over can often take place within it. If the chromosomes are *telocentric* (i.e. with the centromere at one end) and crossing over takes place, one chromatid will be without a centromere (*acentric*) whereas the other chromatid will receive two centromeres (*dicentric*). The dicentric chromatid is drawn in two different directions and becomes like a bridge between the two separating chromosome groups; the acentric chromosome fragment lacks the capacity for orientation and can end up anywhere in the cell (Fig. 9:7). Eventually the bridge breaks and the cell divides. Gametes formed in this way are normal in appearance, but they have incomplete gene sets and cannot therefore give rise to viable progeny. Translocation heterozygotes show similar disturbances in the reduction division. The homologous pair build a cross formation where the homologous parts correspond to each other (Fig. 3:16B), and the result is a relatively large number of gametes with incomplete genetic constitution. In the higher plants this lethal effect operates on the gametes themselves, but in animals the effect is usually first apparent in the zygotes, the result being a high frequency of embryonic deaths. These conditions are of great interest to the animal geneticists. KNUDSEN (1958) has reported on inversion and translocation heterozygosity in bulls which produced apparently normal sperm but of low fertility (p. 247).

Changes in the number of chromosomes have been studied mainly in plants and the lower species of animals. When the chromosome number of an individual or a species is a multiple of the haploid chromosome number (n) of its own species or a closely related species and is greater than the diploid number ($2n$), it is customary to call this *polyploidy* ($3n$ = triploidy, $4n$ = tetraploidy). Some species of wheat have 14 chromosomes, others have 28, and our common wheat has 42 chromosomes. All these chromosome numbers are multiples of 7, which is said to be the basic haploid chromosome number of wheat species. The species with high chromosome numbers have no doubt been derived from the primitive diploid species with 14 chromosomes by genome duplication. Usually an increased chromosome number is associated with increased cell size, and thereby a stronger vegetative development of the plant. However, different species react in different ways to genome duplication. If the new chromosome number exceeds the optimum then the viability is reduced, as, for example, in the 84 chromosome wheat.

Other changes of the chromosome number imply a deviation from the multiple of the basic number, i.e. an increase or decrease takes place of one or more

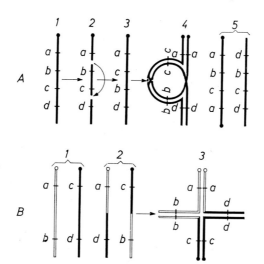

Fig. 3:16. Schematic illustration of chromosomes pairing during meiosis in (A) inversion heterozygotes and (B) translocation heterozygotes.

(A) 1. normal chromosome; 2. two breakages have occurred; 3. the middle piece has reunited with b and c in the reverse order; 4. in conjugation with a normal chromosome a loop has been formed and crossing over takes place within this; 5. the result of the crossing over is that one of the homologues has two centrosomes whereas the other is without.

(B) 1. shows two non-homologous chromosomes; 2. an exchange of segments has taken place between them; 3. shows how these two chromosomes conjugate during the prophase of meiosis. When crossing over takes place, normal or genetically defective gametes may be formed.

chromosomes but not of the complete genome. This is called *aneuploidy*. Alterations in the number of sex chromosomes and their connection with intersexuality in domestic animals have already been dealt with (p. 60). There can also be an excess of autosomes; as a rule, this brings about disturbances in the development of the individual. A number of cases of triplication (*trisomy*) of one of the 22 autosomes of humans have been demonstrated, the effect of which appears to be that the larger the autosome which is triplicated, the more serious is the physical and mental damage. In the so-called Mongolian idiocy (Down's syndrome) one of the smallest autosomes is triplicated.

The spontaneous occurrence of polyploid types has played an important part in evolution in the plant kingdom, and corresponding phenomena have without doubt occurred also in the animal kingdom though very little is known about this as far as the higher animals are concerned. Polyploidy has been demonstrated in the lower vertebrates such as fishes and frogs. The diploid chromosome number of rabbits is 44, but cytological observations indicate that the basic haploid number is possibly 11. According to some investigations the Chinese and European hamster has 22 chromosomes whereas the golden hamster has 44 chromosomes. Whether the golden hamster arose from the European hamster by chromosome doubling is, however, an open question.

Mutation frequency; artificially induced mutations

Mutations can take place both in the somatic cells and in the formation of germ cells. In plants, bud mutations are relatively common, so that mosaics occur, i.e. certain parts of the plant show the mutant trait whereas other parts do not exhibit the mutation. Somatic mutations are also known in domestic animals, an example of which is given in Fig. 3:17. On a Swedish fox farm a platinum fox appeared with a section of one half of its body coloured as in the silver fox. The most likely explanation seems to be that a back mutation occurred in a somatic cell at an early stage in the heterozygous platinum fox ($w^p w$), i.e. w^p mutated to w, with the result that all cells originating from this mutant cell received the gene constitution (ww) characteristic of the silver fox. There are rare cases of a sharply defined black patch appearing in red cattle breeds, and the same explanation is perhaps applicable.

In domestic animals a large number of qualitative characteristics are known which have probably arisen through spontaneous gene mutations. Even when crosses have not taken place, polled animals can appear in horned cattle breeds and this polledness is then inherited by the progeny in the usual way. It has been possible therefore to produce polled types in several of the horned breeds, e.g. Ayrshire and Friesian. During a few decades of fur-animal breeding many mutant colour types have appeared, the majority of which are recessive, but some dominant, to the wild type. About 30 different colour mutations are known in ranch-bred mink; some of these will be discussed later (p. 176). Genetically determined lethal defects have been demonstrated in all types of animals, and it is common for the same defect to appear in different breeds of the same species and even in different species. For example, hairlessness has been observed in nearly all types of domestic mammals; so also have short-leggedness, polydactylism and some other external peculiarities.

With plants and laboratory animals it has been clearly shown that different loci mutate at different rates, some having a high and others a low mutation frequency. This is no doubt also the case with domestic animals. The frequency of spontaneous mutations at different gene loci in domestic animals has not been estimated; but in humans, where accurate registrations have been available, a number of estimates have been made. The normal allele to the recessive, sex-linked gene for haemophilia has been estimated to mutate in about three cases per 100 000 gametes. Other genes mutate only in one case per 1 000 000 gametes. If it is assumed that cattle have 10 000 gene loci and that each of these mutates once per million gametes, then the number of genes which can mutate per individual is 20 000, and 2 per cent of the gametes can be expected to carry a new mutant gene.

A dominant mutant gene can be detected in the first generation, i.e. in the

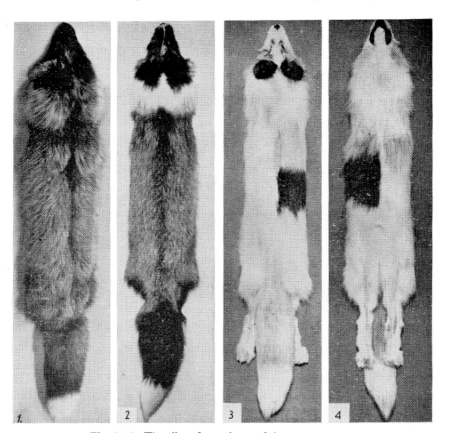

Fig. 3:17. The silver fox and two of the mutant types.
Pelts of (1) a silver fox; (2) a 'white face'; (3) a platinum fox, back view; (4) the same platinum fox, belly view. The dark coloured sector on the platinum pelt is probably due to a somatic mutation. (JOHANSSON 1948)

zygote when a gamete with the mutant gene is fertilised or fertilises. A recessive gene on the other hand can remain hidden in the population for many generations. It is only when two individuals which are carriers of the mutant gene mate with each other that the homozygote can appear and the mutant characteristic be manifested. The mating of close relatives is an effective method of bringing recessive mutant genes to the surface.

Those mutations which have been discussed up to now have been associated with genes having distinguishable effects (oligogenes), manifested in distinct qualitative characteristics. It is very probable that the majority of mutations have so little effect that they are never discovered. The adaptation of domestic animals to their special functions is probably due, not least, to the effect of many 'small mutations' (polygenes) being added by selection.

An increase in the frequency of mutations can be induced in plants and laboratory animals in different ways. In 1927, the American geneticist, H. J. MULLER, reported that he had successfully induced a marked increase in the frequency of mutations in Drosophila by using X-rays. He also developed a special technique for analysing mutations in the X chromosome of Drosophila, and intensive research in this field followed Muller's discovery. The induction of mutations is achieved by *mutagenic agents* such as different types of radiation and certain chemical substances, e.g. mustard gas. Ionising irradiation (X-rays and radio-active radiation) results in chemical reorganisation at the gene loci and/or chromosome breakage with consequent structural changes. It has been shown that the frequency of lethal mutations induced by X-rays is proportional to the radiation dose measured in roentgen units. Tissues with rapid cell division, such as the testicles during the process of spermatogenesis, are especially sensitive to ionising radiation.

It has not been possible to produce mutations at a desired chromosome or locus, but different mutagenic agents have a different effect on different loci, and this makes it possible to direct the mutation process to some extent, at least in plant breeding. Generally, one has to be content with inducing the same mutations as occur spontaneously, and the result is mainly an increased overall frequency of mutations.

Changes in the chromosome number can be induced artificially. Polyploidy can be arrived at either by preventing the reduction division in the formation of gametes or by preventing the completion of cell division after the nuclear division is almost complete. This is achieved by temperature shocks or by certain plant poisons, usually colchicine, that have a paralysing effect on the cell division mechanism. Quite often the viability and fertility of the polyploid types is considerably reduced but these can usually be improved by selection. In plant breeding, good results have already been achieved from the production of polyploids, and it is thought that the method has great prospects in the future.

It may be asked whether polyploid breeding has any possibilities for domestic animals. Some years ago, Swedish research workers reported that they had succeeded in producing triploid rabbits and pigs by inseminating females with sperm to which was added colchicine solution of a certain concentration.

Attempts at several different laboratories to reproduce these results have up to now been unsuccessful. By the use of more advanced methods it has been possible to induce polyploidy of fertilised ova in both rabbits and mice. The polyploid mice embryos developed to about half the foetal stage but no further. No full-term foetuses have yet been obtained.

At present, research is in progress at several institutions in order to test the mutagenic effect of ionising irradiation on domestic animals. An experiment into the genetic effects of paternal irradiation of pigs was started in 1959 at the Iowa State University, U.S.A. The testes of treated males are given a single exposure of 300r X-rays and thereafter the animals are held for at least five months before they are used for breeding. The main object has been to study the effects of sire irradiation on the first generation offspring. The mating plan also provides comparison between progeny of irradiated males with contemporary progeny of untreated males within groups related as double first cousins. At the end of 1964 records were available for about 25 000 pigs from over 2300 litters. No clearly significant effect of irradiation on such quantitative traits as litter size at birth, sex ratio, post-natal mortality or post-weaning growth rate has been found in these comparisons. The effect on genes for red cell antigens (blood groups) and serum proteins has also been studied, but so far no evidence of mutations has been obtained. However, the tests for serum proteins started much later than the other investigations and the data is therefore still somewhat limited. ABPLANALP and co-workers (1964) at the University of California have carried out a long-term experiment with the object of inducing genetic variation in quantitative traits of chickens and then utilising this variation in selection. For twenty years a 'production line' of White Leghorns responded well to selection but appeared to approach a plateau of response (cf. p. 430). One sub-line (X) was irradiated without artificial selection and another sub-line (C) was kept as control. A total of 8000r of X-radiation was administered to mixed semen of the treated line over a period of seven generations with doses of 1000 or 1500r per generation. In each line 40–60 males and a slightly larger number of females were kept. The immediate effect of the irradiation was a pronounced decrease in egg-production and hatchability of the eggs, but egg size was relatively unchanged. After seven generations, the irradiation was discontinued and selection started for higher egg-yield. In two generations egg production and hatchability in the X line were restored to levels only slightly below the C line, suggesting that dominant deleterious genes were eliminated at this stage. Response to selection for egg-production in subsequent generations was about the same in the irradiated and the control lines. It was assumed that the irradiation applied was not sufficient to induce substantial amounts of genetic variation accessible to selection for high egg-yield. The results of full sib matings in the last generation of selection indicated that the irradiated line carried a larger number of recessive deleterious genes than the controls did.

The viability of mutants

Mutant genes, as a rule, lead to a reduction of the viability of the carrier;

structural chromosome changes do so even more. This applies to both spontaneous and artificially induced mutations. If the mutant gene is completely or partially dominant, the effect is apparent in the heterozygote; if it is recessive to its allele, the effect first appears in the homozygote. The degree of impairment varies from complete lethality to an insignificant and hardly discernible reduction of viability. It has already been pointed out that the majority of mutant genes are recessive when they first appear, but that even completely or partly dominant mutants occur. It is possible that dominant lethal genes are responsible for a proportion of the foetal deaths which occur at early stages both in poultry and domestic mammals. A dominant lethal gene eliminates itself automatically in the first generation in which it has a chance to exert its effect.

Fairly often a mutant gene has a dominant effect on traits such as colour or colour pattern with normal viability in the heterozygote, while in the homozygous condition the gene has a lethal or semi-lethal effect.

Mention has already been made of the platinum mutant in the silver fox, which first appeared in Norway in 1933. Another mutant which goes under the name 'white face' was described for the first time in Canada in 1928. Both these mutations are at the same locus and therefore the genes for platinum (w^p), white face (w^f) and silver (w) are alleles. Fig. 3:17 gives an idea of how the three colour types appear. It has been shown that the genes w^p and w^f are not only lethal in the homozygous state but also in the heterozygous combination $w^p w^f$. The platinum gene w^p reduces the viability of the heterozygote (ww^p) to a certain extent, but the white-face gene does not (cf. p. 178). Two similar cases are known in mink.

Some of the recessive colour mutations in mink result in no, or only a very slight, reduction of the viability of the homozygotes, e.g. socklot pastel, royal pastel and silver blue. However, homozygosity for many recessive colour genes result in a distinct lowering of the viability. The dark steel-blue colour mutant 'aleutian' (aa) often shows a disease syndrome ('aleutian disease'), loss of appetite, mucus membrane haemorrhages and anaemia being especially evident. Reproductive disturbances are usual and the death rate of pups is high. Another mutant in the homozygous condition is characterised by white fur but pigmented eyes; it goes under the name 'Hedlund white' with the genetic formula hh. The gene is not completely recessive; for the heterozygotes have white spots of varying area. All Hedlund homozygotes are completely deaf; which results in difficulties for the female taking care of her young. Otherwise the animals have normal viability.

The above examples show that a gene can have more than one effect. When a gene influences two or more traits this is referred to a *pleiotropy*. This phenomenon is also common in the case of 'normal' alleles though it is not manifested quite so clearly.

With regard to the colour types of mink, it has been shown that the effect of lowered vitality due to the mutant genes can, at least partly, be counteracted by several generations of selection for increased viability and relatively good breeding results achieved. The explanation is probably that by selection the mutant genes become associated with an increased number of polygenes which operate in the direction of increased viability and thereby compensate the deleterious effects of the mutations. During the course of generations this has no doubt also taken place in other domestic animals. In a similar way the effects of dominance can also be changed. A dominant mutant gene in an altered gene environment can be recessive or even sub-recessive, as the penetrance is gradually reduced until finally it is zero. In this way, originally deleterious mutant genes, which under certain conditions have a favourable effect, can be protected from elimination from the population by natural selection.

The examples show that even when the immediate effect of a gene mutation is a reduction of the viability of the homozygotes, and in some cases also the heterozygotes, under the prevailing environmental conditions, it is not at all certain that the impairment will be apparent under other environmental conditions or in the long run. The mutant frizzle in poultry (Fig. 3:3) has a loose and open feathering which means that losses of heat from the body at low temperatures or in damp conditions are very much greater than with normal feathering. At relatively high external temperatures, however, this means that the increased heat losses are an advantage. The white breeds of pigs are poorly adapted to warm climatic conditions since in the strong sunlight they are susceptible to sunburn. On the other hand, in regions with little sunshine, the lack of skin pigmentation is an advantage since the pigs are thereby better able to utilise the anti-rachitic effect of sunlight. A mutant gene can have advantageous effects in certain gene combinations in spite of the fact that in other combinations it acts so as to reduce vitality. The process of mutation provides the population with new genetic variation on which natural and artificial selection can operate.

It is to be expected that the majority of mutant genes should have a disadvantageous effect when they first appear. During many generations the gene pool of species and breeds has been balanced so that it is adapted to the prevailing environmental conditions. Changing a gene vital to life, or loss of a block of genes (deletion and deficiency), lead to a disturbance of this balance which easily impairs, or makes impossible, the normal course of the vital processes. By a process of trial and error, evolution is moulding new breeds and species. So far as domestic animals are concerned, it is man himself who, to a relatively large degree, determines the rapidity and direction of the alterations.

GENES AND GENE FUNCTION

When WILHELM JOHANNSEN, about sixty years ago, coined the term gene he maintained that it implied nothing about the nature of the genes but that the term 'is simply a very useful little word which can be used instead of unit character, allelomorph, etc.'. Genotype was defined as the sum of all genes in a gamete or zygote and phenotype as the manifested character which can be observed and measured. About 1930, a definition of the gene concept had been arrived at, primarily due to the work of MORGAN and his co-workers with *Drosophila*. It can be expressed in the following way: 'The gene is that segment of a chromosome which always behaves as a unit in mutation and crossing over.'

It was shown later that this definition was not valid, since both mutations and crossing over can occur at different places within the same gene locus or segment. Biochemical research has thrown new light on the question of the nature and function of the genes.

The real break-through in this line of research came about 1940 with BEADLE and TATUM's investigations with the bread mould, *Neurospora*. Mutations were induced by ultraviolet or X-ray irradiation and it was shown that many mutants were incapable of synthesising either a certain vitamin or a certain amino acid, which therefore had to be added to the nutritive substrate to enable the mould to grow. Some mutants lost the capacity to synthesise the amino acid tryptophan which others can synthesise from sugar and ammonia. A number of different enzymes are required for the synthesis, each enzyme making possible a certain stage of the process. The expression 'one gene, one enzyme' was therefore coined to explain that one gene controls the formation of a specific enzyme which is responsible for a given stage of the synthesis or for the breaking down of more complex organic compounds to simpler substances.

At about the same time investigations into the genetics of bacteria and viruses were commenced. Mutations were induced in different species of bacteria in the same way as with Neurospora and with similar biochemical results; a number of mutants were shown to have lost the capacity to synthesise certain substances. It was also found that a number of mutants had increased, and others had reduced, resistance to antibiotics such as penicillin and to sulphonamides. After the addition of an antibiotic to a culture medium the resistant bacteria grew much faster than the non-resistant and soon became predominant; this applied to both the spontaneously appearing as well as the artificially induced mutations. By mixing strains of bacteria, which contained different biochemical mutants, it was possible to show that recombination of genes took place between the strains and that the genes in each strain constituted a single linkage group. It was therefore deduced that the recombination must be due to 'crossing over' in connection with some kind of sexual process between the different strains of bacteria. With the aid of the electron microscope it was possible to study this process in detail. The nature of the sexual process is that in certain cases bacteria of different genotypes conjugate in pairs when one bacterium (the 'male') transfers to the other bacterium (the 'female') a larger or smaller fragment of a chromosome together with the associated genes. One bacterium can have several cell nuclei, but only one chromosome from a nucleus takes part in the process. This agrees well with the fact that only one linkage group of genes could be demonstrated in the species studied.

The really great discovery was the demonstration that the genetically active substance in bacteria, and as a rule also in viruses, consisted of deoxyribonucleic acid (abbreviated to DNA). If a sample of dead and disrupted ground-up bacteria of a particular strain, or the filtrate of such a sample, is added to a pure culture of a different bacterial strain of known genotype, then the bacteria can be genetically altered by taking up and incorporating DNA from the sample or filtrate. The transfer and incorporation of DNA in the genetic mass of the bacteria has been demonstrated by labelling DNA with radio-active phosphorus.

DNA can be transferred to living bacteria by viruses which attack the bacteria. Virus particles are too small to be observed under a normal microscope but their appearance can be seen with the aid of the electron microscope. The viruses under discussion here consist of a 'head' and a short 'tail'. The head has an outer coating of protein and an inner nucleus consisting of one molecule of DNA. When the virus attacks a bacterium, the tail is bored into the victim and through this DNA is injected. The bacterium is then 'forced' to produce new DNA molecules of the same type as that received and to supply the protein coating and

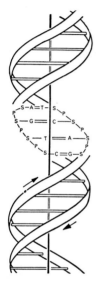

Fig. 3:18. The Watson and Crick model of the structure of DNA.

The two spirally twisted strands consist of 'antiparallel' polynucleotide chains held together by the double hydrogen bond between the constituent bases. P = phosphate group, S = sugar unit (ribose), A = adenine base, T = thymine base, G = guanine base and C = cytosine. The arrows indicate that one polynucleotide chain runs in the opposite direction to the other chain with respect to the sugar-phosphate linkages. (BEADLE 1962, redrawn from WATSON and CRICK 1953)

tail. This procedure continues until the bacterium bursts and the new virus particles are released. These virus particles can, however, carry parts of the original DNA of the bacteria and later they inoculate them into other bacteria which thereby become genetically altered. This process is called *transduction*, the details of which have been elucidated by modern isotope techniques.

Even viruses can mutate, and different virus particles can exchange broken pieces of their DNA molecules with each other (crossing over). It has been possible to make chromosome maps for viruses in the same way as for bacteria and multicellular organisms.

It can be assumed, on very good grounds, that in all higher living organisms the basis of the hereditary material consists of deoxyribonucleic acid (DNA), which always occurs in the nucleus and sometimes also in other parts of the cell. A closely related substance, ribonucleic acid (RNA), is distributed throughout the cell both in the nucleus and in the cytoplasm. The function of RNA will be briefly mentioned later. It has been shown that bull's spermatozoa contain exactly half as much DNA as the diploid body cells, as would be expected on account of the reduction division during spermatogenesis.

It was on the basis of investigations with bacteria and viruses, as well as studies on the structure of DNA, that the American geneticist, JAMES D. WATSON, and the English physicist FRANCIS H. C. CRICK were able, in 1953, to construct their model of the DNA molecule. According to this model the DNA molecule consists of two parallel chains of nucleotides twisted together in a spiral formation (Fig. 3:18). Each nucleotide consists of a phosphate radical, a sugar (ribose) and a nitrogenous base. The corresponding nucleotides in the two chains are held together by double hydrogen bonds between the bases adenine and thymine or guanine and cytosine to form a nucleotide pair. The DNA molecule has been likened to a spiral staircase where the chains of phosphate-ribose make up the supporting sides, and the base pairs, adenine-thymine or guanine-cytosine, form the steps which hold the two spirals together. In Fig. 3:18 the bases are indicated by their initial letter: adenine = A, thymine = T, guanine = G and cytosine = C. Since adenine is always bound to thymine and guanine always bound to cytosine, there are four possible combinations, A-T, T-A, G-C and C-G. The nucleotide pairs can occur in the DNA molecule in all possible sequences and in all possible proportions and, since the number of nucleotide pairs in each molecule is very large, the number of possible combinations is immense. The DNA molecules can have different lengths; it has been estimated that the number of nucleotide pairs occurring in the virus T4 is of the order of 200 000. It is the sequence of the nucleotides which determine the genetic information, and the possibilities for storing information are almost beyond comprehension. The number of nucleotide pairs in a fertilised ovum from humans has been estimated to be of the order of 5 milliards which would correspond to the number of words contained in 1000 printed volumes of 1000 pages each with 500 words per page. This genetic information regulates and directs the development from the fertilised egg to maturity under normal environmental conditions. If this amount of DNA was transformed to a single thread it would have a length of about 1·5 metres but the diameter would be only 2·5 millimicrons.

Watson-Crick's model makes it easier to understand how the genes can reproduce themselves at each mitotic division without losing their identity. Due to the fact that adenine is always bound to thymine and cytosine always bound to guanine both the nucleotide chains in a DNA molecule are complementary with respect to the base sequence. They are separated from each other at the lengthwise division of the chromosomes by breaking of the hydrogen bonds, after which each chain must organise exactly similar partners as they had previously; the substances for this being obtained from metabolic products of the cell. The process is illustrated in Fig. 3:19, which shows a segment of DNA with four nucleotide pairs which split and re-synthesise new partners exactly like those occurring in the molecule before the separation occurred.

The way in which the genetic information is passed on to the cytoplasm has been the object of detailed investigation but only very brief details can be dealt with here.

It has been previously mentioned that ribonucleic acid (RNA) occurs both in the nucleus of the cell and in the cytoplasm. RNA consists of a single chain of nucleotides (phosphate–ribose–base), where the ribose group has one more oxygen atom than DNA and the base thymine is always substituted by uracil which lacks the CH_3 group in thymine. The bases may occur in various sequences, e.g. in a RNA segment A–U–C–G– which corresponds to A–T–C–G– in a DNA segment. Three types of RNA can be distinguished, messenger RNA which is abbreviated to mRNA, transfer RNA abbreviated to sRNA and RNA in the ribosomes, shortened to rRNA. All three are formed by the 'copying' processes of special sections of the DNA of the cell nucleus. Transfer RNA occurs in the cytoplasm with distinctly smaller molecules than mRNA and is often called

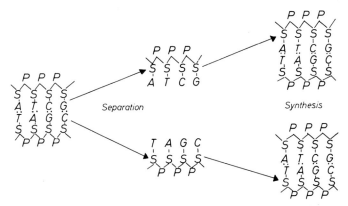

Fig. 3:19. Schematic illustration of the separation of the DNA chains by breaking of the hydrogen bonds between the constituent bases together with the replication by pairing with complementary nucleotides in the reverse order.
P phosphate, S sugar (ribose), ... hydrogen bond between bases.

'soluble RNA' (= sRNA). The ribosomes are small bodies associated with a network of 'threads' in the cytoplasm and can only be observed with the aid of the electron microscope (Fig. 3:20). Ribosómes always consist of two parts, the one being about twice as large as the other. It is on the ribosomes that protein synthesis takes place.

The protein molecules are built up of about 20 different amino acids. Transfer RNA picks up these amino acids in the cytoplasm and carries them to the template of the ribosomes to which coded mRNA is attached. The mRNA is produced by the DNA in the cell nucleus and passes out into the cytoplasm to attach itself to the ribosomes. There is at least one specific sRNA for each amino acid. Each amino acid-specific sRNA has a code which corresponds to a definite code in the mRNA's template so that the right amino acid can be fitted in the right place. This system, which is regulated by the stored information in DNA, is called the genetic code. It has been found that the genetic code is built up of three 'letters' i.e. the base sequence in three successive nucleotides. The code ATT in one of DNA's nucleotide chains corresponds to AUU in mRNA and this has been shown to be the code for the amino acid tyrosine, GGU specifies tryptophan, UGU specifies valine, etc. The code is thought to be the same for all

Fig. 3:20. Electron micrograph of ribosomes in a human submaxillary gland cell. The cytoplasm is pervaded by a membrane system on which the ribosomes appear as small, dark particles. ×70,000. (ALLFREY and MIRSKY, *Scientific American*, Sept. 1961)

organisms from bacteria to mammals. The code system for protein synthesis is illustrated in Fig. 3:21.

The ribosomes are genetically unspecific and can consequently take part in the synthesis of any protein. It is only when the template mRNA is attached to the ribosomes that they receive the order to conduct the synthesis in a particular way. The polypeptide chains are built up in stages by first one amino acid and then another being inserted. When the chain is completed it is released from the ribosomes, which can then begin building a new polypeptide chain of the same type or, if they receive a new mRNA molecule, another type. It is assumed that the ribosomes form groups probably four or five in each group, which is held together by a single mRNA molecule.

According to WAGNER and MITCHELL (1964) the gene concept may be defined as follows: 'A gene is that segment of DNA which determines the base sequence of nucleotides in mRNA which make up the code for a certain biological function, e.g. the synthesis of an enzyme or protein, or, more generally, a segment of DNA which controls a given process in the formation of the individual's phenotype.' The term 'gene locus' can be used in the same way as earlier, but different mutations can take place at the same locus; one refers therefore to 'mutational sites', or 'mutons', within the genes. The number of 'mutational sites' within a given gene can be very large and the gene is no longer regarded as a unit in 'crossing over'. If the same muton mutates repeatedly, but in different ways, so-called homoalleles are obtained; but when different mutons in the same gene mutate, heteroalleles are obtained. The large number of alleles at certain blood group loci may be heteroalleles. The gene is considerably more complex than was first imagined, and it is very often uncertain where the line should be drawn between different genes on the same chromosome. It is comparatively easy to draw the line between genes which regulate the synthesis of a certain amino acid in *Neurospora*, but it is more difficult, for example, with the eye colour of Drosophila. When the term gene is used in the following discussions it will be understood to mean the smallest unit (DNA segment) which shows Mendelian inheritance; whether mutation can take place at different 'sites' within such genes will be left as an open question.

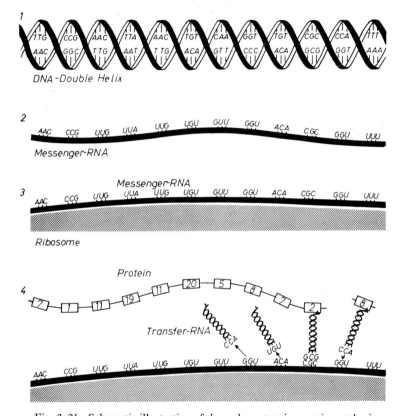

Fig. 3:21. Schematic illustration of the code system in protein synthesis.
1. Double spiral of DNA with the code triplets specified (above).
2. Messenger RNA is formed by the copying of one spiral strand of DNA commencing from the left but in doing so thymine (T) in DNA is replaced by uracil (U) in mRNA.
3. Messenger RNA has moved to a ribosome in the cytoplasm.
4. Coded transfer RNA (the nucleotide chains are bent like a hairpin) has attached itself to a particular amino acid which is conducted to its place on the template of the ribosome (mRNA) on which are built up the polypeptide chains.
In the illustration the code triplets are shown separated from each other for the sake of clarity; actually they form a continuous series with the same distance between the bases. (After NIRENBERG, *Scientific American*, March 1963)

From a biochemical point of view, 'mutation' means an alteration of the sequence of nucleotides within a DNA segment; and the smallest imaginable 'mutational site' is a nucleotide pair. The alteration may consist of additions, subtractions, substitutions, inversions of one or more nucleotide pairs or whole DNA molecules, and even more complicated phenomena. The real (intrachromosomal) mutations probably take place in connection with the separation of the DNA chains and the DNA synthesis which immediately follows at the commencement of meiotic cell division. Mutations have been likened to typographical errors in a book; the information can be essentially altered even with relatively small printing errors.

There is evidence to show that, on the basis of their effect, the genes can be divided into two groups, structural genes and controlling genes. The former determine which enzymes or proteins shall be produced, and the latter regulate the rate of the production at each particular stage—they can initiate, accelerate, retard or stop the production.

Since all the nucleus-bearing cells of the body have the same complement of chromosomes and DNA, the high degree of specialisation of the cells in different tissues and organs of the same individual can appear to be rather puzzling. This specialisation takes place with great precision, mainly during the foetal stages, and in an accurate sequence; more serious disturbances are relatively rare. The sequence of the specialisation is to a great extent a consequence of the fact that not all of the DNA information is operating at the same time. In amphibians, for example, it has been found that different segments of the chromosome are active at different periods of time. The activity sequence of the genes is apparently also genetically determined. The genes that operate during the first months of an individual's life are not necessarily the same as those which are operative later. Congenital defects, which occur now and then in both animals and man, may be due to disturbances in the genetic information which is responsible for regulating the sequence of differentiation of tissues and organs.

BIOCHEMICAL GENETICS

In the first decade of this century the idea had already been suggested that the metabolism of the cells and the development of the individual from the fertilised egg was principally directed by enzymes which are genetically determined not only with regard to specificity but also with respect to the sequence and the qualitative effect of their influence. Among the pioneer work in this field may be mentioned SEWALL WRIGHT'S investigations into the colour inheritance of mammals and Goldschmidt's studies of the intersexuality of the gypsy moth, *Lymantria dispar*. This branch of research was given the name *physiological genetics*. During the past quarter of a century, biochemical methods have been used more and more in the study of the nature and function of genes and the term *biochemical genetics* is now used with about the same meaning as the earlier expression 'physiological genetics'. Some examples will now be given of the application of biochemical genetics in humans and domestic animals.

The amino acid tyrosine is a constituent of proteins and is a precursor in the body's production of both melanin pigment in the skin and hair and the hormone thyroxine in the thyroid gland. Tyrosine can be produced in the body by the oxidation of a closely related amino acid, phenylalanine, which must always be included in the diet. Genetic blocks of different kinds can occur in the metabolism of these amino acids if the enzyme responsible for a definite stage of the metabolism is absent or the amount present is too small. Fig. 3:22 shows examples of four different defects in amino acid metabolism, of which at least three—and probably even the fourth—are due to homozygosity for recessive genes.

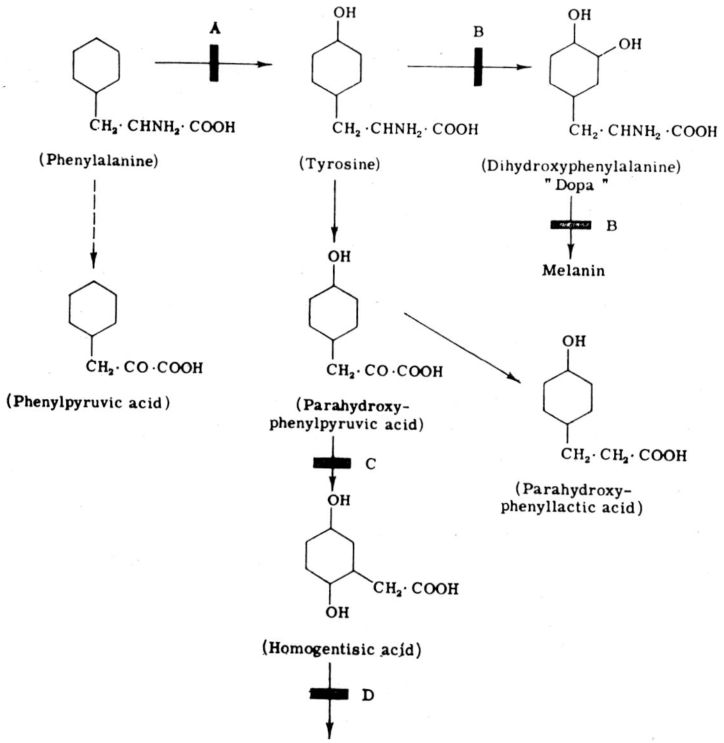

Fig. 3:22. Genetically determined blocking of the metabolism of the amino acids phenylalanine and tyrosine in man.
(A) Blocking of the oxidation of phenylalanine to tyrosine which results in *phenylketonuria* when the diet contains an excess of phenylalanine. (B) Blocking of the oxidation of tyrosine and 'DOPA' which results in *albinism*. (C) and (D) Blocking of the break-down of tyrosine to simpler compounds. If the blocking takes place at (C) *tyrosinosis* occurs; blocking at (D) results in *alkaptonuria*.

As a rule, the food contains more phenylalanine than the body requires for protein synthesis, and the excess is therefore normally oxidised to tyrosine. If the enzyme phenylalanine hydroxylase, which catalyses this process, is absent there is an accumulation of phenylalanine and phenylpyruvic acids in the body. These substances are excreted with the urine and the defect is known as *phenylketonuria*. The concentration in the blood can be fifty times greater than normal. These large amounts of phenylalanine act as a poison to the brain and nervous system, and children with this metabolic disorder therefore will be mentally defective (feeble-minded) unless the content of phenylalanine in their diet from a few weeks of age is reduced to a very low level.

In albinism it is the enzyme tyrosinase that is lacking. This must be present for the amino acid tyrosine to be converted to dihydroxyphenylalanine, abbreviated to DOPA, and also for the next stage in the conversion of DOPA to melanin. A number of other genes can later bring about changes in the melanin synthesis which result in different colour types of individuals. If tyrosinase is lacking, no pigment is formed at all.

Excess tyrosine is broken down into simple compounds, but even here genetically determined blocks can occur. The block which is indicated by C in Fig. 3 : 22 has been demonstrated in only a few cases; the blood becomes overloaded with *parahydroxyphenylpyruvic* acid, and the result is seen in symptoms similar to scurvy (tyrosinosis). Blocking at D is more common. The homozygotes are incapable of breaking the benzene ring in alkapton (homogentisic acid) which is therefore excreted in the urine. If urine containing alkapton is allowed to stand exposed to air for a few hours, it becomes inky black. This defect, which is not particularly harmful, is called *alkaptonuria*. With increasing age the cartilagenous tissues become dark coloured and a form of arthritis usually appears.

It is not only in the decomposition of amino acids, or in their synthesis to more complicated compounds, that a certain stage of the reaction can be blocked by a gene mutation. In some cases a disease of infants occurs which is called 'galactosaemia'. The child cannot convert the simple sugar galactose, a component of milk sugar, into glycogen because of the absence of a specific enzyme. A certain amount of the galactose is excreted with the urine but even so the blood contains a very high content so that the development of the child is affected; he becomes mentally deficient and other defects may also occur. If galactose is eliminated from the diet, the child develops quite normally.

Several genetically determined metabolic disorders are known in domestic animals, though the conditions pertaining to these have not been studied in such detail as with humans. Partial or complete albinism occurs in fowls, turkeys, rabbits, mink and cattle; the inheritance is always recessive to the inheritance for pigmentation. Congenital porphyria occurs in cattle and pigs. Porphyrine is formed in the break-down of chlorophyll and other closely related plant pigments; it is a normal constituent of haemoglobin. In porphyria the excess of porphyrin cannot be broken down and the substance is excreted with the urine, to which it gives a reddish-brown colour. The blood contains a higher than normal content of porphyrin, which accumulates in the skeleton and the teeth; and these become discoloured. In pigs the defect is due to a dominant gene of varying expressivity. The disorder manifests itself in newborn pigs, which are carriers of the gene, as colouration of the eye-teeth. During growth the defect can either disappear or be intensified, in which case the value of the carcass is reduced on account of the discolouring of the skeleton. The growth rate and vitality of the pigs is affected only in exceptional cases. When the same defect occurs in cattle, they show an increased sensitivity to light and, if they are subjected to direct sunlight, serious injury occurs in those parts of the skin which are not protected by hair or pigmentation, such as the skin around the orifices of the body and along the midback where the hair is parted. The defect, which occurs in Shorthorn cattle in Denmark and South Africa, has been shown to be the result of homozygosity for a recessive gene. A similar defect occurs in sheep and is also due to a recessive gene. The sensitivity to light is due to an accumulation of phylloerythrine in the blood.

One of the best examples of the sequence of amino acids in the protein molecule being genetically determined has been obtained from the study of haemoglobin variations in humans. The haemoglobin molecule consists of four polypeptide chains of two different types, usually indicated by α or β. Each chain consists of about 140 amino acids together with a porphyrin ring and an atom of iron. Apart from the normal type, indicated by HbA, some 30 deviant types are known, one of which is the 'sickle-cell' type HbS. Individuals homozygous for the 'sickle-cell' gene suffer from a severe form of anaemia; the haemoglobin has a lower solubility and the red blood cells, when deprived of oxygen, assume a sickle-like shape and are readily haemolysed. Only one-fifth of these individuals reach the age of puberty. Both HbA and HbS are present in heterozygotes but the capacity of the blood to transport oxygen is very little reduced and the viability is normal. The heterozygotes have, however, a distinctly greater resistance to malaria than non-carriers of the sickle-cell gene. This is of importance in areas of Africa and Southern Europe where the sickle-cell gene has reached an appreciable frequency. None of the other haemoglobin types studied show sickle-cell characteristics, but may cause other types of anaemia. A deviant type of haemoglobin, HbC, depends on a gene which is an allele of HbS. In HbC as well as HbS both the β chains are defective. One amino acid is replaced by another at a definite place in the chain as shown in the following:

HbA: valine-histidine-leucine-threonine-proline-*glutamine*-glutamine-lysine
HbS: valine-histidine-leucine-threonine-proline-*valine*-----glutamine-lysine
HbC: valine-histidine-leucine-threonine-proline-*lysine*-----glutamine-lysine

A substitution of a single amino acid in the β chain has thus a tremendous influence on the characteristics of the haemoglobin.

An important factor in the metabolic disturbances just discussed is that the genes, although in general regarded as recessive, have some effect in the heterozygotes and the 'carriers' can in many cases be detected. It has been shown with respect to phenylketonuria that the heterozygotes break down orally administered phenylalanine much more slowly than do persons who are free from the defective gene. This can easily be established by determining the phenylalanine content of the blood at regular intervals after ingestion of the substance; the difference usually is apparent after about four hours. The same technique can be used to detect carriers of the gene for galactosaemia, though the reliability of the test is somewhat less. Heterozygosity for the deviating haemoglobin types can be demonstrated by electrophoretic investigations of the haemoglobin from blood samples. In the cases of other defects use is made of tissue cultures, where the metabolism of the cells can be studied with refined biochemical methods.

Biochemical genetics covers the whole field of the metabolic and immunity reactions of the body. The importance of this branch of genetics to practical animal breeding will undoubtedly increase in the near future. It can be assumed that many metabolic disorders and the capacity to resist infections are genetically determined. In addition, the quality of the fats and proteins in milk and meat depends upon their chemical composition, which is essentially genetically determined. Consequently biochemical genetics will probably be of considerable importance for research into the quality of animal products. During the past ten years great progress has been made in the study of blood groups of domestic animals as well as serum and milk proteins. A review of the results achieved up to now will be given in Chapters 7 and 10.

SELECTED REFERENCES

BEADLE, G. W. 1963. *Genetics and Modern Biology* (73 pp.). Philadelphia.
DONALD, H. P., RUSSEL, W. S. and TAYLOR, ST. C. S. 1962. Birth weights of reciprocally cross-bred calves. *J. Agr. Sci.*, **58**: 405–412.
HUTT, F. B. 1964. *Animal Genetics* (546 pp.). New York.
SRB, A. M., OWEN, R. D. and EDGAR, R. S. 1965. *General Genetics*, 2nd ed. (557 pp.). San Francisco and London.
STENT, G. S. 1963. *Molecular Biology of Bacterial Viruses* (474 pp.). San Francisco.
WAGNER, R. P. and MITCHELL, H. K. 1964. *Genetics and Metabolism*, 2nd ed. (673 pp.). New York.

4 The Principles of Population Genetics

A *population* is a group or collection of individuals with certain traits or characteristics in common, e.g. a breed of cattle or a group within the breed which as a breeding unit is more or less isolated from the breed in general. Population genetics makes use of mathematical-statistical methods in analysing the structure of such breeding units at a particular time as well as the changes which take place under the influence of known causes, i.e. the dynamics of the population are studied.

It has been mentioned previously that one can distinguish between *qualitative* and *quantitative* characters. The difference between the two categories is that typical qualitative traits, such as polled or horned are mutually exclusive, whereas quantitative traits—such as milk yield or withers height—show continuous variation, with all values between the extremes (cf. p. 92). Quantitative characters must be measured in order to study their variation and the results of an analysis depend therefore to a certain extent on the accuracy with which the measurements can be made. The number of pigs born per litter are included in the category of quantitative characters although the variation is in fact discontinuous. The range of the variation is rather large and can be analysed in the same way as, for example, the variation in withers height of horses.

Certain characters have a genetic background comparable with quantitative characters but manifest themselves as alternatives in the same way as qualitative characters. Resistance to certain diseases is an example of this if measured in terms of death or survival, and such characters present special problems in genetic analysis. In presenting the principles of population genetics it is convenient to begin with qualitative characters, for which it is possible to demonstrate the presence of defined genes from generation to generation; but first consider some definitions.

SOME STATISTICAL CONCEPTS

Probability

Practically everything we embark upon is based more or less on a conscious consideration of the probability that something will or will not happen. We can never be absolutely sure about what will happen or why a certain event occurs. We often attribute things to *chance*; which simply means that the causal conditions are so complicated that an individual case cannot be analysed but nevertheless they follow certain laws which can be revealed by studying a large number of random events.

When we spin a coin the chances are equally as great that it will come down heads as tails. The probability of each of these two occurrences is one half. If we throw a die, the probability that we turn up any particular number of the six possible is one-sixth. If we pick a marble out of a box which contains 20 black, 30 white and 50 red marbles, all identical except for the colour, then the probability of picking a white marble is 0·3 and so on. The concept of probability is usually defined in the following way: if an event can occur in n mutually exclusive and equally likely ways of which a have a certain property X, the probability that X will be found is a/n. The sum of the probabilities of an event's happening and of its not happening is always equal to 1.

If two events are independent, the probability that both will occur is the product of their individual probabilities. If a coin is spun twice the probability of getting two heads in succession is therefore $0·5 \times 0·5 = 0·5^2 = 0·25$, getting three heads in succession $0·5^3 = 0·125$, and so on. If two coins are spun simultaneously there are four possible combinations all equally likely, head + head, head + tail, tail + head and tail + tail. Two of these possible combinations are however equivalent and one can say therefore that the combination of head plus tail has twice as great a change of occurring as each of the other two combinations. The probabilities of the different combinations can be calculated in the following way: head + head = $0·5 \times 0·5 = 0·25$; head + tail = $2 \times 0·5 \times 0·5 = 0·5$ and tail + tail = $0·5 \times 0·5 = 0·25$. The probability of obtaining at least one head in spinning two coins simultaneously is obtained by adding the individual probabilities for head + head and head + tail, i.e. $0·25 + 0·50 = 0·75$.

The above examples have a direct application to Mendelian inheritance. It has been previously mentioned that several species of mammals give birth to more males than females: the sex ratio is therefore somewhat higher than 0·5. The deviation is however quite small and can be disregarded here. It can be assumed therefore that the probability of a calf being a heifer is equal to the probability of it being a bull i.e. one-half. If a cow during her lifetime gives birth to 6 calves the probability for each of the 7 possible sex combinations is as follows (F = female, M = male):

6 F	5 F + 1 M	4 F + 2 M	3 F + 3 M	2 F + 4 M	1 F + 5 M	6 M
1/64	6/64	15/64	20/64	15/64	6/64	1/64

These probabilities are obtained by developing the binomial $(\frac{1}{2} + \frac{1}{2})^6$ or in generalised form, $(p+q)^n$, where p is the probability of the one event, q the probability of the other event, and n the total number of events. A distribution of this type is said to be *binomial*. Two more examples may be mentioned. If dominance occurs, e.g. in mating between black animals which are heterozygous for red colour, then the expected proportion of the progeny are 3 black : 1 red. If only 4 progeny are obtained there are 5 different possible combinations and the probability for each of these can be calculated by developing the binomial $(0·75 + 0·25)^4$:

STATISTICAL CONCEPTS 91

4 black	1×0.75^4	$= 0.316$
3 black + 1 red	$4 \times 0.75^3 \times 0.25$	$= 0.422$
2 black + 2 red	$6 \times 0.75^2 \times 0.25^2$	$= 0.211$
1 black + 3 red	$4 \times 0.75 \times 0.25^3$	$= 0.047$
4 red	1×0.25^4	$= 0.004$
	Total	1.000

The probability of obtaining exactly 3 black and 1 red is 0·422 and the probability of obtaining 4 red calves is only 4 in 1000. However, one should be prepared for the latter event happening at some time.

The coefficients for the binomial distribution, i.e. 1, 4, 6, 4 and 1 in the above calculation can be obtained for small numbers quite easily from Pascal's triangle. A particular coefficient is always equal to the sum of the two coefficients immediately above to the right and left. For the first 8 values of n the triangle takes the following form:

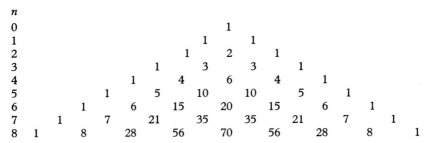

Finally let us assume that we mate two heterozygotes ($Aa \times Aa$) with each other and that all 3 genotypes in the progeny generation can be phenotypically identified. The probability for the different types is thus $0.25\ AA + 0.50\ Aa + 0.25\ aa$. A rather large number of individuals is required if these proportions are to be realised exactly. The probability for each of the possible combinations can be obtained by developing the expression $(0.25 + 0.50 + 0.25)^n$, where n is the number of progeny. If the number of offspring is only 4, then the probability of obtaining exactly the proportion $1\ AA + 2\ Aa + 1\ aa$ is 0·19. This combination of progeny is the most probable of the 15 possible combinations. With a small number of individuals therefore the proportions obtained for a particular type can deviate widely from that expected on the basis of a particular genetic hypothesis without necessarily invalidating the hypothesis. Statistical methods have been developed by which it is possible to test whether the obtained frequencies deviate from the expected frequencies more than can be attributed to chance. Examples of these statistical analyses are given on page 98.

Variation

Most quantitative characters show continuous variation and in unselected biological material the distribution generally follows the *normal curve* quite closely. Measurement of the variation about the mean involves the concept of *standard deviation* which, for populations, is represented by the Greek letter sigma (σ) and for samples of the population by s.

In a normally distributed population $s = \sqrt{\dfrac{(x-\bar{x})^2}{N-1}}$, where $x =$ the variable or variate; this may take any one of a specified set of values, $x_1, x_2, \ldots x_i$, $\bar{x} =$ the mean, $N =$ the number of variates, $N-1 =$ the number of degrees of freedom and $\Sigma =$ summation sign. In a binomial distribution $s = \sqrt{\dfrac{pq}{N}}$, where p, q and N are as defined earlier.

The concept of *degrees of freedom* occurs several times in what follows, and we shall therefore briefly explain its meaning. Let us assume a group of 10 variates

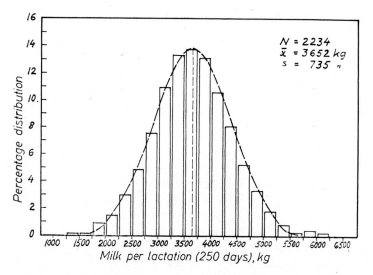

Fig. 4:1. The distribution of butter fat production in the first lactation (250 days) of 2234 cows of the Red Danish breed compared with the normal distribution.

with mean $\bar{x}$ and total $= \Sigma x$. This group is then bound by the condition that $\Sigma x = N\bar{x}$. Nine of the variables $(N-1)$ may take any value but the value of the 10th variate is fixed by Σx and the other 9 variates. We say then that this group of variates has 9 degrees of freedom.

In a normal distribution 68·3 per cent of the variates lie within the limits $\pm \sigma$ from the mean, 95·4 per cent lie within $\pm 2\sigma$ and 99·7 per cent within $\pm 3\sigma$. Fig. 4:1 shows the variation in milk production in the first lactation period (250 days) for 2234 heifers of the Red Danish breed, recorded at heifer testing stations during the years 1948–1952. The mean production ($\bar{x}$) was 3652 kg of milk and, the standard deviation (s) about the mean was 735 kg. The class intervals in the histogram correspond to 250 kg milk. The distribution of the different production classes has been smoothed out into a normal curve and it is apparent that agreement with the normal curve is very good.

The two most usual types of deviation from the normal distribution are excess and skewness. The former case is characterised by an excess of variates in the centre and outer classes. In the asymmetrical, or skewed, distribution the mean and median are different. Skewness is either positive or negative according to whether the median lies above or below the mean. Many of the methods which are used in statistical analysis assume that the material at least approaches normal distribution. Before these methods of analysis are applied, it is necessary to ensure that this assumption is at least approximately true.

For obvious reasons, it is very seldom that we can investigate all the individuals in the population and we have to confine ourselves to *samples* of the population. Every mean of a sample has a certain error, the size of which depends upon the number of individuals in the sample and the way in which the sample has been selected. If we take a large number of random samples each of N variates from a population with a standard deviation σ, we obtain a 'population' of means, the standard deviation of which is $\sigma/\sqrt{N}$. The standard deviation (s) for a particular random sample is an estimate of the population's standard deviation σ. The expression $s/\sqrt{N}$ is a measure of the error with which a sample mean estimates the population mean. It is usually referred to as the standard error of the mean, or simply, the *standard error*. It is usual to give the standard error of a mean after the mean value, e.g. $7\cdot78 \pm 0\cdot285$.

When one investigates certain conditions in a population, it is of the utmost importance that the sample taken is representative of the population. In studying the relative importance of inheritance for the variation of cows' milk yield it is necessary to ensure that the data are as unselected as possible. The investigation should not be confined to the larger well managed herds even if this would make it simpler from a technical point of view. What is needed is a cross section of all recorded herds.

If a particular character is influenced by two independent factors (1) and (2) whose effects can be added together, and each of the factors gives rise to distributions with standard deviations σ_1 and σ_2 respectively, then the standard deviation of the population when both these causes are operative at the same time is $\sqrt{(\sigma_1^2 + \sigma_2^2)}$. If one wishes to partition the variation in a population into its causal components, it is more practical to work with the square of the standard deviation. FISHER, introduced the term variance to denote the square of the standard deviation:

$$\sigma^2_{x_1+x_2} = \sigma^2_{x_1} + \sigma^2_{x_2},$$

where x_1 and x_2 are two independent sources of variation.

On this concept Fisher developed his *analysis of variance*, which has gained wide application in all biological research, not least in animal genetics. With the aid of the analysis of variance it is possible to estimate the proportion of the total variance which is determined by the differences between the group means (e.g. daughter groups from different bulls) and the differences within the groups. Examples of analyses of variance are given later (p. 125).

The different sources of variation are not always independent; when x_1 and x_2 are correlated, the total variance of x_1 and x_2 has the following form:

$$\sigma^2_{x_1+x_2} = \sigma^2_{x_1} + \sigma^2_{x_2} + 2r\sigma_{x_1}\sigma_{x_2},$$

where r equals the correlation between x_1 and x_2; r can have either a positive or negative sign.

The concept of variance is fundamental to every discussion of the inheritance of quantitative characters. The total variance can be divided into (1) directly observable components of variance caused by, for example, the differences between group means and within groups and (2) causal components according to theoretical models of one or another type. It is often practical to distinguish between these two types of variance. The observable variance components in a sample will be represented by the symbol s^2, whereas the causal components will be represented by V, with a subscript indicating which component of variance is referred to.

Covariation

Many characters show covariation, e.g. milk yield and milk fat from the same cow. The degree of *covariation* between two variables is usually measured by the *correlation coefficient* (r) and the *regression coefficient* (b). The correlation coefficient may be defined as the ratio between the variance common to the two variables, co-variance, and the geometric mean of the two variances, i.e.

$$r = \frac{\text{covariance}}{\sqrt{\sigma_x^2 \sigma_y^2}} = \frac{\Sigma(x-\bar{x})(y-\bar{y})}{\sqrt{\Sigma(x-\bar{x})^2 \Sigma(y-\bar{y})^2}}$$

The regression of y on x, i.e. $b_{y/x}$ is given by

$$\frac{\Sigma(x-\bar{x})(y-\bar{y})}{\Sigma(x-\bar{x})^2} = r\frac{\sigma_y}{\sigma_x}.$$

If the variation of the two variables x and y is equally great ($\sigma_x = \sigma_y$), then $r = b$. The variation can however be very different. In a group of cows selected for breeding on the basis of high yield, the variation of yield, as a rule, is considerably less than in an unselected group of daughters of these cows.

The regression coefficient shows how much the dependent variate (y) changes for one unit increase of another (independent) variate (x), i.e. $y = \bar{y} + b_{y/x}(x - \bar{x})$, where $\bar{y}$ is the mean value of the character y. In general it can be said that the regression coefficient measures the slope of the regression line whereas the correlation coefficient measures the spread about the regression line, i.e. the intensity of the relationship of the two variates (Fig. 4:2).

The square of the correlation coefficient (r^2) shows the proportion of the total variation which can be explained by the linear regression; r^2 has therefore been called the *coefficient of determination*.

In many cases the relationship between the variables is curvilinear; the more pronounced the curvilinearity, the less reliable will be the linear regression coefficient as a measure of the covariation. It is then necessary to find another function which better expresses the relationship.

If several variables are correlated with each other and the regression is linear, one can calculate the *partial correlation* between two of the variables, i.e. the correlation between two variables when the other variables are held constant. It is further possible to calculate the correlation between one variable and two or more other variables. This is known as *multiple correlation*.

The measurement of the degree of covariation discussed so far has said nothing about the causes of the covariation. In many cases, the relationship between two variables is directional. The likeness between a father and his

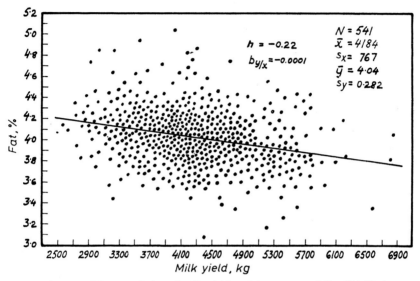

Fig. 4:2. The covariation of milk yield and fat content of Swedish Red and White cattle.

progeny is due, of course, to the fact that the progeny have received a random sample half of the father's genes. The progeny on the other hand have not been able to influence the father genetically. In order to explain such directional biological relationships WRIGHT (1921) introduced his *path coefficients*.

Let us assume a variable X, whose variation is caused by changes in the independent variables A, B and C, thus:

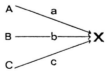

Let σ_x represent the total standard deviation of X and σ_{XA} that part which is due to the variable A (all other factors held constant). The path coefficient from A to X (represented here by a) is thus $= \dfrac{\sigma_{XA}}{\sigma_X}$, similarly $b = \dfrac{\sigma_{XB}}{\sigma_X}$ and $c = \dfrac{\sigma_{XC}}{\sigma_X}$.

The square of the path coefficients which correspond to the previously mentioned coefficients of determination, shows the proportion of the total variance (σ_x^2), which is eliminated if A, B or C are held constant. If the whole of the variation of X is determined by the variables A, B and C then,

$$\sigma_{XA}^2 + \sigma_{XB}^2 + \sigma_{XC}^2 = \sigma_X^2 \text{ and } a^2 + b^2 + c^2 = 1$$

QUALITATIVE CHARACTERS

Gene frequency and genetic equilibrium

In Chapter 3 examples were given of Mendelian segregation in matings between different genotypes. The mating between two heterozygotes of the type *Aa* resulted in progeny in the proportion 1 *AA*:2 *Aa*:1 *aa*. If one gene is dominant, the proportions between the two phenotypes becomes 3:1. There are numerous examples in the animal kingdom of characters which are determined by dominant and recessive genes, e.g. black and red in cattle. In spite of this there are hardly any cattle breeds in which the proportion between animals of the two colour types approach 3:1. It is more often the case that one colour is completely or almost completely predominant in a particular breed, e.g. black in Friesian cattle. In order to find an explanation of this we must introduce a new concept, namely, *gene frequency*.

Let us assume a locus with two alleles, *A* and *a*. The frequency of these two genes can be represented by p and q respectively where $p+q = 1$. The gene frequency must be considered, if one wishes to explain the distribution of the different genotypes in a given population. The frequency of gametes which carry the gene *A* is p, whereas the frequency of *a* gametes is q. If mating is at random within the population with respect to this gene pair, then the distribution of the three genotypes is obtained by random combination of the two types of gametes. The result is illustrated below:

		Sperm	
		pA	qa
Ova	pA	p^2 *AA*	pq *Aa*
	qa	pq *Aa*	q^2 *aa*

The distribution of the zygotic types follows the binomial $(p+q)^2$, i.e.

$$p^2 \, AA : 2pq \, Aa : q^2 \, aa.$$

This is often referred to as the *Hardy-Weinberg* law (cf. p. 10). Populations in which the zygote frequency shows this distribution are said to be in *genetic equilibrium* with respect to the particular locus. In large populations mating at random, the gene and zygote frequencies remain unchanged from generation to generation.

The zygote frequency is a function of the gene frequency. If equilibrium prevails, the proportion of heterozygotes in a gene system with two alleles can never exceed 0·5, a value which is reached when $p = q = 0·5$.

Multiple alleles. The existence of more than two alternative forms of a gene, multiple alleles, appears to be rather common in domestic animals. The frequency of zygotes in a large population mating at random is obtained, in a system with multiple alleles, by developing the expression $(p_1 + p_2 + p_3 + \ldots p_n)^2$ where p_1 and p_2 etc. represent the frequency of the various alleles. Analogous with the conditions in the gene system with two alleles, equilibrium occurs after one generation of random mating.

Combination of two different gene systems. Let us assume two independent loci, one with alleles A and a and the other with B and b. The gene frequencies within the one system are represented by p and q and in the other system by r and s respectively. In Chapter 3 it was shown that four types of gamete and nine different genotypes are possible. If mating is random, the zygote frequencies are obtained in the manner shown in Table 4:1. If $p = 0.4$, $q = 0.6$ and $r = 0.7$, $s = 0.3$, then the frequencies of the different types are as follows: $(p^2 + 2pq + q^2)$ $(r^2 + 2rs + s^2) = 0.0784\ AABB + 0.0672\ AABb + 0.0144\ AAbb + 0.2352\ AaBB + 0.2016\ AaBb + 0.0432\ Aabb + 0.1764\ aaBB + 0.1512\ aaBb + 0.0324\ aabb$.

Table 4:1 The frequency of the possible genotypes when two different gene systems, each with two alleles, are combined

	p^2 $A A$	$2pq$ $A a$	q^2 $a a$
r^2 $B B$	$p^2 r^2\ A A B B$	$2pqr^2\ A a B B$	$q^2 r^2\ a a B B$
$2rs$ $B b$	$2p^2 rs\ A A B b$	$4pqrs\ A a B b$	$2q^2 rs\ a a B b$
s^2 $b b$	$p^2 s^2\ A A b b$	$2pqs^2\ A a b b$	$q^2 s^2\ a a b b$

If two populations with differing frequencies within the two gene systems are crossed, genetic equilibrium is not attained immediately for the nine genotypes, in spite of the fact that equilibrium is attained within the two systems separately. Take the case of a population formed by amalgamating two equally large but previously separated populations. In the one population only the type $AABB$ occurs; in the other only the type $aabb$. In the newly formed population the frequency for each of the four genes will be 0.5. Random mating within this new population gives rise in the next generation to only three of the nine possible genotypes. Only in the next following generation are all possible types present, but the frequencies of these deviate quite markedly from equilibrium. For equilibrium to be attained, the four gametes AB, aB, Ab and ab would have to be produced with equal frequency. In the first generation the gamete types Ab and aB are missing. The difference, d, between the frequency of $AB \times ab$ and of $Ab \times aB$ is a measure of the deviation from equilibrium. It can be shown that d is halved for each succeeding generation when random mating is repeated. The distribution of the different genotypes approaches equilibrium in a relatively short time. Linkage will delay the approach to equilibrium and the length of the delay will depend upon the closeness of the linkage.

Calculation of gene frequencies

Direct gene counting. If all the different genotypes within a locus can be distinguished an estimate of gene frequency can be obtained quite simply by counting the genes in the population or in a random sample. Many such gene systems are known, e.g. that which controls the blood groups F and V in cattle (cf. p. 189). The former group is determined by the gene F^F, whereas group V is determined by the allele F^V. The results of an investigation of blood groups in 630 bulls of the Swedish Red and White breed are summarised in Table 4:2.

Table 4:2 An example of the estimation of gene frequency by direct gene counting and the testing of genetic equilibrium

	Genotypes			
	$F^F F^F$	$F^F F^V$	$F^V F^V$	Total
Observed number	375	218	37	630
Expected number with genetic equilibrium (m)	371·89	224·28	33·83	630
Deviation (d)	+3·11	−6·28	+3·17	—

$\chi^2 = 0.52$; d.f. = 1. The deviation from equilibrium is not statistically significant.

The frequency (p) of the gene F^F is obtained in the following manner:

$$p = \frac{375 + (218/2)}{630} = 0.768,$$

and $q = 1 - 0.768 = 0.232$. By developing the expression $N(p+q)^2$, where N is the total number of individuals, the hypothetical distribution which corresponds to genetic equilibrium is obtained. Table 4:2 shows that the deviation (d) between the observed and expected number is not significant. Whether the deviation can be attributed to chance can be tested in a chi square (χ^2) analysis with one degree of freedom, where χ^2 is equal to $\Sigma(d^2/m)$, where m is the theoretically expected number according to a particular hypothesis. The degrees of freedom in χ^2 analyses of this type are equal to the number of columns (i.e. comparisons) minus 1 and minus the number of parameters estimated from the sample. In this particular case the gene frequency p had been estimated from the sample, whereas q was calculated indirectly since $p + q = 1$. In the case of a tri-allele system, when all six genotypes can be separated the corresponding number of degrees of freedom becomes $6 - 1 - 2 = 3$. Two parameters, p and q, have been estimated from the sample, whereas the frequency of the third allele is calculated indirectly.

The square root method. At a locus with two alleles, one of which is dominant, there can only be two phenotypes, e.g. red or black cattle. In this case one cannot readily distinguish the dominant homozygote from the heterozygote. Estimation of gene frequency must therefore be based on the recessive type. Let us assume that the fraction b of the total is the recessive type (i.e. $b =$ the number of recessives divided by the total number). The frequency of the recessive gene is then $= \sqrt{b}$. In this case, however, there is no possibility of ascertaining whether genetic equilibrium exists. The square root method should therefore be applied only when it can be reasonably assumed that the gene system is in equilibrium; otherwise the calculated gene frequencies may easily be misleading.

The square root method can often be used for calculating gene frequencies at loci with multiple alleles. As an example of this we can take the colour types, full pigmentation, himalaya and albino rabbits which are determined by the alleles C, c^h and c. Gene C is dominant over the other alleles whereas c^h is dominant only over c (see Chapter 6). Let us suppose a rabbit population in genetic equilibrium, where a fraction a of the total are fully pigmented, whereas the fractions b and d are of himalaya and albino type respectively. The frequency of the genes C, c^h and c are represented by p, q and r respectively. In agreement with the situation at a locus with two alleles the frequency of the completely recessive gene is obtained by taking the square root of the frequency of the corresponding phenotype, i.e. $r = \sqrt{d}$. The himalayan type rabbits are of two genotypes $c^h c^h$ and $c^h c$ with frequencies q^2 and $2qr$ respectively. Thus the fraction $b = q^2 + 2qr$. If we increase both sides of this equation by $r^2 = d$ we obtain $(q+r)^2 = b+d$ and $q = \sqrt{b+d} - r$. Since $p+q+r = 1$ then $p = 1 - \sqrt{b+d}$. The general principle which has been applied here can be used also in calculating frequencies in many other multiple allele systems.

In this connection it may be of interest to mention the special problem which applies to genes with *incomplete penetrance*. Let us suppose a locus with two alleles, A and a, and that the genotypes AA and Aa are normal, whereas the fraction x of the genotypes aa show a particular defect. The gene a is then said to have a degree of penetrance x when in the homozygote condition. If d is the number of individuals with the character and N is the total number, then the frequency of the recessive gene a is $\sqrt{d/xN}$. In order to be able to estimate the frequency of a gene with incomplete penetrance, it is therefore necessary first to estimate the degree of penetrance. For a recessive gene this may be done by studying the progeny from mating the types $aa \times aa$, if such matings are possible. The degree of penetrance is then the ratio of the number of progeny with the character to the total number of progeny.

The penetrance in a particular locus often appears to depend upon genes at other loci, and this type of character therefore constitutes an intermediate form to the quantitative characteristics. In many cases therefore it is more correct to use a quantitative approach to the problem and use genetic methods which are applicable to this situation than to postulate genes with incomplete penetrance and attempt to calculate their hypothetical frequency.

Estimation of gene frequencies in material where some of the genotypes are underrepresented. Many genotypes result in reduced viability of the individual; some zygotes die at a very early stage without being observed, others live only a short time. Estimation of the frequency of genes with such effects presents special problems. Only some simple cases will be dealt with here.

1. *Demonstrable effect of the heterozygote when one homozygote is lethal.*

Geno- and phenotype	Number	Estimation of gene frequency
AA	b	$p = \dfrac{b + 0\cdot 5c}{b+c}$
Aa	c	
aa	–	$q = \dfrac{0\cdot 5c}{b+c}$

The type aa is assumed to be lethal and the estimation of the gene frequencies must therefore be based upon the types AA and Aa. The blood disease in humans 'sickle'-cell anaemia, gives rise to a situation very similar to that given above. The heterozygotes have a blood picture which makes it possible to distinguish them from normal individuals whereas those which are homozygotes for the 'sickle'-gene rarely attain adult age.

2. *Recessive defects.* A whole range of defects fall within this group. The gene frequency can be estimated from the frequency of the deformed type (aa). The estimation of this frequency however can be easily misleading, partly on account of the occurrence of certain deformed individuals, dead at birth which is never reported, and partly because the number of defective individuals is often so small that the sampling error is large. Estimation of the frequency can, however, be obtained also by analysis of the result of matings where the male animal happens to be the carrier of the recessive factor. This question will be discussed in more detail in connection with the progeny-testing of males for recessive genes (Chapter 15).

Mendelian gene analysis with population data

Hypotheses about the inheritance of different characters can be tested with data collected at random from a population. If we assume the occurrence of three different phenotypes in a large population mating at random, we can then test whether these types are likely to be determined by two co-dominant alleles at the same locus. The three possible types would be expected to be distributed as $p^2:2pq:q^2$. This expected distribution is then compared with the obtained numbers. The method is thus similar to the example when the FV blood groups of cattle were used to investigate whether the population was in genetic equilibrium (Table 4:2). On the assumption that viability is the same for all three types, a statistically significant difference between the obtained and expected numbers shows that the proposed hypothesis is incorrect, or that the population is not in genetic equilibrium. On the other hand, if a close agreement is found between the obtained and expected, the indications are that the proposed hypothesis may be correct. Conclusive evidence for the correctness of the hypothesis can be achieved only by analysis of family data. This latter method becomes

Table 4:3 The frequency of different mating combinations

Type of mating D = Dominant R = Recessive	Genotypes	Frequency with random mating	Distribution of progeny		
			AA	Aa	aa
D × D	$AA \times AA$	p^4	p^4	—	—
	$AA \times Aa$	$4p^3q$	$2p^3q$	$2p^3q$	—
	$Aa \times Aa$	$4p^2q^2$	p^2q^2	$2p^2q^2$	p^2q^2
D × R	$AA \times aa$	$2p^2q^2$	—	$2p^2q^2$	—
	$Aa \times aa$	$4pq^3$	—	$2pq^3$	$2pq^3$
R × R	$aa \times aa$	q^4	—	—	q^4
Total		1·00	p^2	$2pq$	q^2

especially important when investigating gene systems with one or more dominant genes.

The expected mating combinations within a locus with two alleles is obtained by developing the expression $(p^2 + 2pq + q^2)^2$ as shown in Table 4:3.

It is then possible to compare the expected values with the obtained number of progeny from the different mating combinations. Progeny of the dominant type in matings between the recessives show, without the use of statistical methods, that the proposed hypothesis is incorrect, provided that no other reasonable explanation can be given, e.g. incorrect classification or incorrect pedigrees.

Changes in gene frequencies

From the foregoing section it will be apparent that the gene frequency remains unchanged from generation to generation in large populations mating at random where no selection takes place. It is well known, however, that great changes have taken place within most of the species and breeds of domestic animals during the last fifty or a hundred years. If we compare, for example, pictures of the indigenous unselected pigs at the end of the nineteenth century with the present-day, highly selected Landrace, the difference is seen to be enormous. The difference is partly due, of course, to improved environment, but even with the best feeding it would be impossible to make these indigenous pigs comparable with the present day Landrace. It is evident, therefore, that a change has taken place in the genetic make-up of the individuals and thereby also in the population. We shall now deal with those forces which can bring about a change in gene frequency—*mutation, migration, selection* and *random drift*.

Mutation. As was mentioned in Chapter 3, sudden changes of the genes (mutations) can take place. Mutations can go in both directions, thus $A \rightleftharpoons a$. The frequency is usually much higher in one direction than in the other and the occurrence of reverse mutations will therefore be disregarded. Over short periods of time mutations have very little effect on gene frequencies in a population; over longer periods and in combination with selection, mutations have great importance for evolution.

Migration. This means the interchange of individuals between populations. Moving out (emigration) changes the genetic composition of a population only if the group which moves out deviates from the population in general.

The introduction of genetic material (*immigration*) has been and is still of importance in animal breeding. If the gene frequency in the original population is represented by q_o and that of the introduced animals by q_i, then the change Δ_q in the gene frequency due to immigration is:

$$\Delta_q = bq_i + (1-b)q_o - q_o = b(q_i - q_o),$$

where b denotes that part of the population which consists of the introduced individuals. For practical reasons the factor b often has a very low value, since only a few good breeding animals can be imported to improve an undeveloped breed. The immediate changes in gene frequency are thus very small. By continually choosing for breeding individuals which originate from the introduced animals much greater changes in gene frequency can be achieved. It takes quite a long time, however, before the introduced new genes are distributed throughout the population. For migration to be of practical importance, therefore, it must be combined with selection.

Selection. Not all genotypes have the same capacity for meeting the demands set by nature or man. Those which are least fitted succumb or are permitted to reproduce themselves only in comparatively small numbers, whereas the best fitted have a proportionately large number of progeny; there is a sorting out, or selection, among the genotypes. It is usual to differentiate between natural selection, which is conducted by nature itself, and artificial selection, which is regulated by man. Selection, of course, influences gene frequencies. The size of the change per generation can be calculated if the intensity of selection is known. Let us suppose a locus with two alleles:

Genotype	AA	Aa	aa	Total
Frequency in the original population	p^2	$2pq$	q^2	1
Individuals chosen as parents for the next generation	p^2	$2pq(1-hs)$	$q^2(1-s)$	$1-sq(q+2ph)$

p = Frequency of the desirable gene in the original population
q = Frequency of the undesirable gene in the original population
s = Intensity of selection against aa
hs = Intensity of selection against Aa

The frequency (p_1) of the undesirable gene in the next generation will be:

$$p_1 = \frac{p^2 + pq(1-hs)}{1-sq(q+2ph)},$$

and the change

$$\Delta_p = \frac{pqs[q+h(2p-1)]}{1-sq(q+2ph)}$$

The meaning of the factors s and h will be apparent from the following example. If the genotype Aa leaves 5 per cent fewer and aa 20 per cent fewer progeny than AA, then $s = 0.2$; $hs = 0.05$ and $h = 0.25$. The value for h varies with the degree of dominance. If gene A is completely dominant, $h = 0$. On the other hand, if over-dominance occurs, then h is negative, i.e. the type Aa is more desirable than either AA or aa.

The effect of selection. The change in gene frequency per generation, depends upon the original gene frequency as well as the intensity of selection. Selection for a dominant gene is most effective when the frequency of the gene in the original population is low. When the frequency exceeds 0·3 the change per generation becomes progressively less since the number of recessive homozygotes which can be culled becomes proportionately less. Conversely, the rate of change due to selection for a recessive gene is highest at a frequency of 0·7, while selection for genes without dominance is most effective at intermediate frequencies.

In animal breeding, selection is often for the dominant genes, i.e. against the recessive homozygote type; for example, against recessive defects or against other undesirable characters, such as red in the black-and-white Friesian breed. The effect of selection with complete elimination of the recessive homozygous type can easily be calculated. The values for s and h are 1 and 0 respectively. If the frequency of the recessive gene in generations 0 and n are represented by q_0 and q_n, we obtain: $q_n = \dfrac{q_0}{1+nq_0}$. The lower the frequency of the undesirable gene at the beginning, the slower the process of reducing the frequency. Assuming that no mutations take place, it would take 100 generations to reduce the frequency of the undesirable gene from 0·01 to 0·005, but only 10 generations to change it from 0·1 to 0·05.

Equilibrium between mutation and selection. In every generation a certain number of the genes at a given locus mutate. Suppose the frequency of mutation $A \to a$ is u and that a is a complete recessive. A state of equilibrium between selection and mutation is reached when the number of a genes eliminated is equal to the number of new recessives arising from mutation. By the same reasoning by which the formula for Δ_p was derived, selection leads to a reduction (Δ_q) in the frequency of the recessive undesirable gene according to $\dfrac{sq^2(1-q)}{1-sq^2}$. Equilibrium between selection and mutation occurs therefore when $u(1-q) = \dfrac{sq^2(1-q)}{1-sq^2}$. The denominator deviates only very slightly from 1. The gene frequency at equilibrium can thus be approximately estimated by $\sqrt{u/s}$.

Dominance protects undesirable recessive genes against selection when they occur in the heterozygote condition. The frequency of mutation per generation seems to be, as already mentioned (p. 73), of the order of 10^{-6}. Complete elimination of the recessive homozygotes results therefore in an equilibrium frequency of about 0·001.

Selection for heterozygosity. In some cases the heterozygote is superior to both the homozygote types. In this case the factor h in the formula on p. 102 becomes negative. It is, of course, impossible to achieve constancy of the heterozygote type (Aa), but a certain state of equilibrium can be reached. This occurs when $\Delta_p = 0$. As already explained

$$\Delta_p = \frac{pqs[q+h(2p-1)]}{1-sq(2ph+q)}.$$

When $\Delta_p = 0$, then $p = \dfrac{1-h}{1-2h}$.

Stable equilibrium conditions due to the fact that the heterozygote has greater viability and fertility than both the homozygote types appears to be rather common in nature. BUZZATI-TRAVERSO (1952) studied a recessive mutant in the common fruit fly (*Drosophila melanogaster*) which gave rise to light eye-colour.

He constructed experimental populations with original frequencies for the recessive gene of 0·125, 0·500 and 0·875. Two populations were constructed for each frequency. These were then allowed to reproduce at random for 15 generations. After 7 generations, all the populations had already reached about the same gene frequency—about 0·58—for the recessive mutant, and the frequency remained practically constant during the following 8 generations. The result seems to be due to the fact that the heterozygote was superior to both the homozygous wild type and the homozygous mutant type. The latter, however, was somewhat superior to the wild type homozygote; otherwise the state of equilibrium would have been reached at 0·5.

In the case of the large farm animals there is a lack of experimental evidence to show that stable intermediate gene frequencies are attained due to the selective superiority of the heterozygote. Several investigations indicate, however, that the general degree of heterozygosity is of considerable importance for the animal's viability (Chapter 14). There is no doubt, therefore, that, even in domestic animals stable intermediate gene frequencies occur at certain loci.

The effect of selection on the degree of homozygosity. The changes in homo- or heterozygosity per generation due to selection are relatively small, provided that inbreeding is avoided. If those individuals which are chosen for breeding are mated with each other at random—without inbreeding—then the change in the degree of heterozygosity Δ_F can be calculated as follows:

$$\Delta_F = 2(p + \Delta_p)(q - \Delta_p) - 2pq = 2\Delta_p(q - p - \Delta_p).$$

The changes depend, therefore, on Δ_p and p. There will be a decrease in heterozygosity if p is > 0.5, provided that selection really has some effect. If $p = 0.2$ and $\Delta_p = +0.04$ the selection results in an increase in the degree of heterozygosity from 0·32 to 0·36, and if $p = 0.7$ and $\Delta_p = +0.04$, then the heterozygosity decreases from 0·42 to 0·38 in the progeny generation. The value of $2pq$ alters only slightly when p undergoes small changes on either side of 0·5 but alters rapidly at higher and lower values of p. The effect of selection at low and high values of p however, is, small. Consequently the percentage heterozygosity is only slightly influenced by selection, in spite of the fact that the average for the population can be altered considerably.

Changes in gene and zygote frequencies due to population size and inbreeding

Up to now we have confined our attention to large populations and to the changes which can be brought about by migration, mutation and selection. Every new individual is formed by the union of two gametes whose genes constitute a random sample half of the genes from the two individuals which produced the gametes. When the number of individuals is small, the gene frequency in the new generation may deviate quite widely from the frequency of the parental population, even though the matings were entirely at random. The gene frequency therefore varies from generation to generation.

The influence of the population size on gene frequency has been studied in detail by WRIGHT; some general principles will be dealt with here. Let us suppose a parent population with gene frequency $p = q = 0.5$ and that this produces 40 progeny. The standard deviation for the gene frequency in the new generation will be

$$\sigma_p = \sqrt{\frac{0.5 \times 0.5}{2 \times 40}} = 0.056$$

(cf. p. 92; the denominator becomes $2N$ since each individual has two genes at a given locus.) On the basis of the standard deviation and on the assumption that the possible gene frequencies in the offspring population are normally distributed, the probabilities for the various gene frequencies can be directly calculated. These probabilities are given in Table 4:4.

Table 4:4 *The probability of different gene frequencies (q) in a progeny generation of 40 individuals when the parent population has a gene frequency of 0.5*

	<0.332	0.332–0.387	0.388–0.443	0.444–0.500	0.501–0.556	0.557–0.612	0.613–0.668	>0.668
Probability	0.002	0.021	0.136	0.341	0.341	0.136	0.021	0.02

In approximately 4 per cent of the cases the difference between the new and the original gene frequency will be of the order of 0.11 units or more. It is thus clear that the gene frequency can be pushed up or down as a result of chance occurrences.

Perhaps the picture will become clearer if, instead of taking a single small population of progeny, we imagine the progeny from a very large parent generation, with gene frequency $p = q = 0.5$ at a given locus, divided at random into a number of groups of 40 individuals each. The gene frequency in the different progeny populations can then be expected to have the distribution given in Table 4:4. Similar reasoning can be applied to a number of loci in a single population if the gene frequency for all these loci in the parent generation is $p = q = 0.5$.

The gene frequency may vary up or down; the direction is entirely at random if no selection takes place. If a locus becomes fixed in the homozygous condition due to chance, variation is no longer possible. In a population with the same number of males and females the proportion of loci which become fixed per generation can be shown to approximate $1/2N$ (where N = the number of individuals), if the changes due to migration, mutation and selection are disregarded. Dividing a large population into many sub-groups which reproduce separately results in increased homozygosity within the groups. The average of the gene frequencies in the different sub-groups may be expected to be the same as the gene frequency in the original population. The variation in gene frequency between the sub-groups however, increases, with each generation. The effect of dividing a population into sub-groups has the following main consequences: (1) differentiation between sub-groups, (2) genetic uniformity within groups and (3) an overall increase in homozygosity.

The effective population size. Up to now we have assumed that the number of males and females in the parent generation has been equal or, if this was not the case, that the numbers were so large that the unequal distribution between the sexes was not of importance. In animal breeding males are, as a rule, much fewer than females. Such populations can, in spite of the fact that the total number is rather large, have about the same effect on random variation, fixation and genetic drift as a relatively small population with equal numbers of males and females. In order to be able to compare populations with different structures the concept of *effective population size* (Ne) has been introduced. The actual population is compared with a hypothetical population having the same effect on genetic drift as the actual population and consisting of equal numbers of males and females. The size of this hypothetical population corresponds to the effective size of the actual population. Provided that matings occur at random the effective number can be calculated from the actual number of males and females as follows:

$$Ne = \frac{4N\male \times N\female}{N\male + N\female}$$

If the number of females is very large then $Ne = 4N\male$. The increase in homozygosity per generation becomes about $1/2Ne$.

Suppose we have a breed of cattle consisting of 100 000 cows. With natural mating this requires approximately 2000 breeding bulls. The effective population size if the matings are at random would thus be 8000 animals. However, in improved cattle breeds most bulls are derived from a small number of related élite herds, which will decrease the effective population size. A further decrease may occur when artificial insemination is used.

With the aid of modern deep freezing technique and maximum use of the bulls, a single bull should, without difficulty, be able to serve 10 000 cows per year. Not more than 10 breeding bulls plus a number of test bulls would be required in this breed of 100 000 cows and the effective population size with random mating would be only about 40 individuals. Under these conditions genetic drift and increase of homozygosity would be a factor to reckon with.

Measures of inbreeding and relationship. In the previous section it was shown that a continual increase in homozygosity takes place in small populations. Each individual in the population has two parents; and if the population is small, the probability that these two parents are related and therefore carry genes which have the same origin is increased. Mating between related individuals is called *inbreeding*.

A gene, *A*, which is present in two relatives, can be the result of repeated divisions of the same gene in their common ancestor. Mating between these two relatives can result therefore in homozygotes *AA* where both these genes constitute identical replicates of the ancestor's *A* gene. Inbreeding results in an increase of this type of homozygosity. The *coefficient of inbreeding* can be said to measure the probability that the two genes at a given locus are identical by descent. It is a measure of the increase in homozygosity as a consequence of inbreeding. The size of the coefficient of inbreeding depends upon the degree of relationship between the parents of the individual. The coefficient of inbreeding is usually represented by F.

The degree of relationship between two individuals is measured by the *coefficient of relationship*, (R). In order to avoid confusion with multiple correlation the sign r_g will be used in Chapters 15 and 16. The coefficient of relationship is a numerical index indicating how much more alike the breeding values (see p. 116) in two related individuals are than they would be in two individuals taken at random from the same population. The correlation between two unrelated individuals is 0. The coefficient of relationship between two related individuals (i.e. the correlation between the breeding value of the individuals) can assume any value between 0 and 1.

The principles used in calculating the coefficient of inbreeding will be apparent from the following. Suppose there is an individual X, whose grandparents

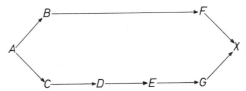

Fig. 4:3. Pedigree of the individual X (see text).

F and G have a common ancestor A, and that a number of generations exist between A and the grandparents F and G (see Fig. 4:3). What is the probability that the individual X inherits two alleles which are identical by descent? Assume any locus in the ancestor A. The probability that B and C inherit the same allele from A is 1/2 and that they inherit different alleles is 1/2. The probability that the latter gene is identical with the former, on account of earlier inbreeding, is, however, $F_A = $ A's inbreeding coefficient. The total probability that B and C inherit genes which are identical by descent is therefore $\frac{1}{2}(1+F_A)$. The probability that B further transmits the gene, which it obtained from A, to F is 1/2 and that F further transmits the gene, obtained from B, to X is similarly 1/2. The same reasoning can be applied for the gene which C passes to D and so on. The probability that X receives two alleles which are identical by descent is thus $\frac{1}{2}(1+F_A)(\frac{1}{2})^{2+4}$. The individual X can however have other common ancestors than A and the general formula for calculation of the coefficient of inbreeding becomes then:

$$F_X = \Sigma 0 \cdot 5^{n_1+n_2+1}(1+F_A)$$

where $n_1 = $ the number of generations from the common ancestor to the father and n_2 the corresponding number for the mother, $\Sigma = $ summation.

The method for calculating the coefficient of inbreeding was developed in 1921 by SEWALL WRIGHT, though from somewhat different premises than have been described here. He also developed methods for calculating the coefficient of relationship. It is necessary to distinguish between two types of relationship; there are relatives that are collateral, e.g. cousins, and relatives that are directly descended, e.g. parent-progeny. The coefficient of relationship for collateral relatives is usually represented by R_{XY} and for relatives directly descended by R_{AO}. The calculation is as follows:

$$R_{XY} = \frac{\Sigma \, 0 \cdot 5^{n_1 + n_2}(1 + F_A)}{\sqrt{(1 + F_X)(1 + F_Y)}}$$

where $n_1 =$ the number of generations from X to A and $n_2 =$ the number of generations from Y to A.

$$R_{AO} = \Sigma \, 0 \cdot 5^n \sqrt{\frac{1 + F_A}{1 + F_O}}$$

where $n =$ the number of generations between the ancestor (A) and the progeny (O).

In a locus with two alleles, inbreeding results in an increase of both homozygote classes. Let us suppose a large population mating at random with gene frequencies p and q. From this population a large number of lines are formed, within which inbreeding is conducted with equal intensity. The resulting changes in the average frequency of zygotes from the different lines are shown below:

Genotype	Frequency in the original population	Frequency in the inbred lines
AA	p^2	$p^2 + pq\,F$
Aa	$2pq$	$2pq - 2pq\,F$
aa	q^2	$q^2 + pq\,F$

An analogous result will naturally be obtained also with polyhybrid segregation. A consequence of inbreeding is that the population is divided up into lines. The genetic uniformity increases within the lines and decreases between lines, as the degree of inbreeding increases. The most effective dividing into lines takes place with generation after generation of full sib mating.

It should be pointed out that R and F are statistical means, and their calculation does not take consideration of selection which may have taken place. Particular individuals and loci naturally exhibit a certain amount of variation from the calculated value. It must be made quite clear that the coefficient of inbreeding measures the increase in homozygosity compared with that existing in a given base population. It is meaningless to say that the coefficient of inbreeding in a certain breed of animals is 0·05, unless one states at the same time that it is calculated on the basis of a given number of generations or related to a certain period of time. The increase in the coefficient of inbreeding per generation is, in general, a good measure of the changes in the degree of heterozygosity which has taken place in the population during a limited number of generations. Admittedly, the coefficient of inbreeding or its increase does not tell us whether a gene pair is homozygous or heterozygous in a given individual, but it is a valuable aid in assessing the average frequency of heterozygosity and the changes within an inbred population.

The meaning of the coefficient of inbreeding can be illustrated by the following hypothetical example. In a population mating at random the animals are assumed to be heterozygous for, on the average, 5000 loci, out of a total 10 000 loci. From this population smaller groups are selected, within which intensive inbreeding is conducted. If the coefficient of inbreeding for one animal is calculated to be 0·6, whereas for another animal it is 0·35, then the former animal can be expected to be heterozygous for about 2000 gene pairs and the other animal for 3250 pairs.

Examples of the calculation of the coefficients of inbreeding and relationship

(1) Calculate the coefficient of inbreeding for an individual Y from the pedigree shown in Fig. 4:4. The ancestors A, B and C are common and the individual C

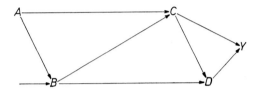

Fig. 4:4. Pedigree, where the inbreeding coefficient of Y and the relationship between C and Y are to be calculated (see text).

is inbred. Applying the formula for calculation of the coefficient of inbreeding (p. 107) we obtain:

Inbreeding through

A	$0·5^{2+1+1}$	$= 0·0625$
B	$0·5^{1+1+1}$	$= 0·1250$
C	$0·5^{1+1}(1+0·25)$	$= 0·3125$
	Total	$0·5000$

The inbreeding coefficient of Y is thus 0·5.

(2) Calculate the coefficient of relationship between the individuals Y and C in the pedigree shown in Fig. 4:4. These are related both directly and collaterally. In order to obtain the relationship between Y and C, the formulae for direct and collateral relationship should be applied (p. 108).

Collateral relationship through A,

$$\frac{0·5^{3+1} \times 1}{\sqrt{(1+0·5)(1+0·25)}} = 0·0456$$

Collateral relationship through B,

$$\frac{0·5^{2+1} \times 1}{\sqrt{(1+0·5)(1+0·25)}} = 0·0913$$

Direct relationship,

$$(0.5^1 + 0.5^2) \sqrt{\frac{1+0.25}{1+0.5}} = 0.6847$$

$$\text{Total } R = 0.8216$$

The relationship between Y and C is thus 0·8216.

The effect of systematic inbreeding. It is of interest to mention something about the increase in the coefficient of inbreeding with certain regular systems of mating. The effect on the degree of inbreeding of full sib mating, half sib mating and repeated back crossing to the sire (or dam) is presented in Table 4:5. With half sib mating, it is assumed that one male is mated to a large number of his half sisters which will also be half sisters themselves. The procedures of full and half sib mating result eventually in complete homozygosity.

Table 4:5 *The inbreeding coefficient with different intensities of inbreeding*

Generation	Full sibs	Half sibs (males × many half sisters)	Repeated back crossing to a sire (or dam) which is not inbred
0	0	0	0
1	0·250	0·125	0·250
2	0·375	0·219	0·375
3	0·500	0·304	0·438
4	0·594	0·380	0·469
5	0·672	0·449	0·484
10	0·886	0·692	0·499
15	0·961	0·829	
∞	1·000	1·000	

Repeated back crossing to the sire results first in an increase in the degree of inbreeding comparable with that which takes place with full sib matings; but after two generations of back crossing the increase in the degree of inbreeding declines quite sharply. This system of inbreeding cannot result in a higher coefficient than 0·5 if the sire (or dam) are not themselves inbred. If, on the other hand, an inbreeding system is applied whereby back crossing is alternate to the sire or dam, i.e. the female progeny are mated to the sire and the male progeny obtained from this mating are mated back to the dam and the female progeny of this mating are mated back to their sire and so on, an increase in the degree of inbreeding is achieved which corresponds to full sib mating.

The increase in homozygosity which is calculated to occur in the above mentioned inbreeding systems can be retarded if the heterozygote is favoured by either natural or artificial selection.

QUANTITATIVE TRAITS

Up to now, we have dealt only with traits determined by one or a very few genes. The most important economic characters in our domestic animals are often polygenic and, as previously mentioned, show continuous variation. Crosses between breeds with different production levels result, in general, in progeny intermediate between the two parent breeds and in the F_2 generation no segregation into discontinuous groups is apparent. The differences in the level of production cannot therefore be traced back to any special gene.

For the moment, we shall leave the influence of environment and, instead, we shall discuss what the effect will be if a trait is wholly determined by a number of independent loci each with two alleles. For the sake of simplicity we shall suppose also that the gene frequency within each system is $p = q = 0.5$. If there is no dominance, the distribution of the different phenotypes is obtained by developing the binomial $(\frac{1}{2} + \frac{1}{2})^{2N}$, where N is the number of gene loci. If we

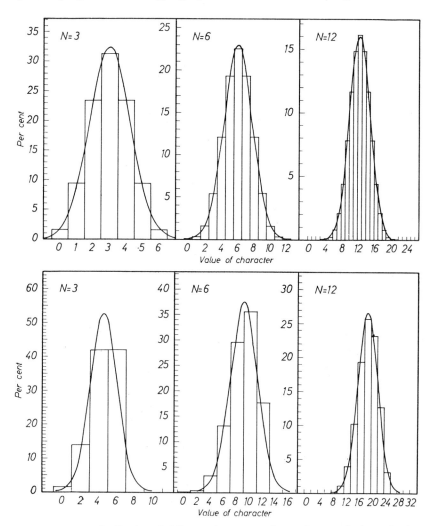

Fig. 4:5. The distribution of different phenotypes when a character is determined by 3, 6 and 12 gene pairs.
Upper part shows cases of no dominance; the lower part shows cases of complete dominance. The normal curves have been superimposed for comparison.

further assume that at each locus one gene in heterozygous condition increases a given character by one unit, and in homozygotes by two units while the other gene is neutral, then the distributions obtained for 3, 6 and 12 loci are as shown in Fig. 4:5.

The greater the number of gene pairs which influence the character, the less will be the difference between the various classes compared with the total variability. The binomial distribution then approaches the normal distribution. For the sake of comparison, the normal curve has been drawn in the diagram and demonstrates the continuous variation typical for the majority of production traits.

The influence of dominance on the variation, with different numbers of involved loci, is also shown in Fig. 4:5, where the presence of one or two dominant genes per locus is assumed to increase the character by 2 units. The distribution is obtained by developing the binomial $(\frac{3}{4}+\frac{1}{4})^N$ where N is the number of loci. With the same number of loci the number of classes is less when dominance is present than with no dominance. When the number of gene loci is small the distribution will be skewed. With increasing numbers of loci concerned the distribution approaches normality.

We have assumed up to now that the variation in the traits is determined solely by inheritance. The environment, however, very often has an important influence on quantitative characters, and this very often causes overlapping between the different classes. For an illustration of the variation in a character determined by a number of genes and environment reference may be made to Fig. 4:1, which shows the distribution in milk yield during the first lactation of 2234 cows. A single individual taken at random from this population of cows can have practically any value within certain limits; but values close to the average have the greatest probability.

The production capacity of an individual is determined by its genotype and by the environment. If all non-genetic factors are attributed to environment, the phenotypic value (P) of the individual can be expressed as $P = G + E'$ where G is the genotypic value of the individuals and E' is the deviation from the genotypic value attributed to environment. The mean phenotype ($\bar{P}$) of the population is thus identical with its mean genotypic value ($\bar{G}$) within the prevailing environment. The only measure one has of the mean genotypic value in a population in a given environment is thus its mean phenotypic value.

The additive effect of genes and deviations due to dominance and gene interaction

In order to illustrate the inheritance of quantitative characters it is usual to set up simple models to show the way in which the genes work. Let us assume a gene system with the two alleles, A_1 and A_2, the former gene having a positive effect on the character in question. The difference between the two homozygous types can, as in Fig. 4:6, be represented by $2a$ and the mid-point between them by O ($= M$ in the text). The genotype A_1A_1 has thus the value $M + a$ and A_2A_2 becomes $M - a$. Without dominance the heterozygote will come to lie exactly between the two homozygotes. Its value will thus be M. If one of the two alleles is dominant

A_1A_2 will deviate from M by the amount d. If A_1 is completely dominant over A_2, then $d = +a$ and if overdominance occurs, d will be greater than a. In the further discussion we shall assume that the quantitative character is determined, apart from the environment, only by the variation at the A_1A_2 locus. The frequency of the genes will be denoted by p and q respectively.

The phenotypic and genotypic mean values of the population can be calculated in the manner shown below (note that $p+q = 1$).

Genotype	Frequency	Deviation from the midpoint of the scale (M)	Frequency × deviation
A_1A_1	p^2	$+a$	p^2a
A_1A_2	$2pq$	d	$2pqd$
A_2A_2	q^2	$-a$	$-q^2a$
		Total	$a(p^2-q^2) + 2dpq$ $= a(p-q) + 2dpq$

The mean value of the population is $\bar{P} = \bar{G} = a(p-q) + 2dpq + M$. The mean value thus depends partly upon the homozygotes which contribute with the term $a(p-q)$, and partly on the heterozygotes which contribute with the term

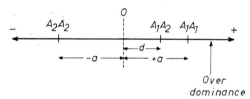

Fig. 4:6. Model for gene effect at a locus which takes part in the control of a quantitative character. In the text M is used instead of O for the average value of A_1A_1 and A_2A_2.

$2dpq$. Without dominance ($d = 0$) the latter term will be 0 and the mean value will be proportional to the gene frequency: $\bar{P} = a(1-2q) + M$. However, if complete dominance occurs, the mean value will be proportional to the square of the gene frequency, $\bar{P} = a(1-2q^2) + M$.

The average (= *additive*) effect of a gene can be defined as the mean deviation from the population average of those individuals which have received the gene. If a number of A_1 gametes unite with a random sample of the gametes of the population, then the deviation of the resulting zygotes from the population mean is a measure of the average effect of the A_1 gene. Let us use the same symbols as in Fig. 4:6 and again confine our attention to a single gene pair. If the A_1 gametes unite at random with the gametes of the population then p^2 zygotes of type A_1A_1 and pqA_1A_2 zygotes will be formed. The proportion of A_1A_1 will therefore be:

$$\frac{p^2}{p^2+pq} = \frac{p}{p+q} = p.$$

Similarly the proportion of A_1A_2 types can be shown to be q. The average value of the zygotes formed from A_1 gametes is $pa + qd + M$. The average effect of the A_1 genes (α_1) is obtained by subtracting the mean value of the population from this phenotypic value, thus:

$$\alpha_1 = pa + qd + M - [a(p-q) + 2dpq + M] = q[a + d(q-p)]$$

In a similar way the average effect of the A_2 genes can be obtained, $\alpha_2 = -p[a + d(q-p)]$. The average effect of the genes is therefore a function of the gene frequency in the population. For characters which are determined by a single locus the additive value can be calculated in the following way:

Genotype	Additive value
A_1A_1	$2\alpha_1 + (\bar{P})$
A_1A_2	$\alpha_1 + \alpha_2 + (\bar{P})$
A_2A_2	$2\alpha_2 + (\bar{P})$

The additive genetic values can be given either in absolute values or as deviations from the population mean. When dominance occurs, these additive values are not identical with the genotypic values from which certain genotypes can differ quite considerably.

Example. At the beginning of the 1950s an inherited form of dwarfism (snorter type) became relatively widespread in Hereford cattle in the U.S.A. There are several different genetically determined types (see Chapter 8). It has been believed that the commonest type is determined by a recessive gene, which will be referred to here as A_2, while the dominant gene will be called A_1. According to WARWICK, A_2 once reached in the United States an overall frequency of nearly 0·2 in the Hereford breed. Later investigations indicate that the inheritance of the snorter type may be more complicated than this, but for the sake of simplicity we shall use the original hypothesis of one recessive gene. GREGORY and co-workers reported that a number of 21-month-old Hereford dwarfs weighed 172 kg, whereas the weight of the normal animals at that age was 381 kg. The weight of the heterozygotes was not given, but we shall assume that the heterozygotes had the same weight as the normal homozygotes. The difference between the two homozygotes is 209 kg; M is thus 276·5 kg and, since complete

Table 4:6 The actual genotypic values together with their estimated additive value and dominance deviations with respect to the locus which results in dwarfism in the Hereford breed. The gene frequency $p = 0·8$ and $q = 0·2$. The population mean $P = 372·64$ kg

Genotype	Frequency f	Genotypic value (G) kg	Additive value (A) kg	Dominance deviations (D) kg	$f(G-P)^2$	$f(A-P)^2$	fD^2
A_1A_1	0·64	381	389·36	− 8·36	44·729	178·917	44·729
A_1A_2	0·32	381	347·56	+ 33·44	22·365	201·282	357·835
A_2A_2	0·04	172	305·76	− 133·76	1 610·256	178·917	715·670
s^2	—	—	—	—	1 677·350	559·116	1 118·234

dominance can be assumed to occur, $a = d = 104.5$ kg. With a frequency of the recessive gene of 0·2 the average effect of the dominant gene (α_1) reaches a value of 8·36 kg (cf. formula above), whereas the average effect of the recessive gene is -33.44 kg and the mean of the population 372·64 kg. On this basis the additive genetic values can also be estimated for the three genotypes (Table 4:6). The dominance deviation is large, especially for the recessive homozygote.

An illustration of the relation between the genotypic and additive values as well as the dominance deviations is presented in Fig. 4:7.

The three right-hand columns in Table 4:6 show also how the variances due to the genotypic values, the additive gene effects and the dominance deviations can be calculated. We shall return to this, however, in a later section.

When a character is influenced by two or more loci the deviations from the 'additive value' can be due also to an interaction effect (I) between the loci

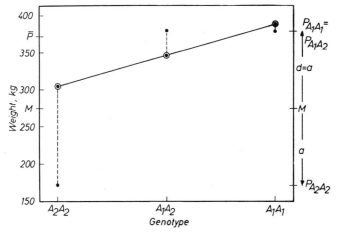

Fig. 4:7. Graphical representation of the genotypic values, the additive genetic values and the dominance deviations for body weight in the gene system which gives rise to dwarfism in Herefords. The recessive gene for dwarfism (A_2) has been assumed to have a frequency of 0·2 (see text and Table 4:6).

(epistasis). The conditions can be extremely complicated and nothing can be gained here by presenting simple models for calculating the effect of such interaction. In addition, comparatively little is known about the importance of this cause of variation with normal breeding in a population. However, with inbreeding and crossing between inbred lines it appears that interaction between loci can be of considerable importance. The interaction deviations can be treated, in principle, in the same way as the dominance deviation. In general it can be said that the total value of a given genotype is determined by its additive value (A) together with the deviations therefrom caused by dominance (D) and interaction (I), $G = A + D + I$. D and I are usually referred to as the *genes' non-additive effects*.

The additive genetic value of a character in an individual can be defined as the additive effect of all genes which influence the character. From the foregoing it will be apparent that this is an average value related to all the different combinations which can possibly be formed from the gene or the genes. The additive genetic value is identical with what will later be called *the breeding value* (i.e. *general* breeding value), which can be estimated according to certain general principles based upon the phenotype of the individual itself and/or its relatives. To this must be added the non-linear dominance and interaction effects which are apparent in certain gene combinations but not in others. The result of these effects is that an individual in certain mating combinations has a *special breeding value*, the size of which it is impossible, as a rule, to estimate in advance. The special breeding value must be experimentally determined for each separate case, not only in mating between individuals but also in crosses between inbred lines or breeds.

In animal breeding the importance of assessing individuals' breeding value is sufficient reason for theoretically illustrating the concept of breeding value. To attempt to assess this value has, however, practical importance only for polygenic quantitative characters and not for simple Mendelian characters. It ought to be pointed out that the approach applied is statistical rather than biological. A gene's effect on a single individual is naturally not altered with altered gene frequency. On the other hand, the gene's influence on the variation of the population as a whole certainly is altered. In population genetics, where one attempts to estimate the components of the total variation, which is due to different causes, it is necessary to use a statistical approach.

The variation of quantitative characters

Since the genotypic value of an individual depends upon the additive value (A), the deviations therefrom caused by dominance (D) and the interaction between gene loci (I), the total genotypic variance V_G has the following composition:

$$V_G = V_A + V_D + V_I.$$

The relative importance of the components of variance which are attributable to additive inheritance and dominance will be discussed first for a single trait determined mainly by a single locus, viz. dwarfism in Hereford cattle. The actual value of the different genotypes, the additive values and dominance deviations are summarised in Table 4:6, which also shows how the different components of variance can be calculated.

The total variance in the population was 1677·350, and of this one-third was due to the gene's additive effect and two-thirds to the dominance deviations. The components of variation can also be directly calculated as follows:

$$V_A = 2pq[a + d(q-p)]^2; \ V_D = (2pqd)^2.$$

The meanings of d and a were defined in Fig. 4:6. The relative importance of V_A and V_D therefore changes with the gene frequency. These changes are illustrated in Fig. 4:8 for a locus with complete dominance. At most gene frequencies the importance of the variance caused by additive inheritance is considerably greater than that attributed to dominance deviations. It is only at the lower frequencies of the recessive gene that the dominance variance is greater. However, for characters influenced by genes with over-dominance, the dominance deviations are of greater importance. With intermediate gene frequencies and strong overdominance, practically all the genetic variation is a result of dominance deviations.

A quantitative character is a result of both inheritance and environment. The phenotypic variation in a population depends therefore upon the genotypic variation, the environmental variation and the possible interaction between them. The phenotypic variance of the population (V_P) can be written:

$$V_P = V_G + V_E' + 2 \operatorname{cov}_{GE}' + V_{GE}',$$

where V_G = the genotypic variance, V_E' = the environmental variance, cov_{GE}'

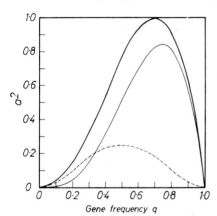

Fig. 4:8. The effect of the gene frequency on the magnitude of the genetic components of the variance caused by a single locus with two alleles and complete dominance. q = the frequency of the recessive gene; for explanation of a see Fig. 4:6.
——— = total genetic variance; ———, additive genetic variance; – – – –, dominance variance.
(Reproduced by permission from FALCONER 1960.)

denotes the covariance of the genotypic value and the environmental deviations and V_{GE}' the variance due to the non-linear interaction between genotype and environment.

Quite often correlation occurs between inheritance and environment. In herds where planned breeding is carried out the feeding is as a rule much better than in herds where no breeding work is conducted. In estimating the relative importance of V_G or $V_{E'}$ it is therefore necessary to eliminate the effect of the covariation, e.g. by making the analysis within herds.

In addition to the additive effect, which the genotype and environment have on the individual's phenotype, a non-linear interaction can occur. Suppose that we have a fairly large number of one-egg, i.e. genetically identical, pairs of heifer twins, that we split the members of each pair between two different environments (feeding, management, hygiene) and accurately measure a number of quantitative characters. If we then find that within both these environments the relative differences between heifers from different pairs is exactly the same, this indicates that the influences of inheritance and environment are entirely additive. On the other hand, if the size of the differences and perhaps the ranking order of the animals is altered, then a non-linear interaction has occurred, the size of which can be statistically estimated. There is no doubt that such interaction effects occur with respect to a number of quantitative characters when the environmental differences are large, e.g. when European breeds are tested in their natural habitat and in tropical countries. The importance of the interaction between genotype and environment, for breeds subject to less extreme environmental conditions or confined to limited areas, is less certain. The investigations that have been carried out on cattle appear to indicate that the interaction effects are relatively small with respect to the milk production of cows on different planes of nutrition, but they can be very important for a management factor such as the length of the milking intervals. It appears, therefore, that the size of the interaction effects depends as much upon the environmental factors as on the character concerned.

Unfortunately, very little is known at present about the importance of the interaction between genotype and environment for different characters and different environmental factors. Therefore we shall leave the subject here and return to it later, in the discussion of practical details of animal breeding.

Environmental influences may be of two different types—those which are common to groups of individuals, e.g. litter mates or whole herds, and those which, with equal probability, can affect any individual in the population. The former type of environmental influence is usually represented by C and tends to increase the phenotypic correlation between individuals within the same group. It is therefore important, in estimating the genotypic variance, to select material and methods in such a way that the variance caused by C will not be completely or partly included in the genetic variance. How this can be done will be explained later. Randomly distributed environmental influences will be denoted here by E. The total environmental effect (E') is therefore $E+C$. If no consideration is taken of the correlation between inheritance and environment or to the non-linear interaction effects between genotype and environment, the phenotypic variance in a population can be divided in the following way:

$$V_P = V_A + V_D + V_I + V_E + V_C$$

The concept of heritability

It is only the genetically determined variation which can be utilised for a permanent improvement of the production characteristics in a population. If all the variation is attributable to environment, selection of the phenotypically superior individuals does not result in any alteration in the next generation. In making breeding plans it is therefore necessary to know the relative importance of the heritable and environmental variation of characters. It is mainly the additive effect of genes which can be utilised by selection. The ratio of the additive genetic variance to the total phenotypic variation is therefore of special interest. This variance ratio (V_A/V_P) is usually called the *heritability* and is denoted by h^2. The reason for choosing the symbol h^2 is that it represents the square of the 'path-coefficient' (h) between heredity and phenotype (cf. p. 95).

From the definition of heritability it is evident that the additive genetic standard deviation for a given character is $h\sigma_p$. With the aid of the heritability and the total phenotypic variation it is easy to obtain an idea of the distribution of the animals' breeding values for a character in a population. Fig. 4:9 shows a comparison between the distribution of the phenotypic and the additive genetic value in respect of milk yield. The heritability of this characteristic was assumed to be 0·3. The same data were used as in Fig. 4:1. The phenotypic

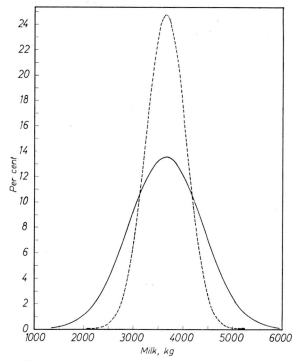

Fig. 4:9. A comparison between the variation of phenotypic and breeding values with respect to milk yield in the first lactation. Same material as in Fig. 4:1, $h^2 = 0·3$ (see text).

standard deviation was 735 kg, whereas the additive genetic standard deviation was estimated to be 402·6 kg.

The coefficient of heritability is an expression for the regression of the breeding value on the phenotypic value; thus $h^2 = b_{A/P}$. Another important aspect of heritability should be mentioned, namely the possibility of predicting the effect of selection in the following generation. If the mean of those individuals selected for breeding deviates from the mean of the population by the amount S (cf. p. 133), then the mean of the progeny generation can be expected to deviate from the original mean by an amount $= h^2 S$.

For this reason it is quite natural that the concept of heritability has come to be one of the corner-stones of quantitative genetics and animal breeding. In assessing the relative value of different breeding methods, consideration must always be taken of the magnitude of the heritability.

Before we go farther into the methods used for estimating heritability, another important concept must be dealt with—the *coefficient of repeatability*. Many characters can be measured several times in the same individual, e.g. the milk yield of successive lactations, or the number of pigs per litter in successive litters from the same sow. The correlation can then be calculated between the repeated results either in a usual correlation analysis or as an intra-class correlation by an analysis of variance. That the result from the same individual repeats itself to a certain degree is rather natural and depends partly on the genotype, which is the same the whole time, even though the effect and activity of some genes may change with age, and partly on certain environmental influences, different for different individuals, so that they result in permanent differences

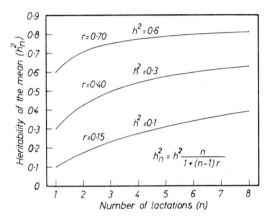

Fig. 4:10. The change in the heritability of the mean (h_n^2) with different values of n, r and h^2 (see text).

between them. The size of the coefficient of repeatability depends upon the following conditions:

$$r = \frac{V_G + V_{Ep}}{V_G + V_{Ep} + V_{Et}} = \frac{V_G + V_{Ep}}{V_P},$$

where V_{Ep} = the variance due to permanent differences between individuals caused by environment, and V_{Et} = variance due to temporary environmental differences which are randomly distributed between and within individuals. The coefficient of repeatability thus includes all the genotypic variation plus a part of the environmental variance and is therefore an upper limit for heritability.

Let us take a character which can be measured several times, e.g. milk yield in successive lactations. Every cow is assumed to have n lactations and $\bar{P}_n$ is the average production for each individual animal from the n lactations.

The total phenotypic variation can thus be expressed in the following way:

$$V_{\bar{P}n} = V_G + V_{Ep} + \frac{V_{Et}}{n}.$$

By the repeated measurements therefore the environmental variance which is due to temporary differences is reduced to V_{Et}/n. As the total variance decreases the heritability is correspondingly increased.

On the assumption that the heritability is the same for every single repeated result, the heritability (h_n^2) for the average of the n replications of the character is obtained as follows:

$$h_n^2 = \frac{nh^2}{1+(n-1)r}$$

The accuracy with which the breeding value of an individual can be assessed increases, therefore, if it is based on the average of several observations. The changes in h^2 for different values of n and r are shown graphically in Fig. 4:10; from which it will be seen that, when the heritability is low, there is most to be gained by using the average from repeated observations. It must be pointed out that the values in Fig. 4:10 are based on the assumption that the heritability for each replication is the same. This assumption does not always apply; the heritability for the milk yield of cows, for example, is lower in the second lactation than in the first; the heritability for the average of the first two lactations may be lower than for the first lactation alone, and repetition of the assessment then brings no gain in accuracy.

Methods of estimating heritability

Heritability can be estimated, in principle, by three different methods, (a) from the phenotypic likeness among relatives (b) from the results of selection experiments and finally (c) by comparisons between the size of the phenotypic variance within isogenic lines and populations mating at random. In the case of the larger farm animals it is, in fact, only the first method that can be used, but we shall briefly mention the other two possibilities.

The estimation of heritability with the aid of isogenic lines. In highly inbred populations, or in crosses between such populations, practically all the variation ought to be due to environmental influences since the animals have the same genotype. An estimate of the size of the genetic variance for a given character in a population mated at random can be obtained by subtracting the variance within inbred lines, or the variance within crosses of inbred lines having the same origin as the randomly mated population, from the total variance of this population. This method has been used on data from experiments with laboratory animals. The assumption made for the method is that the environmental variance in the two groups is the same. Several experiments with laboratory animals indicate however, that the environmental variance is often greater within inbred lines than in populations mated at random. The inbred animals are in fact more sensitive to environmental changes. Estimates of heritability obtained by this method should therefore be regarded with considerable caution. It should also be pointed out that it takes account of all the genetic variance, i.e. even those parts caused by dominance and gene interaction.

Lines so highly inbred that they can be said to be isogenic have not yet been produced in the larger domestic animals, though one-egg twins are isogenic. In cattle the occurrence of one-egg twins is relatively common and it has therefore been possible to conduct experiments on a large scale with such twins. The possibilities of estimating heritability from these experiments will be discussed in Chapter 5.

Calculation of heritability from selection experiments. As already mentioned the selection response (Re) depends upon the heritability and the selection differential (S); thus $Re = h^2 S$. The selection differential is the difference between the mean of the population and the mean of those animals selected for breeding, S being averaged over sexes; Re is that part of S which is regained in the next generation. If S and Re have been calculated in a selection experiment, then the heritability can also be calculated. In order to obtain an estimate of the heritability, selection ought to be carried out for several generations, and one must also maintain a control group, within which no selection takes place, or better still two groups within which selection is carried out in opposite directions. The effect of selection is not necessarily always the same in both directions. Many such selection experiments have been carried out with laboratory animals, and the most important results will be dealt with in Chapter 16. Where the larger farm animals are concerned, it is obviously impossible to conduct selection in an unfavourable direction, or to maintain a control group within which no selection is carried out, simply for the purpose of estimating heritability. Heritability investigations with domestic animals must therefore be based, as a rule, on comparisons of the manifestation of the character in different types of relatives.

Heritability estimates based upon resemblance between relatives. Daughter-dam correlation and regression. Experience shows that full sibs belonging to the same family are more alike than individuals from different families. The degree of resemblance varies, depending upon the character studied. The resemblance is due partly to the fact that, to some extent, relatives have genes in common and partly to the fact that the members of the same family may have a similar type of environment. As a measure of the resemblance between relatives, e.g. dam-daughter, the correlation between them is used or the regression of the progeny on the dam. The causes of phenotypic resemblance between relatives are shown in Table 4:7.

Table 4:7 The genetic and environmental causes of the phenotypic resemblance between different types of relatives

Relation	Regression (b) or Correlation (r)
Parent-progeny	$b = \dfrac{1/2 V_A + 1/4 V_1 + V_C}{V_P}$
Full sib	$r = \dfrac{1/2 V_A + 1/4 V_D + 1/4 V_1 + V_C}{V_P}$
Half sib	$r = \dfrac{1/4 V_A + 1/16 V_1 + V_C}{V_P}$

It should be noted that V_C has not the same value in the three formulae. Generally it contributes much more to the full sib correlation than to the half sib correlation.

The coefficient of relationship between a parent and its progeny is 0·5, if there is no inbreeding (cf. p. 107). In other words the similarity of the breeding values of dam and daughter exceeds by 50 per cent the similarity of the breeding values of individuals paired at random in the population. As a result there will be a covariance of $0·5V_A$ between dams and daughters. That part of the daughter-dam-regression which is due to the genes' additive effect is thus $\frac{1}{2}(V_A/V_P)$. It can also be shown that one-quarter of the variance due to the interaction between two loci, as well as an insignificant part of the interaction variance due to three or more loci, contributes to the resemblance between the dams and their daughters. Finally, there may be a resemblance between them due to similar environments. For the sake of simplicity only the interaction variance due to two loci is considered in Table 4:7.

In the case of dairy cattle, dams and their daughters often remain in the same herd. If dam-daughter pairs from different herds are compared, there automatically occurs a covariation between their milk yield caused by similar environment. This influence can be easily eliminated, however, by making the calculation within herds or, better still, by basing the regression on the animal's deviations from the contemporary mean of the herd. In this way a part of the effect due to differences between years is also removed. The calculation is more difficult if certain families within a herd are given better management than others; but this situation is very rare. The term V_C can quite often be effectively eliminated. The effect of the interaction between different loci is in fact so small that for most purposes it can be ignored. The regression (b) of the daughters' yield (y) on that of their dams (x) becomes thus:

$$b_{yx} = \tfrac{1}{2}\frac{V_A}{V_P} = \tfrac{1}{2}h^2; \quad h^2 = 2b_{yx}$$

The coefficients of correlation and regression will be identical, if the variance among the dams and the daughters is the same. It very seldom is, because usually the dams are a selected group. This selection does not affect the regression coefficient, since the covariance (cf. p. 94) decreases at the same rate as the variance of the dams decreases. The regression coefficient can therefore be used to estimate the heritability, even though the dams are selected. Heritability investigations based on dam-daughter regressions are often made within groups where the daughters are paternal half sibs, i.e. have the same sire. In this way the material is limited in time, since the male animal is used for a relatively short period. Time trends in the environment will therefore have little influence. Furthermore, the possible effect of a correlation between the sire and those females to which he is mated is eliminated.

As an example of heritability investigations based on daughter-dam regressions the following investigation by JOHANSSON is presented (Table 4:8).

He investigated the relationship between the yield of butterfat and the fat percentage of the milk during the first lactation (300 days) of dams and daughters of the Swedish Red and White (SRB) and the Swedish Friesian (SLB) breeds. The basic data were obtained from recording associations. The coefficients of

Table 4:8 The estimation of the heritability of butterfat yield and fat content of the milk in the first lactation (300 days), based on daughter-dam comparisons within herds and bulls (JOHANSSON). x = yield of dams, y = yield of daughters

Source of data	Trait	$\bar{x}$	s_x	$\bar{y}$	s_y	r_{xy}	b_{yx}	$h^2 = 2b_{yx}$
20 SRB herds (60 bulls and 2399 dam-daughter pairs)	Butterfat, kg	140	28·7	147	31·2	0·182	0·196	0·39
	Fat, %	3·94	0·299	4·09	0·285	0·350	0·342	0·68
7 Swedish Friesian herds (61 bulls and 1461 dam-daughter pairs)	Butterfat, kg	156	33·0	154	33·6	0·164	0·174	0·35
	Fat, %	3·58	0·258	3·63	0·262	0·307	0·296	0·59

regression and correlation were calculated within herds and bulls. The yield of the daughters of the SRB animals was somewhat higher than that of their dams, whereas in the SLB material the difference was less pronounced. The variation for the different characters was about the same for dams and daughters, even if there was a slight tendency toward a lower value for the dam group—something which is quite natural if selection has been carried out among the dams. The heritabilities were estimated from $2b_{yx}$. The heritability for butterfat yield in the SRB breed was estimated to be 0·39 and for the fat percentage to be 0·68. The values in the SLB breed were very similar.

Correlation between full sibs. The difference between groups of full sibs is nearly always caused partly by systematic environmental differences between the groups. Animals belonging to the same full sib group have the same dam, and similarities in their early environment can therefore have some influence. The environmental similarities between sibs belonging to the same litter are especially large. The covariation between full sibs is not only caused by environment and additive gene effects, but also includes one-fourth of the variances due to dominance and the interaction between pairs of loci. Heritabilities based upon full sib correlations are therefore often overestimates.

Heritability calculations based upon resemblances between half sibs. The same reasoning can be applied to the causes of correlation between half sibs as to those between dams and daughters. The coefficient of relationship between half sibs is 0·25 in a non-inbred population. The resemblance between half sibs is due to $\frac{1}{2}(V_A/V_P)$, an insignificant part of the interaction variation (V_I) as well as any systematic environmental similarities between members of the same half sib group. The effect of gene interaction on the correlation can be ignored. If no systematic environmental differences occur between the half sib groups, then the heritability is obtained by multiplying the half sib correlation by four. Correlation due to systematic differences of environment between the half sib groups must be avoided, since this error will also be multiplied by four.

Systematic environmental variation can occur in different ways. Take the case, for example, where the feeding of cattle in a district is continually improved, and suppose that we wish to calculate the heritability for a given production character from AI data from this district. If comparisons are made between progeny groups from different AI bulls, some of which were used earlier and others later, then the variation between the groups will be partly due to environmental differences between the periods when the bulls were utilised. When planning heritability investigations, therefore, it is very important to consider possible systematic environmental differences between the half sib groups; otherwise the heritability coefficients can be quite misleading. Such systematic errors can often be avoided if the investigations are based on the animals' deviations from the contemporary mean of the herd.

The correlation between half sibs is usually calculated by an analysis of variance, as an intra-class correlation. It is necessary, therefore, to estimate the variance due to deviations of the mean values for the different half sib groups from the total mean and the variance within the groups. In addition, it is often necessary first to eliminate the influence of the various herds and of year, and other sources of systematic environmental variation. For several reasons the majority of heritability estimates are based on half sib correlations and therefore the principles for the analysis will be dealt with in some detail.

Suppose that we wish to estimate the heritability of butterfat yield and that we have available data from N cows distributed over n herds from a total of k bulls. No bull is represented in more than one herd and all the progeny groups are assumed to be of about equal size. The task then is to calculate the variance caused by the differences between the progeny groups (σ_s^2), after the effect of the different herds has been eliminated (i.e. within herds). The principles for such an analysis of variance are set out in Table 4:9.

Table 4:9 *An example of the estimation of the components of variance due to differences between progeny groups*

Source of variation	Degrees of freedom	Mean square	Components of mean square
Between herds	$n-1$	M_h	$\sigma_w^2 + \dfrac{N\sigma_s^2}{k} + \dfrac{N\sigma_h^2}{n}$
Between progeny groups (within herds)	$k-n$	M_s	$\sigma_w^2 + \dfrac{N\sigma_s^2}{k}$
Within progeny groups	$N-k$	M_w	σ_w^2

The mean square 'within progeny groups' (M_w) represents the best estimate of the variance within progeny groups (σ_w^2). On the other hand, the mean square for 'between progeny groups' (M_s) or between the herds (M_h) does not represent an estimate of the corresponding variance components. The mean square (M_s) includes σ_w^2 and N/k times the variance component 'between progeny groups' (σ_s^2). The components of the mean squares are given in the right-hand column of the table. We have assumed that the number of individuals is exactly the same in each progeny group and that the half sib groups are equal in number within each herd. If the analysis is based on field data, this assumption can seldom be realised. As long as the variation in the number within the groups is not large, a relatively good estimate of the components of variance can be obtained from the mean of the number per group.

The intra-class correlation between half sibs is then calculated as follows:

$$r = \frac{\sigma_s^2}{\sigma_s^2 + \sigma_w^2}.$$

An example of the estimation of heritability, based on half sib correlation, is provided by an investigation into udder proportions of three Swedish breeds of cattle, SRB, SLB and SKB. As a measure of the udder proportions, the amount of milk in the front half of the udder was expressed as a percentage of the total ('front/rear index'). In a similar way the 'left/right index' was computed (cf. Chapter 10, p. 266). The proportions were calculated on the basis of milking trials carried out during two successive days on 591 cows, the progeny of 62 different bulls. The results of the investigation are summarised in Table 4:10. There was a highly significant difference for the 'front/rear index' both between breeds and between groups of paternal half sibs (between bulls) and also between cows. For the left/right index the difference was significant only between cows.

The percentage distribution of the variance among the different causes of variation within breeds is shown in the right-hand part of the table. The

Table 4:10 An analysis of the causes of variation in udder proportions in cattle, see text (from JOHANSSON and KORKMAN, 1952). ***$P < 0.001$

Source of variation	Degrees of freedom	Mean squares		Percentage distribution of variance within breeds	
		Front-rear index	Left-right index	Front-rear index	Left-right index
Total variance	1181	43.5	10.9	—	—
Between breeds	2	2 693.5***	30.5	—	—
Between bulls (within breeds)	59	226.0***	22.0	22.1	1.2
Between cows (within breeds and bulls)	529	59.9***	19.5***	73.5	80.2
Within cows	591	1.8	2.0	4.4	18.6

coefficient of repeatability (within bulls) for the results from the same cow can easily be calculated from these figures. For the front/rear index this is $\frac{73.5}{4.4 + 73.5} = 0.94$, whereas the corresponding value for the left/right index is 0.81. It is quite natural that the animals repeat their result from one day to another. The individual differences in front/rear index were strongly influenced by inheritance. The intra-class correlation between half sibs was 0.221

and the heritability can thus be estimated to be 0·88. There were no heritable differences in left/right index, however, in spite of the fact that the animals showed relatively large individual variation (cf. Chapter 10, p. 266).

Heritability calculations for characters which show alternative variation ('all-or-none characters'). There are a number of important characters in our domestic animals which show alternative variations, i.e. they can assume only one of two values, such as healthy or sick, dead or alive. The distribution is thus binomial, and therefore the earlier methods discussed for estimation of the heritability cannot be used. The genetic background of such characters as resistance to disease seems to be similar as for the majority of quantitative characters, i.e.

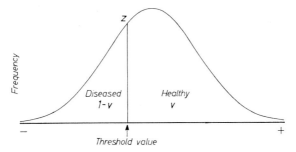

Fig. 4:11. Schematic representation of the genotypic variation in disease resistance. If a certain threshhold value is exceeded with respect to degree of resistance, the animals remain healthy in a given environment (see text).

determined by a large number of genes and the environment. It seems probable therefore that the genotypes in a population show continuous variation with respect to resistance. For an animal to remain healthy in a given environment its genetic resistance must exceed a certain level (*threshold value*). These conditions are shown in Fig. 4:11, which shows the genotypic variation of individuals with regard to resistance to a particular disease. Under certain given external conditions, fraction v of the individuals are resistant, whereas fraction $1-v$ die, or are at least affected by the disease. The dividing line between these two fractions marks the genotypic resistance (threshold value), which must be exceeded if the individual is to be healthy in the given environment. The level of the threshold is determined to a very large extent by environmental conditions, something which is especially pronounced in infectious diseases. If there is no source of infection, all the animals will be healthy regardless of the genotypic variation to resistance, i.e. the threshold value is pushed further to the left in Fig. 4:11. On the other hand a deterioration of environmental conditions, e.g. malnutrition, leads to a decrease in the fraction of animals which can resist an attack of the disease.

As resistance to disease is binomially distributed, parent-offspring correlations cannot be calculated; but regression analysis is possible. LUSH (1950) studied the resistance to mastitis, and determined the regression of the state of health of the daughters on that of their dams. The principle of the analysis was briefly as follows. The dams were divided into two groups, healthy and diseased. The difference between these was set at 100 per cent. The state of health of the daughters was then investigated. If 40 per cent of the daughters of healthy dams showed the disease whereas the daughters of infected dams showed the disease in 60 per cent of the cases, then the difference had been reduced from 100 per cent in the dams to 20 per cent in the progeny generation. If it is assumed that the bulls used on the healthy and infected dams respectively were, on the average, equal, the heritability could then be estimated to be 2 × 20 per cent, or 0·4. A total of 494 dam-daughter pairs were included in Lush's investigation, and by the procedure just outlined the heritability for resistance to mastitis was estimated to be 0·38.

Where large progeny groups are available, as in AI breeding of cattle, the progeny from the same male animal can be divided into several groups of, say, 20 individuals. Within each of these sub-groups the average resistance of the progeny can be calculated, i.e. the percentage of healthy individuals. These percentages can then be used as variables in a normal analysis of variance, and the variance caused by the differences between the sires can be calculated.

Such an analysis was carried out by LUSH, LAMOREUX and HAZEL (1948) for the resistance to different poultry diseases. The study included the resistance to leukosis and to diseases in general. The investigation was carried out on data collected during three years from a large American poultry breeding establishment. The mortality due to leukosis during the first laying year reached a maximum of 22 per cent. The animals were classified in the following way:

1. Birds which survived the first laying year;
2. Birds which died before the end of the first laying year:
   (a) Birds affected by leukosis.
   (b) Birds without leukosis.

In order to get fairly reliable mortality figures, they used only data from cockerels which had such a large number of progeny that these could be divided into several full sib groups of at least 10 birds at the beginning of the laying year. The resistant birds were given the value 1 and those which died were given the value 0, following which the analysis of variance was carried out in the usual way according to the following scheme:

	Variance
Between paternal half sib groups	σ_s^2
Between full sib groups (within cocks)	σ_d^2
Within full sib groups	σ_w^2

The heritability was then calculated from:

$$\frac{4\sigma_s^2}{\sigma_s^2 + \sigma_d^2 + \sigma_w^2}.$$

The variance which is obtained from binomially distributed data is correlated with the mean. The variance becomes therefore small both at low and high frequency of the character (in this case death due to leukosis). It is therefore not possible to compare the heritability from populations with differing mortality rates. A correction was made by Lush and his co-workers for the frequency of surviving birds (v) by multiplying the heritability by the following factor $\dfrac{v(1-v)}{z^2}$, where z is the height of the ordinate of the normal curve at a point on the x-axis where the fraction v intercepts (Fig. 4:11). The result of the investigation carried out by Lush and co-workers (1948) can be summarised as follows:

	Heritability	
	Uncorrected	Corrected
Total mortality	0·083	0·145
Mortality due to leukosis	0·068	0·156
Mortality due to other causes	0·031	0·074

Some general points about heritability, limitations and random variation

The heritability of a given character is the ratio between the additive genetic variance and the total variance in a population. The size of the heritability is therefore influenced by changes in both the numerator and denominator. The additive genetic variance within a population can be changed in many ways. It can be increased through crossing when new genetic material is introduced into the population. Inbreeding, on the other hand, decreases the additive variance within lines, whereas the differences between the lines increases. The inbred animals are in addition often less well adapted to variations in the environment which may result in an increase in the total phenotypic variance.

The reasons for differences in estimates of heritability between different populations can often be found in differences in the environmental variance. The variance caused by errors of measurement is usually reckoned as a part of the environmental variance. In a population where accurate production recording is carried out, the heritability estimate for the character is automatically higher than in an otherwise similar population where the errors of measuring production are large. The environmental variance can be influenced also by the planning of the experiment. In the Danish pig progeny testing (cf. Chapter 11), the pigs are individually fed. The total variance of daily growth rate within the test litters, for example, is considerably less with individual feeding than with group feeding, no doubt due to the fact, that the competition factor at feeding is eliminated. The variance between half sib groups, however, is only slightly affected and, as a result, a much higher heritability for daily growth rate is obtained from individual than from group feeding.

It must be stressed therefore that the heritability for a given character is not in any way a natural constant. It is strictly applicable only to the array of genotypes in a given environment, i.e. to the population from which it was calculated. This limitation, however, is not so serious as would first appear. The estimates of the heritability for a given character within different populations which live and are recorded under similar conditions usually give results which agree fairly closely, at least when the calculations are based on a large amount of data.

In comparisons between different heritability estimates, it is necessary to take into consideration the random variation of the coefficients of heritability. These are affected by the size of the sample and its composition, i.e. the type, number and size of the family groups. The sampling variance of heritability and the optimum structure of the material has been discussed in detail by ALAN ROBERTSON (1959). The error of estimate of the heritability, calculated from daughter-dam regressions is obtained from the error variation of the regression coefficient (b)

$$\sigma_b^2 = \frac{1}{N-2}\left[\frac{\sigma_y^2}{\sigma_x^2} - b^2\right],$$

where N is the number of pairs, y the dependent variable and x the independent.

Since b is usually a small number

$$\sigma_b^2 \approx \frac{1}{N}\frac{\sigma_y^2}{\sigma_x^2} \approx \frac{1}{N},$$

$$\sigma_h^2 = (2)^2\,\sigma_b^2 \approx \frac{4}{N}.$$

When estimating the heritability from the intra-class correlation (t) between half sibs, the sampling variance is obtained from

$$\sigma_t^2 = \frac{2[1+(n-1)t]^2(1-t)^2}{n(n-1)(N-1)}$$

where N = the number of half sib families and n = the number of half sibs per family.

It can be seen from the formula that the error is relatively large when n is small, since n^2 occurs in the numerator. ROBERTSON has shown that with a given total size (T) of the material, and $T = n \times N$, the sampling variance is at a minimum when $n = \frac{1}{t} = \frac{4}{h^2}$. The error variance of the intra-class correlation with this optimum family size is

$$\sigma_t^2 \approx \frac{8t}{T} \approx \frac{8t^2}{N}.$$

From this an approximate estimate can be obtained of the error variation of the heritability calculated from the half sib correlation

$$\sigma_{h^2}^2 \approx \frac{32h^2}{T}.$$

The heritability is seldom known in advance; but the reduction in precision is considerably greater if the number of individuals per half sib group is less than the optimum than if it exceeds the optimum. It is advisable that the half sib group consist of 20–30 individuals.

Robertson has also compared the size of the random error when the heritability is determined from daughter-dam regression, with the error when it is estimated from the correlation between half sibs. With equal numbers of animals and a heritability less than 0·25 the half sib method is subject to less error than the daughter-dam regression, provided that the half sib groups are of optimum size.

Correlation between characters

Many of the production characters of domestic animals are correlated with each other. For instance, big cows give more milk, on the average, than small cows, but the yield is negatively correlated with the fat percentage of the milk

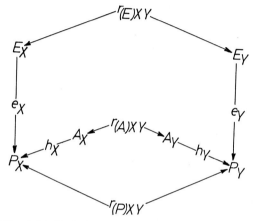

Fig. 4:12. Path diagram showing how the phenotypic or total correlation between two characters X and Y is composed of the genetic correlation and the environmental correlation. (After LERNER 1950)

($r \simeq -0.25$). In pigs there is a correlation of about -0.35 between the back fat thickness and length of carcass when the pigs are slaughtered at 90 kg live weight.

The phenotypic correlation between two characters can be influenced by inheritance, environment, or both. When the correlation is mainly genetic, consideration must be taken of this in the breeding plans. A genetic correlation can be due to pleiotropy, i.e. a single gene influencing both characters at the same time, or to linkage between two gene loci, each of which influences a single character. In the latter case the strength of the relationship depends upon how strongly the two loci are linked. Crossing over between them leads eventually to all combinations being represented in proportions which correspond to the zygote frequencies in the prevailing mating system. As mentioned earlier, it may take many generations before equilibrium occurs between two linked gene systems. If the coupling is so strong that crossing over seldom or never takes place there is, in principle, no difference between pleiotropy and linkage. A temporary genetic association between two characters can occur also when two independent genes, each responsible for its own character, are simultaneously introduced into the breed by a certain breeding animal, which is then used extensively. This relationship is broken down, however, quite rapidly under conditions of random mating (cf. p. 97).

The way in which the phenotypic correlation between two characters X and Y is made up of genetic and environmental correlations is shown in Fig. 4:12. The correlation (path coefficient) between an individual's genotype and phenotype is equal to the square root of the heritability for the character (h). In a similar way the path coefficient between those environmental factors which influence the character and the phenotype can be represented by e. If we further assume that there is a correlation ($r_{(A)XY}$) between the breeding values for the two characters and that the environmental influences are also correlated ($r_{(E)XY}$) with each other, then the total phenotypic correlation (r_{XY}) has the following form:

$$r_{XY} = h_X r_{(A)XY} h_Y + e_X r_{(E)XY} e_Y.$$

The genetic correlation between two characters can be defined as the ratio between their genetic covariance and the product of their genetic variation, thus

$$r_{(A)XY} = \frac{\text{Covariance}_{(A)XY}}{\sigma_{AX}\, \sigma_{AY}}.$$

HAZEL (1943) at Iowa State University, U.S.A. developed general methods, with which it is possible to estimate the genetic correlation between two characters on the basis of the covariation between the characters in related individuals. The principle is the same as in the calculation of the heritability. Instead of calculating that part of the variance which is additively genetically influenced, the corresponding covariance is calculated. If the analysis is based upon groups of half sibs then the covariance between half sib groups is calculated, which represents one quarter of the genetic (mainly additive) covariance. The genetic variances are estimated in the usual way (cf. p. 116), and then, the genetic correlation is calculated from the above formula. The environmentally influenced correlation ($r_{(E)XY}$) can then be calculated from the fact that for any given character $e^2 + h^2 = 1$ and that cov XY = cov $(A)XY$ + cov $(E)XY$ (cf. the discussion on path and determination coefficients, p. 95).

The estimations of the genetic correlation should be treated with great caution since it represents a ratio between values which are themselves estimated with considerable uncertainty. It is important that the related individuals are not influenced by common environment factors, i.e. V_c must be 0. Robertson has shown that methods and material which give optimum precision of heritability estimates are also optimum for estimations of the genetic correlation (cf. p. 30).

In spite of the uncertainties attached to the estimation of genetic correlation it is of great interest in principle. If a genetic relationship exists between two characters then selection for the one character also brings about changes in the other. We shall return to these correlated selection responses in a later section. The genetic correlation of a number of characters has been calculated. FREDEEN and JONSSON (1957) found a phenotypic correlation between the carcass length and back fat thickness of pigs of −0·24, whereas the corresponding genetic correlation was −0·47. MORLEY (1955) estimated the phenotypic and genotypic correlation between body weight and wool production in sheep to be +0·36 and −0·11 respectively. A negative genetic relationship can thus occur between two characters in spite of the fact that the phenotypic correlation is positive. In the case of wool production and body weight of sheep it appeared that the environmentally influenced positive relationship was so strong that it was more than enough to nullify the negative genetic correlation.

Selection

In an earlier section of this chapter it was shown how the gene frequencies in a population can be altered by selection. With quantitative characters, the individual gene effects are so small that they cannot be separated. We can, however, assume that the same principles which applied to the qualitative characters are also valid in selection for quantitative characters. If, for example, selection is made for high body weight then the frequency of genes which have an increasing effect on the weight will increase. The magnitude of the change will depend upon the intensity of selection and the relative effect of the different loci on the variation of the character. As a measure of the strength of the selection, the difference between the mean of those animals selected for breeding and the mean of the population may be used. This difference is called the *selection differential* (S). If, for example, the average yield for all cows in a herd is 4000 kg of milk per lactation period and the mean of cows used for breeding purposes is 4500 kg then $S = 500$ kg milk.

The difference between the mean of the selected characters in the new generation and the mean of the parent generation before selection took place is called the *selection response*. This is denoted by Re and it is a measure of the change in the additive genetic value of the population. It was pointed out in the discussion on heritability that $Re = h^2 S$. The expression $h^2 S$ is a measure of how large a part of the difference between the parent and the population mean can be attributed to additive inheritance and thus transferred to the next generation.

If all the animals in a population are used for breeding and reproduce themselves at the same rate, the selection differential is zero and no alteration in the mean value will take place in a constant environment. The selection differential for mares and cows is of necessity low, since one cannot expect to obtain a large number of female foals, or heifer calves, per female. The selection differential for male animals can be quite large, especially with artificial insemination. In poultry breeding relatively strong selection can be carried out both among hens and cockerels. When the force of selection among males and females is different, the real selection differential is the mean of the selection differentials of the two sexes.

The response to selection per year (Re_y) depends upon the time interval between the birth of the parents and the offspring, i.e. the generation interval. $Re_Y = \dfrac{h^2 S}{y}$, where $y =$ the generation interval measured in years.

It is often desirable to give the selection differential in standard deviation units instead of absolute measure. It is then possible to compare directly the selection differentials in different populations and for different characters, e.g. milk yield, fat content and body weight. This standardised selection differential will henceforth be called the *selection intensity* (represented by i). If it is desired to express the selection response in absolute units the following relationship applies:

$$\frac{Re}{\sigma_p} = \frac{h^2 S}{\sigma_p}; \quad Re = \sigma_p i h^2.$$

The selection intensity can be easily calculated from a knowledge of the percentage animals used for breeding and vice versa. These values can be obtained from tables of ordinates and areas for the normal curve. This relationship will be illustrated in Chapter 16.

The correlated selection response. The results from a number of selection experiments, some of which will be discussed in Chapter 16, show that selection for a given character often leads to changes in other characters. This is because the characters are genetically correlated. Let us suppose two characters X and Y, and that selection is carried out for X. The correlated change in Y is determined by the regression $(b_{(A)YX})$ of the breeding value of Y on the breeding value of X, i.e. ratio of the genetic covariance for the two characters and the genetic variance in X, thus:

$$b_{(A)YX} = \frac{\text{cov}_{(A)YX}}{\sigma^2_{AX}} = r_{(A)XY} \frac{\sigma_{AY}}{\sigma_{AX}}.$$

The conditions are in principle the same as with selection for a single character when the response to selection is influenced by the regression of the breeding value on the phenotype, i.e.

$$Re_X = h^2_X S_X.$$

The correlated selection response in $Y = b_{(A)YX} Re_X$, which can also be written as:

$$i h_X h_Y r_{(A)XY} \sigma_{(P)Y};$$

where $i =$ the selection intensity, $\sigma_{(P)Y} =$ the phenotypic variation of the character Y and $r_{(A)XY} =$ the genetic correlation between X and Y.

Cases can therefore be imagined where it is profitable to select for one character in order to achieve improvement in another character, for example when Y occurs only in the one sex whereas X occurs in both sexes. A condition for such indirect selection is, however, that there is a strong genetic correlation between them.

Selection methods. Selection in practice will be discussed in Chapter 16, after the influence of inheritance on the different production characters has been discussed (Chapters 8–13).

Selected References

Falconer, D. S. 1960. *Introduction to Quantitative Genetics* (365 pp.). Oliver and Boyd, Edinburgh.
Hazel, L. N. 1943. The genetic basis for constructive selection indexes. *Genetics*, **28**: 476–490.
Kempthorne, O. 1960. *Biometrical Genetics* (234 pp.). Pergamon Press, London.
Lerner, I. M. 1958. *The Genetic Basis of Selection* (298 pp.). John Wiley and Sons, New York.
Le Roy, H. L. 1960. *Statistische Methoden der Populationsgenetik*. Birkhäuser, Basel.
Lush, J. L. 1949. Heritability of quantitative characters in farm animals. *Hereditas, Suppl.*, 356–375.
Ostle, B. 1954. *Statistics in Research* (487 pp.). Iowa State University Press.
Pirchner, F. 1964. Populationsgenetik in der Tierzucht (212 pp.). Paul Parey, Hamburg.
Robertson, A. 1959. Experimental design in the evaluation of genetic parameters. *Biometrics*, **15**: 219–226.

5 Multiple Births. Twin Research

MULTIPLE BIRTHS IN NORMALLY MONOTOCOUS ANIMALS

On the basis of the number of offspring per pregnancy, farm animals can be divided into two groups, those which produce a single offspring, known as *monotocous* and those producing several young, known as *polytocous*. Pigs, dogs, cats, mink and rabbits are manifestly polytocous, the birth of only one offspring being exceptional. Horses and cattle are monotocous, though twins and higher degrees of multiple births occur occasionally. Sheep and goats take an intermediate position with single births being about as common as multiple births. Primarily, the number of offspring per pregnancy is determined by the number of ova shed during the heat period and the male with which the female mates has no influence in this respect. The male can, on the other hand, influence the fertilisation percentage and also, by virtue of his genotype, the mortality of the foetuses in the uterus. There are four ways in which multiple births can occur in normally monotocous animals:

1. Two or more ova are shed during the same heat period and these are fertilised by different sperm from the same male. This is the most common way in which multiple births occur in both monotocous and polytocous animals. Twins produced in this way are said to be *two-egg* or *dizygotic* (DZ) twins. The genetic likeness between two-egg twins is on the average the same as for full sibs produced by different pregnancies.

2. If two or more eggs are shed during the same heat period and the female mates with two or more males, the ova can be fertilised by different males. Offspring produced in this way are only half sibs.

3. Sometimes a female which has been on heat, mated and became pregnant may come on heat again and also ovulate. If the female then mates during this new heat period, a new pregnancy can occur alongside the first. This is known as superfoetation and several cases are described in the literature. When such twins are born with several weeks' interval, or if simultaneously they are at different stages of development, this is a sign, but not necessarily proof, of superfoetation. Superfoetation is quite common in mink.

4. When a single fertilised ovum splits at an early stage of development so that two foetuses develop from the same zygote, the offspring are known as *one-egg* or *monozygous* (MZ) *twins*. Such twins are always of the same sex, and since they contain the same set of genes, they are frequently referred to as *identical twins*. Another possibility would be that an ovum divides just prior to fertilisation and

that these two ova are fertilised by different sperm. This could be an explanation of why a number of twins are very much alike without being one-egged twins. This conjecture, however, is not supported by any real evidence.

Apparently the splitting of the fertilised ova can take place at quite different stages of embryonic development, in some cases at the first cleavage, in others during the blastocyst stage or even after gastrulation. When splitting takes place during the later stages, there is a risk that it is not complete and gives rise to so

Fig. 5:1. 'Siamese' twin calves showing mirror image likeness, especially on the front legs. (JOHANSSON 1932)

called 'Siamese', or conjoined, twins. Such partial duplications have a low frequency in all classes of animals. Sometimes it is only the head or some other extremity which is duplicated; in other cases the whole of the front or hind part of the body may be duplicated. These duplications, which seem to occur at random, generally give rise to difficulties at parturition. It is believed that the 'mirror image' likeness of normally developed one-egg twins is due to the fact that the splitting of the zygote takes place at a later stage of development, when to a certain extent the fixation of right and left characteristics has already taken place (Fig. 5:1).

Triplets and quadruplets can be monozygotic, but more usually they represent a combination of mono- and dizygotic foetuses. Several cases of monozygotic triplets and one case of MZ quadruplets have been noted in cattle. The frequency of MZ twins in cattle is relatively high. Only one case of monozygotic twins has been demonstrated in horses, and only one in sheep; the frequency in these species is apparently very low. Study of the foetuses and foetal membranes indicate that in rare cases monozygotic twins can occur even in polytocous animals. To trace living examples is extremely difficult, however, and up to now the search has not met with any success.

The heritability and frequency of multiple births

The frequency of twin pregnancies in *horses* is between one and two per cent of all pregnancies, and only a small number of the twin foetuses are born alive. In *cattle*, the frequency of twins is relatively high in the large breeds such as Simmentaler and Friesians, considerably lower in the Jersey and Guernsey and still lower in the beef breeds such as Hereford and Aberdeen Angus. If the sex combinations of an unselected sample of twins are known, it is possible to estimate the frequency of monozygotic twin pairs. Where dizygotic twins are concerned the following three possible sex combinations can be expected to occur with the following frequencies: $p^2 ♂♂ + 2pq ♂♀ + q^2 ♀♀$, where p = the fraction of bull calves and q = the fraction of heifer calves in the twin material. In a large sample $p = q = 0.5$. If no MZ pairs occur, the same number of like-sexed as unlike-sexed twin pairs can be expected, i.e. $p^2 ♂♂ + q^2 ♀♀ = 2pq ♂♀$. The excess of like-sexed twins over the unlike-sexed can be assumed to be due to the occurrence of MZ pairs. This excess is expressed as a percentage of the total number of like-sexed twin pairs. Table 5:1 shows the frequency of twins and

Table 5:1 *The frequency of twin births and MZ births in some breeds of dairy cattle*

Breed	Total No. of births	% twin births	MZ pairs as % of like-sexed pairs
Simmental	12,625	4·61	6·00 ± 7·43
Holst. Friesian (U.S.A.)	18,736	3·08	8·60 ± 7·62
Swedish Friesian (SLB)	24,670	3·32	6·81 ± 3·19
Swedish Red and White (SRB)	53,554	1·85	11·05 ± 3·31
Swedish Polled (SKB)	3,751	1·81	26·78 ± 10·06
Jersey (New Zealand)	87,926	1·02	16·60 ± 6·06

the estimated frequency of MZ twins in some of the dairy breeds of cattle. The differences between breeds in the frequency of twins is statistically significant and most probably the greater part is genetically determined. The estimated frequency of monozygotic twins, as a percentage of all like-sexed pairs, is on the average, for the breeds studied, $10 \cdot 61 \pm 1 \cdot 84$ per cent. No one breed differs significantly from this average. The figure for the Swedish Polled is very high but the number of twin pairs is small and estimation of the frequency of MZ twins is therefore dubious as indicated by the standard deviation. The conclusion to be drawn from the investigations carried out is that about 10 per cent of all like-sexed twins in dairy cattle are monozygous.

The frequency of triplet births in cattle is very low, probably only about one

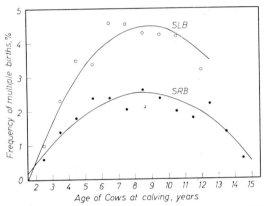

Fig. 5:2. The relationship between multiple births and the age of the dams in the Swedish Friesian and the Red and White breeds. (JOHANSSON 1932)

case per 10 000 births, and the occurrence of quadruplets and higher degrees of multiple births is very rare indeed.

Several different factors influence the frequency of multiple births in a given breed of cattle, e.g. the age of the cow, the time of the year when conception takes place, disturbances in the hormone balance of the cow and, to a slight extent, genetic disposition. The frequency of multiple births in cows under three years of age is very low, but increases thereafter up to 6–10 years of age, after which the frequency decreases again (Fig. 5:2). Seasonal variation is less evident, but it appears as if conception during the autumn (September–October) results in more multiple births than conception at other times of the year. According to American investigations, there is some relationship between cystic ovaries and multiple births.

If a cow gives birth to twins once, the chances of twin births in subsequent calvings is three or four times higher than for the average of the population. If the ovulation of two or more eggs during the same heat period is due to an environmentally induced hormonal disturbance, then it is very likely that it is just this disturbance which remains for several years after. Cystic ovaries show a clear tendency to repetition. In the SRB breed referred to in Table 5:1 the daughters of twin cows had 3·72 per cent twin births compared with 1·77 for other cows. This indicates that the heritability is about 0·04. KORKMAN found, in an analysis of the same data, that there was no correlation between dams and daughters, but the correlation between full sisters was 0·1; which indicates that the genetic variation, which no doubt is present, is of a non-additive nature. It is fairly safe to assume, however, that the influence of inheritance on the variation of twin frequency is very small within breeds.

Polyovulation and multiple births have been obtained in cows of the beef breeds by hormone injections before the heat period. The greatest difficulty lies

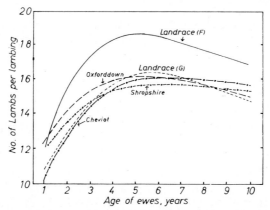

Fig. 5:3. The relationship between the age of ewes and the number of lambs born at each lambing in Swedish sheep breeds: Landrace (F) sheep on the mainland; Landrace (G) sheep on the island of Gotland. (JOHANSSON and HANSSON 1943)

in regulating the hormone dose according to the individuality of the cows. In a large-scale experiment carried out by the Milk Marketing Board in England, 320 cows were injected 2–7 days before the onset of heat with 800–1200 IE of gonadotropic hormones (PMS). Superovulation (two or more eggs) was induced in 37 per cent of the treated cows but only half of these resulted in twin births, i.e. about 18 per cent of the treated cows. No disadvantageous side effects could be attributed to the treatment. It is possible that the method can be further developed and better results achieved. Artificially induced twinning in the beef breeds could be of real economic importance.

Twins and triplets are normal in *sheep*. Table 5:2 shows the results of an investigation into the frequency of multiple births in Swedish breeds of sheep. The data were obtained from registered flocks where the feeding and management were above the average level for the breeds concerned.

The frequency of multiple births is highest in the Landrace (F) on the mainland and lowest for the Landrace (G) on the island of Gotland. Triplets are relatively common in the Landrace (F). To some extent, but not entirely, the difference between these two Swedish Landrace sheep is due to different environmental conditions. In the data presented in Table 5:2 the fre-

Table 5:2 The frequency of multiple births in Swedish breeds of sheep (JOHANSSON and HANSSON, 1943)

	No. of births	No. of lambs per lambing	Percentage multiple births			
			Twins	Triplets	Quads	Quins
Oxford Down	6,838	1·53	45·8	3·4	0·06	—
Shropshire	16,765	1·48	43·0	2·4	0·10	0·01
Cheviot	25,779	1·46	43·6	1·3	0·06	—
Landrace (F)	7,277	1·66	54·8	5·2	0·33	0·03
Landrace (G)	1,722	1·43	42·2	0·5	—	—

quency of like-sexed twins was lower than the frequency of unlike-sexed; which is evidence for the fact that monozygotic twins must be comparatively rare in sheep.

The influence of the age of the ewes on the frequency of multiple births is nearly as pronounced in sheep as in cattle (Fig. 5:3). The time of mating also has a clear effect. The mating season for sheep is in the autumn and the frequency of multiple births is highest for ewes mated about the middle of the season when the sexual activity is at a peak. Within breeds the number of lambs per lambing increases with increasing body size of the ewes. The influence of the plane of nutrition has been demonstrated in English and American experiments. When the ewes are in comparatively low condition the frequency of multiple births can be increased by 'flushing', which is the term given to the practice of moving the ewes onto better grazing and possibly giving supplementary concentrate feed a few weeks before the tupping season. When the ewes are in a 'rising condition' immediately before and during the tupping season more eggs are shed and the number of lambs per ewe is increased.

In the previously mentioned investigation into lamb production in Swedish breeds of sheep the repeatability of number of lambs per lambing for the same ewe was estimated to be 0·18 and the heritability to 0·10, both figures within breeds, flocks and age groups. There is thus a certain amount of genetic variation within breeds, but it is relatively small.

Disadvantages of multiple births

As a general rule, the more pronounced the tendency to single births in a species of animal, then the less is the viability of offspring in multiple births. Twin births in horses are always detrimental. An investigation into the German half-blood (East Prussian) showed that both twin foetuses were aborted in 65·7 and one of the foetuses in 9·1 per cent of all twin pregnancies. Furthermore, only 29 per cent of twin foals born alive survived to maturity.

Twin calves in *cattle* have, on the average, a birth weight 20–30 per cent lower than single born calves, and the death rate is 3–4 times higher. In addition, the calving interval is prolonged after the birth of twins, as the cows are more difficult to get in-calf. Furthermore, about 90 per cent of all heifers born twins to bull calves will be sterile, because of defective development of the genital organs. About fifty years ago, KELLER and TANDLER in Austria and LILLIE in the U.S.A. demonstrated, independently of each other, that in the majority of cases the foetal membranes of bovine twins fuse at an early stage of embryonic development and that vascular anastomosis is established between the foetuses (Fig. 5:4). The explanation of the intersexuality of the female twin was believed to be that, because of the vascular anastomosis, male hormones enter the female twin and inhibit the normal sexual differentiation. Recently, however, it has been found that primordial germ cells are transported with the blood stream from one twin, or triplet, to another thus producing a 'germ cell chimerism' in the foetuses. This has been found to take place already when the foetuses have attained a crown-rump length of 10–12 mm. At present there is

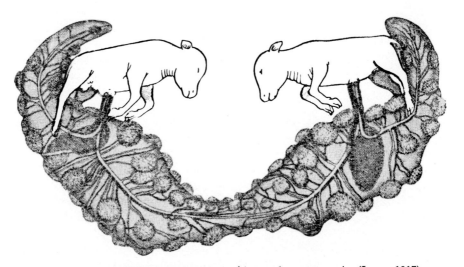

Fig. 5:4. Fusion of the foetal membranes with vascular anastomosis. (LILLIE 1917)

no genetic evidence that functional gametes are produced by transplanted germ cells, and it has not been explained why the sexual development of only the female foetuses is affected by the germ cell chimerism. The genital development and fertility of males born co-twin to females is perfectly normal.

Intersexual female cattle twins are usually called 'freemartins'. The ovaries remain very small, the uterus and vagina are underdeveloped and there is often no communication between them, the vagina being merely a blind end. The vulva is small but the clitoris is usually large. The teats are also smaller than normal. At maturity, freemartin heifers have a steer-like appearance and never come on heat. If the foetal membranes do not fuse, or the fusion takes place at a later stage so that no circulatory anastomosis is established, then the heifer is just as likely to be fertile as a single born heifer. However this is the case in only about 10 per cent of heifers born twin to a bull.

Fusion of the foetal membranes also takes place in sheep when several embryos develop simultaneously in the uterus, but it is very seldom that anastomoses occur and it can generally be reckoned that unlike-sexed twins are not affected. However, live weight at birth decreases, and death rate of lambs increases as shown in an analysis of data obtained from Swedish breeds of sheep (Table 5:3).

Table 5:3 The mortality, weight at birth, and at 5 months, of single and multiple birth lambs (JOHANSSON and HANSSON, 1943)

	Oxford Down	Shropshire	Cheviot	Landrace (F)
Mortality, per cent				
Single born	9·2	9·4	9·1	4·3
Twins	15·4	16·0	13·8	6·2
Triplets	31·0	31·5	27·7	18·1
Birth weight, kg				
Single born	6·0	4·9	4·0	2·9
Twins	5·1	4·1	3·4	2·7
Triplets	4·2	3·6	3·1	2·3
Weight at 5 months, kg				
Single born	42·1	35·1	30·9	28·4
Twins	39·2	31·6	27·6	24·5
Triplets	39·6	33·4	26·3	23·0

Twin lambs are an advantage when feed availability is average to good; for the annual weight of lambs produced per ewe increases considerably compared with single born lambs. There is hardly any advantage with triplets or quadruplets on account of the high death rate of lambs. The weight of multiple born lambs at five months of age is on the average lower than that of single born lambs, but in conditions of good feeding the difference evens out.

VARIATION IN LITTER SIZE IN POLYTOCOUS ANIMALS

A typical example of a polytocous animal is the pig. What can be said about the pig applies in principle to other polytocous animals though the number of young per litter varies between species and between breeds of the same species. In comparing different rabbit breeds, for example, there is a strong relationship between body size and number of young per litter. The Belgian Giant gives birth, on the average, to 9 or 10 young, the medium large White 8 and the small Polish rabbit only 4 young per litter.

Distinct breed differences are found in pigs in respect of size of litter at birth. At the Beltsville Research Centre, U.S.A., the following average sizes of litter were obtained for different breeds under practically identical conditions: Yorkshire 11·0, Duroc 9·5, and Poland China 6·9. The long slim bacon breeds have a higher fertility than the short, chunky pork and lard breeds. The Swedish Landrace and the English Large White (Yorkshire) both have an average of very nearly 11 pigs per litter.

The variation in litter size of herdbook recorded sows of the Large White breed are shown in Fig. 5:5. The litter size varied between 2 and 21. Single births can occur in rare cases and litters are sometimes born with 25 pigs or more. Experiments have shown that sows generally shed considerably more eggs during heat than correspond to the number of pigs born. Foetal deaths in the uterus increase with an increasing number of ova released and is, on the average, estimated to be about 40 per cent.

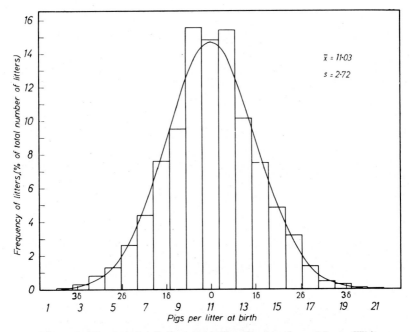

Fig. 5:5. The variation in litter size of herdbook registered Large White sows in Sweden.

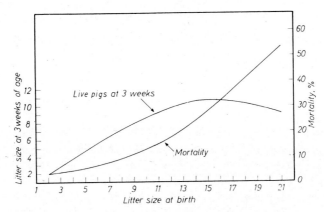

Fig. 5:6. The relationship between the number of pigs born per litter and the litter size at 3 weeks old, together with piglet mortality up to that age. (AXELSON 1934)

One reason for the variation in litter size is the age of the sows. An analysis carried out by KORKMAN on records from breeds of pigs in Sweden showed that litter size increased from 9·7 in the first litter to 11·0 in the fourth to sixth litter, after which the number decreased slightly. The quality and intensity of feeding is of great importance, whereas time of year has very little influence. If the sows are over-fat, the number of eggs released is decreased. Over-fatness during pregnancy results in increased foetal deaths. Deficiency of minerals and certain vitamins, as well as unsuitable protein mixtures, have the same effect.

That litter size is largely a matter of chance is illustrated by the fact that litter size varies within very wide limits from one farrowing to another for the same sow. Investigations into the repeatability of the same litter size show that the individuality of the sow has some influence on litter size. The coefficients of repeatability obtained vary between 0·10 and 0·20. Heritability has been estimated by a number of authors, and values from 0·03 to 0·12 have been obtained.

As the number of pigs per litter increases, so the birth weight decreases and death rate during the first week of life increases. It has been shown in several investigations that within litters of a given size the risk of death is greatest for the smallest pigs in the litter. Fig. 5:6 (after AXELSSON) shows that the number of pigs surviving at three weeks was at a maximum for litters with 14–16 at birth. A limiting factor is the number of functional teats of the sow which generally is not greater than 14–16. With large litters the amount of milk produced by the sow is often insufficient. The weaker pigs have to be content with those teats which give the least milk and thus are progressively stunted in their development. The variation in the weight of pigs at weaning at 7–8 weeks of age increases with increasing litter size. The average number of pigs per litter surviving at three weeks of age is a little over 9 in the Swedish breeds of pigs. On account of the low heritability for number of pigs born per litter, the possibilities for increasing litter size by selection are rather limited. Any gain in this direction would in any event be very small, unless the number of teats on the sow increased simultaneously and there was a general improvement in the milking capacity and mothering ability.

THE USE OF MONOZYGOUS TWINS IN RESEARCH

From the foregoing it will be clear that it is only with cattle that monozygous twins occur in such numbers that they can be of importance as experimental animals. In the 1920s there were already descriptions in the literature of cases of cattle twins which were so alike in appearance that it was assumed that the twins were monozygous. In 1930 C. KRONACHER of Germany published an article on the use of such twins in animal research, and in 1932 he published a larger work on the diagnosis of monozygosity. In the same year JOHANSSON, working with Swedish material, published the first estimate of the frequency of monozygosity in cattle twins, according to which 11·4 per cent of all like-sexed twins could be assumed to be monozygous.

When studying the influence of a particular environmental factor, e.g. plane of nutrition, milking interval or housing, on one or several characters, e.g. milk yield or milk composition, it is obviously a tremendous advantage if the animals which are to be subjected to different treatments are genetically identical throughout the experiment. In order that this advantage can be utilised, it is first necessary to distinguish accurately between mono- and dizygotic twins. The error in diagnosis ought not to exceed one, or at the most two, per cent.

Kronacher and his co-workers developed a method of diagnosis, which was based on a study of the similarities and differences between the twins in respect of a large number of traits. Consideration was given to muzzle print, hair whorls, coat colour and pattern as well as a number of body measurements. The degree of similarity for each character was scored according to a point scale 1–3, and an average score, called the 'similarity index', was calculated for each pair. However, no clear dividing line between mono- and dizygotic twins was possible on the basis of this index, and later researchers have not used the index calculation. Kronacher and his co-workers also tried blood grouping, but the method was far too undeveloped in the 1930s to give any result of real value. Since then the situation has changed radically, as far as blood groups are concerned.

Valuable contributions to the improvement of similarity diagnosis have been made primarily by the Ruakura Research Station in New Zealand and also by Bonnier and co-workers at the Wiad Experimental Station, Sweden. Similarity of muzzle prints and hair whorls are no longer given the importance they were given earlier, though some consideration is still given. In breeds which have a large variation in the intensity of pigmentation of different parts of the body, as is the case with the Jersey and, to a lesser degree, also with the Ayrshire and Swedish Red and White cattle (SRB), great importance is attached to colour similarities. It should be recognised, however, that colour shade and distribution can be quite considerably modified by environmental influences. With good health and proper feeding the coat becomes smoother and more shiny and red pigmented animals have a darker tone of the colour than those in poor health or with inadequate feeding. Colour pattern and level of pigmentation are also important, but the details with regard to the extent of the white pattern are subject to rather large non-genetic variation (Fig. 5:7). Head and body conformation are highly heritable and marked differences in this respect are attributed therefore to dizygotic origins. Unfortunately, such morphological characters are indistinct in suckling calves, which are always more difficult to diagnose than older, more developed animals. As examples of other characteristics of diagnostic interest, mention may be made of 'mirror image', escutcheon, tail length, and appearance of the switch (Fig. 5:8), the pigmentation of the skin around the body orifices and the size and placement of the teats.

When the twins are to be used for investigations into the influence of certain environmental factors on growth rate and conformation, the diagnosis must be carried out at an early stage, preferably at 2–4 weeks of age, so that the animals can then be placed in their respective experimental groups. Those investigations which have been carried out with this early diagnosis, based entirely on morphological characteristics, show that it must be reckoned that about 15 per cent of the twins classified as MZ are in fact dizygous. Some months later, when the dizygosity is more clearly reflected, approximately half of the DZ pairs could be rejected; but even a trained person has little hope of reducing the error of the similarity diagnosis below 6–8 per cent, since even dizygous twins can be

Fig. 5:7. Swedish Ayrshire identical twins. (Department of Animal Breeding, Agric. Coll., Uppsala)

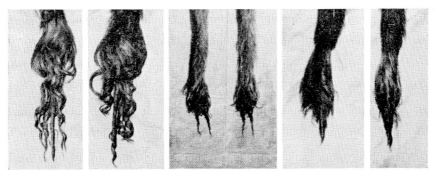

Fig. 5:8. Switch similarities in three pairs of Swedish Red and White monozygous twins. (BRÄNNANG and RENDEL 1958)

remarkably alike. With the aid of blood group determinations and other serological tests, it is now possible to diagnose monozygotic twins with greater accuracy than was possible with morphological likenesses alone; a more detailed discussion of this will be given later (p. 205). By a combination of laboratory tests and similarity diagnoses it is possible to reduce the error to about one per cent, which may be considered as fully acceptable.

Monozygous cattle twins are used for research on (1) the modifying effect of environmental factors on various characters, (2) the effect of the interaction between heredity and environment, and (3) the relative importance of heredity and environment for the total variation of quantitative characters (their heritability).

1. The twin method has, as already indicated, been used extensively for investigations into the influence of certain environmental factors on a number of quantitative characters, e.g. growth rate and milk yield. BONNIER began investigations of this type at the Wiad Experimental Station, Sweden, in 1937, and was the pioneer in this field. The Wiad experiments have been in progress since then and have been primarily designed to elucidate the influence on body development and the future production capacity of rearing at different planes of nutrition. In addition, the influence of differing intervals between milking on milk yield and composition have been studied. In analyses of the results, the interaction between heredity and environment have been taken into consideration, even if the material for this purpose is comparatively small. Similar investigations have been carried out in a number of countries, especially New Zealand, Great Britain and the U.S.A.

Bonnier and his co-workers carried out studies into the efficiency of the twin method compared with the conventional group method, which makes use of genetically different individuals. This type of study was later continued at the Ruakura Research Station in New Zealand where twin experiments could be conducted on a larger scale than anywhere else. The twins are reared and produce milk under conditions which are as alike as possible (*uniformity trials*), a comparison is then made of the variance between twin pairs (V_B) and the variance within twin pairs (V_W). On the assumption that the twin pairs are a random sample from the population, and that the environmental influences are equally distributed between and within pairs, then the ratio $V_B/V_W = Ef$ ('the twin efficiency value'), indicates the number of animals, in each of two experimental groups, which can be replaced by a MZ pair without reducing the statistical precision of the experiment, i.e. without loss of efficiency. Naturally, such uniformity trials must be conducted with a relatively large number of twin pairs if the random variation in the twin efficiency value is not to be too large. At Ruakura, 10 MZ pairs have usually been used in these trials. However, even a large number of twin pairs cannot obviate the systematic errors, which must always be reckoned with.

The variance component V_B will always be more or less inflated as an expression for the genetic variation between MZ pairs (cf. Table 5:2), and because of this the estimates of Ef exaggerate the efficiency of the twin method. BRUMBY (1958) points out that 'the important point of these efficiency estimates lies not so much in the actual figures put on Ef but in their relative magnitude; that is to say, their value lies in the extent to which they can indicate the type of study for which MZ twins appear to be particularly useful'. However, even the relative value of Ef may be misleading, especially when the MZ pairs enter the uniformity trials at different times, as they usually do because of the difficulties involved in securing the necessary number of pairs. As an example, the following Ef values may be quoted from the Ruakura uniformity trials (HANCOCK, 1954):

Milk yield per lactation	22	Haemoglobin content of blood	13
Butterfat content of milk	15	Number of sperm per	
Growth rate	11	ejaculate from bulls	8

It would be expected that the efficiency of the twin method increases with increasing genetic variance (and heritability) of the character. The high Ef value obtained for milk yield, compared to butterfat content, is probably due to environmental differences between the pairs (cf. Table 5:4). A trial with only a few MZ pairs will never be a reliable basis for conclusions but may serve as a preliminary orientation.

The object of the MZ trials is usually to compare the effect of different environmental factors, or environmental levels, for example levels of feeding and management. In investigations involving several environmental factors or levels, the relative efficiency of the twin method decreases; how much it decreases depends on the design of the experiment.

The twin method is especially useful for long-term trials, covering for example the whole growth period of the animals, whole lactations or several lactations. When the control group and the experimental group are genetically identical they will be comparable throughout the experiment.

Monozygous twins have been used also to assess the genetic differences between dairy herds of the same breed but with different production levels. An experiment of this type has been carried out at the Ruakura Research Station. Twenty high and twenty low producing herds, each with at least 50 milking cows, were located and 120 MZ pairs of heifer calves were split between these two groups, each herd receiving 6 calves. The twin calves were reared and treated in exactly the same manner as the other animals in the host herd; they calved at two years of age and during the lactations the feeding and treatment were identical. At the end of the first lactation a sample of the twin pairs were returned to the Ruakura Station where the cows were managed as a uniform herd. Seventy-five MZ pairs completed normal lactations on the participating farms. The results indicate that the difference in milk yield between the two groups of farms, on an average 850 kg milk, was wholly environmental, whereas about half the difference in the average fat content of the milk (0·41 per cent fat) was genetically determined.

2. The method of splitting monozygous twin pairs between different environments may be used also for a study of the interaction between heredity and environment. Knowledge about the importance of such interactions is needed for a rational planning of the breeding procedures. The existence of a pronounced genotype-environment interaction would mean that the best genotype in one environment, e.g. a high level of feeding and management, may not be the best in another environment where feeding conditions are poor; different strains should therefore be bred for different environments. When genotype and environment act additively, or nearly so, no such adaptation of strains is necessary.

FALCONER (1960) pointed out that when a character of the same, or related, animals is measured in two different environments it may be regarded as two characters with a genetic correlation between them. 'The physiological mechanisms are to some extent different, and consequently the genes required for high performance also are different to some extent. For example, growth rate on a low plane of nutrition may be principally a matter of efficiency of food utilisation,

Table 5:4 Components of variance between and within pairs of twins and single-born animals

Type of pair	Variance between pairs: V_B	Variance within pairs: V_W
MZ	$V_A + V_N + V_{HE} + V_M + V_C = {}_{MZ}V_B$	$V_E = {}_{MZ}V_W$
DZ	$0.5V_A + 0.25V_N + 0.5V_{HE} + V_M + V_C = {}_{DZ}V_B$	$0.5V_A + 0.75V_N + 0.5V_{HE} + V_E = {}_{DZ}V_W$
FS (full sibs, not twins)	$0.5V_A + 0.25V_N + 0.5V_{HE} + V_C$	$0.5V_A + 0.75V_N + 0.5V_{HE} + V_M + V_E$
HS (half sibs)	$0.25V_A + 0.25V_{HE} + V_C$	$0.75V_A + V_N + 0.75V_{HE} + V_M + V_E$
U (unrelated)	$V_C = {}_U V_B$	$V_A + V_N + V_{HE} + V_M + V_E = {}_U V_W$

V_A = additive genetic variance.
V_N = non-additive genetic variance (dominance and epistasis).
V_{HE} = variance due to interaction of heredity and environment.
V_E = variance due to randomly distributed environmental factors.
V_C = variance due to systematic, non-genetic differences between pairs.
V_M = additional variance within pairs of single-born animals due to differences in the pre-natal environment.

whereas on a high plane of nutrition it may be principally a matter of appetite'. If the genetic correlation between the performance in the two different environments is high, then the performance must be determined by very nearly the same set of genes; if it is low, high performance requires a different set of genes in the two environments.

In the Ruakura split-pair experiment referred to above, it was found that the genotype-environment interaction had a distinct effect on the variation in milk yield; and in the Wiad experiments on the effect of the milking interval on yield, the different pairs reacted quite differently. When one of the twins in each of the nine pairs was milked only once per day during 280 days of the first lactation the milk yield fell by about 50 per cent, on the average, compared with the co-twins which were milked twice per day with 15·5 and 8·5 hours interval. The decrease in yield varied from 85 to 25 per cent in the different pairs.

3. Twin data has been used also for estimating the heritability of quantitative traits, especially milk yield and milk composition. In order to study the results and to make comparisons with field data it is necessary to know how the various estimates have been calculated.

HANCOCK and others have estimated the heritability directly from the intrapair correlation of monozygous twins, i.e. $h^2 = \dfrac{V_B}{V_B + V_W}$. However, the variance between MZ pairs (V_B) depends not only on additive gene effects (A) but also on non-additive effects (N) which BRUMBY (1960) estimated to account for about 10 per cent of the total variance in lactation yield. In addition, the interaction of heredity and environment (HE) is included as well as the effect of pre-natal (M) and post-natal (C) differences between pairs (Table 5:4). DONALD (1958), has shown that the pre-natal maternal effect (M) is quite important with respect to the liveweight at an early age but that it gradually disappears, to become negligible at about 18 months of age. The effect on milk yield is probably small, but according to Brumby (1960) it may play some part in determining the variation between pairs. The effect of the genotype-environment interaction depends on the kind and magnitude of the environmental differences. In the Ruakura split twin-pair experiment it seemed to be rather important; the observed correlation between the deviation of the co-twins from their respective high and low herd means was only 0·11 (BRUMBY, 1961). However, when the twins are reared and produce in the same herd, the most important source of systematic errors would seem to be the post-natal environmental differences between pairs (C). The members of the same pair are contemporary but the different pairs are not contemporary.

The relative magnitude of the variance components within the same population varies between different characters, as well as with the age of the animals and the environment to which they are exposed.

An interesting experiment was started by the Animal Breeding Research Organisation (ABRO), Edinburgh, in 1949. MZ and DZ pairs are kept under the same environmental conditions as pairs of paternal half-sisters (HS) and unrelated pairs (U). The age difference between members of the HS and U pairs is only a few days. Table 5:4 shows the variance components within and between pairs which may be expected to add up to the total variance of the groups.

DONALD and WATSON (1960) estimated the heritability of milk yield on the basis of the intra-pair variance within MZ and DZ pairs according to the formula $h^2{}_{MZ/DZ} = \dfrac{2({}_{DZ}V_W - {}_{MZ}V_W)}{2({}_{DZ}V_W - {}_{MZ}V_W) + {}_{MZ}V_W}$, and they also calculated intra-pair correlations for the different types of pairs. The main results are summarised in Table 5:5.

Table 5:5 Intra-pair correlation between milk yields of twins and single-born animals in the first lactation and heritability estimates according to the formula stated (DONALD and WATSON, 1960)

Intra-pair correlation between milk yields of twins and single-born animals				Heritability estimates according to the formula stated above: 51 MZ and 39 DZ pairs. $h^2{}_{MZ/DZ}$	
Type of pairs	d.f.	305 days	70 days		
MZ	45	0·84	0·87	Body weight at 18 months of age	0·81
DZ	35	0·68	0·57	Milk yield in 70 days	0·72
HS	21	0·16	0·12	Butterfat percentage (305 days)	0·91
U	8	0·15	0·03	Solids-not-fat (305 days)	0·32

The intra-pair correlation for milk yield of MZ twins in the same herd is very high; which is in good agreement with the results from other twin experiments. The corresponding correlations for the DZ pairs are also high, but for the half sib pairs they are low. The number of unrelated pairs is small and the coefficients for these pairs have therefore large sampling errors.

The heritability estimate, $h^2{}_{MZ/DZ}$, for milk yield during 70 days of the first lactation is lower than the corresponding intra-pair correlation of MZ twins; but it is still at least twice as high as estimates based on dam-daughter or paternal half sib correlations obtained from analyses of field records (cf. page 289). There are several reasons for this discrepancy. First, it must be remembered that when h^2 estimates are based on intra-pair variances of MZ and DZ twins, a somewhat exaggerated estimate of 'heritability in the broad sense' is obtained $[2({}_{DZ}V_W - {}_{MZ}V_W) = V_A + 1·5V_N + V_{HE}$ according to Table 5:4], whereas estimates from dam-daughter or paternal half-sib correlations are very little influenced by non-additive gene effects (cf. p. 122). Several other causes may also contribute to the divergence between the estimates. One cause may be that the data from the twin investigations are obtained at research stations where the milk recording is much more accurate than under field conditions. WATSON (1960) found that when an analysis of the experimental twin data was made using only one butterfat test every four weeks, instead of the weekly tests, the intra-pair correlation fell from 0·88 to 0·79 in MZ twins and from 0·67 to 0·49 in DZ twins. However, the recording of milk yield monthly instead of daily had no measurable influence on the variance components. Another reason for the high intra-pair correlation between the twins at the research stations may be that practically no culling is made on the basis of production, the aim being to keep all the twin pairs for one or several complete lactations. In commercial herds a considerable amount of culling takes place, and when, for example, the regression of the daughters on dams is analysed, not only the dams but also the daughters may be selected to a certain extent. It is not likely, however, that the selection among the daughters is so intense as to have pronounced effect on the results obtained. The most important cause of the high values obtained for intra-pair correlation and heritability in twin experiments would seem to be contemporaneity of the two members of each twin pair.

Watson made a comparative study of data obtained for twins, full sisters and half sisters in the field and in the ABRO experimental herd. All animals in the field belonged to the Ayrshire breed. The intra-pair correlations which were obtained in the analysis of the data are shown in Table 5:6.

Table 5:6 Intra-pair correlation between lactation yield (305 days) of twins and sisters in the field and at the experimental station (WATSON, 1960)

	No. of pairs	Intra-pair correlations Milk yield	Butterfat %
Dairy herds: pair members on same farm:			
Twins*	234	0·37	0·40
Full sisters	207	0·20	0·25
Half sisters	502	0·31	0·35
Experimental station (ABRO):			
MZ twins	64	0·88	0·88
DZ twins	56	0·67	0·65

* No diagnosis of zygosity was made. The sample of twin pairs probably contains at most 10% MZ pairs.

In the field data, the intra-pair correlation is higher for half-sisters than for full sisters, and this is most certainly due to the fact that the former were selected to calve within a month of each other, whereas the intra-pair differences in calving time of the full sisters must have been at least one year. The correlation between the members of the DZ pairs at the experimental station is about twice as high as that for twins in the field; this may be partly due to the absence of culling in the experimental herd.

In another paper WATSON (1961) analysed the effect of the contemporaneity on the intra-pair correlations and variances. The members of the pairs were considered contemporary if they calved within two months of each other, and non-contemporary if they calved more than twelve months apart. It was estimated that about 10 per cent of the total intra-pair variance in milk yield was due to contemporaneity in time of calving, and that a further 10 per cent was due to contemporaneity with regard to lactation number. Other sources of environmental variation contributed to an additional 25 per cent. Further, the analysis indicated that an appreciable part of the genetic variance may be due to non-additive effect of the genes.

A recent publication by BARKER and ROBERTSON (1966) is of interest in this connection. They analysed production records from the progeny of 65 British and 24 imported Friesian bulls, each one with at least 45 daughters which complied with certain requirements; a total of 10 967 cows which had completed at least one lactation (305 days). Distinction was made between contemporary lactations in the same herd (cows calving in the same recording year (cf. Chapter 10, p. 287), and lactations made in an 'analogous' environment (calving in the same herd, in the same month and at the same age in months). It was found that the variance between 'analogous' lactations was only about 2/3 of that between 'contemporary' lactations, both defined as above. Furthermore, the analysis showed that the month (season) of calving had a much greater effect on the variance than the age of the cow at calving.

Another investigation of field data has been made by RENDEL and JOHANSSON (1966). Twin pairs were located in 170 herds of Swedish Red and White cattle and diagnosed for zygosity. When available, pairs of unrelated cows (U) were drawn from the same herds, according to certain rules as to contemporaneity within pairs. The first lactations of pair members were considered contemporary if the cows calved less than four months apart within the same calving season (September–February), or else less than one month apart, and if the age at calving differed by less than 4·5 months. A total of 22 MZ, 162 DZ and 117 U pairs fulfilled these requirements. Comparisons within the same herd could be made with respect to 70 DZ and 70 contemporary U pairs, denoted DZ_c and U_c respectively. In addition, 48 DZ pairs were available which did not fulfil the requirements as to contemporaneity. These pairs will be denoted DZ_n. Further, non-contemporary U pairs, denoted U_y and U_n were drawn from the 60 herds which contained at least one DZ and one U_c pair. The U_y pairs were formed by pairing unrelated first calvers at random within herds, and recording years. The U_n pairs were formed in the same way, but the partners were required to have calved at an interval of at least 950 days—the average was in fact 1094 days—corresponding to the interval between the first calving of a dam and her daughter. The intra-pair correlations, as well as the variance in milk yield within U_n pairs, are based on the deviation of the first lactation record from the contemporary herd average. The MZ intra-pair correlation for milk yield in the first lactation is about the same as that obtained at ABRO's experimental station (Tables 5:5 and 5:6), but the corresponding figure for the DZ pairs is somewhat lower, although higher than that for Watson's field data. The significant correlations for the U pairs are a result of the intra-pair contemporaneity; for the non-contemporary U_n pairs the correlation is zero. The intra-pair correlation for the fat content of the milk is very little affected by the contemporaneity. By comparing the intra-pair variance in milk of the U_c pairs with that of the non-contemporary U_n (or U_y) pairs in the same herds it is found that the former is only about 70 per cent of the latter; the 30 per cent reduction due to contemporaneity is three times greater than in Watson's investigation, where no U pairs were available for comparison; but it agrees very well with the results obtained by BARKER and ROBERTSON (1966).

Four different estimates of the heritability of milk yield in the first lactation are presented in Table 5:7. The highest estimate (0·88) is obtained from the intra-pair variances of contemporary DZ_c and U_c pairs in the same herds. Here, contemporaneity exerts a considerable effect. The estimate of h^2 (in the broad

Table 5:7 *Intra-pair correlation and variance in milk yield of MZ and DZ twins and of unrelated cows (U). The intra-pair correlations for fat percentage are stated for comparison* (RENDEL and JOHANSSON, 1966)

Type of pairs	No. of pairs	Diff. between calving dates, days Variation	Diff. between calving dates, days Average	Intra-pair correlation* Milk	Intra-pair correlation* Fat %	Intra-pair variance in milk yield	Heritability according to diff. estimates
Contemporary lactations:							
170 herds:							
MZ	22	0–82	26	0·88³	0·84³	92,814	$h^2_{MZ/DZ} = 0.67$
DZ	162	0–120	25	0·56³	0·45³	189,788	$h^2_{DZ/U} = 0.72$
U	117	0–117	24	0·23²	0·07⁰	272,440	
60 herds:							
DZ$_c$	70	0–117	25	0·68³	0·33²	150,722	$h^2_{DZ_c/U_c} = 0.88$
U$_c$	70	0–103	25	0·20¹	0·00	253,948	
Non-contemporary lactations							
DZ$_n$	48	33–321	152	0·29¹	0·19⁰	315,415	$h^2_{DZ_n/U_{ys}} = 0.34$
U$_{ys}$	48	34–328	152	—	—	379,118	
U$_y$	1049	0–360	58	0·00	0·00	361,317	
U$_n$	831	952–1474	1094	0·00	0·00	361,339*	

⁰P > 0·05; ¹P < 0·05; ²P < 0·01; ³P < 0·001

* Based on deviations from herd average. All other variances (not marked*) refer to the actual production records.

sense) from the variances within MZ and DZ pairs, without consideration of whether or not they belonged to the same herd, is a bit lower (0·67) and the estimate from the DZ and U pairs is about the same (0·72). A fourth estimate was obtained from the non-contemporary DZ$_n$ pairs and a comparable group of U$_{ys}$ pairs which were selected by matching the DZ$_n$ pairs with those U$_y$ pairs in which the members had calved in the same month of the year, and with approximately the same interval between their calving dates, as the corresponding DZ$_n$ pair. The DZ$_n$ and U$_{ys}$ groups should therefore be fairly well comparable. The h^2 estimate based on these two groups is 0·34, which corresponds very closely to estimates for the first lactation yield obtained from daughter-dam regression or paternal half-sib correlations using field data (cf. Table 10:3). It should be pointed out, however, that the age at calving was not considered; and this probably had the effect of slightly elevating the intra-pair variance of the DZ$_n$ group compared with the U$_{ys}$ group.

All four estimates in Table 5:7 have large sampling errors because the number of pairs is limited, but nevertheless the results indicate that contemporaneity has a much higher effect on intra-pair variance in twin research than is generally realised. One heritability estimate may be as true as another under the environmental conditions to which it is applicable. For a right interpretation of the estimates it is important that the conditions are specified, and especially that the degree of contemporaneity is defined. In principle, the same would apply to other characters which are sensitive to environmental influences. Further, heritability in the narrow sense cannot be estimated from twin data, since the non-additive gene effects cannot be eliminated.

When heritability estimates are applied in breeding practice, they must correspond to the field conditions where they are used; otherwise they will be misleading. In the field neither the lactation of dams and daughters nor the progeny groups of sires are contemporary.

Our conclusion is that the twin method is valuable for investigations into the influence of certain environmental factors, or the total effect of environment, on growth and milk production in cattle, as well as for analyses of non-linear interactions between heredity and environment; but that heritability estimates based on twin data are not applicable in general breeding practice.

The use of monozygous twins entails with certain disadvantages even in investigations in which they are otherwise excellent experimental animals:

1. It is generally difficult to collect within a limited period of time, a sufficiently large number of MZ pairs for the commencement of a large scale investigation. The different MZ pairs consequently commence a given experiment at varying times, and this means a loss of precision. Because of this, several research stations have begun using DZ twins, which are easier to collect in sufficient numbers. It was pointed out earlier that the intra-pair correlation for DZ twins is relatively high.

2. The cost of obtaining MZ twins is high, and as a rule it is necessary to purchase many more twins than the number which eventually, after accurate diagnosis, are shown to be one-egg twins.

3. When one member of a pair is eliminated from the experiment on account of accident, disease or sterility, there is generally no further use for the partner. The result is that the number of animals excluded from an experiment is much higher than the number that would be excluded from a conventional group experiment during a corresponding period. In the Ruakura split-pair experiment only about 60 per cent of all twins sent out to the host herds as calves successfully completed first lactations (BRUMBY, 1961).

In spite of these practical disadvantages the twin method has many advantages for certain types of investigations, and will therefore probably be of great importance to animal research in the future.

SELECTED REFERENCES

EUROPEAN ASSOCIATION FOR ANIMAL PRODUCTION, 1960. Research work with monozygotic cattle twins. *Publication No. 9* (189 pp.).

HANCOCK, J. 1954. Studies in monozygotic cattle twins. *Publ. No. 63, Animal Research Division, N.Z. Dept. Agriculture.*

JOHANSSON, I. and VENGE, V. 1951. Studies on the value of various morphological characters for the diagnosis of monozygosity of cattle twins. *Z. Tierzücht. ZüchtBiol.*, **59**: 389–424.

PERRY, J. S. 1960. The incidence of embryonic mortality as a characteristic of the individual sow. *J. Reprod. Fert.*, **1**: 71–83.

POMEROY, R. W. 1960. Infertility and neonatal mortality in the sow. *J. Agr. Sci.*, **54**: 1–66.

RENDEL, J. 1958. Studies of cattle blood groups. III. Investigation of twins with special reference to diagnosis of zygosity. *Acta Agric. scand.*, **8**: 162–190.

—— and JOHANSSON, I. 1966. A study of the variation in cattle twins and pairs of single born animals. IV. The effect of contemporaneity and some other factors on heritability estimates for milk yield and fat percentage during the first lactation. *Z. Tierz. ZüchtBiol.*, **83**: 56–71.

6   The Inheritance of External Traits

Many of the external attributes of domestic animals can be assigned to the category of qualitative traits and, in general, they show relatively simple inheritance. Coat colour pattern is an example of this, as are horns or polledness in cattle and sheep.

COLOUR INHERITANCE

From time immemorial man has been interested in the colour and colour pattern of animals, chiefly perhaps in horses and sheep, where symbolic or religious importance was attached to the colour. One can read in the Revelation of St. John the Divine, Ch. 6, v.2, 'And I saw, and beheld a white horse: and he that sat on him had a bow; and a crown was given unto him: and he went forth conquering and to conquer.' The belief that the white horse was a token of good luck persisted well into modern times. The pale (dun-coloured) horse is used in the Revelation of St. John to symbolise death and evil.

    Attention has also been given to the colour of cattle, but for completely different reasons. In this case it was believed that colour had a direct relationship to the capacity to produce milk and meat. This idea has been shown to be groundless, except in some special cases connected with the sexual development of heifers (cf. pp. 240 and 242). Colour is, however, a breed recognition or 'trade mark' not only in cattle but also in sheep, pigs and poultry. The coat colour of fur animals, including sheep in some cases, is of great economic importance since the price of the pelt is determined, to a high degree, by the colour.

    Colour of the same type has been shown to occur in widely different species of animals. As far as possible, therefore, the same gene symbols will be used for these common colours in the different species. This does not mean that genes which are given the same notation in different species are necessarily homologues, i.e. that they have the same origin. The recessive gene, which in guinea-pigs, rabbits, pigs, poultry and other animals gives rise to red colour, is given the notation e without any assumption that it is exactly the same gene. It is quite probable, though, that this gene acts at about the same stage of the biochemical chain of reactions which determines colour in the various species. The inheritance of colour has been very thoroughly worked out in a number of rodents (guinea-pigs, mice and rabbits), perhaps primarily due to the extensive work of SEWALL WRIGHT, who has also made outstanding contributions to population genetics. The gene notations first introduced by Wright will be used here.

Pigment formation

All skin and hair pigments in mammals consist of a group of substances which are included under the common name of *melanins*. They are formed by the oxidation of the amino acid, tyrosine, or closely related compounds. In addition to melanin, a yellow coloured substance called *xanthophyll* occurs in birds. This

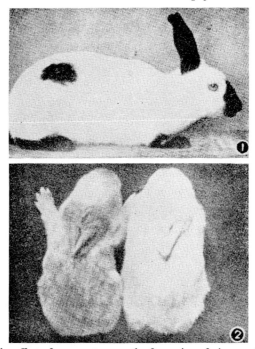

Fig. 6:1. The effect of temperature on the formation of pigment in the Himalayan rabbit.
1. A patch of hair has been shaved off, after which the rabbit was exposed to low temperature. 2. The young on the left was subjected to cold water treatment for several days after birth; its growing hair became grey. The young on the right is an untreated litter mate. (PICKARD and CREWE 1931)

cannot be synthesised in the animal organism but is abundant in the green parts of higher plants. Birds obtain xanthophyll from their food and it is then stored in the epidermal layers and in the body fat. Plant pigments also are stored in the fat of many species of mammals.

Two types of qualitatively different melanin can be distinguished: (1) *phaeomelanin* — round, red pigment particles which are soluble in alkalis; (2) *eumelanin*, which includes two pigment types, black and brown, both of which are much less soluble than phaeomelanin. The formation of melanin in the body takes place under the influence of different oxidising enzymes (cf. p. 85) and the oxidising process is, among other things, sensitive to temperature. This can be demonstrated by the colour of the Himalayan rabbit, which is white except for the feet and ears where the body temperature is lower than in other parts of the body. If the Himalayan rabbit is exposed to extreme cold, the extent of the black colour increases. Fig. 6:1 shows the result of such treatment. The low temperature apparently influences the enzyme system which controls the production of black pigment.

Detailed studies have been made into the type and distribution of hair pigments in the various colour types of rodents; both black and red pigments appear to always occur in the form of granules of varying size. Some researchers have claimed that the red pigment in larger domestic animals exists in a diffused form, but it is difficult to decide whether this is a real difference from the situation in rodents.

The outer layer of the hair consists of an extremely thin layer of cells, the cuticle, laid in the manner of roofing tiles, one overlapping the other (cf. p. 328). There is no pigment in the cuticle. The next layer, the cortex, consists of cornified cells which in coloured individuals contain pigment, with the exception of hairs forming the white flecks in speckled individuals. The central part of the hair is called the medulla, or pith, and consists of loosely arranged cells with air spaces between them. The hair of cattle and horses has a continuous medulla which extends from the root of the hair to just below the tip. The wool hair of fine-woolled sheep is completely without a medulla; where a medulla does occur, it is only very slightly developed (cf. p. 330). The cuticle and cortex cover the whole of the hair.

The amount and distribution of pigment granules vary with the different colour types. A recessive blue-grey colour type occurs in mice, rabbits and mink, in which the reduction in colour has been shown to be due to an aggregation of the pigment granules. In rabbits it has been shown that exposure to low temperatures leads to an aggregation of pigment granules and that this alteration in granule size gives the fur a darker shade of grey. It is likely, therefore, that the grey colour observed in wild animals during winter is connected with the aggregation of pigment granules similar to that occurring in rabbits.

The genetics of colour in rodents. Gene symbols

WRIGHT studied the genetics of the colour of guinea-pigs for almost half a century. The amount of pigment in the different colour types has been determined with the aid of gravimetric and colormetric methods as well as by the classification of colour intensity by comparison with selected standard pelts. It has been shown that colour type is due to a large number of genes and the interactions between them. The interactions are often very intricate and we shall therefore deal only with the main points of the genetics of colour in guinea-pigs. Table 6:1 summarises the more important gene loci and environmental factors influencing colour.

Table 6:1 Factors influencing coat colour in guinea-pigs
Major genes

E, e^p, e. Eumelanin; tortoise pattern and phaeomelanin.
C, c^k, c^d, c^r, c^a. Albino series. The alleles determine the amount of pigment.
P, p^r, p. Dilution of the eumelanin.
F, f. The amount of phaeomelanin.

Modifying genes

A, a. Agouti (wild type) and non-agouti hair.
S, s. Self-coloured versus spotting.
Si, si. Silver.
Dm, dm. Reduction in the amount of pigment.
Gr, gr. Grizzling, white hair, progressive with age.

Other modifying factors

Age of individual and age of mother;
Amount of androgen;
Temperature and other factors.

Wright distinguishes between a number of major genes, essential for the type and amount of pigment, and a number of modifying factors which influence the extent and intensity of pigment or the localisation of pigmentation to certain sections of the coat. The so called *E*-locus determines whether the coat contains eumelanin, phaeomelanin or both. The dominant *E* gene (extension) distributes black or brown throughout the whole coat, provided it is not influenced by modifying genes; *ee* produces only phaeomelanin, whereas $e^p e^p$ and $e^p e$ give rise to the tortoise pattern, i.e. yellowish-red patches of the coat on a black or brown background. There is reason to believe that the *E*-series of alleles influences the differentiation of the melanocytes into eumelanin- or phaeomelanin-producing cells.

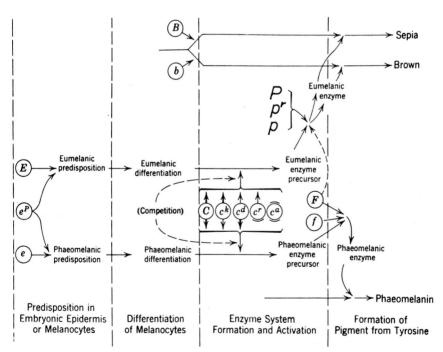

Fig. 6:2. Wright's explanation for the action of the different kinds of factors controlling melanin formation in the hair of guinea-pigs. (From Wagner and Mitchell 1964)

The albino series includes five different alleles, C, c^k, c^d, c^r and c^a. Homozygosity for the latter always gives albino. The genes in the albino series can be detected only in animals which are also carriers of the gene E; in them they give rise to pigment amounts which reduce in the order $C\ c^k\ c^d\ c^r\ c^a$. In the homozygote ee only three groups of alleles can be identified, C, $c^k + c^d$ and $c^r + c^a$, the latter two resulting in the albino. The C-alleles are believed to be concerned with the production of an enzyme with varying specificity and efficiency in the eumelanin- and phaeomelanin-producing processes. The way in which the different gene systems are believed to interact with one another and build reaction chains which finally result in one or the other kind of pigment is illustrated in Fig. 6:2.

The enzymes which are formed by the different alleles in the C-series are in turn affected by the genes in the P and F loci. P operates only in the reaction chain which gives eumelanin and F mainly in the phaeomelanin chain. The recessive alleles in the respective systems reduce the amount of pigment. Individuals of the genotype $E\ C\ pp\ F$ have more eumelanin than $E\ C\ ppff$, since the enzyme which is determined by the gene F also has some effect on eumelanin formation. The enzymes found in the final phase of the two reaction chains are thought to operate on tyrosine or closely related compounds, whereby eumelanin or phaeomelanin is formed. The genes in the B locus are believed to influence the substrate on which the enzymes act so that B gives black (sepia) and bb brown eumelanin. This locus has no effect on the red-yellow pigment.

The effect of the major genes described above can be modified by a number of gene systems with more or less drastic results and also by environmental factors such as age, the concentration of androgens and temperature. The agouti or wild-colour gene (A) influences the extent of the two types of pigment in individual hairs. The genotype, $E\ C\ P\ A$ is recognised by the hair on the back having a yellowish-red band of varying width just below the dark tip of the hairs. The hairs on the abdomen are practically yellow, except for isolated hairs which have black pigment at the base of the hair. The allele a in the homozygous condition forms uniformly black or, with bb, uniformly brown hair. The A gene has no recognisable effect on individuals of the ee type. The part played by the agouti gene in the chain reactions which determine colour is not known. It should, however, come into play at the differentiation of the melanocytes.

The distribution of pigmented and pigment-free hair and skin segments is controlled primarily by the S locus. The recessive gene, s, gives rise to patches, whereas the dominant allele, S, extends the colour over the whole body (self colour). The gene, S, is however, not completely dominant, and a certain amount of patchiness can occur in Ss individuals. The extent and location of the white flecks are caused by genes with small effect, together with non-genetic causes such as sex and age of the mother, and by developmental irregularities which are not the same even for litter mates.

The recessive genes in the Si, Dm and Gr loci give rise to varying amounts of white hairs. Some types are completely white. The gene, gr, is interesting in that it first manifests itself in older individuals, so that the number of white hairs increases with age. Colour intensity is also influenced by age. Brown and black show a distinct reduction in intensity with age, whereas the effect on the red-yellow pigment is much less.

The genetics of colour in rabbits and other rodents shows great similarity to that of guinea-pigs. In the rabbit, at least five alleles in the albino series can be distinguished, C, c^{ch} (chinchilla), c^m (sable), c^h (himalaya) and c (albino) (Figs. 6:1

Fig. 6:3. Four colour types in the albino series.
Top left: all black. *Top right*: Chinchilla. *Bottom left*: Sable. *Bottom right*: Albino.
For another colour type in the albino series, Himalayan, see Fig. 6:1.
(*Handbuch der Tierzüchtung*, II)

and 6:3). The dominance relationship is as follows: $C \to c^{ch} \to c^m \to c^h \to c$. The gene system ($E$), which determines the differentiation of black and red-yellow pigment, includes besides the genes E, e^p, and e, the allele E^D, called dominant black. In the homozygous condition this gene gives black colour even in the presence of the agouti gene. When individuals are heterozygous for E^D, the gene does not completely suppress the agouti gene (A), the result being a dark grey type, which was first described in the Belgian Giant rabbit. The E^D gene is dominant over all genes at the E locus.

At the locus which controls the wild colour or agouti type there occurs an interesting allele, a^t, which produces a type 'black and tan', recognised by a dark (black or brown) back fur and a light yellow belly. The a^t gene is mainly recessive to A but dominant over a. Homozygotes for the latter allele are, as in guinea-pigs, completely black or brown haired.

The colour of farm animals

Horses. There is very little colour formalism in horse breeding. The division into breeds has taken place mainly on the basis of other and more rational grounds. The importance of colour type is that it serves as an aid to the identification of the animals. It is usual to denote the three types, black, bay and chestnut as the three *basic colours* of horses since these are the commonest in the majority of horse breeds. There are, in addition, many different colour types which can be regarded as modifications of these basic colours. Many gene symbols have been introduced; but we shall confine ourselves here to the nomenclature used by, among others, the American, CASTLE (1954, 1961) and the Norwegian, BERGE (1963). Much of the information on the inheritance of colour in horses traces back to the extensive work of the Spaniard, ODRIOZOLA (1951).

Basic colours. Horses which have completely black bodies, black mane and tail are collectively designated as *black*. The colour of the hair in a number of animals in this group becomes lighter with age, especially in sunlight, when it takes on a reddish tinge. Others do not bleach at all but remain jet black throughout life.

Animals with black manes, tails and fetlocks but bodies of varying shades of brown are designated collectively as *bay*. The body colour can vary from blackish-brown or seal-brown to yellowish-brown. The black-brown (seal-brown) type can be so dark as to be difficult to tell apart from black. However, the nose, the area around the eyes, the belly, and especially the median site of the thighs, have a lighter brown colour than the rest of the body. The hair on the body of the seal-brown type often has alternating zones of black and yellowish-red, similar to the agouti pattern.

Animals with a body colour of red-brown and tail with the same or somewhat lighter colour are designated as *chestnut*. The colour varies from liver chestnut to light red (sorrel).

Before going further with the genetics of colour it is necessary to describe still another colour type, i.e. wild colour. The so called *Przewalski* horses, which even in modern times roamed the plains between Siberia and Mongolia, constitute the last of the wild horses. They have sand coloured, yellow-brown body, black mane, and tail with a dark stripe along the back. This colour type resembles in some ways the bay and the dun. According to GREMMEL, the hair on the body has alternating lighter and darker zones. There is also a dilution of the colour intensity in that the fibres have less pigment on the side towards the body than on the outer side. The same type of colour dilution and zone divisions occurs in the dun bay Fiord horses. Both Castle and Berge maintain that the colour in the Przewalski horse is controlled by a gene A^+, comparable with the agouti gene in rodents. On account of the similarity of colour type in the wild horse, the dun and the seal-brown, all these colours are assigned to the same group. Wild colour, seal-brown and bay are considered to be controlled by the alleles A^+, A^t and A, whereas aa gives the absence of agouti pattern. Whether the dun-bay colour of the Fiord horse is determined by the same gene as the agouti pattern in the wild horse is not clear. Berge believes that it is a question of a separate allele, A^D.

Black occurs in animals which carry the B gene, whereas individuals which are homozygous for bb are chestnut. The effect of the genes in the B system are modified by a number of genes. Bay colour occurs in horses, which in addition to the B gene also carry the gene A. Results from a large number of matings of the

Fig. 6:4. *Left*: Dun North Swedish horse. *Right*: Norwegian Fiord horse of the dilute dun-bay type.

Fig. 6:5. *Left*: Arabian half-blood with constant roaning and dappling. *Right*: Mare of the progressive whitening type, with foal.

type black × black, black × bay and bay × bay, show that black is, in general, hypostatic to bay. The frequency of bay animals from the mating black × black, however, is so great that it can hardly be explained by classification errors or incorrect pedigrees. CASTLE maintains that there are two types of black (dominant and recessive) and that the former, as in rabbits, is influenced by the gene E^D, which has the ability to suppress the effect of the agouti gene and thereby distribute black pigment over the whole body. The E^D gene is assumed to occur in animals with jet black colour. The other two alleles of the E system (E and e) are assumed to have a slight modifying effect only on the basic horse colours. This is in marked contrast to the situation in rodents, where the E and e genes apparently control the differentiation of the melanocytes into cells which produce black and cells which produce red-yellow pigment. Horses of the type B aa ee

are thus black, whereas the corresponding formula in rodents gives red colour.

Dominant colour dilution. In horses there is no colour type directly comparable with albino in the rodents. Completely white horses with unpigmented skin are homozygous for an incompletely dominant gene, usually denoted by D. The effect of the gene in the heterozygotes depends upon the rest of the genotype. *Isabelle* coloured horses, more commonly called *palomino*, are recognised by a pale yellow body and white mane. The colour appears in heterozygotes for D, which are at the same time homozygous for bb. The most favoured type has a cream-coloured body and the genotype $A\ bb\ ee\ D$. The gene E is thought to give a sooty-yellow colour. The D gene, in the single dose, gives rise to a more or less clear pale yellow colour in genetically bay (AB) horses; the mane and tail on the other hand, are black. This colour type is known as dun or buckskin. In the Norwegian Vestland, or Fiord, breed there is a similar colour type which in addition to the gene for black and heterozygosity D, also carries the agouti gene typical for the breed. The result is a horse with beautiful yellowish-white coloured body, a dorsal band down the length of the mid-back and a black mane flanked with lighter hairs (Fig. 6:4). When the mane is clipped (hogged), there appears a clearly defined difference between the darker inner hairs and the lighter outer hairs.

Roan. Roan occurs in two forms, both of which are controlled by dominant and probably independent genes. The first form is known as 'constant roan' since it is constant throughout the life of the horse. The other form is called 'progressive whitening' (Fig. 6:5). Foals of the latter type are born with only a few white hairs but, as they grow older, the number of white hairs increases, so that at the age of about twelve years the animal is almost completely white. Progressive whitening is often accompanied by dappling.

In addition to the two white types described earlier there is a completely dominant white type. These horses have white coats but coloured eyes. The muzzle can be either pigmented or unpigmented.

Cattle

Albinism. In several herds of cattle, animals have been encountered which completely or partially lack pigment in the hair and skin. One of the best described examples comes from a herd of Friesians in Minnesota, U.S.A. For a considerable time the herd included a number of albino-like animals (Fig. 6:6). The calves were born completely without pigment in the eyes, skin and hair. Eventually small amounts of pigment developed in the eyes and, later still, also in those parts of the coat which correspond to the black markings of Friesian animals. The animals appeared as a consequence with a peculiar form of marking called 'ghost pattern'. The trait is determined by a recessive gene. Partial albinism with the 'ghost pattern' has recently been demonstrated in the Hereford breed by PADGETT *et al.* (1964). Padgett and his co-workers made the interesting observation that their partial albinos had abnormal granulations in the leucocytes which were similar to those which occur in certain forms of partial albinism in man and also in the Aleutian colour type of mink.

Varying degrees of pigment reduction have been demonstrated in a number of other cattle breeds, e.g. the Red Danish cattle. All these albino-like types exhibit photophobia.

Black and red colour. These two colours are the most commonly occurring in cattle. The black colour is determined by a dominant gene E, whereas red animals are homozygous for the recessive allele e. Crosses between black and red breeds result, therefore, in black progeny though it is possible to observe a

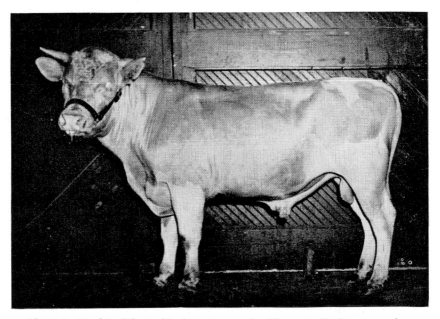

Fig. 6:6. Partial albino with 'ghost pattern'. (Courtesy Professor emeritus W. E. Petersen, University of Minnesota)

reddish tinge to the coat in some animals. In the majority of black cattle breeds the allele e is also present, but at a very low frequency. Now and then in the Friesian cattle red calves are born but these animals differ from the other Friesian animals only in that they are homozygous for the gene e.

Modifications of black and red colour. As in rodents, the black and red colour can be altered due to the influence of other gene systems. A number of cattle breeds are grey or include variations of grey. Berge has studied the inheritance of the grey colour of the Norwegian Westland, or Fiord, breed. Histological examination showed that the tips of the hairs were for the most part free of pigment and that other parts of the hair had a varying number of black and yellow zones. The similarity to the wild type in rodents is apparent and in the same way the grey colour is controlled by a dominant gene, which, to emphasise the likeness, is denoted by A. Red colour is apparently not influenced by the gene A, and therefore grey individuals can appear after mating between red and black animals, when the former are carriers of the *A* gene. The grey colour of the Brown Swiss cattle and many individuals of the Jersey breed, is probably also influenced by a dominant gene. The variation in the intensity of the grey colour is often considerable, and is due most likely to modifying genes and to sex. The bulls are in fact consistently darker than the females.

A similar sex dimorphism also occurs in Ayrshires, but in this case it is a question of an extension of the dark brown pigment in otherwise red animals. The bulls are darker and also begin development of the darker colour earlier than the females. GILMORE and FECHHEIMER (1965) conducted an interesting experiment at the University in Columbus, Ohio, in which eighteen bulls with blackish body markings were castrated. After castration the blackish colour markings disappeared from the flanks but remained on the head, neck and extremities. Administration of testesterone restored the dark brown colour, though not to exactly the same extent as before castration. The dark brown colour is considered to be due to a dominant gene, *Bs*, blackish. The effect of *Bs* can be modified by the sex and also by other genes.

BERGE (1956) has investigated the inheritance of the brown colour of the Norwegian Döla breed. The brown colour appears first after the first change of hair and apparently, as in the Ayrshires, is caused by a dominant gene. Another colour type, so called 'brindling', occurs in the Döla breed (Fig. 6:7). The dark pigment is concentrated into stripes which contrast against the yellow brown or red ground colour. This pattern also appears in the Jersey and in the French Normandy breed. The extent of the black stripes varies considerably, but this variation is not necessarily all genetic. According to the Norwegian investigations, brindling is determined by a dominant gene, *Br*, which is neither an allele to *E* nor to *Bs*. In order that *Br* shall express itself, the presence of the gene for dark brown, *Bs*, is necessary and also homozygosity for *e*. If the *E* gene is present, the animal is black and thus the stripes are not visible.

Colour pattern and white markings. In cattle, there are all gradations from completely white to self coloured animals. If an animal from the Friesian breed is mated with a self coloured animal, either black or red, an almost completely black offspring is obtained. If F_1 progeny are mated together a segregation takes place into self coloured animals with very little white markings and animals with large white patches in the ratio of approximately 3:1. The self colour type depends, therefore, on an incompletely dominant gene, usually designated *S*.

COLE and JOHANSSON (1948) studied the distribution of the white markings in the F_1 generation from crosses between Friesians and Aberdeen Angus. None of the crossbreds were completely lacking in white markings in the inguinal region, some had much white on the belly, and in some isolated cases the white extended up the sides. It was also shown that the appearance of white markings on different parts of the body had a certain relationship to one another. Some of the F_2 segregates showed the typical Friesian pattern of white marking.

The expression of the colour pattern seems to be only partly genetically determined as is apparent from the variation which can arise between one-egg twins (cf. Fig. 5:7), but the general level of pigmentation appears to be under strong genetic influence. LUSH and co-workers studied the Friesian breed in the U.S.A. and came to the conclusion the heritability for percentage of white probably exceeded 0·90.

Dominant white markings. The characteristic white markings of the Hereford

Fig. 6:7. Brindling or tiger pattern on a cow of the Norwegian Döla breed. (BERGE and MIDTLID 1949)

Fig. 6:8. *Top left*: Hereford. *Top right*: Groningen. *Bottom left*: Normandy. *Bottom right*: Simmental.

and Groningen cattle is determined by a dominant gene. The Hereford is recognised by the white head (white faced), white feet, a white tip of the switch, white underline as well as white brisket and crest. In the Groningen animals the white markings are similar but have considerably less spread. The eyes are ringed by black pigment which gives the animals the appearance of having spectacles. The white face also occurs in the Simmental cattle and the Normandy breeds (Fig. 6:8). In crosses with other breeds it has been shown that

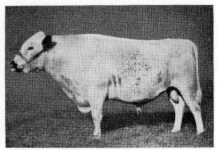

Fig. 6:9. Side markings of Nordic cattle breeds. *Left*: East Finnish. *Right*: Swedish Polled (Highland). In this breed the white is often very extensive.

Fig. 6:10. Roan Shorthorn.

this colour type is dominant at least with regard to the colour of the head. Where the Hereford is concerned, this is true even in crosses with the American bison.

The colour-sided type of marking (Fig. 6:9), which occurs in several Nordic breeds, is apparently due to a dominant gene. The extent of the colouring is also influenced by modifying genes. HILDEMAN studied the colour in the F_1 generation from crosses between Swedish Red Polls, which is solid red, and animals of the old highland breed, which is basically colour-sided although usually the pigmented parts are very small. Of the crossbreds, 108 were colour-sided, 24 self coloured and 9 were spotted. The side-colouring varied from very little white along the back and on the belly to an almost complete lack of coloured spots on the body.

White cattle. White animals in the Swedish Highland cattle nearly always have pigmented ears similar to the Park cattle in England. The white colour type of the Swedish Highland cattle is thought to represent the extreme expression of the colour-sided type. However, in the English Shorthorn cattle there is an all white type due to homozygosity for a gene W. The heterozygotes are roan (Fig. 6:10), and the homozygotes, ww, are red (cf. page 44). Animals which carry the gene for black but are at the same time heterozygous for Ww, are blue-grey (blue roan). In fact the coat consists of a mixture of black and white hair.

Sheep

Two types of hair fibre occur in sheep, wool hairs and the ordinary hair fibres known as guard hairs. These have not always the same colour in one and the same individual. Wool fibres are as a rule lighter than guard hairs. In the wool-producing breeds the fleece consists almost entirely of wool fibres, whereas in the more primitive breeds there is a mixture of wool and guard hair. The head and legs are often covered with short hair-like fibres called *kemp*. This difference in the distribution of the two hair types in different parts of the body partially explains why the head and extremities are often pigmented in breeds which otherwise have white or yellowish-white fleeces.

Black and brown. The black and brown colours may be divided into dominant and recessive types according to how they behave in crosses with white animals. The dominant types occur mainly in the Karakul breed. The black and brown colour in this breed is, however, modified by the presence of one or more dominant genes for progressive whitening. Analogous with the situation for progressive whitening in horses, Karakul lambs are born either all black or all brown. The roan colour makes its appearance when the animals are about one year old. The gradual increase in the number of white hairs gives the full-grown animals a grey colour although the ground colour remains black or brown. Brown lambs can segregate out from matings between black Karakul sheep, but matings between the brown sheep never produce black progeny. The gene for black is thus dominant over the gene for brown as well as over the gene for white.

In North European sheep breeds the black and brown colour types are usually recessive to white colour. Matings between white sheep can therefore produce black lambs and the expression 'black sheep of the family' is, no doubt, based upon this well known fact. Brown is now very rare in the North European breeds; it is recessive to black.

Grey colour. A number of grey colour types occur in sheep. As already mentioned the full grown Karakul sheep are grey, whereas the lambs are all-black or all-brown. A certain type of Karakul lamb, though, are grey at birth with a mixture of black and white hairs. According to a number of concurring investigations the trait is controlled by a dominant gene. The heterozygotes are quite vigorous, whereas the homozygote lambs seldom reach the age of maturity. The gene for grey in Karakul lambs can thus be regarded as a recessive lethal but the grey colour is manifested in the heterozygote (cf. page 220).

Varying shades of grey also occur in the North European, short-tailed 'Landrace' breeds, e.g. the Gotland sheep, the Icelandic sheep and the old Norwegian 'Spaelsau' breed. The colour of Gotland sheep is of great economic importance, since the pelts are used in the fur trade. Pelts with an even grey colour fetch the best prices. The inheritance of grey colour, and the various shades thereof, is still far from clear. According to Swedish as well as Norwegian and Icelandic investigations, it appears that the white colour is, in general determined by a gene which is dominant over, or epistatic to, the genes for grey and black. In a similar way the grey colour shows a tendency to dominate over black. Studies on

Fig. 6:11. Gotland indigenous sheep. Light grey.

the inheritance of the grey colour are complicated by the fact that the dividing line between the different colour types is not clear, since all shades, from black to almost white can occur (Fig. 6:11). Of those animals which have a white fleece, some have white extremities and head, others are pigmented in these parts of the body.

Since the middle of the 1950s SKÅRMAN has been conducting investigations into the influence of age, sex and inheritance on the colour type in the Gotland sheep. Up to now over 4000 lambs have been examined. The colour of the fleece is classified at birth and again at 2 and at 5 months of age. At the latter two ages the classification is carried out by the same person. No effect of sex on colour type has been established but changes with increasing age were very pronounced. At birth, over 40 per cent of the lambs were classified as being black but only 5 per cent at 5 months of age. The lambs became uniformly lighter in colour with increasing age. Several of the lambs, which at birth were classified as light grey, were considered at 5 months of age to have white fleeces. Those parts of the extremities and head which were covered with short kemp hairs remained pigmented.

The distribution of the different colour types of 4009 lambs of the Gotland breed at different ages (after SKÅRMAN 1961) is shown in the following table:

Colour	Percentage distribution of colours at:		
	Birth (base of fleece)	2 months	5 months
No data	3·6	—	—
White	0·9	1·2	1·9
White grey	0·4	8·9	15·7
Light grey	4·1	16·3	21·3
Grey	26·0	39·6	34·4
Dark grey	13·1	14·1	12·1
Black grey	7·4	10·2	9·7
Black	43·2	9·6	4·9
Motley, brown, etc.	1·3	0·3	—

Skårman found, as in earlier investigations, that white colour is dominant over grey or black. The intensity of the grey colour of lambs was, on the average, darker, the darker the grey colour of the parents. The matings often gave rise to transgressive segregations which is illustrated by the mating combination grey × dark grey, where 149 lambs at 5 months of age had the following percentage distribution for the different colour types:

White	White grey	Light grey	Grey	Dark grey	Black grey	Black
0·7	10·7	14·1	32·2	19·5	17·4	5·4

The variation was similar in other mating combinations and the results indicate that the intensity of the grey colour is under the influence of many genes. The dropping of white lambs from grey parents may possibly have been due to the presence of genes for dominant black or grey. Skårman is of the opinion that it is more likely that the white fleece is caused by an accumulation of polymeric factors which operate in the direction to white colour. The smoothness of the variation curve for the intensity of colour is evidence in favour of this, as is also the fact that certain lambs classed as grey at birth later became lighter, so that at 5 months of age they were considered to have white fleeces.

Pigs

During the past twenty years a number of crossing experiments have been carried out between different breeds of pigs. The object has been to test the production capacity of the crossbreds or to evolve new breeds which combine the best qualities of the original breeds. In a number of these experiments the inheritance of colour also has been studied and is therefore fairly well worked out. The most detailed study of colour has been carried out by H. O. HETZER (1945–1948), at the Beltsville Agricultural Research Centre, U.S.A.

Black and red colour. The Scandinavian Landrace and the English Yorkshire or Large White are examples of white breeds of pigs. The English Large Black is—as the name implies—black, whereas the Tamworth and Duroc from England and the U.S.A. respectively are red (Fig. 6:12). The result from crossing between red and black breeds appears to depend entirely upon which black breed is used. Crosses between the Hampshire (black with a white belt) and Tamworth or Duroc breeds show that black is completely dominant over red, and in the F_2 generation segregation is 3 black to 1 red. On the other hand, if the Hampshire is replaced by Berkshire or Poland China, both of which are black apart from small white markings, the progeny are not all-black but spotted

Fig. 6:12. *Left*: Berkshire. *Right*: Tamworth.

Fig. 6:13. Pig with black spots on a red-brown ground coat. This pattern occurs after crossing between Berkshire and Tamworth. (Courtesy of Dr H. O. Hetzer, U.S.D.A., Beltsville Md.)

red and black (Fig. 6:13). A 3 to 1 segregation of red and black spotted and solid red individuals takes place in the F_2 generation. Hetzer is of the opinion that the inheritance of black and red colour in pigs is genetically similar to that in rodents. In the Hampshire, for example, the black is determined by a dominant gene, E; the almost black Berkshire and Poland China are assumed to be homozygous for the gene, e^p, but the spotted red and black colouring has been altered by an accumulation of modifying genes. The effect of the e^p gene, however, is clearly seen in the progeny of matings between the Berkshire and the different red breeds.

White colour and markings. The all-white colour of the Scandinavian Landrace and Yorkshire is controlled by a dominant gene, usually denoted by *I*. Crosses between the Scandinavian Landrace and black, spotted or red breeds give progeny which are all-white. The Landrace colour is dominant over even the wild pig colour.

The commonest type of white marking in pigs is characterised by a white belt, which extends from the front feet and encircles the body over the shoulders and back. The width of the belt varies greatly. The English Wessex breed has a narrow white belt whereas individuals typical of the Hanover-Braunschweig breed only the head, hams and possibly the back legs are black, the rest of the

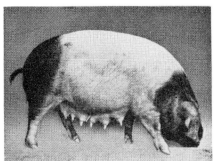 

Fig. 6:14. *Left*: Sow of the Hanover-Braunschweig breed. *Right*: A Wessex sow.

Fig. 6:15. Wild pig with young. (Photo: The Nordic Museum)

body being white (Fig. 6:14). Variation within breeds is, however, very great. In crosses between breeds with and without the belt, the belt marking shows a tendency to dominance. It is very likely that several different genes must be present to produce belts typical of the various breeds.

Wild pig colour. Adult wild pigs are recognised by a dark greyish-brown colour; but the piglets, up to 4 or 5 months of age, have a reddish colour with longitudinal creamy-white stripes on each side of the body (Fig. 6:15). The wild colour is dominant in matings with the majority of other breeds, e.g. Berkshire, but not with the all-white breeds. The difference between the wild pig type and black colour of the Berkshire is apparently due to a dominant gene in the wild pig. This gene is probably a homologue of the agouti gene in rodents.

Fur animals

The mink and fox are kept exclusively for the sake of the pelt. The colour of the hair is of great importance for the value of the pelt, and several extensive investigations have been carried out to elucidate the inheritance of the various colour types. In mink, more than 20 colour-determining chromosomal loci have been demonstrated; the number of colour types in the fox is somewhat less. In Chapter 3 the inheritance of colour type in fur animals was used to illustrate Mendelian inheritance, and therefore only some complementary points will be dealt with here.

Grey or grey-blue colour type in mink. There are a number of grey or so called 'blue' colour types in mink. The type first studied was called platinum or silver

Fig. 6:16. *Left*: White mink (Hedlund white). *Right*: Sapphire mink.

blue and is determined by a recessive gene p. The variation in intensity of colour is rather large and a 'pepper and salt' pattern often occurs. SHACKELFORD (1948) has shown that platinum colour is due to an aggregation of pigment granules in the hair cells. The type is thus similar to that in the blue Viennese rabbit and mice with diluted colour. There is another allele, at the same locus as the gene for platinum, which is recessive to the standard type but dominant over p. This gene, p^s, gives the colour type steel-blue, which usually is somewhat darker than platinum, although the colour shade may vary considerably. Another platinum type, imperial platinum, cannot be distinguished phenotypically from the common platinum. Crosses between the two types give standard mink; which shows that imperial platinum and common platinum are determined by genes at different loci. A fourth grey type, called aleutian (gunmetal), (gene symbol a) is distinctly more steel-blue than the other platinum types. According to Shackelford, this is due, not to an aggregation of pigment granules, but rather to a reduction in the amount of pigment. As mentioned previously, the aleutian colour type is associated with a disease complex called *aleutian disease* (cf. page 76). The greatest importance of the aleutian gene is that in combination with other recessive genes it gives rise to much favoured colours, for example the Sapphires ($a\ a\ p\ p$) (Fig. 6:16).

Brown colour types in mink. The colour of the modern standard mink is such a dark shade of brown that it appears to be almost black. The wild mink is considerably lighter. Apparently a number of modifying genes have been accumulated through selection in the ranch-bred mink.

In the ranch-bred mink a number of light brown, so called 'pastel' types, have appeared, which are all determined by recessive genes. The royal or brown-eyed pastel (gene symbol b) is recognised by the eyes, nose and fur being decidedly lighter in shade than standard mink. Other, but genetically different colour types, are green-eyed, imperial and amber-gold pastel. A fifth pastel type, called *socklot*, is also determined by a gene recessive to the wild type.

Socklot appeared for the first time in 1945 on a mink ranch in Finland. Animals of this type are often large and have good fertility, but they tend to have a somewhat coarser fur. The socklot mink has shown itself to be valuable particularly in certain combinations with other mutants in the brown series.

The locus which controls the socklot type includes, apart from the standard and socklot genes, three alleles which give the colours Swedish-palomino, Finn-white and Nordic-buff. The first of these has buff-coloured fur and dark eyes, whereas the two latter types are cream-coloured and have pink eyes. A number of other brown colour types have been identified.

In mink there are several gene systems that give white colour and markings. Pure albino, however, has not been identified. The so called American albino, which is controlled by a recessive gene, is not completely white, the nose and tail being more or less pigmented, similar to the himalayan type of pattern in rabbits. The eyes are pink. The Hedlund-white colour type has a completely white fur, but the eyes are pigmented. It has been previously mentioned that animals of this type are completely deaf; this reduces their value, since the females are poor mothers. The gene for Hedlund-White is not completely recessive, as the heterozygotes have large white patches.

Several other genes in mink give white markings and increased extent of white hairs in the fur; only two will be dealt with here, the bluefrost and stewart. The former is determined by a gene F in the heterozygote condition. Apart from the white markings and interspersed white guard hairs, the bluefrost type is recognised by a distinctly lighter underfur than the standard type. The F gene is lethal in the homozygous condition. The stewart type closely resembles the bluefrost, and at first was thought to be controlled by the same gene. The pelt of the stewart colour is more in demand because it has less white markings, but at the same time the under-fur is of a lighter shade than the bluefrost. That it is not the same gene in both types follows from the fact that the homozygotes for the stewart factor are viable. These so called 'homos' are distinctly lighter than the heterozygotes. The pelts of homo pastel and other colour combinations have commanded high prices on the fur market. The stewart factor has therefore achieved a certain degree of distribution in spite of the fact that the homozygotes have a reduced vigour and fertility. The males lack the ability to produce sperm and are thus completely sterile. There is nothing wrong with their sexual behaviour, however; the males mate with great vigour whenever the opportunity presents itself. The females may be fertile but the degree of fertility is considerably reduced.

The genes for colour in mink can be combined in many different ways. A number of combinations have given rise to valuable commercial types, e.g. the sapphire, in which the animal is homozygous both for silver blue and aleutian; and topaz, in which the animal is homozygous for two mutant genes which individually give brown of the palomino type or darker. Classical genetics has, without doubt, received more application in mink breeding than any other farm animal. Fashions change and the fur farmer tries therefore to produce new colours by combining known, or new, colour genes. Each new colour which becomes popular can mean a great deal of money if the right breeding stock and knowledge of how to combine them are available.

Fox

Two different species of fox, the red fox (*Vulpes vulpes*) and the Arctic fox (*Alopex lagopus*) have been used as fur animals. The guard hair of the red fox has typical agouti markings with a black tip, grey hair-base and reddish-yellow band. Two recessive mutant genes individually bring about the disappearance of the reddish-yellow band and produce black colour. These black foxes have a varying number of white hairs. The silver fox colour type has been obtained by selection for an increased number of white hairs. A number of colour mutants occur, among which are those described on page 73, for platinum and 'whiteface' types.

In the case of the Arctic or blue fox, two types are distinguished, the Alaska blue fox, which is blue-grey both in winter and summer, and the somewhat lighter Greenland blue fox. From the latter type the so-called 'white' fox, occasionally segregates; it is recognised by the fact that in winter the fur is white, but in summer it is grey-brown. According to Norwegian and Swedish investigations, the white colour is due to homozygosity for an incompletely recessive autosomal gene. The heterozygotes are consistently lighter than individuals which are homozygous for the dominant gene in the double dose.

Poultry

As already mentioned, the colour of poultry is due, not only to melanin, but also to the pigment, xanthophyll. In addition, certain colour types are influenced by the way in which the pigment is stored in the cells and the way in which light is refracted in the tissue layers. Examples of such a structurally determined colour nuance are the blue-green colour of wing quills and the green colour of the toes in some pigmented breeds. In the latter case the green colour is due to the fact that the distribution of the melanin is confined to the dermis, whereas the

Fig. 6:17. *Left*: Black Minorca. *Right*: Blue Andalusian.

epidermis contains the xanthophyll. The black colour of the dermis is thus seen through the yellow outer layer which produces a green nuance.

The ability to store xanthophyll in the skin is determined by homozygosity for a recessive gene, the dominant allele of which limits the storage of the yellow pigment to the body fat and blood. A prerequisite for yellow skin of the recessive homozygote is that xanthophyll is present in the food. If it is not, then the skin will be white and the two genetically different types cannot be identified.

White plumage in the fowl is controlled by several different genes. There are, for instance, two different albino types both of which are due to recessive genes; one is autosomal and the other is sex-linked. The white colour of the White Leghorn dominates, by and large, in crosses with black breeds, e.g. Black Minorca and is therefore called dominant white and given the gene symbol *I*. The white colour of Leghorns was one of the first traits of farm animals which could be shown to be genetically determined (cf. page 49). The white colour of the Wyandotte and Silkie breeds, unlike that of the Leghorn, is recessive in crosses with brown and black breeds. The recessive genes in Wyandotte and Silkie are, however, not identical since crosses between them produce pigmented progeny.

Black and red colour. Black colour occurs in several breeds of fowl and is under the control of a dominant gene which, as in mammals, is usually given the symbol E. For the E gene to manifest itself, the genotype of the bird must otherwise be such as to allow pigmentation in the feathers. Red colour is determined by homozygosity for the recessive gene e. The extent of the red colour is influenced by a number of different gene systems.

Black colour is altered to blue-grey when the gene Bl is present in a single dose. The blue-grey colour is due to the aggregation of black pigment granules. The feathers are, however, not entirely blue but black spots also occur. The Blue Andalusians are heterozygous for the Bl gene (Fig. 6:17). Homozygotes are predominantly white with the exception of a number of black spots.

In the fowl, there are many different colour patterns and several of these have been genetically investigated. The barred pattern of the Plymouth Rock and the silver of the Light Sussex are, as already mentioned (page 62), due to sex-linked dominant genes. The barred pattern of the Hamburger breed, on the other hand, is due to an autosomal recessive gene. Other colour patterns which occur are spangling, mottling and lacing. Some of these colour patterns are due to dominant genes, others to recessives.

OTHER EXTERNAL TRAITS

Horns

Cattle, sheep and goats all belong to the family *Bovidae* or *Cavicornia*, i.e. the hollow-horned ruminants. The horns in this family consist of an inner boney core and an outer sheath of horn. The core is an outcrop from the frontal bone of the skull. At the base of the horn the sheath is thinnest and continues over into, and becomes part of, the surrounding skin. In cattle and sheep there are breeds as well as animals within breeds, with and without horns. Some individuals have incomplete horn development, which can vary from barely visible 'buds' in the skin of the skull to almost fully developed horns. In the latter case, though, there is no solid connection with the bone of the skull. The difference between real horns and horn-like growths, such as lumps, scurs or loose horns, is that in the former case the bone core is an actual outgrowth from the skull, while the horn-like formations lack this solid connection with the skull.

The inheritance of horns and polledness has been investigated in both cattle and sheep. BATESON and SAUNDERS concluded in 1902 that polledness in cattle was dominant over hornedness. In the main, this conclusion has stood the test of time. However, the dominance is not complete. In the F_1 generation of a cross between horned and polled breeds some of the male animals have horn buds or even loose horns, whereas the females often, or most often, lack these altogether. In matings between the polled Angus breed and the Zebu type of cattle, the male progeny have, without exception, these horn-like formations. When the Friesian is used as the horned part of this cross the buds are much less developed. When the results of the crosses carried out between horned and polled breeds are summarised, it is found that in the F_2 generation the ratio of polled (completely hornless or with 'buds' or scurs) to horned animals is approximately 3:1. This is sufficient evidence to conclude that polledness is controlled by an incompletely dominant gene (usually indicated by P), whereas proper horns occur in the homozygote pp. The degree of dominance exercised by the P gene is influenced by the sex and a number of modifying genes which vary in different breeds.

The inheritance of horns in sheep is even more complicated. In some breeds,

Fig. 6:18. *Top left*: Dorset Horn ewes; *top right*: Dorset Horn ram. *Bottom left*: Merino ewe; *bottom right*: Merino ram.

Fig. 6:19. Syrian ram with four horns.

e.g. the British Dorset Horn and Scottish Blackface both the rams and ewes have horns. In the Rambouillet and Merino breeds only the rams are horned (Fig. 6:18), but the ewes can have small horn buds. Both rams and ewes are hornless in the Suffolk, Shropshire and several other breeds. When the Dorset Horn and the Suffolk are crossed, the F_1 ewes are hornless and the rams horned. The sex difference in this cross is generally used to illustrate sex influenced dominance. The result agrees closely with that of the Aberdeen Angus × African cattle though the sex difference here is less pronounced.

The castration experiment of HAMMOND and MARSHALL with an English breed of sheep, where the rams were horned and the ewes hornless, clearly showed the importance of sex for horn development. As soon as the ram lambs were castrated, horn growth ceased. Removal of the ovaries of the ewe lambs did not, however, result in the ewes developing horns. In conclusion, it can be stated that, as with cattle, it appears as though polledness in sheep shows a tendency to dominate over horns. The degree to which the gene for polledness dominates in the heterozygote is determined by the sex, by modifying genes, and possibly also by the other allele. The gene for horns, e.g. in the Dorset Horn, apparently has such an overriding effect that the influence of the gene for polledness is partially neutralised.

In Icelandic and some other primitive sheep breeds, animals with four horns occur (Fig. 6:19). The type is more common in rams than ewes. The four horn characteristic apparently dominates over two horn but both are recessive to polledness.

Hair characteristics

In the majority of farm animal species there are individual differences in hair length, diameter and general appearance—such as waviness, curliness, etc. A part of this variation is clearly genetic. In the rabbit, there are types with normal fur, types with long soft fur (Angora) and those with extremely short hair. The Angora type is due to a recessive gene. The name is taken from

Fig. 6:20. Rabbit and mink of angora type. Angora (ll) mink; this individual is also a Black Cross (Sa).

the woolly Angora goat which originates from the Ankara (or Angora) region of the Middle East. The Angora type is found in a number of other species, e.g. cat, dog, mink, and guinea-pigs (Fig. 6:20). In those species where the trait has been genetically investigated, it has been shown to be due to homozygosity for a recessive gene. The extremely short haired, so-called Rex rabbit was first described in 1919 in France. The guard hairs are drastically shortened and insufficient to cover the under-coat. Macroscopic examination gives the impression that the fur consists only of under-coat. Even this is shortened but not as much as the guard hairs. At first it was thought that the pelt of the Rex type would be popular and utilised by the fur trade and the Rex rabbit was in considerable demand. It was soon shown, however, that the pelt lacked durability and wearing capacity, and the dream of a new and important fur rabbit collapsed. The Rex mutation is thought to give rise to metabolic disturbances; the birth weight of the young is lower than normal and early mortality increases. At least two other recessive short haired rabbit types have been identified. With regard to hair type in sheep see page 328.

Plumage in fowls

Silkie. The appearance of feathers in birds, like the hair of mammals, can deviate from the normal. The Silkie feathering is an example of this (Fig. 6:21). The name describes the silky soft nature of the feathers which are almost like the down of chicks. The shaft of the feathers is soft and the barbules lack hooks

Fig. 6:21. Silkie fowl.

and are therefore spread out in all directions. Otherwise the bird appears to be quite normal. The change in the structure of the feather is caused by homozygosity for a recessive, autosomal gene. The frizzle fowl described on page 45 is another example of genetically determined change in feather structure.

Many other deviations from the normal type have been identified in the various species of farm animals. Many of these, however, lead to reduced viability, such as hairlessness and dwarfism. These and similar traits will therefore be dealt with in Chapter 8.

SELECTED REFERENCES

BERGE, S. 1965. Haarfarbe und Haarzeichnung bei Kreuzung von gelbenweissen Charaolaise mit roten und schwarzen Kühen. *Z. Tierzücht. ZüchtBiol.*, **81**: 46–54.

CASTLE, W. E. and SINGLETON, W. R. 1961. The palomino horse. *Genetics, Princeton,* **46**: 1143–1150.
HETZER, H. O. 1948. Inheritance of coat color in swine. VII. Results of Landrace by Hampshire crosses. *J. Hered.,* **39**: 123–128.
HUTT, F. B. 1949. *Genetics of the Fowl* (590 pp.). New York.
LAUVERGNE, J. J. 1966. Génétique de la couleur du pelage des bovins domestiques (Bos taurus, Linné). *Bibl. Genetica.,* **20**: 1–68.
RAE, A. L. 1956. The genetics of the sheep. *Adv. Genet.,* **8**: 189–265.
RENDEL, J. 1959. Farbe und Zeichnung. *Handbuch der Tierzüchtung,* Vol. 2 Chap. 4; 105–141. Edited by Hammond, Johansson and Haring. Verlag Paul Parey, Hamburg.
WAGNER, R. and MITCHELL, H. 1964. *Biochemical Genetics,* Chapter 10: 427–438.
JOHANSSON, I. 1965. Studies on the genetics of ranch mink. II. Fecundity, viability and body size of various colour mutants. *Z. Tierzücht. ZüchtBiol.* **81**: 55–72.

7 Inheritance of Blood Characteristics. Basic Results and Practical Applications

The characteristics of the blood of men and animals have long attracted the interest of the scientist and at the present time more is known about the genetic variations of blood components than of any other animal tissue or fluid. Differences between the blood of animals from different species had already been reported by LANDOIS at the end of the nineteenth century, who found that agglutination or haemolysis occurred when human blood was mixed with that of higher animals. That there were differences between the blood of individuals from the same species was established by LANDSTEINER in 1900, when he made his fundamental discovery of the A, B and O groups of human blood. This detection was made in the same year that the Mendelian principles of heredity were rediscovered and it was not long before a fruitful co-operation was established between immunology and human genetics. *Immunogenetics* is now one of the most prolific branches of human genetics.

Investigations into the blood groups of farm animals also began in 1900, when EHRLICH and MORGENROTH demonstrated differences between the blood of different goats. They introduced the immunisation technique in which the blood of one individual is injected into other individuals. Antibodies may then be produced against the injected blood and the antibodies can be used for investigations of the blood of other subjects. This method has been widely used in animal blood group research and will be discussed in more detail later on. There was not a great deal of work done on the blood groups of farm animals until the beginning of the 1940s when a research team at the University of Wisconsin, U.S.A., led by M. R. IRWIN, began systematic studies of the blood groups of cattle, sheep and poultry.

The investigations mentioned so far are all based on the reaction between the antibodies in the serum and the antigens on the red blood cells (*erythrocytes*). In recent years it has been possible with the aid of biochemical methods, mainly electrophoresis, to demonstrate genetic differences in blood proteins both in humans and animals.

The genetic classification of the various constituents of blood is based mainly on these immunological and biochemical methods. The immunological approach is by far the oldest and consequently the term 'blood groups' has tended to become more or less synonymous with blood characteristics detectable by immunological techniques. However, the term 'blood group' is sometimes used more broadly to include other inherited blood characters. When distinction is needed more specific terms, as 'blood antigens' or 'erythrocyte antigens', will therefore be applied to those characteristics which are determined by immunological methods, while other blood characteristics will be referred to in a way which indicates their biochemical nature or function. We shall deal first with the erythrocyte antigens and then with the biochemical characteristics. The final part of the chapter will be devoted to a discussion of practical applications which generally involves both these main classes of blood characters.

BLOOD ANTIGENS

Definitions and technique

An *antigen* is usually defined as a substance, which when injected into an individual, evokes the production of *antibodies* which react with the antigen in question. In animal blood group research, the term antigen is used in a more restricted sense and indicates a genetically determined entity with antigenic specificities. An antigen can have more than one antigenic specificity, i.e. it can give rise to several distinct antibody types. These 'parts' of the antigen which evoke and react with these different types of antibodies are called *antigenic factors* or simply *blood factors*.

The antigens are distinguished with the aid of *antigen-antibody reactions*. One or two drops of a serum containing antibodies, known as *anti-serum*, are mixed together with a dilute solution of erythrocytes. If the reactive sites of the antibody and the erythrocyte antigen correspond, a reaction will take place which can be made visible in different ways. Antibodies of a certain type may react simultaneously with antigens on two different erythrocytes. These are drawn together to form clusters or clumps which can be easily observed in a test tube. The erythrocytes are then said to *agglutinate*. The agglutination method is used, for example, in investigations concerning humans and fowls but is less suitable in studying cattle blood antigens. Instead, *haemolysis* reactions are used, in which the red blood corpuscles are disrupted with the liberation of the blood pigment, *haemoglobin*, and colouration of the surrounding fluid. For haemolysis to take place, the presence of a thermolabile serum factor, *complement*, is required. Sera of different animals, such as rabbits and guinea-pigs, are utilised as complement sources.

In tests for blood antigens use is made of either normally occurring anti-sera, e.g. anti-A and anti-B in humans, or the antibodies are produced by immunisation. In the latter case, blood is repeatedly transferred from one individual to another. The recipient may then produce antibodies against the antigens which are present on the donor's erythrocytes but absent in the recipient. The greater the difference between the two individuals, the more types of antibodies the recipient can be expected to produce. When an antiserum is obtained it therefore often needs purification by an absorption technique so that it contains only one type of antibody. Anti-sera with only one antibody type are known as *blood group reagents*. Fig. 7:1 shows schematically the method of immunising cattle. Ten ml of blood are injected intramuscularly into the recipient at weekly intervals. In this hypothetical case it is assumed that the donor carries the antigens A, B

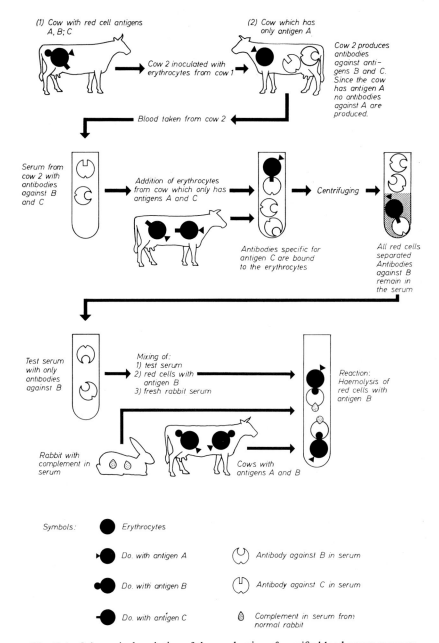

Fig. 7:1. Schematic description of the production of specific blood group reagents by immunisations. (STORMONT and CUMLEY 1943)

and C, whilst the recipient has only A. The recipient is assumed to produce antibodies against the antigens B and C. In order to obtain a B reagent it is necessary to mix the B, C anti-serum with erythrocytes from an individual with the C antigen, whereby the C antibodies become bound to the erythrocytes and can be separated by centrifuging. The serum obtained after centrifuging contains only B antibodies and can thus be used as a reagent in studies of animals with unknown blood type.

Erythrocyte antigens of cattle

No other species of farm animal has been the object of such extensive studies of blood antigens as cattle. Stimulated and influenced by the discovery of the human ABO system, the first attempts to study the blood antigens of cattle were made with the help of normally occurring anti-sera and agglutination. A few antigens could be demonstrated in this way, but the work was associated with technical difficulties since cow anti-sera lose their agglutinating power after only a few days storage in a refrigerator. In order that a cow anti-serum may be able to agglutinate the erythrocytes, the presence of a thermolabile serum component is required. This technical problem was overcome by the previously mentioned group of researchers at the University of Wisconsin, who began using immune sera instead of normal sera and haemolysis instead of agglutination.

From the immunisations, reagents could be produced which gave clear and distinct haemolysis and by carrying out a large number of immunisations of the type described in Fig. 7:1 it has been possible to produce some 70 different reagents. This in turn has made it possible to demonstrate a corresponding number of antigenic factors. These have been designated alphabetically in the order in which they were discovered, i.e. A, B, C, etc., and when the letters of the alphabet were used up superscripts to the letters were employed, A', B', and so on. No antigenic similarity was assumed to exist between A and A'. Serologically related antigenic factors have instead been given the same letter with different numerical subscripts, e.g. O_1, O_2 and O_3. The majority of antigenic factors were first demonstrated by STORMONT and FERGUSON (U.S.A.).

At first it was believed that each of the different antigenic factors was controlled by a separate gene system. For factor A it was assumed that there was a corresponding gene *A*, whereas the absence of the antigen A would be due to homozygosity for the gene *a*. The same reasoning was applied to B, C and etc. It soon became apparent, however, that this could not be the case. STORMONT and co-workers (1951) in a study of large progeny groups from different bulls were able to show that certain antigenic factors were always inherited together. This may be illustrated by a study of the progeny of a bull with the factors B, O_2, O_3, A', E_3', J' and K'. No less than 66 offspring were obtained from matings with cows which lacked all or most of these factors, and of these progeny 35 had the combination $BO_2A'E_3'$ and 31 had the combination $O_3J'K'$. There was no other combination of factors. The result indicated, therefore, that the bull was heterozygous for two genes of which one gave an antigen with the factors $BO_2A'E_3'$ and the other an antigen characterised by the factors $O_3J'K'$. The fact that some blood factors occur in combinations or *phenogroups*, which are transmitted as units from generation to generation, has been verified in several studies of family data. Continued investigations have shown that the known blood group antigens in cattle are controlled by genes at no less than 11 different loci.

Table 7:1 A summary of the genetic systems controlling blood antigens in cattle

System	Antigenic factors	Number of alleles	Special comments
A	A_1, A_2, H, D_1, Z'	10	
B	B, G, K, I_1, I_2, O_1, O_2, O_3, P, Q, T_1, T_2, Y_1, Y_2, A', B', D', E_1', E_2', E_3', F', G', I', J', K', O', Y' and others	>300	Genotypic classification directly from the phenotype often possible
C	C_1, C_2, E, R_1, R_2, X_1, X_2, L', W, and others	>35	Genotypic classification sometimes possible
F–V	F_1, F_2, V_1, V_2	4	Genotypic classification generally possible
J	J	>4	J is a soluble plasma constituent. Anti-J in genotype j/j, which lacks J
L	L	2	
M	M_1, M_2	3	
SU	S_1, S_2, U_1, U_2, U', and others	>10	Genotypic classification often possible
Z	Z_1, Z_2	3	Genotypic classification generally possible
R'–S'	R', S'	2	Genotypic classification possible

Table 7:1 gives a summary of cattle blood group antigens within 10 of these gene systems. The eleventh was recently discovered independently by LIND-STRÖM and MAIJALA (1965) in Finland and GROSCLAUDE (1965) in France. The system has not yet been given a definitive name. We shall first consider the L system, which contains only one factor, L. This is caused by a gene, also called *L*, when present in single or double dose. The $R'S'$ system also contains only two alleles, but in this case it is a question of co-dominant alleles, i.e. both the genes are manifested in a single dose. The nature of the FV system is similar in all the north-west European breeds, but in southern Europe, Asia and Africa there are additional alleles. All the other blood group systems carry multiple alleles. Before describing any of the really complicated systems we shall first deal with the J system, which from many points of view has a unique position among the cattle blood groups. The J factor was discovered with the help of antibodies which normally occur in the serum of J-negative animals. STORMONT (1949) showed that the J factor is not originally a blood group antigen, but a serum factor. Individuals of certain genotypes have a mucopolysaccharide in their serum—known as the J substance—which reacts with J antibodies. This sub-

stance is present in the serum of new-born calves but is not present on the erythrocytes. These take up the J substance from the surrounding plasma when the calves are about one week old. It is also possible to make J-negative erythrocytes-positive by incubating the J-negative cells in blood plasma from J-positive animals. There is a great deal of individual variation in the concentration of J substance in the serum from positive animals. Those with the lowest concentration generally lack J substance on the erythrocytes. The individual differences are apparently determined by multiple alleles with a certain amount of dominance.

The B system is the most complex genetic system that has been identified in any species of animals. It includes at least 300 alleles which control as many antigens. The majority of these are characterised by several antigenic specificities. In different laboratories about 40 reagents have been prepared which react with an equal number of antigenic factors in the B system. Some B antigens (or B-phenogroups) are shown in Table 7:2. The first B allele, $B^{I'}$ occurs with

Table 7:2 Example of four different B antigens in cattle

Antigen	Allele	Code number
I'	$_BI'$	89
GY_2E_1'	$_BGY_2E_1'$	39
$BO_3Y_1A'E_3'G'$	$_BBO_3Y_1A'E_3'G'$	22
$BGKO_2Y_1A'B'E_3'G'K'O'Y'$	$_BBGKO_2Y_1A'B'E_3'G'K'O'Y'$	28

low to medium-low frequency in several of the cattle breeds. The corresponding antigen is represented only by a single factor, I'. The next allele, $B^{GY_2E'1}$, gives rise to an antigen which reacts with the reagents, G, Y_2, E_1', E_2' and E_3'. This allele is extremely common in the American Holstein-Friesian breed, but is also found in several other breeds. The allele, $_BBO_3Y_1A'E'_3G'$, is often called 'the Shorthorn allele' as it has a frequency of nearly 40 per cent in the Shorthorn breed in the U.S.A. It has also a very high frequency in the Swedish Red and White cattle, which were built up from crosses between Shorthorn, Ayrshire and Swedish Landrace. Finally, the allele which has been given the code number 28 (Table 7:2) gives rise to an antigen with no less than 14 different antigenic factors. This antigen is common in Jerseys (allele frequency about 25 per cent) and is also found in the Guernsey breed (10 per cent).

The other blood group systems in cattle are, in principle, of the same type as the B system, but are less complex.

Since the complex B antigens are inherited as units it has been assumed that they are determined by alleles. Similar, though less complex systems, occur in several species, e.g. the Rh system in man. One alternative explanation of the inheritance of the Rh antigens in humans is based on the assumption of closely linked genes instead of alleles. A pertinent question is whether there are any results which would support such a hypothesis in the case of farm animals. During recent years several research workers have reported at least 12 cases of changes taking place in a particular B antigen from one generation to another which can be an indication of recombination.

The majority of these changes have involved a loss of one or several specificities of the antigen concerned. Two of the cases have resulted in additional antigenic factors and another case was mainly a duplication. Three of the changes observed have occurred in the genes the individual received from the mother, and can hardly be ascribed, in these cases, to errors in pedigree. STORMONT reported on the mating ♂PY_2/E_1' × ♀BO_1Y_2D'/BO_1B' in which an offspring of the type PY_2/BO_1 was produced. A loss of factor B' had thus occurred on the antigen which represented the gene transmitted from the dam. BOUW reported on a Dutch sire, which by virtue of numerous progeny had been classified genotypically at the B locus as $BGKO_xY_2A'O'/O_xY_2D'E_1'O'$. However, one of his male offspring had the odd genotype $BGKO_xY_2A'D'O'/I'$. A change constituting a gain of factor D' had evidently occurred in the antigen transmitted to the son. This B antigen, which was of a previously unrecognised type, was thereafter transmitted to approximately half of the son's progeny.

A particularly interesting family studied by LINDSTRÖM of Finland should also be mentioned. She typed twenty-nine offspring of an Ayrshire bull with the B locus genotype PI'/E_3'. Fourteen offspring had received the allele for PI' and thirteen the allele for E_3'. However, two had inherited neither but what would appear as a recombination of the two, namely $E_3'I'$. These two offspring were males and both of them were shown to transmit the recombinant gene to some of their offspring. The probability of obtaining two identical mutations or recombinations within a comparatively small family is very low. It seems therefore likely that a change had occurred during the development of the germinative epithelium of the sire's testicle so that a small segment was of type $E_3'I'/P$ rather than PI'/E_3'. The occurrence of an offspring which had received P only from this sire would almost constitute a proof of this hypothesis; but in spite of further testing, no such progeny was encountered.

Blood antigens in cattle twins

As discussed in Chapter 5, approximately 10 per cent of the like-sexed twins are monozygous. The majority of twins are thus dizygous and as a result genetically no more alike than ordinary full sibs. The probability that two dizygous cattle twins will inherit exactly the same blood antigens in each of the numerous antigen systems is therefore very small. Contrary to expectation, however, the two members of most twin pairs have exactly the same blood antigens. This remarkable discovery was made by OWEN over twenty years ago. After painstaking research, Owen was able to demonstrate that most of the twin sets with identical blood types had an admixture of two different erythrocyte types. This admixture was referred to as *erythrocyte mosaicism*. The investigations on the freemartin problem in the early part of this century had disclosed that there are vascular anastomoses between approximately 90 per cent of all fraternal twin foetuses.

Owen suggested that the erythrocyte mosaicism was a result of the vascular anastomoses, which led to an exchange of primordial blood cells between the twins. These primordial cells would then become established in the blood-forming tissues of the new hosts. Such twins would therefore, during subsequent periods of life, produce two different erythrocyte types, one corresponding to its own genotype and the other to that of its co-twin (see Fig. 7:2).

There is one exception to the rule that twins with erythrocyte mosaicism have exactly the same antigen type. As shown by STORMONT (1949), twins genetically different for the J factor will remain phenotypically different. The J substance is, as already mentioned, primarily a soluble plasma constituent, which is not present on the erythrocytes at birth. In twins with erythrocyte

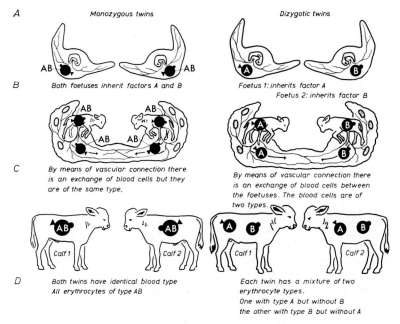

Fig. 7:2. Schematic illustration of how vascular anastomosis in dizygous twins (*right*) leads to erythrocyte mosaicism, whereas monozygous twins (*left*) always inherit the same type and are therefore not affected by the anastomosis. (TOLLE 1960)

mosaicism both the erythrocyte types will become coated with the J substance if this occurs in the plasma. In mosaic pairs all the red cells of one member may therefore very well carry the J antigen, while those in the other member are entirely negative.

The two fractions in the blood admixture can be isolated and the proportions determined in a so-called differential haemolytic test. In general, one of the erythrocyte types will be more frequent in both twins. In some cases the proportions are greatly displaced so that one erythrocyte type may hardly be detectable. RENDEL and GAHNE (1961) reported a case in which a routine check had been made on the stated parentage of a young Friesian bull which would be used in A.I. service. The young bull carried the blood factor W, which was absent in both its parents. The dam had the B type, $BGKYA'E_1'O'$ which strongly indicates the genotype $B^{BGKYA'O'}/B^{GYE_1'}$ or $B^{BGKYA'O'}/B^{E_1'}$. The young bull was entirely lacking in B factors, i.e. its genotype was $B^b B/^b$. It was therefore concluded that the stated parentage was erroneous and that it was unlikely that the young bull was assigned to the right dam. However, further inquiry gave the information that the dam was born twin to a dead calf. By refined haemolytic tests and absorption experiments it was possible to prove that the dam had an admixture of red cells and that one of the two types made up less than 10 per cent of the total number of erythrocytes. This rare type, which evidently represented the genotype of the dam, carried W and had a B factor combination which was consistent with the type of the young bull.

Early studies on the proportions of the two erythrocyte types in twins with blood admixture indicated rather stable cell proportions, at least over periods of one or two years. However, a recent report by STONE (1964) and co-workers shows that this is not always the case. A number of twins were followed over a period of several years, in some cases as long as eleven years. Drastic changes in the cell proportions were encountered in some of the pairs. In brief, the results show that if a change occurs the direction of change seems to be the same in both twins of a pair; furthermore, the direction of the shift with respect to the initially predominant type seems to be random. The same group of workers described an interesting case in which a twin bull with two erythrocyte types developed a third recombinant erythrocyte fraction. At eight years of age the proportions of his blood types had changed from 90 per cent of the type called I and 10 per cent of type II to approximately 2 per cent of each type. The remaining 96 per cent of the cells in this mosaicism were of a new type containing antigens on the same cells which were previously on different cells in the mixture. How this new cell type came about is not known. It was assumed, however, that it might have resulted from somatic cell mating in the admixed haematopoietic tissues.

Erythrocyte mosaicism may also occur in higher orders of multiple births. In a Finnish set of probably five-egg cattle quintuplets, no less than four different erythrocyte types could be distinguished in each animal. However, the proportions of the cell types varied greatly between the quintuplets. The set was composed of four females and one male. The latter developed normally while all the four females were sexually retarded.

Erythrocyte mosaicism is not limited to cattle; a few cases have been reported also in man, sheep and mink. The detection of erythrocyte mosaicism in cattle has been of great importance for the development of the modern concept of immunological tolerance. Normally, tissue grafted from one individual to another is rejected completely within a few days. However, the cattle twins constitute examples of natural successful grafts, where cells giving rise to haematopoietic tissue have been exchanged between the co-twins. Owen's discovery gave the immunologist a lead as to what factors might influence immunological tolerance. The time factor is of great importance. If animals, as for example dizygotic twins, are exposed to foreign tissue at a very early age, they will not later on be able to distinguish the foreign tissue from that of their own. They will become tolerant. The degree of tolerance varies depending on species and on the age at which the animal was first exposed to the foreign tissue or antigen.

Blood antigens in other farm animals

The blood antigens in cattle were dealt with in some detail. In principle the same technique and genetic pattern apply also to the blood antigens in other farm animals. These will therefore be discussed more briefly.

Horses

At the beginning of the 1930s, SCHERMER and KAEMPFFER described 6 different agglutinogens (i.e. antigens detected by virtue of agglutinating anti-sera). Their work was disrupted during World War II, but about 1950, studies of horse blood antigens were taken up at the Pasteur Institute in Paris, by Dr PODLIACHOUK (1957). She detected 10 antigenic factors on horse erythrocytes. Very recently STORMONT and SUZUKI (1964) published a series of papers in which they described 16 antigenic blood factors, 10 of which were 'new', i.e. they were not previously described by Podliachouk and co-workers. Stormont's factors were shown to be controlled by 8 different loci, some of which contained series of multiple alleles. Taken together all these studies indicate that there are at least 21 antigenic factors in the horse, controlled by at least 13 loci.

Sheep

Seven blood antigen systems have been described in sheep and six of these were first worked out by RASMUSEN (1958, 1960), now at Urbana, Illinois. The serological technique is the same as that used in cattle. Several of the antigenic systems comprise multiple alleles and some of them are serologically related to certain antigenic systems in cattle. We shall here mention only the R–O–i and M systems which are interesting in several respects. In the latter system there are three alleles, M, M^x and m. The homozygotes mm lack blood factors of the M system, while the two alleles M and M^x control two related M-antigens. RASMUSEN and HALL (1966) recently made an interesting discovery that the M locus apparently is identical with the K locus of EVANS and KING (1955) which controls the potassium level of the red cells. It is not yet known how the M gene exerts its two effects; the M antigen on the red cell membrane and the elevated potassium content (see also p. 208).

The R antigen of sheep is closely related to cattle J and human A. It is detected by use of cattle anti-J sera or by normal antibodies from sheep lacking the R antigen. Lambs do not develop R on the erythrocytes until they reach an age of about three weeks. The R substance is primarily a secretor substance occurring in the plasma of certain genotypes. It is passively acquired by the cells from the surrounding plasma. R-negative cells may therefore be converted into R-positive ones by incubation in suitable plasma samples.

STORMONT detected in certain cattle sera a normally occurring antibody which reacted selectively with all sheep lacking R. He suggested that this new antigen, O, was determined by a gene recessive to that of R. These suggestions were at least partly confirmed in studies by RENDEL and co-workers, who, however, found an additional group, i, which lacked the R as well as the O substance. The group i was determined by a recessive gene at a second locus which suppressed the development of the R-substance. Mating between i and O individuals might therefore produce offspring of group R, thus constituting an exception to the rule that a cellular antigen is present in an individual only if present in one or both of its parents. Similar examples of epistatic gene action on blood antigens have now been found also in man (the Bombay groups) and in swine (the AO groups).

The R–O–i groups were described in some detail, partly because of their exceptional nature and inheritance and partly because they were recently found to be associated with the occurrence of certain alkaline phosphatase enzymes. More will be said about this later.

Pigs

The blood antigens of the pig have been subject to intensive study during the last decade, particularly by E. ANDRESEN (1963). At least 12 blood antigen systems are known and some of these carry extensive series of multiple alleles.

Andresen and Baker recently described a system, C, which was shown to be linked to one of the already known systems, J. There was a crossing over frequency between them of 5·3 per cent. As will be apparent from the preceding pages, the blood antigens of farm animals constitute good examples of several classical situations in genetics as dominance, co-dominance, and epistasis. We can now add linkage to this group of events. The observations of Andresen and Baker provided the first unequivocal evidence of linkage between genetic traits in larger farm animals.

Fowls

The genetics of the blood antigens of the fowl have been worked out primarily by W. E. BRILES in U.S.A. and D. G. GILMOUR in England. By use of isoimmune sera and agglutination, 12 antigenic systems have been identified, some of which comprise large numbers of multiple alleles. The B system is particularly complex and shows similarities, in principle, to the B system of cattle. Additional blood antigens have been identified by various other methods. The discussion here will be limited to the Hi antigen described by SCHEINBERG and RECKEL (1962). They found that extracts from certain legumes as *Lathyrus cicera* or *Pisum arvense* agglutinated the red cells of certain egg-laying hens. However, when egg production ceased, the antigen gradually disappeared. Although the red cells of chicks and cocks were not agglutinated by the legume extracts, injections of oestrogen resulted in a transformation of the red cells in some of these birds so that they became agglutinable.

Family data indicate that the Hi antigen is under the control of a single autosomal gene, but its actual manifestation is a function both of heredity and hormonal influences. The oestrogens do not seem to act directly on the red cells. When the chicks have been injected with oestrogen, it takes some time before the agglutinogen appears and the red cells in the bone marrow show it before the mature red cells.

BIOCHEMICAL BLOOD CHARACTERISTICS

Most of the biochemical characteristics which will be described here are distinguished with the aid of *electrophoresis*, a method based on the fact that different proteins have different electrical charges. If a solution of the proteins, e.g. animal serum, is subjected to an electric potential, the proteins will migrate towards the anode or cathode depending on their molecular charge. In certain cases the electrophoresis can be performed on filter paper soaked in a suitable buffer solution. In the middle of the 1950s Dr OLIVER SMITHIES, working at Toronto University in Canada, showed that a better separation could often be obtained if the electrophoresis was carried out on a starch gel. The pores of the

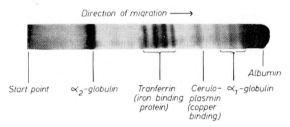

Fig. 7:3. The result of a starch gel electrophoresis of a cow serum. The proteins are stained with amido black. The different protein fractions are shown in the picture. The transferrins belong to the β-globulins. (Photo, GAHNE)

starch gel are so fine, that the migration speed of the protein molecules becomes proportional not only to their electrical charge, but also to the molecular size. With starch gel electrophoresis it is often possible to detect differences between proteins which with the traditional electrophoresis would appear identical. A number of these differences have been shown to be genetically determined. The result of a gel electrophoresis of cow serum is shown in Fig. 7:3. The various protein constituents are shown below the figure. The variation in the β-globulin region is genetically determined (cf. Fig. 7:4). By altering the conditions of the electrophoresis with respect to time, voltage and buffer solution it is possible to vary the degree of separation of the different proteins.

The first genetic variation in proteins was demonstrated by ordinary paper electrophoresis. The 'double' Nobel Prize winner, PAULING, and his co-workers showed in 1949 that humans with the 'sickle-cell' anaemia (cf. p. 87) had a haemoglobin type different from that which occurs in normal individuals. This discovery led to intensive research in the haemoglobin field, and several new heritable haemoglobin variants have been found in humans and the chemical differences between several of these types are now accurately determined (cf. p. 87). In 1955 the Frenchmen, CABANNES and SERAIN were able to demonstrate three haemoglobin types in Algerian cattle. The variation was shown to be due to two co-dominant alleles. The commonest type, now called A, was recognised by a single slowly migrating haemoglobin zone. Type B also had only one zone, but it migrated more rapidly than the A zone. The heterozygotes AB had both the zones.

Haemoglobin B occurs with relatively high frequency in Zebu cattle and with medium-high to low frequency in south and central European breeds of cattle. With the exception of the Channel Island and S. Devon breeds, B haemoglobin is not found at all in the north-west European breeds. Continued research has resulted in the discovery of at least a further three genes which control haemoglobin variation in cattle. So far they have been found only in African or

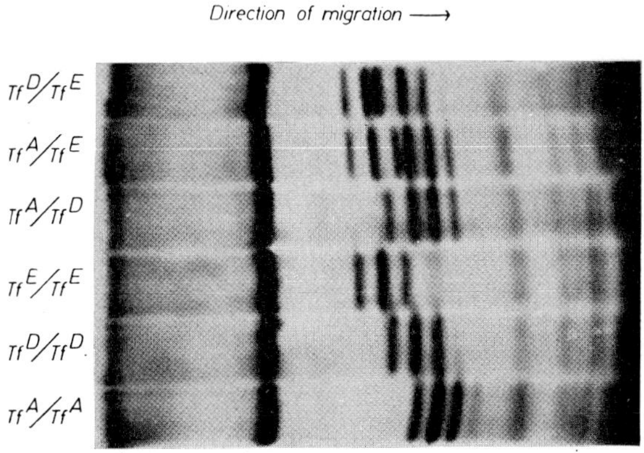

Fig. 7:4. Photograph of six different transferrin types in cattle.
(Photo, GAHNE)

Indian cattle. Heritable haemoglobin variations, similar to those of cattle, also occur in sheep. LARSEN (1966) recently showed that the haemoglobin locus and the red cell antigen locus A in cattle are closely linked. This appears to be the first good case of linkage reported in cattle.

The introduction of Smithies' starch gel method has led to the discovery of a number of genetic differences in blood serum. In 1957, SMITHIES and HICKMAN of Toronto, and ASHTON, working independently in England, showed definite genetic differences in certain β-globulins in cattle sera. Closer study showed that these β-globulins were identical with the proteins which facilitate the serum transport of iron ions and which are usually referred to as *transferrins*. In the west European breeds there are three main alleles in the gene system which controls the variation in the transferrins, and each of these genes gives rise to 3 or 4 protein zones, depending on the technique used. Both alleles are manifested in the heterozygotes. The variation in the transferrin pattern is shown in Fig. 7:4. The transferrins have been the object of investigation in a number of different breeds, and in addition to the three alleles just mentioned a further two occur in west European breeds. The variation in Zebu cattle is even greater than that in European cattle. No less than seven different alleles have been identified and of these, four appear to be identical to those in European cattle. In spite of the fact that the transferrins have been subjected to extensive genetic studies there is still very little known about the chemical differences between the various types. It is an open question why each transferrin allele corresponds to more than one protein zone.

There is no reason here to describe all the genetic variants which have been demonstrated in the various serum proteins. Only a few examples will be given. GAHNE (1963) at Uppsala showed a genetic variation in cattle for those proteins which, at pH 8·5, migrate immediately after the albumins. These are called post-albumins and the variation is determined by two co-dominant alleles. ASHTON and BRAEND (1964) have, independently of each other, demonstrated heritable differences in the serum albumins and ASHTON and GAHNE have found genetic variation in the slow migrating α_2-globulins. This variation is interesting in that individuals homozygous for a particular recessive gene completely lack the slow migrating α_2-globulins. In spite of this, the animals are perfectly healthy. By injecting serum from a normal animal into an animal which lacked α_2-globulin, Gahne was able to stimulate the latter to produce antibodies which reacted specifically with α_2-globulin. In laboratory animals such as mice and rabbits, and also in humans, heritable protein differences have been demonstrated with the aid of iso-precipitins. Rabbit γ-globulin variants detected in this way are referred to as allo-types. Variations in cattle α_2-globulins are of interest as they can be demonstrated both by electrophoresis and by the usual precipitin test.

A summary of the genetic systems which determine the variations in blood proteins in farm animals is presented in Table 7:3. We shall deal first with

Table 7:3 *A summary of the genetic variation of blood proteins in farm animals. The figures indicate the number of alleles and asterisks that genetic variation is described.*

Protein	Locus Symbol	Cattle European	Cattle Zebu	Sheep	Swine	Horse	Fowl
Haemoglobin	Hb	2	5	3	–	–	–
Pre-albumin	Pr	–	–	–	2	4	–
Albumin	Al	2	3	2	3	2	2
Post-albumin	Pa	2	2	–	–	–	–
Transferrin	Tf	5	7	10	3	7	2
Ceruloplasmin	Cp	–	–	–	2	–	–
Haptoglobin	Hp	–	–	–	4	–	–
α_2-globulin	Sα	2	–	–	3	–	–
Alkaline phosphatase	F	2	–	*	–	–	–
Amylase	Am	2	3	–	4	–	–
Esterase	Es	–	–	–	*	>3	–

the upper two-thirds of the table. This shows that haemoglobin differences have been identified in cattle and sheep. Pre-albumin types have been demonstrated in pigs by KRISTJANSSON (1963) in Canada and in horses by GAHNE (1966) in Sweden. Albumin variations occur in cattle, sheep, horses and chicken, and post-albumin variations in cattle and possibly in horses. Differences in transferrin are found in all the larger farm animals. According to investigations by IMLAH in Edinburgh, the copper binding protein, ceruloplasmin, occurs in two genetically determined types in pigs. Haptoglobin is a protein which facilitates the transport in the serum of haemoglobin to the liver. KRISTJANSSON, and others, have found at least four haptoglobin variants in pigs. Recent studies indicate that these proteins are not true haptoglobins, as they do not bind fresh haemoglobin but degraded haemoglobin only. The term 'haem-binding globulins' has therefore been suggested instead.

GENETIC VARIATION OF ENZYMES

By combining the starch gel electrophoresis technique and various histochemical methods it has recently been possible to detect genetic differences in a number of blood enzymes. The term *isoenzyme* has been given to these multiple molecular forms of enzymes with essentially the same substrate specificity. Isoenzyme variation has also been detected by immunologic methods and by measuring enzyme activity in relatives. We shall first discuss a study in which the latter approach was used. AUGUSTINSSON and H. B. OLSSON (1961) (Stockholm) studied the activity of two esterases in pigs and detected individual variation in the acetylarylesterase. The activity was measured by the amount of carbon dioxide released from a given amount of plasma per unit of time. The individual values ranged from 0 to 200 units. Parents and offspring in 72 matings with a total of 713 pigs were studied. Newborn pigs had no esterase activity, but after a few days piglets with the appropriate genotype started to show activity and full activity was reached at about six weeks of age. Augustinsson and Olsson postulated that the enzyme variation was due to 5 multiple alleles with additive effects. Gene a was presumed to give no activity whatsoever, while A_1 gave the value 25, A_2 50, A_3 75, and A_4 100. An esterase activity of 100 would occur, for example, in the genotypes A_4a, A_1A_3 or A_1A_2. The results from the various matings supported this hypothesis, but the boundaries between the classes were in some cases diffuse, and the values in the sexually mature boars were somewhat variable. It is possible, therefore, that the hypothesis needs some modification, but whether or not, the esterase variation was under close genetic control.

A similar quantitative variation was recently described by HOPKINS and co-workers (1963) in the acid phosphatase of human red cells. They were able to prove also that the difference in activity could be ascribed to particular enzymes which could be distinguished by starch gel electrophoresis. This variation in enzyme activity is of particular interest because it can serve as a model of gene action in quantitative characters. These traits show continuous variation, but usually no effects of single genes can be isolated. In the case of human red cell acid phosphatase, however, the quantitative variation in activity could be broken down into additive effects of separate isoenzymes, i.e. the genes underlying the quantitative variation could be distinguished.

Genetic differences in the enzyme alkaline phosphatase have been detected in several animal species by means of starch gel electrophoresis. The alkaline phosphatases are hydrolytic enzymes responsible for the break-down of phosphate esters. Their most important *in vivo* activity is concerned with the transfer of phosphate from one alcohol radical to another. The alkaline phosphatases are thought to play an important part in bone formation as well as in the resorption and synthesis of fat.

GAHNE (1963) detected marked differences between the phosphatase pattern in monozygous pairs of cattle twins. One fast moving enzyme zone, designated A, was found to be caused by a dominant gene with a frequency of 0·15 in the Swedish Red and White breed. Using Gahne's technique, Beckman and co-workers were able to show a genetic variation in the alkaline phosphatases in humans. They made the interesting observation that the phosphatase type was associated with the ABO blood antigens. In individuals of blood group A a particular phosphatase zone was almost entirely lacking but had a relatively high frequency among persons of blood group O. It should be remembered that the human blood antigen A is closely related to cattle J and sheep R. It is now evident that all these blood antigens are associated with variations in the alkaline

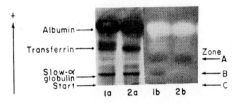

Fig. 7:5. Phosphatase zymograms of two sheep (1b and 2b). Animal No. 1 is blood group O and has the A as well as the B phosphatase while No. 2 is blood group R and has only the A phosphatase. To the left are the two samples (1a and 2a) stained with ordinary protein stain for comparison. (RENDEL and STORMONT 1964)

phosphatases in their respective species. The association seems to be particularly strong in sheep. These questions were recently studied by RENDEL, STORMONT and AALUND (1964). Fig. 7:5 shows a picture of the phosphatase variation in sheep. There are two phosphatase zones A and B in the region between the transferrins and the α_2 globulin. The fastest zone A appears to occur in all individuals while B occurs only in some. More than 300 mature sheep were tested both for the alkaline phosphatases and for the blood antigens within the R-O-i system. Among the 121 sheep of group O, 108 (i.e. 89 per cent) had a distinct and usually strong B phosphatase zone, while the corresponding percentages for the 193 and 23 sheep in group R and i were 11 and 15 per cent respectively. The B phosphatase zones in the two latter classes were usually very faint. The total phosphatase activity was measured in about half the material, and sheep of group O were found to have about 60–70 per cent higher activity than the R and i sheep. Injection of fluids containing the soluble blood group substance O into sheep belonging to blood group R and lacking B phosphatase leads to a rapid appearance of B phosphatase in the recipient's blood. Further studies indicated that the B phosphatase was of intestinal origin and that the O substance was not a part of the phosphatase. The results suggest that the O substance influenced the release of the B phosphatase into the blood circulation. Genetic variation has also been found in other serum enzymes. Amylase isoenzymes occur in cattle and pigs and esterase differences are found in pigs and horses (cf. Table 7:3).

BLOOD GROUPS AS GENETIC MARKERS IN BREED STUDIES

Human blood groups have become one of the most important aids of the anthropologist. By studying the frequency of blood group genes in different populations, attempts have been made to elucidate the relationship between races and to study the migrations of populations which have taken place through the centuries.

As far as blood group investigations of farm animals are concerned, these have been made mostly on American and west European breeds of cattle. However, some recent studies have included also south and east European, as well as African breeds. Quite a number of interesting results have already been reported. The breed differences in the frequency of the various blood group genes is, in general, large, particularly in the complex B system. The Jersey breed has been included in several of these investigations and in certain respects it appears to have a unique position among the west European breeds of cattle. It has certain special B alleles not found in the other breeds, except possibly the Guernsey. There is also a relatively high frequency of the blood factors V_2 and Z', as well as haemoglobin type B in the Jersey. These characteristics are extremely rare in the west European breeds but have a high frequency in Zebu cattle in Asia and Africa and also in cattle from south-east Europe. There are conflicting explanations about the origin of the Jersey breed but according to one of these it was formed from animals introduced into Europe from the Indus valley in Asia via north Africa. The results of blood group investigations give a certain amount of support to this idea.

The distribution of the common haemoglobin types, A and B, in cattle is interesting, as is illustrated by a large-scale survey of the haemoglobins in a number of breeds in Africa and Europe recently carried out by OSTERHOFF (1966). It has already been mentioned that haemoglobin B occurs in S. Devon and the Channel Island breeds, but it is also present in several French breeds, in the Simmental, in several Hungarian and Yugoslavian breeds, and in most of the African breeds studied. The dividing line between breeds with and breeds without the B haemoglobin extends from south-west England across northern France and through the south of Germany. The reason for this relatively sharp line of division is unknown but it is worth noting that the occurrence of the B haemoglobin in Europe bears some relation to the boundaries of the ancient Roman empire.

A study by BRAEND and co-workers (1962) on the blood groups of Icelandic cattle is worthy of special note. The Icelandic cattle probably originate from cattle brought to Iceland by the Nordic settlers over a thousand years ago, since when the cattle have been more isolated than most other breeds. Blood group studies were carried out on 971 individual cattle and, in spite of the long isolation, segregation was observed in 9 of the 10 loci studied. In the B system 20 alleles were detected; which pointed to a high degree of heterozygosity in the breed. Breed comparisons showed that the blood groups of Icelandic cattle were much more similar to those of Norwegian native cattle than to any of the numerous other breeds studied. This was most pronounced in the B system, and no fewer than 16 of the 20 Icelandic B alleles also occur in the cattle in Norway. This is particularly interesting, because the frequencies of the blood groups of the Icelanders of today are more like those of people in the northern parts of the British Isles than in Norway. It has therefore been suggested by human anthropologists that either the majority of Icelanders came from the British Isles with a small number of rulers from Norway or the modern Norwegians are different from those who colonised Iceland. Regardless of where the Icelanders came from, the blood groups of their cattle strongly indicate that these originated from Norway.

Differences in blood groups occur not only between breeds but may also exist between subgroups or strains within a breed. RENDEL (1958) found, in a study of the Swedish Red and White breed, significant differences in the gene frequency of the main strain and the 'Ayrshire' strain at 9 blood group loci. The difference was especially marked within the B locus. Several B alleles were present in one strain only, and even when the alleles were present in both strains the frequencies differed widely. Similar results were obtained in studies of several strains of Swedish Friesian cattle. This blood group variation between strains is very probably an indication of other genetical differences. As any advantage of strain crosses will depend on the genetic diversity between them, blood grouping is useful in checking on the existence of such diversity.

PRACTICAL APPLICATIONS

In the majority of farm animal species it has been possible to identify genetically determined blood groups. These characteristics have been employed in studying a number of practical and theoretical problems in connection with animal breeding and animal health. The discussion of the practical applications of animal blood groups will be confined to cattle and fowls, the two species which have so far been the most thoroughly investigated. In many cases, however, the principles and results will be applicable to other species.

Determination of parentage

The various blood antigens and biochemical groups, with the exceptions mentioned previously, are already developed in the new-born offspring and remain unaltered throughout the life of the individual. In the process of reproduction, those genes which determine the blood groups can *combine* in many ways; so the probability that two individuals have exactly the same blood type is very small indeed. The blood type can therefore be used for verifying an animal's identity. This method of identity control is especially useful in species where individual animals may be very valuable and other methods of identification are uncertain, e.g. in horses.

In animal breeding practice it often happens that a female is served or inseminated by different males in the same or in two successive heat periods. The parentage of the offspring may then be uncertain and the uncertainty may perhaps result in refusal of herdbook registration. Blood grouping has been used extensively in solving such cases of uncertain parentage, especially in cattle breeding.

Parentage tests by blood grouping is based on the so-called *exclusion principle*. All blood characters which are present in the offspring (with certain exceptions, discussed earlier) must also be present in one or both of the parents. If they are not found in either of the parents, then the stated parentage is incorrect. In certain blood group systems, the genotype of the individual can be directly ascertained from the blood type. From these systems it is possible to state, in certain cases at least, that an animal is definitely excluded as the sire (or dam), regardless of the blood type of the other parent. The principles of parentage determination are further explained in Table 7:4.

In case (a) the calf has the blood factors W and Z as well as the transferrin D, all of which are lacking in the dam. None of these factors occurs in one of the bulls (35), which is therefore excluded as the sire. Bull (36), on the other hand, has a blood type which is consistent with paternity of calf (83). Provided that only these two bulls were used and that the dam is correctly stated, it is correct to conclude that bull (36) is the sire of the calf (83). In case (b) the dam of the calf was slaughtered and no blood grouping was available. The calf reacted only with the V reagent at the FV locus and is therefore assumed to be F^V/F^V. Bull (488) is homozygous F^F/F^F and is thereby excluded as the sire of calf (280), irrespective of which cow is the dam. The same thing applies to the Tf locus where the calf has the genotype Tf^E/Tf^D and bull (488) is Tf^A/Tf^A. On the other hand there is nothing to indicate that the bull (833) is not the sire.

Exclusions in so called incomplete parentage cases e.g. when the dam has not been grouped, can be made only in those genetic systems where the genotype can be determined directly from the phenotype. As far as cattle are concerned, this is in most of the protein systems, in the antigen systems B, FV, SU, R'S' and sometimes also in the A and C systems.

At the Department of Animal Breeding of the Agricultural College of Sweden

Table 7:4 *Example of parentage investigation with the aid of blood antigens and transferrins (see text*)*

	\multicolumn{10}{c}{Locus}									
(a)	A	B	C	FV	J	L	M	SU	Z	Tf
Calf 83	–/–	BGKY$_2$A'O'/	C$_1$WX$_2$	F/F	–/–	L/	–/–	S$_2$/	Z/–	A/D
Dam 118	A/–	BGKY$_2$A'O'/	C$_1$X$_2$	F/F	–/–	L/–	M/	S$_1$/	–/–	A/A
Bull 35	A/	BO$_1$Y$_2$D'/BGKY$_1$A'O'	C$_1$	F/F	–/–	–/–	–/–	S$_2$/	–/–	A/E
Bull 36	–/–	BGKY$_2$A'O'/	C$_1$W	F/F	J/–	–/–	–/–	S$_2$/	Z/–	A/D
(b)										
Calf 280	A/	BO$_3$Y$_1$A'E$_3$'/	C$_1$W	V/V	–/–	–/–	–/–	S$_1$/	Z/–	D/E
Dam 28 Slaughtered										
Bull 488	A/	BO$_1$Y$_2$D'/BO$_3$YA'E$_3$'	C$_1$	F/F	–/–	–/–	–/–	S$_2$/	Z/Z	A/A
Bull 833	A/	GY$_2$E$_1$'/	C$_1$W	F/V	J/	–/–	M/	S$_1$/	Z/–	A/E

* For the sake of simplicity gene symbols are not written in full, e.g. $F^V F^V$ is represented by V/V and Tf^A/Tf^D by A/D etc.

over 2000 parentage investigations have been carried out with the aid of blood grouping. Half of these have been studied with respect only to the erythrocyte antigens. Of the complete cases involving two bulls, i.e. when the offspring, dam and the two possible sires have been investigated, parentage was established in 81–82 per cent of the cases. In 1–2 per cent of the complete cases both the possible sires were excluded and in the remainder no exclusions could be made. By making use of the transferrin groups it has been possible to increase the number of solved cases to about 90 per cent. In the incomplete cases the number of solutions has naturally been lower, about 40–50 per cent.

It is of course not always necessary to resort to blood grouping in order to determine the parentage of an animal when the dam has been served by two bulls. In many cases the pedigree can be ascertained with a satisfactory degree of certainty by a comparison of the expected and observed time of calving in relation to the two services. According to our experience and certain theoretical considerations, blood grouping should be resorted to for herdbook purposes if the dam of the animal was served by different bulls, (a) within an interval of 15 days and (b) with a normal service interval of 16–26 days between the bulls, provided that the gestation period after the service by the last-but-one bull is less than a certain time. The Swedish Red and White cattle have an average gestation period of 281 days and it is unlikely that many mistakes will be made in the latter case (b) if blood group investigations are limited to cases where the gestation period after service by the last-but-one bull is less than 295 days. The corresponding figure for the Friesians would be 293 days since their average length of gestation is 279 days.

The combined results of parentage determinations in cattle carried out in Holland, Sweden and Germany show that only a few per·cent of all cows, which are served twice with a normal interval of about three weeks, are fertilised and become pregnant as a result of the first service. Those cows which calve earlier than expected in relation to the second service, naturally have a higher percentage of conceptions attributable to the first service.

It was mentioned above that, in some of the complete paternity cases, both the possible bulls could be eliminated as sires. These are cases of incorrectly stated parentage and in a study of about 600 sire-dam-progeny groups in a number of larger, recorded herds in Sweden where several breeding bulls were kept, it was found that the stated parentage was incorrect in about 4 per cent of the cases. A similar figure was obtained in a study of about 200 such groups in an A.I. association in southern Sweden, and approximately the same results have been noted in Norway and Denmark. The reason for the errors have not always been established but one of the most common causes appears to be interchanging of calves.

Diagnosis of monozygosity in cattle

In Chapter 5 the possibility of using blood groups for the diagnosis of one-egg twins was mentioned. Naturally, all twins which have different blood types are two-egg, and similarly all twins with erythrocyte mosaicism are two-egg.

Investigations by GAHNE, among others, has shown that the transferrins are not influenced by vascular anastomosis between the twin foetuses. If the twins have different genotypes for the transferrins, then they will also be phenotypically different. The same applies to the J antigen as well as to the phosphatases and post-albumins in adult animals and also probably to other biochemical groups. A relatively large number of two-egg twins can be identified with the aid of these systems. In the Swedish Red and White breed it is possible, using the transferrins as the only diagnostic method, to distinguish something over half of the dizygous twins.

Extensive studies have been carried out in Sweden into the possibility of demonstrating erythrocyte mosaicism in dizygous twins. Data on 441 unlike-sexed twins have been analysed. Apart from the system J, the members in 12 per cent of the pairs had completely different blood types, i.e. there had been no vascular anastomosis between them; 81 per cent had erythrocyte mosaicism. In one per cent of the cases no mosaic was detected but the animals were unlike for J. It was not possible to demonstrate dizygosity in the remaining 6 per cent by means of erythrocyte antigens in spite of the fact that unlike-sexed pairs are definitely dizygous. In a number of these pairs the animals had probably inherited identical blood types and in others there was probably such a strong shift in the proportion of the cell types that mosaicism could not be detected. No investigation of the transferrins was carried out on this material. In a study of some other twin material which was also tested for transferrins, post-albumins and alkaline phosphatase it was estimated that, at most, 2 per cent of the dizygous twins would pass undetected when tests for these biochemical characteristics and blood antigens were applied.

Diagnosis of fertility in freemartin heifers

About 90 per cent of the heifers born twin to bulls will be intersexes, due to an influence of hormones from the male foetus via the vascular anastomosis (cf. Chapter 5). The presence of erythrocyte mosaicism shows that the foetuses have been in vascular communication with each other. Blood grouping can therefore be used at an early stage to decide whether a heifer which is co-twin to a bull will be fertile or not. If the heifer has erythrocyte mosaicism, she will certainly be infertile. If blood antigens (apart from J) are different in the twins, then the heifer can be assumed to be as fertile as if she had been born a single.

As already mentioned, erythrocyte mosaicism may occur, in rare cases, in humans and sheep. In the latter, the vascular anastomosis between unlike-sexed twins appears to have the same effect as with cattle. In the human cases described so far, however, there was no evidence of any detrimental effects on the sexual development of the females.

BLOOD GROUPS AND DISEASE

The discovery by LEVINE and his associates in 1941 that erythroblastosis foetalis in man was directly caused by Rh blood group incompatibility between mother and child was a milestone in medical science. The disease was shown to result from iso-immunisation of the pregnant female against dominant hereditary blood factors present in the foetus. The antibodies produced by the mother will then pass through the placenta to the foetus, whose erythrocytes might be more or less seriously affected. Later studies have shown that blood-group systems other than that of Rh may also be involved in the production of erythroblastosis foetalis.

A similar disease usually referred to as haemolytic icterus, or simply haemolytic disease of the newborn, has been reported in the horse and mule by CAROLI, BRUNER and others, and in the pig by BUXTON and BROOKSBANK. In the development of the disease, however, there is a fundamental difference between these farm animals and man. In man the young are already affected at birth, whereas new-born foals or piglets are quite normal and first develop the anaemia after suckling. The reason for this difference is the complexity of the placenta in these farm animals. Bruner and co-workers have further shown that foals lost the ability to resorb antibodies from horse serum introduced into their stomachs when more than 24–36 hours old. Mares expected to have antibodies against their foals should therefore be milked by hand for the first 36 hours, after which the foals should be able to suckle without any danger.

Studies by GOODWIN and co-workers (1955) indicate that haemolytic icterus in the pig occurs mainly among offspring of sows vaccinated with crystal violet swine vaccine. This contains blood from infected pigs. The injected blood evidently evokes the production of antibodies of sufficient strength to cause a haemolytic disease of the newborn.

Numerous investigations in humans have shown relations between blood antigens and certain diseases. AIRD (1953) demonstrated that the frequency of group A in patients suffering from stomach cancer was significantly higher than in the local population to which they belonged. These results have been verified in several studies, in some of which relationships have been found also between other cancer syndromes and the frequencies of the A and O groups. A particularly strong association has been found between the frequency of blood group O and duodenal ulcers. This disease is about 40 per cent commoner in people of group O than in people of the other groups.

BLOOD GROUPS AND PRODUCTION PERFORMANCE

All the species of farm animals which have been studied sufficiently have shown considerable genetic variation in respect of blood antigens and other blood characteristics. The presence of two or more genetic variants of a certain trait within the same population is usually referred to as *genetic polymorphism*. FISHER showed long ago that genes controlling stable polymorphism are highly unlikely to be neutral with regard to fitness. The most simple mechanism for maintaining polymorphism would be the average superiority of heterozygotes. The causes of the very extensive polymorphism in several blood characters are, as yet, mainly unknown. However, the fact that it exists suggests that the blood antigens and electrophoretic variants are, or have been, associated with fitness. Consequently it is natural that attempts have been made to relate these polymorphic blood traits to quantitative traits such as milk production, growth rate, fertility, etc. Some of the more pertinent results from these studies are discussed below.

How relationships between blood characters and production may arise

Genetic relationship between blood groups and production may arise principally in three different ways (Fig. 7:6), by:

1. pleiotropy, i.e. when a gene controls a blood character and influences production as well;
2. linkage between two loci, one of which determines a blood characteristic while the other one influences production;
3. heterozygosity at a blood group locus, which may have a positive effect on genetic viability and therefore also on production traits.

Fertility may, in addition, be influenced by the interaction between characters in dam and foetus. Pleiotropy will give rise to a stable association between the blood character and production, while linkage produces a positive relation to production in some families and a negative one in others. If the linkage is not very close, it is difficult to detect and even more difficult to utilise. Linkage will therefore not be considered here.

As was stated earlier, the majority of production traits are determined by a large number of genes and by the environment. It is therefore not to be expected that any relationships between blood groups and production traits will be especially strong.

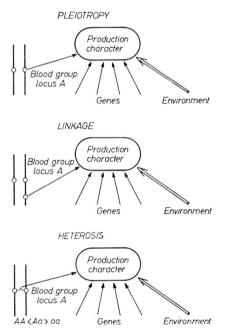

Fig. 7:6. Possible genetic causes of the relationship between blood groups and production characteristics.

The biochemical blood characters are concerned with substances such as haemoglobin and enzymes which have important physiological functions. It would appear possible then that haemoglobin, for example, can reflect differences in the capacity of the erythrocytes to transport oxygen and that they can thereby influence the production performance of the animal. A group of Dutch research workers have been able to show that as far as sheep are concerned, the haemoglobin type A has a greater oxygen-binding capacity than type B. Within certain limits the heterozygotes appear to be able to vary the proportions of the two haemoglobin types according to the availability of oxygen. The rate of oxygen-dissociation appears to be modified by another genetically controlled erythrocyte character, namely the concentration of potassium. This is determined by a pair or alleles, K^L and K^H which give rise to low and high levels of potassium respectively. K^L is dominant to K^H.

New knowledge about the haemoglobin types in sheep and their relations to cell potassium levels and other erythrocyte characters is accumulating rapidly. EVANS and his associates (1962) in Australia and KHATTAB and co-workers (1964) in England have shown that the volume of packed red cells (haematocrit value) is greatly influenced by the haemoglobin type. The Hb^A homozygotes have significantly higher values than the H_b^B homozygotes, the heterozygotes being intermediate. The genes for potassium level have a similar effect, K^L giving an increase and K^H a decrease in the haematocrit value. The differences in the volumes of packed cells seem to be due to an increase in the number of erythrocytes rather than in the size of the individual cells.

KHATTAB and co-workers found in a study of 1777 Welsh mountain sheep a direct association between haemoglobin type and the potassium level of whole blood. At both high ($K^H K^H$) and low ($K^L K^L$ or $K^L K^H$) levels of potassium in individuals of haemoglobin type A had a significantly higher level of potassium in their blood than haemoglobin B animals, while AB heterozygotes had intermediate values. However, these differences could be accounted for by the effect of the various genotypes on the red cell number. Khattab and co-workers also made the interesting observation that the increasing effect of haemoglobin type A on cell number was associated with increased fragility of the cells. According to very recent studies by Evans the volume of packed red cells and their fragility are influenced by haemoglobin type also in cattle.

The rate of oxygen dissociation, erythrocyte number and cell fragility are all factors of great importance in connection with the animal's ability to resist environmental stresses of various kinds. Attempts have therefore been made, quite naturally, to relate the haemoglobin and potassium types to viability, fertility and other production traits. Evans and associates observed that the frequency of the Hb^B was consistently higher than the Hb^A gene in the lowland types of sheep breeds whereas the situation was reversed in the mountain breeds. At higher altitudes the oxygen content of the air is low and the stronger oxygen-binding capacity of haemoglobin A ought therefore to be an advantage. On lower land with its higher oxygen content haemoglobin B would appear to be more advantageous since the oxygen dissociates more easily. It is very possible that the differences in gene frequency between lowland and mountain sheep are the result of a natural selection for the haemoglobin type best suited to the environmental conditions.

KING and his associates (1958) have carried out a study of the possible relationships between the haemoglobin type and the production traits such as fertility, growth rate and wool production of 1500 Scottish Blackface sheep. No significant effect of the haemoglobin type could be established for any of these traits, nor could they find any effect of the genes for low or high potassium. This result does not necessarily mean that the haemoglobin type is entirely unimportant. Even a very slight positive effect on the animal's fertility and vigour by a certain gene in a special environment could, by natural selection, lead to gene frequency differences of the kind that appear to exist between the British lowland and mountain breeds. It is clear, on the other hand, that the effect of the individual haemoglobin types is so small that there is nothing to be gained in attempting to use them as guides in artificial selection.

As far as the erythrocyte antigens and transferrins are concerned, a large number of investigations have been conducted into finding possible connections with production traits. The most comprehensive of these have been carried out on fowls and cattle. For several reasons it is convenient to deal with these investigations separately for each species.

Fowls

The investigations have been mainly concerned with the effect on the various production traits of heterozygosity for genes which control the blood antigens. This complex problem has been on the whole approached in two different ways, partly by comparing the production results of homozygous and heterozygous birds and partly by studying the remaining degree of heterozygosity in highly inbred lines. GILMOUR (1959) followed the genetic changes in four inbred lines in which the inbreeding was calculated to have ultimately reached between 98.6 and 99.4 per cent. Six different blood group systems were studied and heterozygosity was found for 2 – 4 systems in each of the lines. Similar results have been obtained by BRILES (1957) in U.S.A. in studies of 32 lines with over 50 per cent inbreeding. In all these lines selection was made for various production traits (mainly egg production) and the results indicate that the heterozygotes for blood groups were favoured by the selection.

In several American investigations the production with respect to various traits has been compared in birds which were heterozygous or homozygous in the B system. Fertilised eggs from heterozygous cocks showed, on average, a hatchability of 73.2 per cent, whereas the corresponding figure for the homozygotes was 61.3 per cent. The investigations also showed that the embryo's own heterozygosity was of great importance (Table 7:5).

In matings which gave no heterozygosity at the B locus the hatchability was 57.4 per cent but rose to 74 per cent when all the fertilised eggs were heterozygous.

Briles and his co-workers have further shown that heterozygosity at the B locus has a positive effect not only on the growth rate of the chicks but also on

Table 7:5 Influence of the chicken's own heterozygosity in the B system on the hatchability and vigour. Summarised result from three years within one line (after BRILES, 1956)

Expected heterozygosity %	Number fertilised eggs	Number hatched	Hatched Fertilised %	Chicks which lived to 9 weeks %
0	148	85	57·4	44·7
50	643	436	67·8	52·3
75	162	122	75·3	59·8
100	104	77	74·0	63·6

the egg production and general vigour of the hens. This latter trait was measured under two different periods, from the birth to 150 days of age (i.e. during the rearing stage) and during the period 150–450 days of age (primarily the first year of laying). In the majority of lines only two alleles occurred at the B locus. No significant differences could be found between the best homozygote type and the heterozygotes during either of the periods. However, the homozygote which showed the best vigour during the first period was in many cases worst during the second period. Taking both periods together, the heterozygotes were therefore on the average superior to the homozygotes.

The American investigations have not been confined to the B system. Other blood group loci, however, appear to have less or no effect on production characteristics. The heterozygotes on the D locus were in no way superior to the homozygotes in respect of egg production or egg hatchability. That the heterozygotes on the B locus had a positive effect on many production traits does not appear, therefore, to be solely a reflection of the general heterozygosity of the bird. Rather, the indications are that the alleles at the B locus have an influence on some fundamental physiological process and that this is better balanced in the heterozygotes. Continued investigations by Briles and co-workers and also by Gilmour, indicate that where egg production and vigour are concerned, birds with certain B alleles are consistently superior to birds with other alleles, irrespective of whether they occur in the homo- or heterozygous condition. These alleles appear, therefore, to have a pleiotropic effect on the production traits, a situation which has been utilised in breeding work at some of the larger poultry breeding concerns in the U.S.A.

Most of the studies on the effect of chicken blood antigens on production have been carried out on inbred lines or crosses between such lines. Inbreeding will tend to preserve those alleles which in the heterozygous condition have the most marked heterotic effect on viability and production. Gilmour has pointed out that the process of inbreeding may have eliminated most blood antigens other than those exhibiting a strong effect on fitness, thus giving an overestimate of the effect of heterozygosity for blood antigens in non-inbred populations. Accordingly, Gilmour and his co-workers investigated the effect of B blood type on hatchability in a relatively non-inbred Light Sussex strain having four alleles in the B locus. Hatchability was found to be associated with the blood antigens of the zygote but not with those of the dam. The B locus was generally over-dominant in its effect. Of the six possible heterozygotes four were of superior fitness and none of the four homozygotes was as viable as the mean of the whole population. However, the effect of the B locus on hatchability was not sufficient in itself to explain the maintenance of the polymorphism within the strain.

Cattle

A considerable number of studies have now been carried out in an effort to discover pleiotropic relationships between genes for blood groups and milk production. One of the first studies of this kind was made by MITSCHERLICH and co-workers (1959) with Friesian heifers at some German progeny-testing stations. The comparisons were made between heifers within groups of paternal half sibs for the presence or absence of certain antigenic factors belonging to 8 different genetic systems. Heifers carrying the M gene were found to produce 332 kg less milk than their contemporary paternal half sibs lacking this gene. The difference was statistically significant.

In Swedish studies comprising three independent sets of data a relatively strong positive association was found between the B allele $B^{BO_1Y_2D'}$ and the fat content of the milk. Heifers carrying this allele had 0·19 percentage units higher fat content than their paternal half sibs. Essentially the same result was recently obtained in American Holstein-Friesians.

The most comprehensive study made so far of the relationship between blood groups and production is that of NEIMANN-SØRENSEN and ROBERTSON (1961), who analysed data from 2378 heifers belonging to 185 progeny groups at the Danish progeny testing stations. Characters with high heritability (fat content) as well as intermediate (milk yield) and low heritability (fertility) were included.

Nearly all the results were insignificant, and when the multiplicity of the comparisons was taken into account, only two blood genes had a significant effect on production. In Red Danish cattle the B allele $B^{BO_1Y_1D'}$ increased fat content by $0·064 \pm 0·013$ percentage units and in the Jerseys the B allele $B^{O_1TE_3'K'}$ decreased fat content by 0·200 units. The $B^{BO_1Y_1D'}$ gene is serologically related though not identical with the allele $B^{BO_1Y_2D'}$, which, as just mentioned above, appeared to increase fat content in two Swedish breeds and the American Holstein-Friesian.

The studies on production relationships have not been confined to blood antigens. Ashton and co-workers compared the effect of two transferrin genes on milk production in British and Australian cattle. These comparisons, in contrast to most of the studies on blood antigens, were not made within progeny groups. One set of data comprised progeny-tested A.I. bulls in England and Wales. The progeny index for milk yield was 26 gallons higher for the $Tf^D Tf^D$ bulls than for $Tf^A Tf^A$, the heterozygotes being intermediate. A positive effect of the Tf^D allele on milk yield was also found in the Australian set of data.

The studies mentioned above were designed to detect pleiotropic effects of the blood group genes. The influence of heterozygosity has also been investigated in a few cases, but, contrary to the results with fowls, no heterotic effects of the blood group genes have yet been found in cattle.

Independent studies on cattle, sheep and pigs indicate that the transferrin type of the dam and sire may influence fertility. The segregation ratios have been disturbed in specific mating combinations and there have been some differences in fertility. The situation is still far from clear, however, and some conflicting results have been obtained.

In order to be of practical use the relationships between production characters and blood groups should be so strong that prediction is possible as to the future production of a young animal or of a bull's progeny. This matter was discussed in detail by NEIMANN–SØRENSEN and ROBERTSON and they came to the conclusion that the predictive value of blood groups was rather low. In Danish Red cattle, at the most 8 per cent of the genetic variation in fat percentage could be ascribed to blood antigens. The average proportion of the variation for milk yield caused by blood groups was estimated to be 5 per cent, but the influences on milk yield and fat percentage were in opposite directions, so that the net effect on butterfat yield was even less.

Even if blood groups are of limited value in selection programmes it would seem to be of fundamental importance to explore and explain the forces which maintain the very pronounced polymorphism in blood antigens and other blood characters. The very large number of alleles which occur in some of the blood group systems, for instance the B locus, indicates that heterozygosity for this system may be particularly advantageous.

CONCLUDING REMARKS

During the past fifteen years a large number of genetically determined blood groups have been demonstrated in the different species of farm animals. These groups can be used in the study of various genetic and breeding problems. In cattle breeding parentage determination with the aid of blood group investigations has achieved widespread application. This application is in itself important, but is by no means the most important field of application. It has been shown that the blood groups can be utilised in studies of similarities and differences between separate populations (lines and breeds) as well as for studying the changes that take place as a result of breeding systems and selection. Certain of the blood substances, for which genetic differences have been established, have fundamental physiological functions in the animal. Continued studies of the effects of these qualitative differences can therefore most probably contribute substantially to explaining the causes of the variation in many quantitative characteristics of farm animals.

SELECTED REFERENCES

ASHTON, G. C. 1965. Cattle serum transferrins: a balanced polymorphism. *Genetics*, **5**: 983–997.

COHEN, C. (editor). 1962. Blood groups in infrahuman species. *Ann. N.Y. Acad. Sci.*, **97**: 1.

SELECTED REFERENCES

EVANS, J. V. and TURNER, H. G. 1965. Interrelationships of erythrocyte characters and other characters of British and Zebu crossbred beef cattle. *Aust. J. biol. Sci.*, **18**: 124–139.

KIDDY, A. C. 1964. Inherited Differences in specific blood and milk proteins in cattle; a review. *J. Dairy Sci.*, **47** (5): 510–515.

MATOUSEK, J. (editor). 1965. Blood groups of Animals. *Proc. 9th European Animal Blood group conference.* Publ. House, Czechoslovak Academy of Sciences, Prague.

RENDEL, J. 1967. Studies of blood groups and protein variants as a means of revealing similarities and differences between animal populations. *Anim. Breed. Abstr.*, **35**: 371–383.

RENDEL, J., AALUND, O., FREEDLAND, R. A. and MØLLER, F. 1964. The relationship between the alkaline phosphatase polymorphism and blood group O in sheep. *Genetics*, **50** (5): 973–986.

SHAW, C. R. 1965. Electrophoretic variation in enzymes. *Science, N.Y.*, **149**: 936–943.

STORMONT, C. 1965. Mammalian immunogenetics. *Genetics Today. Proc. 11th Int. Congr. Genetics*, **3**: 715–722.

8 Hereditary Defects and Disease Resistance

The process of differentiation and development which every new individual undergoes from the fertilised egg to birth follows practically the same course in all the higher animals. The normal development of a particular organ requires that other tissue systems have attained a certain stage of development. Differentiation is continuous, and at any time or period it is dependent on the development which has already taken place. Deviations from the normal pattern of development can result in disruption of the whole process of differentiation, with the result that the individual succumbs perhaps at a very early stage of embryonic development; lesser deviations result in less serious defects. Every imaginable variation is found, from directly lethal malformations to defects that only slightly reduce the fitness of the animal. The genetic background of anatomical malformations and metabolic disorders is often rather simple, and it has been possible to determine the inheritance of many of these defects.

During the life of an animal it is subjected to various stresses in the form of attacks by microorganisms, nutritive deficiencies, unsuitable climate, and so on. In farm animals, the extent to which these stresses are overcome is of utmost importance for production. Even though the resistance to these disturbances often has a genetic background, the interaction between heredity and environment is so complicated that the classical gene analysis is unsuitable in most cases. Infectious diseases, for example, are first manifested when particular pathogens gain entry into the body and nutritional diseases when the balance of essential nutrients is upset.

DEFECTS AND ANATOMICAL MALFORMATIONS

It is very difficult to make a satisfactory classification of the numerous defects which occur in all species and breeds of farm animals. Sometimes *hereditary defects* are spoken of in contrast to *acquired*, or *environmentally induced*, *defects*, but no dividing line can be drawn between the groups. The same phenotypic defect may in one case be due to a mutant gene in single or double dose, or a chromosomal aberration, while in another case it may be produced by environmental agencies, e.g. certain chemicals, acting on the mother during pregnancy. A classification into *morphological and physiological (biochemical) defects* may be useful as a broad description of the type of manifestation, but it should not imply fundamental differences in the hereditary basis of the defects. Many morphological malformations are observable in the new-born animal and therefore called *congenital defects*, whereas others manifest themselves at later stages of life. For example, various types of genital malformation, or of functional sterility, are usually not diagnosed until the age of puberty. Even with such defects, transitional cases make the classification difficult.

Experimental induction of congenital defects

Since the turn of the century a considerable amount of work has been done on the artificial induction of congenital malformations in laboratory mammals (mice, rats, guinea-pigs, hamsters and rabbits), and also in poultry and pigs. Interesting results have been obtained. The initial experiments involved the administration of X-rays to pregnant mothers and these have been followed by investigations into the effect of chemical agents, pathogens and dietary deficiencies. By injecting pregnant laboratory rodents with such drugs as insulin, cortisone or sulphanilamide, as well as pituitary hormones, oestrogens and various antimetabolites, it has been possible to produce in the young practically the whole array of congenital malformations which are known to appear spontaneously in the same species as a result of homozygosity for recessive genes. Substances which, when administered to the pregnant female, give rise to malformations of the young, are usually referred to as *teratogenic agents*. Essentially the same malformations may appear when the diet of pregnant females is deficient in certain vitamins or amino acids. In rats and pigs, for example, lack of vitamin A has been shown to produce eye anomalies, cleft palate, harelip and leg deformities.

The sensitivity of the foetuses to various teratogenic agents shows considerable variation at different stages of pregnancy. As a rule, the sensitivity is greatest at the time when an organ is in the stage of most rapid differentiation, i.e. during the early stages of pregnancy immediately after implantation in the uterus. The same teratogenic agent may have different effects at different stages of foetal development. LANDAUER (1947) found that insulin injected into fowl's eggs 72 hours after the beginning of incubation produced rumplessness, similar to the hereditary types of this defect, in about 5 per cent of the embryos surviving to the 17th day of incubation. Injections made 120–135 hours after the beginning of incubation, caused a pronounced increase in the frequency of beak abnormalities, short legs and abnormal eyes, but no cases of rumplessness. By injecting eggs with 2·5 mg of a 40 per cent solution of nicotine sulphate several defects were produced in more than 50 per cent of the embryos, the most characteristic being a shortening and twisting of the neck. When the dose was doubled, this defect appeared in practically all the embryos and it was accompanied by dwarfing, shortening of the upper beak, muscular hypoplasia and abdominal hydrops. An interesting result in Landauer's experiments was that the severity of the induced defects seemed to differ between breeds.

The great majority of congenital malformations, which appear spontaneously from time to time in all species of farm animals, can be produced experimentally by various teratogenic agents. However, this does not mean that these defects are non-genetic when they appear spontaneously. On the contrary, evidence from breeding experiments with laboratory rodents shows quite clearly that the spontaneous occurrence of these induced defects is usually due to homozygosity for recessive genes. More or less perfect *phenocopies* may, however, be produced by various environmental influences acting on the pregnant mother or the fertilised egg. Among these influences can be mentioned drugs, dietary deficiencies, virus infections, or hypothermia, which blocks the action of the genes that control foetal development. The important lesson from experimental teratology is that spontaneous cases of congenital malformations, especially when several malformations occur in the same individual, should be carefully scrutinised before they are accepted as hereditary.

Severity of action of mutant genes

Mutational changes of genes are likely to upset the balance in the carefully adjusted system of gene co-operation and in the reaction of the genotype to the environment; therefore most mutant genes are unfavourable under the environ-

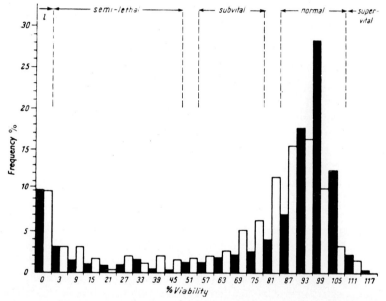

Fig. 8:1. Viability spectra for factors from wild populations of *Drosophila pseudoobscura*. Black columns: distribution of relative viabilities of homozygotes for 326 2nd chromosomes; white columns: the same for 352 4th chromosomes, *l* = lethal. (HADORN 1961)

mental conditions to which the individuals are adapted. Under altered environmental conditions, however, the situation may be different.

HADORN (1955) proposed that genes with unfavourable effect be classified into *lethal*, *semi-lethal*, and *sub-vital*. When a gene in an effective dose always causes death of the carrier before the attainment of puberty, it is said to be lethal; when the percentage of survivors is greater than zero but less than 50, the gene is said to be semi-lethal; and when more than 50 but less than 100 per cent of the carriers survive, the gene is called sub-vital. Categories, especially the last two, may overlap depending on the external environment and the internal genic constitution. The terms have some justification, if only for classification purposes. The sub-vital factors are of especial interest, since some of them can aid in explaining the causes of the variation which exists between individuals, lines and breeds in respect of resistance to infectious diseases and external stresses of various kinds.

Very little is known about the frequency of gene mutations in farm animals, or about the frequency of genes with lethal or semi-lethal effects. Experiments with laboratory animals and analyses of human data show that the frequency of mutations can be very different at different loci. The results of an investigation by DOBZHANSKY and co-workers (1942) with *Drosophila pseudoobscura* are of principal interest. The investigation was concerned with the viability spectrum of the genes on different chromosomes in flies from wild populations. By a special technique, flies were produced which were homozygous for the 2nd and 4th chromosomes occurring in the wild populations. A comparison was made between the viability of the 'chromosome homozygotes' and a random sample of wild flies from the same area. The viability spectrum of the two chromosomes are shown in Fig. 8:1. About 10 per cent of the chromosomes studied carried lethal genes, whereas 10 per cent of the 2nd and 15 per cent of the 4th chromosomes carried semi-lethals. Approximately half of the homozygotes for the two chromosomes had normal viability. Chromosomes which carried genes with sub-vital effects were quite common, but genes with an increasing effect on viability were rare. The relative frequency of genes with lethal, semi-lethal or sub-vital effects can, of course, vary from species to species. Fig. 8:1, however, gives an indication of a situation which probably has some validity for all species of animals.

Defects of definite genetic origin

Only some typical cases will be dealt with here; for a more complete description of the different hereditary defects reference may be made to the monographs listed at the end of this chapter.

Lethal factors with demonstrable effect in the heterozygotes

There is no evidence of any dominant lethal factors in farm animals. There are, on the other hand, a number of lethal factors with demonstrable effect in the heterozygous state but usually with little effect on the viability. An interesting blood characteristic occurs in humans and rabbits. It is known as *Pelger's nuclear anomaly*, after its discoverer, Pelger, and is recognised by the polymorphonuclear neutrophil leucocytes having only two lobules instead of the normal three or more. The individuals do not appear to be affected by the fact that the nuclei are less lobulated. NACHTSHEIM (1950) has carried out extensive research into the inheritance of the Pelger anomaly in rabbits and has shown that all Pelger animals are carriers of a gene P, which in the homozygous condition leads to serious malformations in the young. In matings between normal rabbits and the Pelger type, half the progeny were normal and half Pelger (Table 8:1). When the Pelger animals were mated with each other they produced

Table 8:1 Results from mating Pelger type rabbits

Mating	Litter size	Number of young obtained			Expected number on assumption of equal viability in all classes		
		pp	Pp	PP	pp	Pp	PP
Pelger (Pp) × normal (pp)	5·4	237	217	—	227	227	—
Pelger × Pelger	4·7	223	439	39	220·7	441·3	220·7

smaller litters and the ratio of normal to Pelger type offspring was 1:2. In addition, 39 deformed offspring were born, the majority of which died shortly after birth.

Nachtsheim succeeded in rearing some of the deformed young and in a mating experiment one deformed male was shown to be homozygous for the Pelger factor. Due to the phenotypic likeness, all deformed young were assumed to be of the type PP. The expected number of progeny of the three genotypes from the mating Pelger × Pelger in Table 8:1 was calculated entirely from the number of normal and Pelger progeny obtained, on the assumption that all genotypes were equally viable. Since the number of malformed offspring was only 17–18 per cent of the expected number, the conclusion can be drawn that most of the PP homozygotes die in the uterus. The reduced litter size in the matings between Pelger animals is also evidence of this.

The white blood cells in individuals homozygous for the Pelger factor are of a type previously unknown in mammals (Fig. 8:2); the nuclei are unsegmented and pyknotic (filled with clumps of undifferentiated chromatin). The animal has a severely stunted development, the birth weight is low and the extremities are bent and shortened. The ribs are also short and thickened so that the thoracic cavity cannot function normally, and this leads to difficulties in breathing. These difficulties are often the direct cause of death in those animals which survive to birth.

Several examples of defects with a similar inheritance are known in the larger farm animals. Perhaps the best known is bulldog calves in the Dexter cattle. Cattle of the Dexter breed can be recognised by their extremely short legs and compact bodies, *achondroplasia*. When Dexter animals are mated with each other there is a segregation of 1/4 normal individuals (Kerry type), 2/4 Dexter and 1/4 bulldog calves (Fig. 8:3). The latter are usually aborted during the 4th to 8th month of pregnancy. Thus the Dexter gene acts as a recessive with regard to lethality, but with regard to length of legs and body type it is incompletely dominant. Since animals of the Kerry type and bulldog calves consistently segregate out, the Dexter breed cannot be fixed by selection. Bulldog types, determined by recessive genes, occur in several other breeds of cattle. These will be dealt with in a later section.

Dwarfism, caused by dominant genes, has been demonstrated in several breeds of cattle. A case is reported in the Swedish Red and White cattle, where a non-inbred bull produced 25 deformed and 28 normal calves. The deformed animals were recognised by a comparatively short head, an arched forehead and undershot upper jaw. The extremities were shorter than normal, especially below the knee and hock. The bull calves showed more pronounced achondroplasia than the heifers. The numerical ratio between deformed and normal calves was in close agreement with that which would be expected if the sire was heterozygous for a dominant gene. The bull itself was normal and it therefore appears probable that a mutation had taken place at an early stage of gonad development.

Extreme shortening of the limbs is a breed characteristic in a number of poultry breeds, e.g. the American Creeper and the Japanese Bantam. As with the Dexter cattle, it has proved impossible to get these short legged poultry to breed true by selection. Individuals with normal length of shanks segregate out consistently. Comprehensive studies by LANDAUER and co-workers, have shown that the Creeper fowl is heterozygous for a gene Cp, which is lethal in the homozygous condition. The homozygotes never hatch, the majority dying

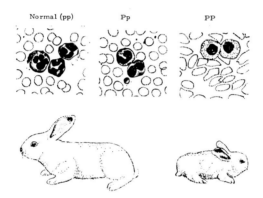

Fig. 8:2. The blood picture of the three genotypes at the Pelger locus, and the body development of a normal young rabbit compared with one of the same age which is homozygous for the Pelger factor. (NACHTSHEIM 1950)

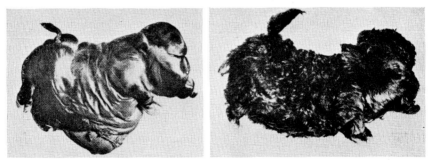

Fig. 8:3. 'Bull dog' calf foetus with ventral hernia: *Left* 6–7 months old; *Right*: nearly full-time. Both were produced in matings of Dexter cattle. (CREW 1923)

before the 4th day of incubation. In exceptional cases some embryos survive until the last week of incubation and can be recognised by their severely stunted development, the legs and wings often being only rudimentary and the femur and humerus bones missing altogether.

Even if the Creeper fowl appears to be normal except for the shortened legs, the Cp gene in the single dose anyway exerts a negative effect on viability. The Creeper fowls are very susceptible to rickets if the feed is only slightly lacking in vitamin D, and the hatchability of the eggs carrying the Cp gene is lower than for normal eggs.

Dominant colour mutations which reduce the vitality of the homozygotes

In addition to the platinum and 'white-faced' types in the fox and a number of colour mutations in mink discussed earlier (p. 76), there are a number of dominant colour genes which are lethal in homozygotes. The dominant yellow colour in mice is probably the most thoroughly investigated of these types. It is determined by a gene which belongs to the agouti series and is dominant over the wild type genes. The homozygotes die at an early stage of embryonic development. The heterozygotes are affected by metabolic disturbances, and are listless with a tendency to be over-fat.

Animals that are homozygous for the gene for dominant grey colour in the Karakul and closely related sheep breeds suffer from digestive disturbances. In matings between Karakul sheep, which are grey even as lambs, 1/4–1/3 of the progeny are black, the remainder grey. The grey progeny includes homozygotes as well as heterozygotes, but the former appear to be very seldom fertile. About 75 per cent of the homozygotes die from digestive difficulties before they reach 9 months of age.

Colour genes which have a depressive effect on vigour and fertility are also known in the larger farm animals. Some workers are of the opinion that the dominant white colour in horses has a lethal effect in the homozygotes. In two breeds of cattle, the English Shorthorn and the Swedish Polled, a relationship has been demonstrated between the occurrence and extent of white colour and certain reproductive disorders. These defects will be discussed in more detail in the next chapter (pp. 239–242).

It is thus quite clear that several of the genes which influence the synthesis of melanin are also concerned with the health of the animal. All these colour factors, except the dominant yellow in mice, result in an increase in the extent of unpigmented hairs.

Recessive defects

Cattle. The great majority of known defects in farm animals are due to recessive genes. A comprehensive investigation by MOHR and WRIEDT (1928) into hairlessness in Swedish Friesians will be taken as a typical example of how the *inheritance of a recessive lethal trait* can be elucidated. The hairless calves are born after a normal term of pregnancy and have no hair on the body except for varying amounts around the muzzle, eyes, ears, the tip of the switch, pasterns and the external genitalia (Fig. 8:4). The calves die immediately after birth. Both parents appear fully normal.

Mohr and Wriedt studied the results obtained from mating three bulls, which had produced hairless progeny, with normal cows out of bulls which had also produced hairless progeny. The matings gave the following results:

Bull	Number and type of calves	
	Normal	Hairless
I	32	4
II	37	2
III	29	6
Total	98	12

On the assumption that the trait is determined by a recessive autosomal gene, half of the daughters of the heterozygous bulls ought to be carriers of the gene for hairlessness. If the daughters are then mated with bulls which are themselves carriers of the trait, one in eight of the progeny ought to be hairless. One would expect, therefore, that of the 110 calves included in the investigation 14 would be hairless and 96 normal. Agreement between the observed and expected values was thus very good, and Mohr and Wriedt drew the conclusion that hairlessness was controlled by a completely recessive lethal factor. By a study of the pedigrees of known carriers it could be further shown that the gene for hairlessness was

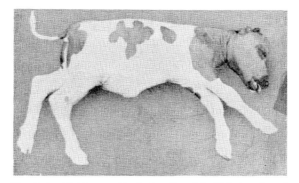

Fig. 8:4. Hairless calf of the Swedish Friesian breed. (MOHR and WRIEDT, 1928)

probably introduced into the Swedish Friesians with the bull Prins Adolf, which was imported from Dutch Friesland in 1902. For a time this bull was one of the most popular in the Swedish Friesian breed and no fewer than 80 of his sons were sold for breeding so that the gene for hairlessness rapidly reached a high frequency throughout the breed.

Several other recessive defects are known in Friesian cattle but we shall deal with only three of them here, namely 'amputated', congenital dropsy, and bulldog. The first two have been rather common in Swedish Friesians at some period of time.

Amputated calves (Fig. 8:5) are born after a normal pregnancy, but the degree

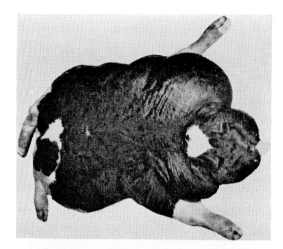
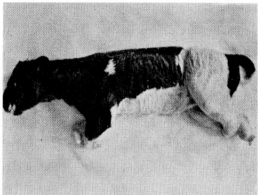
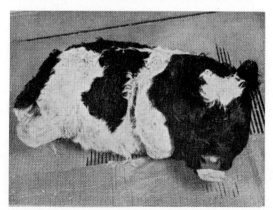

Fig. 8:5. Lethals occurring in the Swedish Friesian breed: hydrops (above), 'amputated' (middle) and bulldog calf (lower).

of deformity may vary considerably. In some cases the head is almost normally developed and the legs amputated below the front knee and the hock; in other cases the head is very deformed and shortened. The upper jaw is curved downwards and the lower jaw is lacking, or partly so, causing the tongue to hang out. In a number of calves the free ends of the limbs are absent altogether. There appears to be some connection between the degree of amputation and the deformity of the head in the same animal. The more amputated the legs are the

Fig. 8:6. Hereford 'snorter' type dwarf.

more deformed is, usually, the head. Calves which are very deformed are usually dead at birth, whereas less affected animals can live for several days or weeks. Animals with differing degrees of deformity seem to be homozygous for the same gene and the variation in expressivity is probably due to modifying genes.

The type of bulldog calf that occurs in Swedish Friesians is completely different from the Dexter bulldog calf. In the former case, the pregnancy is normal and although the calves are generally born alive they die shortly after. The malformation is characterised by a shortening of the jaw bones so that the head resembles that of a bulldog. The legs are also shortened (Fig. 8:5). Similarly malformed calves have been observed in other breeds, e.g. Telemark cattle in Norway.

Congenital dropsy has been observed in several other cattle breeds besides the Swedish Friesian, and is recognised by a large accumulation of fluid in the body cavities and subcutaneous tissues, so that the calf is extremely swollen. The degree of abnormality is very variable and some calves can weigh up to 100 kg i.e. two-and-a-half times the weight of a normal Friesian calf (Fig. 8:5). The dropsical foetuses are normally aborted, and if this takes place early there are no difficulties. If the foetus is full term and abnormal, the parturition may be difficult and endanger the life of the mother.

There are several different types of recessive dwarfism in American beef cattle, especially Herefords. The commonest type is called 'snorter' dwarf because of the snoring or wheezing sound it makes when breathing. It is recognised by a short, stunted body and a short, almost bulldog head with a bulging forehead and protruding eyes (Fig. 8:6). The lower jaw overshoots the upper. The dwarfs are not as viable as normal animals and they are especially

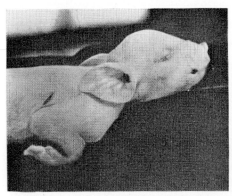 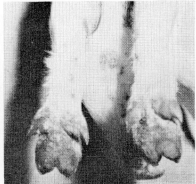

Fig. 8:7. *Left*: pig with bent legs. (HALLQVIST 1933). *Right*: pig with club-feet. (FLATLA 1956)

susceptible to digestive disturbances. Fertility is reduced; matings between dwarfs produce only dwarfs. A number of other dwarf types in the Hereford are probably caused by alleles of the 'snorter' gene or by modifying genes which influence the expression of the homozygotes for the snorter gene.

Several other defects, most probably recessive, have been observed in cattle, e.g. paralysis of the hind legs in Red Danish, non-lethal hairlessness in the Guernsey and Swedish Red and White, prolonged gestation in Holsteins and Swedish Red and White in addition to metabolic disturbances such as porphyrinuria (cf. p. 86).

Sheep. There are several homologous defects in cattle and sheep. Paralysis of the hind legs of the young occurs in both species and a defect characterised by amputated legs has been observed in sheep in Holland. Metabolic disturbances caused by recessive genes also occur in sheep, for example, the inability to break down phylloerythrin, a defect which leads to severe photosensitization (p. 87).

Pigs. There are quite a large number of hereditary malformations in pigs. In the past, a rather common defect in the Swedish Yorkshire was that of bent legs. The front legs are bent and fixed at the front knee; which can lead to difficulties at farrowing. The deformed piglets are generally stillborn. Breeding experiments by HALLQVIST showed that the defect was determined by a recessive autosomal gene.

Another defect, club-feet (*Dermatosis vegetans*), occurs in the Swedish and Norwegian Landrace, and is of interest since it can appear with three rather different symptoms which can vary in degree and severity in different individuals. The lower part of the front feet is already more or less thickened at birth and the hooves are deformed; hence the name club-feet (Fig. 8:7). The thickening is due to oedema in the connective tissue. Associated with club-feet is an eczema which starts on the belly and then generally spreads over the greater part of the body. It may begin immediately at birth, but usually it first appears during the following weeks; the epidermis becomes thickened due to extreme keratinisation and assumes a papillate appearance. The maximum spread of the eczema is normally reached when the pigs are 5–8 weeks old; several die, but a number can survive though they show reduced vigour and growth rate. Only a very few club-feet pigs have reached a greater age than 5–6 months. A third symptom is pathological changes in the lungs with many 'giant cells' in the alveoli, resulting in breathing difficulties and a specific pneumonia. FLATLA and co-workers (1961) in their breeding experiments succeeded in producing three litters from matings between a carrier boar and a sow which recovered from the defect; carrier boars and carrier sows have also been mated together. The results of this and other experiments indicate that the defects are due to a recessive gene in the homozygous condition.

In Germany, a lethal defect in pigs has been described which is characterised by thickened fore-limbs. The deformed piglets are full-term and alive at birth, but as a rule they die within a few hours. The defect is due to a recessive gene. In the American Poland China breed, pigs born without any limbs have been reported. Since the pigs are incapable of movement they soon succumb. The defect is determined by a recessive gene.

Fowls. More heritable defects are known in fowls than any other species of farm livestock. The short review to be presented here is based mainly on the extensive work of HUTT. Reference to two of his monographs is given at the end of this chapter.

The hen's egg can be examined at different stages during incubation and in several cases it has been possible to study in detail the biological development of defects. Many lethals exert their effect before hatching. The defect 'stickiness' causes the embryo to die by the fourth day of incubation. The affected embryos have oedema and are small by comparison with normal embryos at a corresponding stage of development. The defect has derived its name from the fact that at the time of normal hatching the foetal fluids are 'sticky'. It is determined by a recessive, autosomal gene.

Another lethal defect is crooked-neck dwarf. Development appears normal up to the eleventh day of incubation, after which the embryo develops oedema, with the result that growth slows down and the embryo dies during the 20th–21st day. The breast and thigh bone muscles are dystrophied and the neck is crooked. The trait is due to a recessive, autosomal gene.

226 HEREDITARY DEFECTS AND DISEASE RESISTANCE

Mortality during the embryonic stages is highly concentrated in short periods around the fourth and twentieth days of incubation, as shown in Fig. 8:8.

Several types of nervous disorders, caused by recessive genes, occur in newly hatched chicks. One of them, *congenital loco*, is characterised by the chicks being unable to stand immediately after hatching, the head is drawn back and

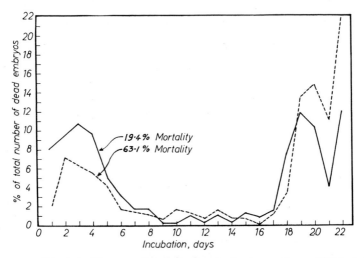

Fig. 8:8. The mortality during incubation in two lines of White Leghorn with 19·4 and 63·1 per cent total mortality before hatching. (LANDAUER 1948)

Fig. 8:9. Newly hatched chicks with the hereditary balance disorder, congenital loco. (HUTT 1958)

the chicks fall over (Fig. 8:9). When they get up the same procedure is repeated. Attempts to get these chicks to survive have been unsuccessful, and after a few days they die from starvation and thirst. Congenital shaking is another nervous disorder though the expressivity is rather variable. Some chicks shake so much that they are unable to stand, while others appear more normal. About 90 per cent die before they are one month old. Individual birds can attain maturity but they are invariably smaller and generally retarded compared with their normal full sibs.

Nakedness can be mentioned as an example of a sex-linked recessive defect. The chicks exhibit varying degrees of nakedness already at hatching but the condition is most pronounced at four weeks of age. In the adult birds the feathering is often much better developed. There is a high mortality both during incubation and after hatching; nearly 50 per cent of the homozygotes and hemizygotes die in the final three days of incubation.

In addition to the above hereditary defects there are a number of others, e.g. lethal and semi-lethal skeletal deformities and at least two types of hereditary blindness.

Defects with hereditary disposition but mode of inheritance unknown

A characteristic feature of many morphological malformations is that they give the impression of an arrest at a certain stage of the foetal development. Harelip, cleft palate, *atresia ani*, *cryptorchidism* and amputated or deformed legs may be mentioned as examples. The expression of *atresia ani* (imperforate anus) varies within and between sexes. The blind end of the rectum is closer to the surface

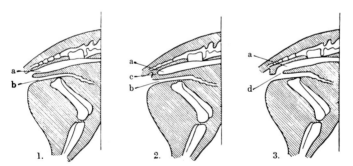

Fig. 8:10. Schematic picture of *atresia ani* in female piglets.
(1) Normal female piglet. (2) Pig with closed rectum. (*atresia ani*). (3) A fistula has developed between rectum and vagina. (a) rectum, (b) vagina, (c) point of closure of the rectum, (d) fistula between rectum and vagina. (CARSTENS, WENZLER and DURR 1937.)

of the skin in the female piglet and the rectum may very often open into the vagina, forming a cloaca (Fig. 8:10). In the male piglet the defect is lethal if surgery is not resorted to to facilitate defecation. KINZELBACH (1931) and HENRICSON (1963) have operated on male pigs with *atresia ani* and mated them when full grown to defective sows. These matings produced 13 litters with a total of 115 pigs of which 59, or 51·3 per cent, were afflicted with *atresia ani*.

Superficially, the hypothesis that *atresia ani* is caused by a recessive gene with about 50 per cent penetrance might seem to fit the data fairly well, but the situation is probably even more complicated. For example, when a defective boar was mated to unrelated defective sows, only 47·7 per cent of the offspring were defective; but when the same boar was mated to his own defective daughters, the percentage of defective offspring rose to 66·7 per cent. The difference between these two percentages is not statistically significant but other observations point in the same direction. Inbreeding has a depressive effect on the vitality and growth rate of the progeny, and when the mother is inbred she provides a less favourable intra-uterine environment for the foetuses than when she is not inbred. It is possible therefore, that inbreeding increases the frequency of foetal malformations, not only because it increases the segregation of lethal and semi-lethal genes but also because of its general depressing effect on metabolism and growth rate.

Homologous defects in different species

The above review of various hereditary defects shows that similar deformities often occur in different species. Nervous disorders which lead to paralysis of the extremities, especially the hind legs, are encountered in cattle, sheep and rabbits and different degrees of amputation of the legs have been observed in cattle, sheep and pigs. Hairlessness or nakedness occurs in cattle, pigs, rabbits and fowl. The list could be continued, but at present it is possible only to speculate as to the reasons for these homologies. It is possible that differentiation of one and the same organ in two closely related species is, at least in part, under the control of genes with a common origin, and that one or some of these have a high frequency of mutation in both the species. It is more probable, however, that mutations in completely independent gene systems can result in similar defects, e.g. hairlessness in cattle and rabbits as well as nakedness in fowls. As already pointed out, the processes of differentiation follow roughly the same course in all the higher animals, and when the development of the embryo is 'derailed' it is done in a similar way in all the different species. The end result is consequently a similarity in the deformities.

'Accidents' in development

In all species of farm animals a number of defects appear with low frequency and have never been observed to aggregate in particular lines or strains. Examples of these are symmetrical and asymmetrical *duplications* (conjoined twins, duplication of a hind or fore leg, etc.), *acrania* (absent or incomplete skull), *ectocardia* (exposure of the heart), *exencephaly* (exposure of the brain), and *situs inversus* (lateral transposition of visceral organs). It seems most likely that these defects are the result of 'accidents' in development (intangible causes). An alternative explanation would be chromosomal aberrations, or dominant mutations which eliminate themselves as soon as their effect has been manifested.

The analysis of the genetic basis of congenital malformations is more complicated than has been generally appreciated during the past few decades. What is hereditary in one case may be a phenocopy in another; furthermore, a hereditary disposition for malformations may have an observable effect only under certain environmental conditions.

DISEASE RESISTANCE

Under the wide range of conditions in which animal husbandry is conducted the animals are subjected to a multitude of stresses such as dietary deficiencies, unsuitable climate, and attacks by insects and pathogenic microorganisms. It is well known that to withstand these stresses is easier for some animals and breeds than for others. These differences in resistance have many causes, a number of which are heritable. The importance of inheritance is evident from the fact that some species of farm animals have complete resistance to certain diseases, whereas others are easily infected. An example of this is foot-and-mouth disease which attacks cattle and pigs but not horses.

Resistance to nutritional deficiencies

For satisfactory production, farm animals require sufficient quantities of feed which is well balanced with regard to vitamins, minerals and certain essential amino acids. Requirements in this respect are very different. A vitamin or mineral content which is adequate for one animal may be directly fatal to

Fig. 8:11. Chick with perosis. (HUTT 1958)

another. In poultry, a deficiency of manganese results in perosis, also known as slipped tendon or hock disease (Fig. 8:11). The legs are shortened and thickened, and in the more advanced stages there is a slipping of the Achilles tendon from its condyles, with a lateral rotation of the metatarsus so that the bird has difficulty in moving. The Leghorn breeds are less sensitive to manganese deficiency than the heavier breeds like the Rhode Island Red. An American study showed that 14 per cent of the chickens in the latter breed developed the disease but the frequency in the Leghorns on the same feed was only 0·7 per cent. In order to prevent perosis in Rhode Island Reds almost twice as much manganese is required in their feed as in the feed for Leghorns. It is possible to increase or decrease the requirement of manganese in different poultry lines by selection; which is good evidence that there is a genetic variation in the tolerance to manganese deficiency.

Deficiency of thiamine (vitamin B_1) in poultry feeds results in serious nervous disorders of the birds, characterised by convulsions, unsteady gait and cramp. The disease is known as polyneuritis. The Leghorn is better able to withstand thiamine deficiency than, for example, the Rhode Island Red, and it also produces eggs with about 60 per cent more thiamine than the latter breed, when both breeds are kept on the same feed. In general the Leghorn breed is less demanding in its vitamin requirements than the heavy American breeds.

Deficiency of riboflavin results in lowered hatchability, the growth rate of the chicks is low and the mortality high. LAMOREUX and HUTT (1948) have been able to show in selection experiments that resistance to riboflavin deficiency is genetically determined. In one line, selection was made for a low requirement of riboflavin; this line had considerably better growth rate and vigour than another line, which had been selected for a high requirement of riboflavin, when both the lines were fed on a ration with low riboflavin content. Swedish investigations have established that Rhode Island Red chicks are much more susceptible to rickets than Leghorns when the feed is low in vitamin D.

The vitamins play a very important role as catalysts in the metabolism. It is clear that in some individuals the utilisation of these substances is better than in others, and the requirement is accordingly less. The differences which exist between the different poultry breeds in this respect are due, at least in part, to heredity. Whether similar differences occur in the larger farm animals has not been investigated, but there seems to be every reason for believing that this is the case.

Climatic sensitivity

Attempts to introduce the improved European breeds, e.g. Friesian cattle, to tropical or sub-tropical regions have often given disappointing results. The European cattle do not tolerate the high temperatures. Their heat regulating mechanism is poorly developed in comparison with, for example that of the Zebu cattle. American experiments have shown that the respiration rate of Friesians increases much more rapidly with increasing temperature than does that of the Zebu. Crosses between these two breed types are about intermediate. The reason for the greater heat tolerance of the Zebu cattle is still only partially explained. One of the contributing causes would appear to be the large dewlap and the hump, which give the Zebu a relatively larger surface area per kg liveweight than the European breeds. Another important factor is the short-haired, shiny coat of the Zebu, which is an effective reflector of sunlight.

Resistance to infectious diseases

A large number of experiments have been carried out with laboratory animals and poultry to determine the causes of the individual differences in resistance to infectious diseases. In the experiments the animals have been exposed to, as near as possible, a constant amount of the infectious agent, after which the variation in individual resistance has been observed. If mice, or other experimental animals, are given a suitable amount of paratyphus bacteria in the feed, it can happen that some animals are quite unaffected, some die, and others develop a mild form of the disease and survive. This last group produces antibodies against the strain of bacteria and acquires thereby an active immunity against infection. The acquired immunity can be transferred, at least temporarily, from the mothers to their young.

An acquired passive immunity passed to the offspring through the placenta or the milk cannot be regarded as heritable, but it can easily be mistaken as such. By the use of suitable experimental techniques it is possible to avoid confusion between acquired, passive immunity and genetically determined resistance. Several experiments have been performed in which the disease resistance has been changed by selection. GOWEN and SCHOTT (1933) selected strains of mice with varying degrees of resistance to a particular type of paratyphus. Females of a susceptible strain were mated in the same heat period with males of both susceptible and resistant types. The males also carried different colour genes. The resultant litter consisted of both young with a resistant sire and young with a sire from a susceptible strain. The two types of offspring could be distinguished by the colour. When the young were exposed to a standard dose of paratyphus bacteria, all the young from the susceptible father died, while 47 per cent of the offspring with the resistant sire survived. With this experimental design the difference in resistance could not be due to the passive immunity but must be genetic in origin.

Pullorum disease, also called bacillary white diarrhoea (B.W.D.), is caused by *Salmonella pullorum*, a bacterium belonging to the paratyphus group. Infection may take place in several ways, for instance, an infected hen may transmit the disease to chicks via the egg. The disease takes the form of an acute general infection which spreads rapidly and results in a high mortality. The resistance to pullorum disease has been considerably increased in American experiments by continuous selection. In one Leghorn line, selected for high resistance for nine years, 70 per cent of the chicks survived a standard dose of the infection introduced by way of the feed. In the unselected line only 28 per cent survived. It has also been observed that Leghorn chicks have a higher resistance than chicks of the heavier breeds.

Undoubtedly the disease responsible for the greatest losses in poultry breeding in the past two or three decades is leukosis. It occurs mainly in two forms, neural lymphomatosis, or fowl paralysis, and visceral lymphomatosis. Both forms of the disease are characterised by an abnormal increase in the number of primitive undifferentiated blood cells (lymphocytes), which accumulate along nerve tissue, in the liver, and in other internal organs. The leukosis is caused by one or several viruses. Newly hatched chicks are especially susceptible to leukosis infections.

The heritability for resistance to leukosis has been estimated to be about 0·08–0·15 (p. 128). For over twenty years, HUTT and his co-workers (1965) at Cornell University in U.S.A. have carried out selection for high and low mortality to leukosis. As the disease is caused by a virus it is very difficult to give each individual a standard dose of infection, and the method chosen has therefore

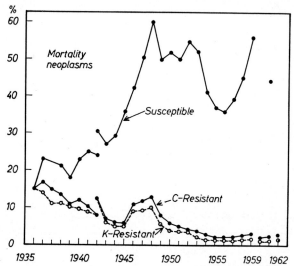

Fig. 8:12. Per cent mortality in neoplasms after selection for high and low susceptibility to fowl leukosis. (HUTT 1965)

been to expose the birds to natural infection, by mixing diseased and healthy birds together. In addition the chicks have been reared as close as possible to the adult birds. The mortality in different half and full sib groups has been determined and each individual bird which died has been examined. By selecting birds for breeding from families with low mortality, two highly resistant lines of Leghorns have been produced. In another line, selection was for high mortality.

At the commencement of the selection experiments the mortality from leukosis was about 15 per cent. After 15–20 years of selection for high resistance the leukosis mortality was down to 2–3 per cent but in the line selected for susceptibility it had risen to 40–60 per cent (Fig. 8:12). At the same time as selection was being made for high resistance to leukosis, selection was also being made for high egg production and low mortality from other causes. The results of this breeding work have been very good, as is shown by the production figures obtained when random samples of the resistant stock are compared with leading commercial poultry lines in the New York Official Random Sample Poultry Test. For this test, a number of eggs are selected at random from the poultry lines which are to be compared. Hatching is carried out by the testing stations

(p. 472). The results of three years' testing are summarised in Table 8:2. As far as mortality was concerned, the Cornell line was superior to all the other tested strains and second best in respect of income per chick started.

Table 8:2 A comparison of the leukosis-resistant Cornell lines and other stocks in respect of mortality and production, in testing according to the 'random sample' method (Hutt, 1958)

	Highest	Production Mean of 25 entries	Cornell line
Mortality (1–500 days)			
All causes, %	39.7	25.0	13.3
Leukosis, %	21.3	10.6	4.0
No. eggs per pullet housed	212.0	191.7	208.3
Income per chick started, dollars	2.46	2.03	2.40

By selection it has also been possible to increase the resistance to other bacterial and virus infections as well as to various internal parasites. There has been evidence of genetic variation in the resistance to various diseases in all the poultry breeds and strains which have been studied. However, resistance to one disease does not necessarily mean a bird is resistant to other diseases. Hutt's selection experiment has clearly shown, however, that by systematic breeding poultry lines can be produced with a high degree of general resistance and at the same time good production performance.

As far as the larger farm animals are concerned, it has not been possible to carry out any comprehensive selection experiments for increased disease resistance. On the other hand, a number of interesting observations have been made. In cattle, susceptibility to mastitis appears to be fairly highly heritable. The heritability has been estimated to be $0.2-0.4$ (p. 127). Large breed differences have been established for resistance to bovine tuberculosis, and Zebu cattle are much more resistant than the European breeds. The Zebu also has a greater resistance to piroplasmosis (redwater) and many tropical diseases transmitted by parasites.

Heritable differences have been found in the resistance of pigs to brucellosis and also to atrophic rhinitis. The latter disease causes atrophy or shrinkage of the scroll-like turbinate bones in the anterior part of the nose, and there is a distortion of the nasal septum which may also involve the upper jawbones. The causes of the disease are obscure. Several workers, among them BRAEND and FLATLA (1954, Oslo) have presented evidence that the disease can be transmitted experimentally; most authorities agree on this point. However, the same symptoms can be produced by several types of bacteria, and so far no single causative agent has been isolated. BENDIXEN and LUDVIGSEN (1958, Copenhagen) are of the opinion that nutritional deficiencies and genetic factors are primary causes, while infections and parasites are secondary causes of atrophic rhinitis. Infections first become important when the dystrophic changes have provided a suitable breeding ground for them. KROOK (1965, Cornell) is of a somewhat similar opinion and claims that atrophic rhinitis may be produced experimentally by alterations in the Ca–P ratio of the feed.

JONSSON (1965) at the Danish National Research Institute of Animal Husbandry, working in co-operation with the Danish Veterinary College, carried out extensive investigations to determine whether the deformation of pig snouts has any genetic background. Since 1956, approximately 14 000 pigs have been examined by veterinary experts to determine the degree and occurrence of atrophic rhinitis. Pathological changes were allotted points on a scale of 1–5, where 1 represents a normal snout and 5 a snout with pronounced changes. The average score for both gilts and barrows was about 2·0. In order to estimate the influence of heredity on the variation, the variance caused by differences between half-sib groups was calculated, eliminating differences between stations, years and sex (p. 125). The following result was obtained:

Cause of variation	Variance, %
Herd origin	4·0
Additive genetic factors	26·0
Litter environment	21·0
Within litters and sex	49·0

The herd environment in which the pigs were reared prior to being moved to the testing station at about 20 kg, had very little effect on the severity of the disease. It must be pointed out, however, that most herds participating in the official breeding programme have relatively good environmental conditions. The special litter environment — which includes, among other things, the sow's ability to provide the piglets with a sufficient and adequately balanced nourishment — accounted for 18 per cent of the variation. The importance of the nutrition of the piglets is emphasised by the fact that the pathological changes in the nose, are more frequent and more severe in litters from young gilts than from full grown sows. It is well known that milk production is lower in gilts. The age of the sows can also have some influence, due to the larger amount of specific antibodies in the milk of older sows.

Heritable causes accounted for 26 per cent of the variation in points for nose deformity.

There was a weak negative correlation ($-0·1$, approx.) between the degree of nose deformation and the growth rate. The decline in growth rate was, however, so small as to be of very little practical importance. When the disease condition score increased by 1 point, the daily growth rate declined by about 4 g.

Atrophic rhinitis is regarded in many countries as a serious problem. The results from the Danish Pig Progeny Testing Stations, indicate that it should be possible to increase the resistance to rhinitis by selection. Whether the results from the Danish investigations are applicable to atrophic rhinitis under the widely varying conditions of practical pig rearing must, for the time being, be left an open question. In two smaller investigations, carried out in the U.S.A. and Canada, it was found that the daily growth rate of pigs with atrophic rhinitis was about 5 per cent less than that of litter mates which were free of the disease. This reduction was much greater than that observed in the Danish investigation.

THE BIOLOGICAL BASIS OF RESISTANCE

The biological causes behind the hereditary resistance are still only partially explained. Several recent investigations, however, have provided a certain amount of information as to how the genes can influence the defence mechanism of the body.

A defect called congenital agammaglobulinaemia occurs in humans. The content of gamma globulin in the blood is then less than 25 per cent of the normal value. The defect is probably caused by a recessive gene. Since the antibodies are gamma globulins, it is quite natural that individuals afflicted with the disorder are highly susceptible to infections. The mortality from infections will therefore become high as soon as the passive immunity conferred through the mother's milk begins to decline. Differences in the ability to produce antibodies, though less drastic, have been discovered in several other species. SCHEIBEL (1943) found, for example, that guinea-pigs have differing capacities to produce diphtheria antitoxin. By selection it was possible to separate two strains, one with a good capacity to produce antibodies (only 2·5 per cent non-producers after one generation of selection), the other one in which the proportion of animals that did not produce the antitoxin was 88 per cent even after 5 generations of selection.

In the larger farm animals there are several examples of differences in the ability to produce antibodies. Horses are used for the mass production of antisera against various disease-causing organisms, but individual horses vary in their effectiveness to produce antibodies. Cattle also vary in their effectiveness to produce antibodies used in blood grouping (cf. previous chapter). Whether this individual variation is heritable is not known, but there is every reason to believe that it is. These differences in the ability to produce antibodies, as well as the differences in the efficiency of the complement, may explain part of the genetic variation in the resistance of farm animals to various diseases.

HUTT and CRAWFORD (1960) have shown that resistance to pullorum disease in chicks is associated with the time the chicks require to raise the body temperature to that of the normal in adult birds, 41–43°C. Prior to hatching, the chicks are entirely dependent on the temperature of the incubator. The day after hatching, the temperature of the chicks is, on average 39·4°C. White Leghorn chicks can increase their temperature much more rapidly than chicks of the heavy breeds, ·e.g. Rhode Island Red. Two lines of White Leghorn were differentiated by selection based on average chick temperature in three readings during the first six days after hatching. The stock were completely free from S. pullorum. The average temperature in the two lines differed by 0·24°C in the first selected generation and by 0·33°C in the second. When four different samples of these two lines were experimentally inoculated with S. pullorum, subsequent mortality to three weeks of age was consistently lower in the high-temperature line than in the other. With the smallest dose the mortality was only 8·6 per cent in the high-temperature line compared with 40·7 per cent in the low-temperature line.

The reasons for the Zebu cattle's superior power of resistance to a number of parasitic diseases, compared with that of the European breeds, has been the subject of extensive study in South Africa. Partially at least, the differences appear to have a purely mechanial explanation. The Zebu has a shiny coat with short hairs in which it is difficult for the parasite carriers to obtain a hold. Furthermore, the hide is thicker than that of the European breeds. The number of ticks per unit body surface area in the European cattle was found to be 2 to 7 times larger than in the Zebu cattle when both groups were mixed together on pasture.

CONCLUDING COMMENTS

The way in which breeding can assist in reducing the spread of various hereditary defects will be discussed in more detail in Chapter 15. As far as resistance to disease is concerned, this is mainly a quantitative trait, influenced both by heredity and environment. In order to increase resistance, use should therefore be made of those methods which are suitable for other quantitative traits with corresponding heritabilities. Selection may be made for individuals with high resistance or those which are from families with good resistance (cf. Chapters 15 and 16). So far, very little has been done to improve the resistance of animals to infections or dietary deficiencies by breeding methods. Instead, efforts have been directed towards controlling the sources of infection by hygienic and bacteriological methods. When the aetiology of a disease is known exactly and the carriers of the disease can be traced with objective methods, the spread of the disease can often be effectively controlled by a slaughter policy combined with protective measures for infection-free areas. If the animal disease is harmful also to humans, as is the case with bovine tuberculosis and brucellosis, then the slaughter policy is the only one justified. This practice has been used in many countries with great success in connection with these two diseases.

Many other diseases, such as mastitis in dairy cows, owe their existence to the ubiquitous bacteria which occur in the housing and on the pasture. It will hardly be possible to consider eliminating all these sources of infection. The aetiology of many of the virus diseases is incompletely known and simple methods of tracing the carriers of the viruses are lacking. This applies, for example, to leukosis in poultry and to many of the common pig diseases. Resistance to leukosis has been shown to be genetically determined to some extent and the experiments carried out have shown that the resistance can be appreciably improved by breeding. The genetic methods of increasing resistance to disease deserve more attention and application than has hitherto been given to them.

SELECTED REFERENCES

HADORN, E. 1961. *Developmental Genetics and Lethal Factors.* John Wiley and Sons, New York.
HUTT, F. B. 1965. The utilization of genetic resistance to disease in domestic animals. *Genetics to-day. Proc. 11th Int. Congr. Genetics,* **3**: 775–782. Pergamon Press, London.

——, 1958. *Genetic Resistance to Disease in Domestic Animals* (198 pp.). Constable and Co. Ltd., London.

KOCH, P., FISCHER, H. and SCHUMANN, H. 1957. *Erbpathologie der landwirtschaftlichen Haustiere* (436 pp.). Verlag Paul Parey, Hamburg.

JOHANSSON, I. 1965. Hereditary defects in farm animals. *Wld. Rev. Animal Production*, **1**: 19–30.

STORMONT, C. 1958. Genetics and disease. *Adv. Vet. Sci.*, **4**: 137–162.

9 Sterility and Low Fertility

Normal fertility in animals used for multiplication is of tremendous importance for the economic aspects of animal production. This is true, not least in cattle breeding, where reproductive disturbances have reached a rather high frequency. LAGERLÖF (1950) studied available data on 13 000 bulls all insured by a Swedish insurance company. The company paid compensation relating to 10 per cent of these bulls and about 1/3 of the claims were due to the inability of the bulls to serve. The frequency of fertility disturbances was especially high among young bulls, 1–2 years old, when newly brought into service. Similar results have also been obtained in the U.S.A.

According to data from the Swedish Milk Recording Societies, reproductive disorders in the female constitute the most important reason for culling. About 40 per cent of the cows are culled for this reason and the figure is especially high in the age class 4–7 years, which is the time when cows attain their highest production capacity. It can perhaps be argued that the carcass value of the young cows covers the rearing costs, so that this early rejection does not result in a direct financial loss to the owner. A serious consequence, on the other hand, is that the possibilities for selection of the females on the basis of milk production capacity is considerably reduced. In addition, the effect of early rejection due to fertility disturbances is similar to that of prolonged calving intervals, in that there is a reduction in the average production of the herd. On top of this come increased costs for veterinary visits and repeat inseminations.

Sterility and reduced fertility occur also in other species of animals, though the frequency is generally much lower than in dairy cattle. Of great importance for the profitability of sheep and pigs is the production per breeding animal of viable lambs or piglets respectively.

The prime requirement for fertility is the same for both sexes, viz. the production of viable gametes. Normal sexual drive is of course necessary, as is also the ability of the females to provide a suitable uterine environment for the development of the foetus and its delivery at full term.

Reduced fertility may be congenital or acquired, permanent or temporary. It is impossible to make a general classification between heritable and non-heritable disturbances in the reproductive functions. All such attempts at systematisation give misleading results. Some malformations of the genital organs are clearly genetically determined, but others are due to external effects on the foetus. Reproductive disorders may often be due to infections, though resistance to these infections varies between individuals, and the possibility that this variation is to some extent genetically determined cannot be entirely excluded. 'Acquired' reproductive disturbances are often due to a chain of interacting causes, on which it is not possible to decide from individual cases the comparative importance of genetic and environmental influences. It is perhaps possible to obtain an idea of this from the population as a whole. In order to facilitate the discussion, reproductive disturbances will be dealt with under the following three headings: (1) malformation of the genital organs, (2) gametic sterility, and (3) hereditary disposition to reproductive disturbances of different kinds. These three groups are fairly well demarcated from each other.

MALFORMATIONS OF THE GENITALIA

Malformations and hypoplasia (underdevelopment) of the genital organs may not be apparent until the animal has reached the age of puberty. As an example of a genital defect caused by the uterine environment, mention may be made of the intersexuality of heifers born twin to a bull. Whether or not the foetal membranes fuse giving rise to vascular anastomosis is apparently a matter of chance. If, however, relatively closely related species are compared with each other, then differences can be established which must be assumed to be due to genetic influences. In cattle, vascular anastomosis occurs in 90 per cent of all twin pregnancies; but in sheep and goats it is rare, in spite of the fact that the foetal membranes fuse at a relatively early stage (cf. Chapter 5).

A number of malformations of the genital organs are caused by inhibitions in development, which take place at a certain stage of the foetal period. The degree of malformation generally shows a rather large variation, therefore, according to the intensity of the check and the time when it first exercised its effect (cf. Fig. 2:1). Developmental inhibition may be genetically determined, but it can also be due to conditions in the uterus during the earlier stages of foetal life. Several genital malformations have been studied in detail from a genetic point of view.

Gonad hypoplasia

In the 1930s it was shown that hypoplasia of gonads had reached a very high frequency in the polled cattle in the north of Sweden. The degree of hypoplasia varied, and during the investigation a distinction was made between definite hypoplasia and more or less uncertain cases. In a survey undertaken in 1936 by LAGERÖF it was found that 17·5 per cent of the females examined showed clear ovarian hypoplasia, and a further 16 per cent were classified as uncertain cases. The frequency of testicular hypoplasia in the bulls was, in the unselected material approximately as high. In a number of cases only one gonad (left) was hypoplastic and the other normal, but in some cases both gonads were underdeveloped. The relative frequency of left-sided, right-sided and double-sided hypoplasia was approximately the same for males and females, viz. 82·1, 3·4 and 15·5 per cent respectively (Fig. 9:1).

ERIKSSON (1943) studied the segregation of hypoplastics in different mating combinations and from the results came to the conclusion that the defect is caused by a recessive gene with incomplete penetrance. In 125 cases single-sided hypoplastics were mated with each other, and of the progeny obtained 49·6 per cent were hypoplastic, 20·0 per cent were uncertain and 30·4 were normal. Matings between normal animals produced 12·2 per cent hypoplastic and 14·2 per cent uncertain cases. The penetrance of the gene in homozygotes was estimated to be 43 per cent in males and 57 per cent in females, the overall average being 50 per cent.

It has been shown later that gonad hypoplasia in Polled Swedish cattle has a definite relationship to the degree of pigmentation, since the defect occurs only

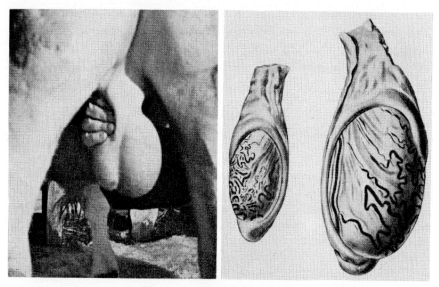

Fig. 9:1. Testicular hypoplasia.
Left: Left-sided hypoplasia. The scrotum is pronouncedly asymmetric. *Right*: Hypoplastic and normal testicle, weighing 69 and 283 grams respectively. (LAGERLÖF 1936)

in almost-white animals. In an examination of 882 females, SETTERGREN (1954) found that in the category $W^9 = (9/10$ white), 76·2 per cent were normal, 13·5 hypoplastic and 10·3 were uncertain. No hypoplasia was observed in the category W^0–W^5 and only 2 cases of uncertain hypoplasia were found in the category W^6–W^8. Perhaps the reason is that the penetrance is complete or almost complete in the all-white animals. Whatever the circumstances, the defect is clearly genetically determined and in one way or another is associated with the pigmentation of the animals.

Testicular hypoplasia has been demonstrated in other cattle breeds and in boars, but the frequency is generally comparatively low. When the defect is bilateral natural selection would be expected to keep the frequency at a low level. A relation between colour and genital defects has been shown in laboratory mammals, e.g. mice and guinea pigs. In the mink a partially dominant gene for reduced pigmentation (Stewart type) is known, which in the double dose makes the males completely sterile (aspermic).

Other genital malformations in males

Bulls of the Red Danish dairy breed have been shown to lack segments of the epididymis, the ductus deferens and the vesicular glands, i.e. the parts of the genitalia arising from the Wolffian ducts (Fig. 2:1); the testicles were found to be normally developed. The defect was often single-sided (Fig. 9:2). If it is double-sided, the bull is naturally completely sterile, since the sperm produced have no way out from the testicles. Examination of the ancestry of the bulls gave some reason to assume that the defect was hereditary.

A comparatively common defect in billy-goats is spermiostasis, usually double-sided; it is believed to be caused by developmental irregularities in the epididymis. The defect has been shown also in rams and bulls. An accumulation of sperm takes place in the efferent ducts of the testicles, and this brings about a swelling in the head of the epididymis (Fig. 9:3) when the animal reaches the age of puberty and spermatogenesis. The defect occurs with varying degrees of severity in bulls; it is usually single-sided, but even double-sided cases have been shown to occur. When the defect is double-sided it leads, sooner or later, to sterility. BLOM (1950) studied a large number of young bulls in the Danish breeds and found that spermiostasis occurred in 2·1 per cent of the Danish Friesian and 3·9 per cent of the Danish Red bulls. The defect is considered to be heritable in he-goats, and Blom assumes that the same is true of bulls also.

Further examples of inhibitory malformations may be mentioned: hypospadia, in which the male urethra opens towards the base of the penis; cryptorchism, in which one or both testicles are retained in the abdominal cavity; and scrotal hernia, in which the connection between the abdominal cavity and the scrotal sac (the inguinal canal) is so wide that the intestines protrude into the scrotum. By injecting relatively large doses of oestrogen into pregnant rats it has been possible to reproduce both hypospadia and cryptorchism in the progeny, which indicates that an excess of the female sex hormone is inhibitory to male sex differentiation.

Cryptorchism is more common in horses, goats and pigs than in cattle; and it has long been considered a hereditary defect. From investigation with goats and pigs it has been concluded that cryptorchism is due to a recessive gene in a double dose.

In the U.S.A. scrotal hernia has been studied in the Poland China and Danish Landrace breeds. It was not possible to demonstrate simple Mendelian inheritance, but the heritability was estimated to be 0·15. In addition, there was a maternal influence of about the same order. Scrotal hernia in boars is thus dependent partly upon the boar's own genetic constitution and partly on the mother as an environmental factor. Whether this maternal influence is mainly operative before, during or after birth of the pigs is not known.

Underdevelopment of the female genitalia

Underdevelopment of the uterus and vagina in Shorthorn heifers has long been known in England, called the 'white heifer disease' since the defect occurs mainly in all-white animals. The ovaries and Fallopian tubes are normally developed, but the uterus and anterior part of the vagina, formed from the Mullerian ducts, show more or less arrested development. In general, there is no communication between the vagina and the uterus, the cervix region consisting of a solid cord without a lumen. RENDEL (1952) investigated an English Shorthorn herd consisting of 232 daughters from seven bulls, all of which had defective progeny. Of 23 all-white daughters 9 were defective, of 115 roan 4 were defective, and of 94 all-red daughters only one was defective. The relationship between the pigmentation of the animals and the frequency of the defect was thus very evident.

A similar defect has been described in Friesian cattle in the U.S.A. and in south Sweden. An American Holstein-Friesian bull was mated with his own daughters and of the 23 grand-daughters reared to sexual maturity 13 were defective and 10 normal. It has been assumed that the bull was homozygous for a recessive gene which prevented the development of the Mullerian ducts to normal uterus, cervix and vagina. According to this hypothesis it can be expected that matings with heterozygote daughters would give approximately 50 per cent defective and 50 per cent normal progeny. NORDLUND (1956) studied the progeny of 410 daughters inseminated from one bull of the Swedish Friesian breed and found that 12 (2·9 per cent) were similarly defective. If the hypothesis of recessive inheritance is to be applied here, then it must be assumed that bulls used earlier had distributed the gene throughout the population so that about 12 per cent of the cows were carriers. Segregation of the recessive homozygotes following mating with a heterozygous bull would then be expected to be 2·9 per cent (p. 98). There is no evidence of a relationship between the extent of white and frequency of the defect in the Friesians. The American Holstein-Friesians are predominantly white, many of them almost entirely white, whereas in the Swedish Friesian black is, as a rule, more extensive than white.

It should be noted that hypoplasia of the gonads affects both sexes with about the same frequency, whereas underdevelopment of the epididymis, ductus deferens, uterus and vagina as well as hypospadia, cryptorchism and scrotal hernia are sex limited; they are inherited in the same way by both sexes but are manifested by only one sex.

Intersexuality

Intersexuality (hermaphroditism) in varying degrees occurs in all species of domestic animals. It is relatively common in goats and pigs but rare, at least in the more pronounced form, in horses, cattle and sheep. The frequency of intersexuality in pigs varies both between breeds and between lines within breeds; on the average, for the Large White and improved Landrace breeds the frequency is probably of the order of 0·01 – 0·02 per cent. At the Agricultural Research Station at Beltsville, U.S.A. it was found that a pedigree line of pigs had an unusually high number of intersexes. The segregation agreed with simple Mendelian segregation, i.e. 3 normal : 1 intersex, and it was therefore assumed that the defect was due to homozygosity for a recessive gene. In a number of the intersexes there was only testicular tissue but in others there was both testicular

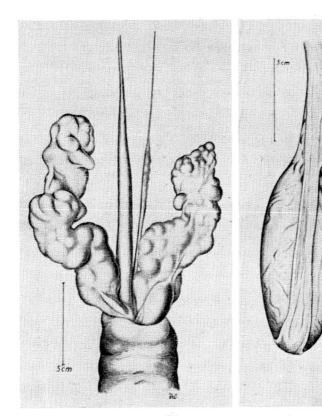

Fig. 9:2. *Left*: Underdevelopment of right vas deferens in a Danish Red bull. *Right*: Underdevelopment of the right epididymis of the same bull. (BLOM and CHRISTENSEN 1951)

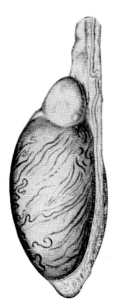

Fig. 9:3. Spermiostasis in an 18-month-old bull with double-sided defect. The head of the epididymis is very swollen. (BLOM and CHRISTENSEN 1951)

and ovarian tissue. No gametogenesis could be identified in either type. Cytological studies showed that all intersexes had the female chromosome constitution XX. The external genitalia of intersexual pigs are markedly of the female type, but the clitoris is often so strongly developed that it protrudes from the vulva (Fig. 9:4). Male sex characteristics generally appear as the age of puberty is attained. This means that the intersexes have a reduced value as slaughter animals unless slaughter takes place fairly early.

Intersexuality is common in hornless but rare in horned goats. It has therefore been assumed that a recessive gene for intersexuality is closely linked to a

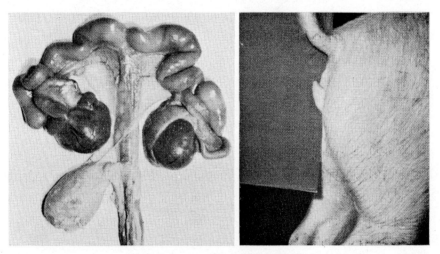

Fig. 9:4. Intersexuality in pigs.
Left: Gonads with both ovarian and testicular tissue (ANDERSON 1956). *Right*: Vulva with abnormally enlarged clitoris. (H. SKOGEN 1961)

dominant gene for hornlessness. It was found that in an experimental herd of Saanen and Toggenburg goats at Beltsville 11·1 and 6·0 per cent respectively, were intersexes. The sex ratio of kids born indicated that the interesexes were genetic females. Asdell assumes that this applies to mammals in general.

It has already been mentioned that the sex differentiation depends on a balance between sex chromosomes and autosomes (Chapter 3). Perhaps the intersexuality of goats and pigs, which according to several research workers show simple Mendelian inheritance, can be explained by the assumption that an autosomal male determining gene with a relatively high 'strength' exerts its effect at a certain stage of foetal development, thereby influencing sex differentiation in the male direction in zygotes with the chromosome constitution $2X + 2A$, i.e. genetic females. The large variation in expressivity can be due to the residual genotype and perhaps also on the intra-uterine environment. The variation in inhibitory malformations, discussed earlier, can be explained in the same way.

GAMETIC STERILITY

Gametic sterility occurs when the individual apparently produces gametes but these are not capable of fertilising or being fertilised. This type of sterility, which may be complete or partial, has been studied mainly in bulls. Most probably it occurs in both sexes and in all types of animal.

BLOM (1950) distinguishes between two main types of sperm abnormality: primary abnormalities (e.g. abnormal head, middle piece and tail), which are due to degenerative changes taking place at the spermatogonia stage, and secondary (e.g. loose sperm heads, loose 'cap' (*galea capitis*), or coiled tails (Fig. 9:5)). It has been shown that the last-mentioned defect can be produced in the ejaculated sperm by hypotonic diluents or rapid cooling.

KNUDSEN (1954, 1958), who has carried out extensive cytological studies of the spermatogenesis of bulls, is of the opinion that it is possible to distinguish between acquired and congenital disturbances in this process.

Acquired disturbances may be due to disease associated with high fever or other temporary or long term depressions of the animal's condition. Degenerative changes of uniform type take place in the spermatogenic epithelium, and cell division proceeds more or less abnormally. If all the spermatogonia degenerate, the result is permanent sterility; but in some cases the disturbances can be of a temporary nature. The frequency of abnormal sperm and the extent of the abnormality can show considerable variation. The sensitivity to unfavourable environmental conditions may be different in different individuals.

Congenital abnormalities mean, in general, a certain type of abnormality that can be diagnosed by suitable techniques. There is always reason to suspect that such sperm defects are genetically determined.

In Friesian cattle in England and Holland a type of gametic sterility has been demonstrated which probably depends on a recessive autosomal gene. Bulls which are homozygous for this gene produce a large number of deformed sperm (80–95 per cent of the total) and they are completely sterile. The defect, called 'knobbed', is visible in living spermatozoa under high magnification, as an eccentrically placed thickening of the anterior borderline of the sperm head, the acrosome (Fig. 9:6). A similar defect has been shown in boars but as yet nothing is known about the inheritance.

KNUDSEN (1958) has studied two types of sterility in bulls, both presumably genetically determined, where spermatogenesis goes 'off the rails' at an early stage. In the one case the chromosomes stick together, usually when the spermatocytes begin their division. The semen is thin and watery and contains no properly formed sperm but only rounded cells with pycnotic nuclei (chromosome clumps). In the second case several spindles are formed in the cell during meiosis, the chromosomes divide but the cytoplasm does not, with the result that giant cells are formed. Both these types can be diagnosed fairly easily with an ordinary microscope.

With the aid of the electron microscope, Knudsen has shown that sperm of apparently normal appearance may have an unbalanced gene constitution as a

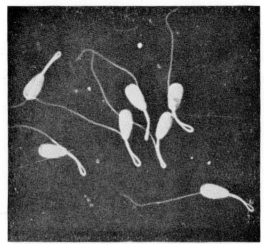

Fig. 9:5. Sperms with coiled tails, a defect which, as a rule, is not genetically determined. (BISHOP *et al.* 1954)

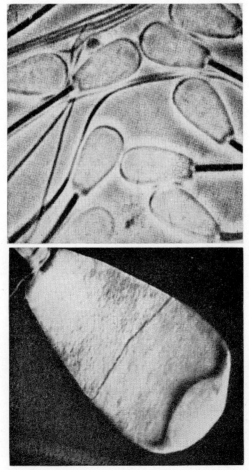

Fig. 9:6. *Top*: Normal bull sperm magnified ×2000. (*Bishop et al.* 1954).
Bottom: Sperm with defective head—'knobbed'. (HANCOCK 1953)

result of chromosomal aberrations. The material for this study was a number of bulls, which according to conventional methods of assessment had excellent sperm but nevertheless exhibited low fertility. Knudsen's findings and interpretations were as follows. Four of the bulls were found to be inversion-heterozygotes and three bulls were found to be translocation-heterozygotes. In both cases the disturbance takes place at the reduction division during spermatogenesis. With each meiotic cell division of the inversion-heterozygote, the

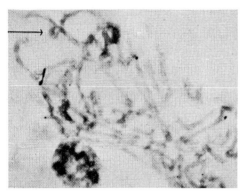
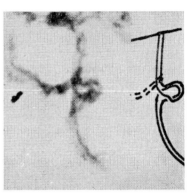

Fig. 9:7. Aberration in the division of the primary spermatocytes of a bull which is an inversion heterozygote. *Above left*: The homologous chromosomes have arranged themselves in pairs; the arrow shows the inversion loop (Mag. × 2700). *Above right*: The same inversion loop magnified × 6000 and alongside a drawing of the loop. *Below*: A chromosome with two centromeres between the separated chromosome groups (bridge); drawn from a microscopic picture. (KNUDSEN 1958)

chromosomes of the inversion pair form a loop, as described earlier and if crossing over takes place within the loop the result is one chromosome with two centromeres and one chromosome without a centromere. The latter chromosome lacks orientation during cell division and can pass over into either of the two daughter cells, whereas the chromosome with two centromeres lies like a bridge between the two groups of chromosomes (Fig. 9:7). The bridge eventually breaks and the cell divides, the result being two, in respect of chromosome complement, unbalanced cells. Spermatozoa formed in this way can probably fertilise ova, but the zygotes will lack the capacity to develop and die at an early stage. The result of translocation is, in principle, the same though the orientation of the chromosomes on the spindle will be different (p. 71). The chromosome aberrations lead, therefore, to reduced fertility in the bulls. Sperm which contain an inversion chromosome where no crossing over has taken place, or two non-homologous chromosomes with translocated parts, have complete gene complements and may transmit the defect to the next generation.

HENRICSON and BÄCKSTRÖM (1964) have made a cytological investigation of testicular tissues from normal, as well as from sterile or subfertile, bulls and boars. The bulls had been used in A.I. and their conception rates were known. At least 200 meiotic nuclei were recorded for each male. The duration of metaphase and anaphase of the first and second meiotic divisions varied very little between normal males but in the sterile and subfertile males there was a pronounced time displacement between the divisions, metaphase I being prolonged. A similar time displacement was found also in two completely sterile boars with the acrosome defect of the spermatozoa. The proportion of nuclei which 'fail' to undergo normal meiosis was found to be as high as 8 or 9 per cent, and anaphase bridges were found to occur in normal animals with a surprisingly high frequency. The authors suggest that the main cause of the formation of defective gametes is to be found in the nonchromosomal structures, i.e. the division apparatus, rather than in the chromosomes themselves. In another publication (1964) the same authors report a case where a Swedish Landrace boar when mated to 51 sows produced litters of only about half the normal number. In litters with other boars 21 of these sows farrowed 12·7 pigs per litter, but with the boar in question only 5·6 pigs per litter. A cytological study was made of the boar's blood leucocytes, cultured *in vitro*, and it was found that 3/4 of one chromosome of the pair 4 (or 3) had been translocated to one of the pair 14. This was a constant finding in all the metaphases suitable for chromosomal identification.

Apparently there is a need for further studies in this field. Cytogenetic investigations based on tissue cultures will probably throw more light on the subject. Such investigations can be applied also to female animals, where knowledge on the frequency of chromosomal aberrations is lacking.

HEREDITARY DISPOSITION FOR REDUCED FERTILITY

The result of a mating depends on the fertility both of the male and of the female, and it may also depend on the gene combinations of the zygote. In individual cases of sterile mating it may be difficult to decide where the fault lies. All variations are found, from normal fertility to complete sterility; fertility shows continuous variation and can consequently be regarded as a quantitative trait. Attempts made to study the relative importance of heredity and environment in the total variation in fertility, however, have yielded only slightly informative results. In the first place, it is very difficult to measure the degree of fertility; and secondly, the environmental variations are large and of many different kinds. In genetic analyses it is particularly desirable to diagnose the nature of reproductive disturbances, since different kinds of disturbances apparently have different degrees of heritability; attempting to measure them collectively can easily lead to erroneous results.

Sterility and reduced fertility often have endocrinal causes. The hormone balance of the body is disturbed and this has repercussions, to a greater or lesser degree, on the reproductive functions. In the male, this can manifest itself as lack of libido or disturbances in spermatogenesis; in the female, weak or irregular heat or inability to carry the foetus to full term may be the consequence. Hereditary predisposition to these disturbances can be found, but the manifestation is usually brought about by environmental stresses of various kinds. To a certain extent the animal is buffered against unfavourable environmental factors, but if the resistance threshold is exceeded the disturbance is phenotypically manifested. The age of the animal is generally a contributory cause in this respect (cf. p. 254).

Reproductive disturbances resulting from infections of the genital organs or inadequate nutrition, such as deficiency of certain vitamins or minerals, will not be dealt with here. The consequences of high or low plane of nutrition will be briefly mentioned, together with the relation between the intensity of production and fertility as well as the risks that are associated with intensive inbreeding.

Variations in the fertility of male animals

Quantity and quality of semen. Several investigations have been carried out with identical twin bulls in order to elucidate the question of the influence of heredity on breeding capability. BANE (1954) studied six pairs of identical twin bulls, reared on different planes of nutrition at the Wiad station until they were 18 months, and then transferred to the Veterinary College of Sweden for detailed study of mating behaviour, sperm production and quality. During the rearing period one bull of each twin pair was allocated to one of two differing planes of nutrition, which were, high-low (H−L), medium/high (M−H) and medium/low (M−L). The high plane of nutrition corresponded closely to that employed in herds raising pedigree bulls for breeding. The medium plane bulls consumed 82 per cent and the low plane bulls 68 per cent of this amount calculated in Scandinavian feed units. From 18 months of age onward, all the bulls were fed at the same level, which was slightly above the medium plane. Semen was collected once a week by two successive services, with accurate checks on volume of ejaculate, sperm concentration, sperm motility, the frequency of abnormal sperm types, etc.

There was a rather large variation in the volume of ejaculate and the sperm quality between different ejaculates from the same bull. This was assumed to be due to differences in the ejaculation process. In spite of this variation, it was possible to show significant differences between twin pairs with respect to volume of ejaculate, sperm concentration, sperm motility and a number of sperm defects, though not for 'coiled tail'. The differences demonstrated were assumed to be genetically determined. No influence of the rearing intensity on semen quality could be shown. According to American investigations the plane of nutrition of bulls during active service has little effect on the semen quality, but the volume of semen produced is lower on a low plane than on a high plane of nutrition.

Twin experiments carried out at the Ruakura station, as well as in the U.S.A. and Holland, have produced results which are in more or less close agreement. DE GROOT (1961) published a report of an experimental series including 6 MZ and 6 DZ twin pairs. He found clear differences between twin pairs with regard to a number of semen characteristics, and also that the variation was greater in DZ than in MZ pairs. He was able also to show a rather strong correlation between the frequency of various sperm defects, e.g. between coiled tail and head abnormalities ($r = 0.44$) and between coiled tails and abnormal middle piece ($r = 0.84$).

There are a number of reports in the literature that low sperm quality and unsatisfactory conception rates have occurred with several bulls from the same pedigree line, which indicates that the disturbances were genetically influenced.

In an investigation of 353 American A.I. bulls, all of which had been in service for at least three years, it was found that the repeatability from one year to another of 60–90 day 'non-return' was 0.44 on the average for three dairy breeds. 'Non-return' means that within the given interval the cows have not been reported for re-insemination. It is thus an approximate expression of conception rate. In Sweden a similar investigation was carried out on young SRB bulls used by the A.I. associations during a period of 9–12 months. The time that the bulls were in service was divided for each bull into two equal halves, after which the conception rate was corrected for the influence of time of year. The repeatability of conception results within the insemination centres was estimated to be 0.60 and the heritability to 0.30, the latter figure being based on the correlation between half sibs. The investigation included no bulls which, according to normal routine testing, had low quality semen. It is probable, therefore, that estimates for an unselected population of bulls would give higher figures. MAIJALA (1965) obtained similar results in analyses of data from Finnish A.I. associations.

In summary, it can be said that all the investigations carried out agree in showing that, under essentially similar conditions, there is considerable genetic variation in semen production and quality between bulls of the same age and breed. The non-genetic variation, however, is apparently considerably larger. The variation is especially large between different ejaculates from the same bull, and it is therefore necessary to have a large amount of material in order to show differences between individuals. This applies, of course, even to experiments with identical twins, where as a rule, the material is such a small sample from the existing population that the results can hardly be generalised.

Mating behaviour and potency

The desire and ability to carry out a complete act of mating is typical of sexually mature males. Mating behaviour is exhibited long before puberty and is also common among females which have been sexually stimulated. When some of the group are on heat cows ride each other, with the rhythmic back movements typical of the bull. This behaviour is no doubt instinctive and present in both sexes. The complete act of mating by the male, on the other hand, must to a certain extent be learnt, even if only a few acts of mating are required for this. The rapidity and smoothness with which the mating act is carried out depends mainly on three factors: (1) the sexual drive of the male, (2) the stimulation he experiences from his mate, normally a female on heat, and (3) environmental influences. Especially in collecting semen with an artificial vagina, but even with natural mating, a number of conditioned reflexes can be established which determine the service ability of the male. A bull gets used to a certain attendant and a certain environment, and quite often he will refuse to serve under altered circumstances.

Copulatory impotence and lack of sexual drive are in the majority of cases psychologically and/or hormonally influenced, and it is natural that sensitivity to external circumstances varies widely between individuals. The experience of pain in connection with mating, e.g. injuries of the joints of the hind legs, can be another cause, though this will not be discussed further here. BANE (1954) found in his investigations with identical twin bulls an example of a twin pair which served normally all its life in spite of deformed joints, whereas the members of another pair, which were free from any affliction in the skeleton or joints, became impotent at $3\frac{1}{2}$ years of age. Ossification of the sacro-iliac joints in Bane's material did not appear to have any influence on copulatory ability.

When the male is brought together with a female on heat he shows a particular pattern of behaviour, which is subject to individual variation. The fore-play, prior to service, varies in length depending upon the sexual drive of the male, on the degree of attractiveness he finds in the object of his attentions, and on strange or inhibitory external factors which may be present.

In a study of about 2100 Swedish Red and White bulls, 15–26 months old, HULTNÄS (1959) differentiated between libido, i.e. the willingness and eagerness to mount and serve, and mating technique, i.e. the ability to perform a complete service. Libido was scored according to a 7-point scale and notes were made on mating technique. The bulls were tested both with natural service and with artificial semen collection. The results showed that 5·25 per cent of the bulls received a score of less than 3, the majority of these being under 21 months of age. Natural mating was carried out without comment by 83·7 per cent of all bulls, the corresponding figure for artificial semen collected being 74·3 per cent. No influence of the time elapsing since the previous service could be detected.

Hultnäs carried out a number of analyses of variance in order to study the relation between sires and sons with respect to libido and a number of semen characteristics. He found significant differences in libido between paternal half sib groups of bulls. If the intra-class correlation for these groups is calculated from Hultnäs's tables, then, after eliminating herd differences, a figure of 0·096 is obtained; which corresponds to a heritability of 0·38. The results are uncertain, however, on account of the large variation in the size of the sib groups (1–41 bulls). When the investigation was limited to 10 sib groups in 4 herds, no significant difference was obtained between groups within herds. Taken at face value, the figure of 0·38 would indicate that libido is a rather highly heritable trait; but further investigations must be made before any definite conclusions can be drawn.

Hultnäs also studied the relationship between the degree of inbreeding of the bulls and their libido and semen characteristics. The degree of inbreeding was consistently low. There seemed to be a slight tendency to reduced sexual functions with increasing degree of inbreeding.

In his study of monozygotic twin bulls, Bane found striking similarities within pairs and pronounced differences between pairs, in the matter of mating behaviour; the plane of nutrition up to 18 months was apparently without effect. For each bull a record was kept of all attempts to serve and their result. The percentage of unsuccessful (i.e. without erection, without thrust, and refusal) attempts was very similar for bulls belonging to the same pair, but different between pairs. The bulls of two pairs were very good servers and another two pairs of bulls were poor servers. Two pairs of bulls became impotent at 42 months of age and a third pair showed the beginnings of impotence at the same age.

Similar observations have been made at other institutions (Ruakura and U.S.A.) that the behaviour pattern at service is much alike for identical twin bulls. DE GROOT (1961) studied the length of the interval between two jumps at the same service for 6 DZ and 6 MZ bulls; this interval was used as a measure of the strength of sexual drive. The statistical analysis was based on 20 tests for each bull. It was shown that the difference between MZ pairs was significant, but the variance between DZ pairs was exceptionally large in relation to that between MZ pairs. It is difficult to draw any conclusions from this study as to the influence of heredity on libido.

Another Dutch investigation included 22 cases of the inability of Friesian bulls to copulate. According to the diagnosis the bulls lacked the ability to straighten out the sigmoid curve of the penis during erection and it was concluded that the defect was due to an autosomal recessive gene. To check the spreading of the gene in the population, the previous method of treating the defect by resectioning the retractor penis muscles was discontinued.

ERIKSSON (1949) carried out a genetic study of Swedish Polled (SKB) and Swedish Red and White (SRB) cattle. He found that in the SKB material 45·8 per cent of the sons of bulls with reduced potency showed the same defect as their sires, whereas the corresponding figure for the sons of normal sires was only 28·0 per cent. He found, too, that many of the bulls used in the founding of the SRB breed had insufficient sexual drive. It was further established that paternal ancestors with low copulatory efficiency were over-represented in the pedigrees of SRB bulls which showed reduced potency. The environmental differences which most certainly occurred between bull groups were not taken into consideration in this study.

The investigations which have been carried out show rather close agreement that copulatory inability (*impotentia coeundi*) in bulls is genetically influenced, and that vigilance should be observed for this defect in selection of bulls for breeding. Lagerlöf has suggested that the breeding ability of the bulls decreased as a result of selection for a feminine, gentle type in preference to the robust masculine type. Many are sceptical of this hypothesis, but there may be something in it.

Copulatory inability occurs also in other species of animals than cattle. A form of copulatory inability has been demonstrated in Swedish Landrace boars, apparently caused by pathological changes in the joints of the hind legs. The condition is assumed to be genetically influenced. Even comparatively young boars show the defect when they are brought into service.

Variations in female fertility

Endocrine imbalance is a relatively common cause of reduced fertility or sterility in females. The background to these disturbances may vary, but a certain degree of hereditary predisposition is most likely an underlying factor. The time at which the disturbance manifests itself during the life cycle of the animal depends largely on environmental conditions. Two different types of endocrine imbalance have been studied from a genetic point of view.

GARM (1949) made a comprehensive clinical and histological study of cystic ovaries in cows. He distinguishes between nymphomania and adrenal virilism depending on whether the erotomania is manifested in the female or male direction. Nymphomania is usually characterised by return to heat at short and irregular intervals due to an increased production of the follicle stimulating

Fig. 9:8. A $3\frac{1}{2}$-year-old nymphomaniac SRB heifer, showing the raised sacrum and tail head with a depression between the sacrum and thurl in addition to a pronounced brisket and enlarged dewlap. (GARM 1949)

hormone (FSH) and a reduced production of luteinising hormone (LH). The follicles do not rupture but are enlarged and remain as cysts. According to Garm, the clinical symptoms of nymphomania vary within rather wide limits; it can also happen that the heat symptoms are absent or weaker than normal. A nymphomanic cow undergoes particular changes in temperament and external appearance. She becomes restless and has a 'wild look'; the pelvic ligaments are relaxed which means that the tail head becomes more prominent and the sacrum is tilted forward (Fig. 9:8).

In adrenal virilism, when the cow is known as a 'buller', Garm found that the adrenal cortex was hypertrophied; an increased production of androgenic substances was therefore assumed to be the cause of the bullish behaviour. Many veterinary gynaecologists maintain, however, that the 'buller' behaviour is often accompanied by nymphomania in the advanced stages. Garm found a high frequency of nymphomania among daughters of cows which manifested nymphomania, and he therefore assumed that the defect is hereditary. Infections of the genital organs did not appear to have any part in the aetiology of this defect.

In the United States several investigations have been carried out on these defects. CASIDA and CHAPMAN (1951) studied a herd of Holstein cattle during a period of ten years. The material comprised 341 cows with 1280 service periods. Cystic ovaries were found in 18·8 per cent of the cows and during 7 per cent of the service periods. Comparison between cows in the same age groups showed that cystic ovaries occurred during 31·0 per cent of all service periods for daughters of dams which had, at some time, manifested the disturbance, whereas the corresponding figure for daughters out of normal dams was only 9·4 per cent. The heritability was estimated, accordingly, to be 43 per cent (cf. p. 128). In another study a significant difference was found between daughter groups of different bulls. ERB and his co-workers studied one of the best-known Holstein herds in the U.S.A. where comparisons could be made between 2076 daughter-dam pairs. The difference in frequency of cystic ovaries in daughters from 'positive' and 'negative' dams was in this case only 2·6 per cent; which indicates a very low heritability. A rather clear relation could, however, be established between several different sexual disturbances in the cows, primarily between the tendency to cystic ovaries, twinning and retained placenta.

HENRICSON (1956) studied the frequency of cystic ovaries in a population of about 10 000 cows of the Swedish Red and White breed in two A.I. associations. All the herds involved in these data had, in connection with A.I., been under continuous veterinary control for at least 10 years. The average frequency of cystic ovaries in one association was 18·9 per cent and in the other 16·2 per cent with an average age of the cows of about 4 years. As in the investigations previously referred to, the frequency was low in young cows, increasing up to about 5 years of age; after which the frequency was stable. It was decidedly lower in the summer than in the winter. Henricson calculated the risk of a cow repeating the defect, having once manifested the condition, to be 0·23 on the average for the two populations and the heritability to be 0·15, but preferred to regard the defect as a qualitative trait. On the hypothesis that the defect, as a rule, is due to a recessive gene, Henricson comes to the conclusion that the frequency of this gene in the data studied was about 0·7, when only definitely diagnosed cases were included, and that the penetrance (for aa) varied with the age of the cows; at $2\frac{1}{2}$ years of age it was only 2 per cent; for cows $5\frac{1}{2}$ years old, 20 per cent; and for $8\frac{1}{2}$ years old cows, about 50 per cent.

However, the predisposition to this trait should probably be regarded as a quantitative one depending on a large number of genes (polygenic inheritance); even the phenotypic manifestation shows a continuous variation from uncertain cases to repeated cases in the same individual. Henricson's figures, referred to above, for the chance of repetition and for the heritability, would seem to give a good indication of the importance of individuality and the genetic component in the occurrence of the defect in the population studied.

The regularity of the oestrous cycle in cows has been the subject of several investigations. CHAPMAN and CASIDA (1937) calculated the repeatability of the length of the cycles between 17 and 27 days to be 0·41 in a large Holstein herd. At the Beltsville Research Station a repeatability coefficient of 0·18 and a heritability of 0·05 was obtained from an analysis of data on 834 cows, correction being made for seasonal variation and time trends. The regularity of the oestrous cycle was studied at the Minnesota Agr. Expt. Station with 9 pairs of identical twins and it was found that the intra-pair correlation for intervals between 14 and 25 days was 0·61. On the other hand, at the Animal Breeding Research Organisation, Edinburgh no influence of individuality on the length of the oestrous cycle could be found for 15 monozygous, 15 dizygous and 15 half-sister pairs.

Weak heat would appear to be due to a low production of oestrogenic hormone. According to ASDELL (1944), failure to detect signs of heat is one of the most common causes of prolonged calving intervals. In the Swedish A.I. associations heat symptoms are classified as distinct, indistinct or very indistinct. In an analysis of data from approximately 230 000 cows the percentage of diagnosed pregnancies in these three groups was 57, 45 and 26 respectively. Weak heat makes the timing of insemination in relation to ovulation very difficult. LAGERLÖF (1957) states that heat symptoms are more pronounced in cows of the Swedish Polled than the Swedish Red and White breed. The occurrence of weak heat is common in the latter breed, especially in winter.

From experiments with identical twins at the University of Minnesota the general statement is made that, 'Behaviour during oestrous and intensity of demonstration was markedly similar in heifers of a set'. At the Danish Bull Progeny Testing Stations the intensity of heat symptoms of the heifers have been scored on a scale of 1–4 points. ROTTENSTEN and TOUCHBERRY (1957) analysed these data and found that the repeatability of the score for the same animal, calculated within progeny groups, was 0·29 and the heritability 0·21. Heifers which conceived at the first service showed more distinct heat symptoms than those which required two inseminations, and the latter showed more distinct heat symptoms than those which required three or more inseminations. The studies which have been made give a rather clear indication that intensity of heat symptoms are to a certain extent hereditary.

Various measures of female fertility

Investigations have been carried out into the repeatability and heritability of a number of more or less comprehensive measures of the fertility of cows, e.g. length of calving interval and number of inseminations per conception. Both these measure can be influenced not only by the fertility of the cow but also by the fertility of the bull and the viability of the foetus. In addition the length of the calving interval can be regulated, to a certain extent, by the breeder, for example when it is desired to concentrate calving to a particular time of the year; the accuracy with which the heat symptoms are observed is also of importance. It is obvious that the length of the calving interval and the number of inseminations per conception reflect, only in a very broad sense, the fertility status of the cows. In order to calculate the repeatability of the length of calving interval, a cow must have calved at least three times, so that there are data on at least two intervals. Cows culled before the third calving on account of fertility disturbances are not usually included in the investigation.

According to several investigations the repeatability of the length of the calving interval is very low, about 5–10 per cent, and the heritability is not significantly different from zero. The same is true of the number of inseminations per conception. In the twin experiments at Edinburgh no significant difference was found between monozygous and dizygous pairs with respect to the within-pair variation of the number of inseminations per conception.

Investigation into the variation of the interval between calving and the next following heat has indicated a repeatability of 15–29 per cent, which is somewhat higher than the figure for the previously mentioned trait.

KORKMAN (1947) studied the fertility in 38 Finnish Ayrshire herds including over 4000 cows and about 13 000 calvings. He came to the result that the chance for a cow to conceive at the first service was 5 per cent hereditary, and the corresponding figure for the chance that a cow becomes pregnant at all, or is culled as sterile, was 10 per cent. In a comprehensive analysis of data from Finnish and Swedish milk recording and A.I. associations, MAIJALA (1964) estimated the heritability for a number of traits, e.g. prolonged calving intervals, services per conception, non-return rates etc. The h^2 estimates were low (1–3%) but significant due to the large number of animals included in the data.

The fact that all these studies show low repeatability and heritability coefficients should not be interpreted to mean that the predisposition for fertility disturbances is of so little importance that it can be ignored. However, as with the investigations into endocrine imbalances, the indications are that the main causes of these disturbances must be sought among the many environmental influences which can affect sexual functions.

Foetal mortality

When possible, a distinction should be made between fertilisation percentage and foetal mortality, calculated on the number of fertilised ova. Another concept which has been introduced is, perinatal mortality, i.e. the mortality immediately before, after and during parturition. Determination of the number of ova shed can either be made directly or, with a certain amount of error, by counting the number of yellow bodies which develop in the ovaries after ovulation. Fertilised ova are assumed to be those which complete a number of divisions without signs of degeneration.

If both the male and female are fully fertile, and if the insemination takes place at a suitable time in relation to the ovulation, it can be assumed that practically all ova shed will be fertilised. Deviations from this, in most cases, are probably due to unfavourable conditions in the female genital organs either for the sperm, or the ova shed, or both. Naturally, the deviations can be due to gametic sterility in the male or female.

In the U.S.A. CASIDA (1961) and his co-workers have made comprehensive studies into the fertilisation rate and foetal mortality in cattle. In an investigation with 74 young heifers, which were slaughtered 3–4 days after heat and insemination, it was found that when mating took place with a bull known to have a high fertilisation rate, all ovulated ova were fertilised; but when mating was made to a bull with a low fertilisation rate, only 72 per cent of the ova shed were fertilised. The experimental heifers had been allotted at random to one of the two groups and the difference in fertilisation percentage ought, therefore, to be attributable to the bulls. In another experiment with 55 heifers, fertilisation percentages of 97 and 77 were obtained for the bulls with high and low fertilisation rates, respectively. It has also been shown that foetal mortality before 33 days' pregnancy was 10·6 and 19·6 per cent for the two groups of bulls respectively; there is thus a relationship to be found between low fertilisation rate and high foetal mortality.

Investigations have been carried out by the same group of research workers into the causes of the failure of return breeders to become pregnant after insemination. In material comprising 96 heifers, of which half were slaughtered 3 days after insemination and the remainder after a further 27 days, it was found that in 13·5 per cent of the heifers the state of the genitalia was such that pregnancy could not be expected, and that in 11·3 per cent of the heifers it could not be shown that ova had been shed in connection with heat. In the other heifers slaughtered 3 days after insemination, 33·3 per cent of the ova either had not been fertilised or had died within two days after fertilisation. Live embryos were found in only 30·6 per cent of the heifers slaughtered after 30 days. In a similar investigation involving 104 cows, live fertilised ova were found in 66·1 per cent of the cows slaughtered 3 days after insemination, but in only 23·1 per cent of the cows slaughtered after 34 days; embryonic mortality during the period between 3 and 34 days after insemination was therefore estimated to be 65 per cent.

In another investigation with 694 cows, in which pregnancy was diagnosed by rectal palpation, it was found that embryonic mortality between diagnosis and the full term of the pregnancy was 6·4 per cent. The great majority of embryonic deaths appear to occur prior to the implantation of the embryo in the uterus. After implantation, embryonic mortality in cows in normal health is relatively small. The perinatal mortality, however, is probably of the order of at least 5 per cent on the average.

The Milk Marketing Board made a study of the fertility of over 1000 bulls, each of which had at least 500 inseminations in the same year. The conclusion was that the best result that could be expected from a fully fertile bull, used on a normal population of cows, was 75 per cent pregnancies. That the wastage is at least 25 per cent depends, most probably, on the inseminated cows and on the genetic constitution of the embryos.

The foetal mortality in polytocous animals is much greater than in cattle. The loss of ova and embryos in sheep has been estimated to be 30–40 per cent and in

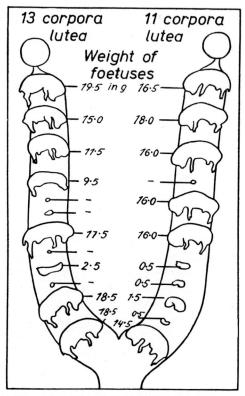

Fig. 9:9. Example of foetal mortality in pigs (mid-pregnancy). In one ovary were found 13 and in the other 11 yellow bodies, together 24. There were 12 normal foetuses and 11 dead or degenerated foetuses, one ovum was missing. The weight of the foetuses is in grams. (HAMMOND 1914)

pigs to 40–45 per cent (Fig. 9:9). It has been clearly shown that in pigs embryonic mortality increases with an increased number of embryos. Approximately 3/4 of the embryonic deaths in pigs take place 10–20 days after fertilisation; after implantation has been completed, the death rate is relatively small. The frequency of pigs born dead is given, on the average, as 5–6 per cent of all births; the figure is decidedly lower for medium-sized litters than for small and large litters. A number of investigations have shown that the feeding of the sows, especially the nutritive quality of the feed, during the gestation period has a marked influence on the foetal mortality.

BISHOP (1964) has advanced the hypothesis that an unexpectedly large part of the foetal mortality is attributable to genetic causes and that the majority of the genetic factors involved probably arise *de novo* in each parent generation. This hypothesis would account for the nature of unexplained differences between males in the conception rate. It is suggested that a considerable part of the embryonic mortality is unavoidable and that it should be regarded as a normal way of eliminating unfit genotypes in each generation. However, it must be remembered that in many cases when, for example, sows return to service, their genitalia are not in the healthy condition required for fertilisation and normal foetal development. When such faults are corrected, the normal fertility of the animals may be restored.

INBREEDING AND FERTILITY

It is well known that close inbreeding is nearly always accompanied by lowered fertility, reduction of libido, an increase in gametic sterility and embryonic mortality, as well as a reduction in the viability of newborn. There is abundant evidence that this is the case from the inbreeding experiments carried out with pigs and cattle in the U.S.A. during the past decades (cf. p. 373). The crossing of inbred lines has resulted in improvement in the breeding results.

The results of inbreeding experiments with pigs at the Animal Breeding Research Organisation, Edinburgh are, in the main, in agreement with American experience. Inbred boars and sows have been tested in crosses with non-inbred animals. The degree of inbreeding was equivalent to rather more than two generations of full sib mating ($F = 0.4$). It was shown that inbred sows shed 2.9 fewer ova during heat than non-inbred sows and that the embryonic mortality up to 28 days was, on the average, higher by 0.37 foetuses after mating with inbred than with non-inbred boars. The degree of inbreeding of the dam had thus a greater influence on the size of the litter at birth than did the degree of inbreeding of the sire. The same applies to mink.

Mention may also be made of the breeding experiment with Holstein-Friesian cattle, now in progress at the University of Wisconsin, U.S.A. In it relatively cautious inbreeding combined with selection and crossings between inbred lines is being tested. Comparisons have been made (cf. Chapter 14) in respect of fertility percentage and embryonic mortality in inbred cows inseminated with semen from related and unrelated bulls. In the first case, when the dam and the foetus are both inbred, the embryonic mortality was higher than when only the dam was inbred. In comparisons between inbred and non-inbred cows, it was shown that embryonic mortality was 15 per cent higher for the former than for the latter. When inbred bulls were mated to related inbred cows the percentage of diagnosed pregnancies was only 36.8 compared with 65.7 for matings between unrelated and non-inbred animals. The influence of inbreeding on pregnancy results was thus quite considerable. From inbreeding experiments with cattle which have been conducted at Beltsville and at Davis, California, a reduction in fertility and an increase in calf mortality have been reported in connection with increasing degree of inbreeding (cf. p. 378).

The evidence that inbreeding carries with it reduced fertility and a lowering of viability in the offspring, especially in the first few weeks after birth, is so clear and concurrent that there exists hardly any doubt. Naturally, the results vary from case to case depending on the genetic constitution of the animals used for inbreeding. On the average, it can be reckoned that a definite inbreeding depression begins as soon as the degree of inbreeding corresponds to half sib mating or closer relationship. It is also clearly demonstrated that crosses between different inbred lines, within which degeneration symptoms have been evident, result in increased fertility and viability of the progeny. Such increases in vitality (heterosis) are often seen in crosses between different pedigree lines or between breeds, even if no inbreeding of importance has taken place previously. This is in complete agreement with the results of experiments with cross-fertilised plants and laboratory animals.

Whatever the genetic explanation may be, the heterozygotes seem to be better 'buffered' against various physiological disturbances than the homozygotes. Apart from genital malformations and congenital gametic sterility, which at least in some cases can be regarded as qualitative traits, the resistance to fertility disturbances ought to be regarded as quantitative and under the influence of polygenes, in addition to the fact that a considerable part of the genetic variation is probably of a non-additive nature.

'STRESS' AS A CAUSE OF REPRODUCTIVE DISTURBANCES

The individuals in a population have varying resistance to unfavourable external influences. When the environmental strains exceed the resistance threshold, more or less serious disturbances occur in the physiological processes which must proceed in harmony with each other if the animal body is to remain in good health and efficiency. The reproductive functions are sensitive to such disturbances, which throw the endocrine balance out of equilibrium. Defining the relevant strains by the commonly used term 'stress' has the advantage that it says nothing about the nature of the strain. The animal can be subjected to stress by harsh treatment, which brings about nervousness and fear, by uncomfortable stable conditions, or by very intensive feeding given to achieve the quickest possible growth during rearing, without any compensatory exercise, or, in the case of cows, maximum milk yield.

Natural selection leads in the long run to an improved genetic acclimatisation of the population to the prevailing environmental conditions, mainly with regard to fertility and viability. Man can exert an influence on this acclimatisation process in two different ways, partly by artificial selection for particular purposes —whereby the viability in the prevailing environmental conditions is perhaps not given sufficient consideration—and partly by more or less drastic changes in the environment. During the past half century very great changes have taken place in both these respects. It is not unlikely that the interplay of genotype and environment has been less favoured by such changes. However, this interplay would probably have been still worse if environmental changes alone had taken place, without simultaneously planned selection. When the one variable is altered, the other variable must be also altered in such a way that the harmonic interplay is maintained.

In the veterinary literature it is often stated that our present day milk cows have an 'unphysiologically' high production and that this is an important cause of reproductive disturbances, e.g. cystic ovaries and lactation diseases, such as puerperal paresis. The 'unphysiology' can hardly be said to mean anything more than that the right balance between genotype and environment has not been successfully achieved. A high milk yield, as such, should not be regarded as anything unphysiological. Under comparable environmental conditions, apart from lower feed consumption, the fertility and viability is no higher in the beef breeds than in the dairy breeds, in spite of the fact that the cows in the former produce no more milk than the calves require. By no means all, or even the majority, of high yielding milk cows exhibit cystic ovaries or puerperal paresis.

That overfeeding of heifers and cows with a poor inheritance for milk yield is often accompanied by reproductive disturbances is well known from experience. Both in England and the U.S.A. it has been shown that the frequency of sterility is much higher in heifers and cows of the beef breeds which have been specially prepared for show purposes. In practice, it is probable that the level of feeding and the balancing of the ration are less adequate for high-yield than for low-yield cows. This is, however, no justification for discontinuing breeding for high production in dairy cows. But the more intensive the selection for production traits, the more important becomes the environmental adaptation. This would apply to all sections of animal production.

Several investigations have shown that the animals' resistance to reproductive disturbances is partly due to environmental factors, e.g. the system of rearing has a distinct influence. The results of the twin research at Wiad point in this direction. Still clearer evidence can be found in Danish and American investigations. A comprehensively planned experiment was carried out in Denmark where heifers of the Red Danish breed were reared on three different planes of nutrition up to the time of first calving, whereafter they were all fed according to accepted scales. In the groups that were reared on the high plane, 50 per cent were culled, sooner or later, on account of reproductive disturbances; in the groups on medium and low plane of nutrition, the corresponding culling percentages were 46 and 28 respectively. A similar experiment with Holstein-Friesians was carried out at Cornell University. In the groups on the high plane (140 per cent of 'normal' feeding), 2·14 inseminations were required per pregnancy, the corresponding figures for the normal (= 100 per cent) and the low plane (65 per cent of the normal plane) were 1·43 and 1·25 respectively. In the part of the experiment already completed, all the heifers were served at about the same age, though sexual maturity was considerably later in groups on the low than on the high plane. A drawback with the restricted rearing was the increased frequency of parturition difficulties at first calving. The milk yield was lowest in the intensively reared groups.

Similar rearing experiments have also been conducted with pigs. With them, the intensity of feeding had no influence on the onset of puberty. Heat symptoms were most distinct and the number of services per pregnancy lowest in those sows which were reared on a sparse diet. With a few exceptions weaknesses occurred in the legs of the sows which had received the high plane rearing.

Several studies have been made into the relationship between milk yield and fertility in cows. In the analysis of data from two Swedish A.I. associations, HENRICSON (1956) found that cows which manifested cystic ovaries had, on the average, a higher milk yield than contemporary stall mates which did not have the defect. The differences were, however, relatively small and do not allow any definite conclusions. More suitable material is available from the Danish Bull Testing Stations. These data have been analysed by TOUCHBERRY (1958) who could not show any relationship between butterfat yield during the first lactation and the number of inseminations for the second pregnancy. Casida and Chapman, investigating an American Holstein-Friesian herd, found no relationship between the occurrence of cystic ovaries and milk yield. On the other hand, it was shown that, when cows on the so called 'Official Test' were milked three or four times a day and specially prepared for high production, the frequency of cystic ovaries did increase. Here it may be appropriate to talk about 'cows under stress'.

Only if it is demonstrated that there exists a genetic correlation between high milk yield capacity and predisposition to reproductive disturbances and/or puerperal paresis is there any real reason for taking notice of this correlation in selection. There is, however, no well established evidence. To the extent that a phenotypic correlation exists, it may quite well be environmentally determined. In such a case, the most important action would be to improve the environment, including nutrition, until it corresponds to the genetic potential of the production capacity of the animals.

LENGTH OF GESTATION

Extreme variations in the length of gestation lead to a reduction in the viability of the offspring produced; and for this reason the genetic background of these varations will now be briefly dealt with.

Table 2:2 shows the mean and the variation of the length of gestation for different species and breeds of animals. It must be expected that outside the limits of the mean ± 2, or 3 times the standard deviation, the viability of the foetus at parturition is seriously reduced, the reduction of viability increasing with increasing deviation from the mean. It should be noted that the gestation period for single births is normally somewhat longer for male than for female foetuses, and is somewhat shorter for primiparous than for older mothers. There is also a seasonal variation; in cattle the period of gestation is somewhat shorter for calves born in the summer than those born in the winter. All these variations are, however, relatively small and of little importance. Because twin foetuses in cattle and horses are usually carried for a shorter time than single foetuses, there is some influence on their viability. There is also a tendency in multiple-birth animals for the period of gestation to be somewhat reduced with an increase in the number of foetuses.

The distinct differences in the gestation length of different breeds within the same species indicates that the variation is to some extent hereditary (Table 2:2). Studies carried out with data from cattle have shown that the differences in length of gestation within one and the same breed, and even in reciprocal crosses between different breeds, depends more on the genotype of the foetus than on the genotype of the dam. RENDEL (1959) made a study of approximately 5000 single pregnancies in the Swedish Red and White cattle and came to the result that 26–36 per cent of the total variation was determined by the genotype of the foetus whereas permanent differences between dams accounted for 11 per cent of the variation. The repeatability for successive pregnancies of the same cow was 17·2 per cent. It was further established that the period of gestation was, on average, about three days longer for cows with pure Ayrshire pedigree than for the closely related Swedish Red and White cattle (SRB). When SRB cows were mated with SRB bulls, the average gestation period was 1·7 days shorter than after mating to an Ayrshire bull.

Abnormally prolonged gestation in cattle has been reported by several authors, and the general conclusion has been that the defect is hereditary and depends on the genotype of the foetus. In a study of American Guernsey cattle, KENNEDY and co-workers (1957) found a number of cases where the length of gestation varied between 292 and 526 days. It was shown that in all the foetuses examined the hypophysis was underdeveloped and the defect was assumed to be under the control of a recessive gene in the homozygous condition. The foetuses were relatively small and stillborn. JASPAR (1959) made a study of 30 cases in Holstein-Friesian cattle in which the length of gestation was 20–28 days longer than normal and the foetuses had weights of up to 75 kg. Parturition was not possible without surgical help. In four cases Caesarian operations were carried out, but the calves were not viable. Hallgren has described 9 cases in which the period of gestation was 367–473 days and parturition was impossible without artificial means. The pedigree of the calves indicated that the defect was heritable.

SELECTED REFERENCES

HAFEZ, E. S. E. and JAINUDEEN, M. R. 1966. Intersexuality in farm animals. *Anim. Breed. Abstr.*, **34**: 1–15.
JOHANSSON, I. 1960. Genetic causes of faulty germ cells and low fertility. *Suppl. J. Dairy Sci.*, **43**: 1–30.
MAJALA, K. 1964. Fertility as a breeding problem in artificially bred populations of dairy cattle. I. Registration and heritability of female sterility. *Ann. Agric. Fenniae. Suppl.* 1. **3**: 1–94.
REID, J. T. 1960. Effect of energy intake upon reproduction in farm animals. *Suppl. J. Dairy Sci.*, **43**: 103–122.

10 Udder Development, Milking Rate, Yield and Composition of Milk

It is well known that, even under the same environmental conditions, different cows yield different quantities of milk. It has been possible by selection to specialise some cattle breeds for milk production and others for beef production. Among the dairy breeds there is great variation in the composition of milk; the Jersey for example gives a very high butterfat content, whereas the black-and-white Friesian has been known for its relatively low fat milk. Since the introduction of milk recording, however, there has been considerable success in increasing the butterfat of Friesians. All this indicates that the capacity for milk yield and composition are heritable traits.

In the early days of animal genetics, attempts were made to explain the variation in milk yield and butterfat content as being due to Mendelian segregation of a few pairs of genes. In Germany, VON PATOW carried out a comprehensive exercise with gene symbols in this respect, and at several American research stations crossbreeding experiments were carried out in order to study how these traits were inherited. This work is only of historical interest and will not be reviewed here. Milking capacity is one of the most typical quantitative characteristics of our farm animals. It is influenced by such a large number of genes and external factors that application of the classical gene analysis must be regarded as quite meaningless. There is only one possibility available at present, and that is to try to estimate the importance of different causes of variation by statistical analysis of available data. The aim must be to estimate the genetic part of the variation between individuals in a population (the heritability) instead of finding the number of contributing genes.

During the past twenty years increasing attention has been paid to the milking rate of cows, i.e. how quickly they can be milked, and the development of the udder, including the size and placement of the teats — traits which are of great practical importance in machine milking. In many places milking rate, udder development and teats are now measured in connection with milk recording in order to obtain a better basis for selection. The logical order would seem to be to start with a review of what is known about the variation and inheritance of these traits, although, of course, the yield of milk and milk solids is of still greater importance.

UDDER AND TEATS

Only 80–90 per cent of the milk present in the udder at the commencement of milking is extracted at any one milking. The amount of milk formed in the

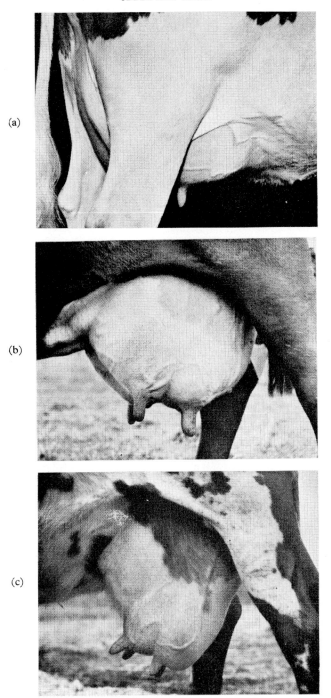

Fig. 10:1. Different types of udders: (*a*) the ideal 'dish-shaped' udder, (*b*) a rounded udder and (*c*) a pendulous udder. (JOHANSSON 1957)

udder during milking is so small that it can be ignored. It is not unusual for a high-yielding cow to give 20 kg of milk, or more, at a single milking, and only a large udder can contain such a large amount of milk. Together with the weight of the udder tissue itself ('empty weight'), the total weight before milking can be something like 40–50 kg. It is obvious that under such conditions the suspension system of the udder is subject to considerable strain.

The distinction may be made between a flat dish-shaped udder, a rounded udder and a pendulous udder. These are illustrated in Fig. 10:1. The dish-shaped udder is attached to the body over a large area, reaching well forward and well back. It is characterised by a moderate and fairly constant depth and can be likened to a rectangular dish or tureen. The rounded udder is shorter and rounder and has a greater depth than the dish-shaped udder. The pendulous udder is a faulty type of udder which normally does not become apparent until the third or fourth lactation, but thereafter it becomes progressively worse with increasing age. A short and tilted ('rear-heavy') udder often becomes pendulous, but seldom does a dish-shaped udder. When the rear half of the udder is much more developed than the front half, it usually takes longer to milk the rear quarters. This means either that the rear quarters are incompletely milked or that the milking machine is working on empty fore quarters for some time until the rear quarters are evacuated. Injury to the teats, which predisposes the udder to infections, may easily be a consequence. Pronounced pendulous udders present difficulties with machine milking and the udder and teats are continually liable to being trodden on or otherwise injured.

The shape of the udder is, to some extent, a breed characteristic though the variation is considerable, even within breeds. The dish-shaped udder is typical of the Ayrshire breed and fairly common in the Jerseys. The rounded udder is predominant in most other breeds, though Friesians have a relatively high frequency of pendulous udders.

To assess the proportions of the udder by eye alone is extremely difficult. It is true that it is possible to observe large deviations from the desirable type, but this does not give a satisfactory measure of the quantitative variation. Such a measure can, however, be obtained by collecting and weighing the milk from each quarter separately. From the results it is possible to calculate two udder indexes. The first, called the *front-to-rear index*, shows the part of the total milk produced from the two front quarters, and is thus a way of expressing the proportions of the front and rear of the udder. The other, called the *left-to-right index*, gives the contribution of the left half of the udder to the total milk, and is an expression of the symmetry of the udder. Investigations with several different breeds have shown that, on the average for a fair number of cows, the left and right halves of the udder yield practically the same amount of milk, whereas the rear half yields considerably more milk than the front half.. It has been shown further that the standard deviation of the front-to-rear index is almost double that of the left-to-right index, 6·2 compared with 3·3 (Fig. 10:2). A considerable variation can be found between the left and right front quarters or between the left and right rear quarters of individual cows, but it appears to be completely random whether it is the left or right quarter that yields the smaller amount of milk. Differences may be due partly to 'accidents' in the development of the mammary tissue and partly to infections of individual quarters; the latter reason is common

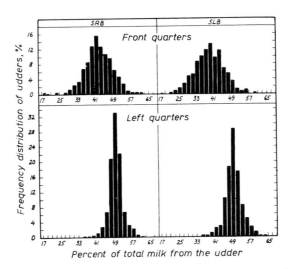

Fig. 10:2. Frequency distribution of the front-to-rear index (above) and the left-to-right index (below) of the udders of 569 cows of the Swedish Red and White breed (SRB) and 488 cows of the Swedish Friesian breed (SLB). (JOHANSSON 1957)

in older cows. The heritability of the left-to-right index is only slightly more than zero, but is quite high for the front-to-rear index, as the figures in Table 10:1 show.

Table 10:1 *Front-to-rear udder index; average and heritability within five dairy breeds*

Breed	No. of bulls	No. of daughters	Front-to rear index, av.	Heritability	Author	
Swedish Red and White	80	569	42.8	0.76	JOHANSSON	1957
Red Danish	201	2870	44.4	0.47	JOHANSSON	1957*
Danish Jersey	54	898	46.8	0.31	JOHANSSON	1957*
Danish Black and White	94	1418	43.5	0.49	JOHANSSON	1957*
Dutch Friesian	54	1122	44.5	0.50	POLITIEK	1961

* The heritability estimates for the Danish breeds are based on data from the progeny testing stations.

The following figures are an example of the inaccuracy with which even an experienced judge of dairy cattle assesses the udder proportions by scoring, compared with the results of quarter milking. The assessments and test milking were made at five-week intervals. The repeatability of the front-to-rear index was 0.90 and of scores 0.47. The correlation between indices and scores obtained at the same milking was 0.33. Because of the high repeatability of the front-to-rear index within the same lactation, it is not necessary to make more than one test milking in the first lactation in order to assess the udder proportions. In one experiment a study was made of the repeatability of the two udder indices when measured two lactations apart; the repeatability was found to be 0·6 for the front-to-rear index and 0·15 for the left-to-right index. The instability of the latter might be expected, as the 'original' variation is much less than for the front-to-rear index.

On average, the heritability of the front-to-rear index is about 0·5 (Table 10:1). There may, however, be some differences between breeds with respect to the

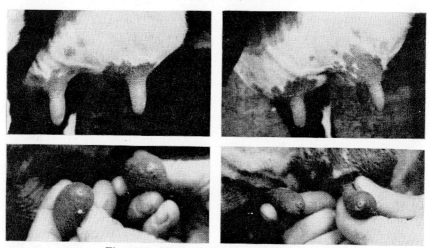

Fig. 10:3. Udders of Friesian identical twins.
Note the similarities in the shape of the udders and teats as well as the teat tips.
(JOHANSSON 1957)

genetic variation and heritability of this trait depending on the selection which has taken place within the breeds.

Observation on identical twins at Wiad, the Ruakura Experiment Station and in Germany show that not only the proportions between front and rear quarters but also the length of the teats, their shape and placement are highly heritable (Fig. 10:3). The similarities within pairs of identical twins and the differences between pairs is striking. Fig. 10:4 shows some of the deviations from the normal udder shape which in greater or lesser degree, occur relatively often. The same Fig. shows also a very poorly developed udder with pronounced conical rear teats.

Certain anomalies in the development of the udder and teats, such as the complete absence of a quarter, underdevelopment of both the front quarters, teats which have grown together etc., have been described in the literature. Sometimes several defective individuals have appeared in the same progeny groups which has been taken as an indication that the defect was due to monofactorial inheritance. There are also some defects which occur sporadically, e.g. 'blind' teats which have no orifice. It is likely that this particular defect is due to

(a)

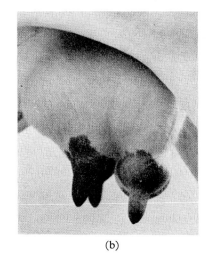
(b)

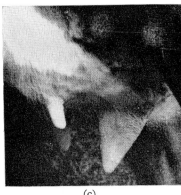
(c)

Fig. 10:4. Undesirable shape of udder and teats. (a) a rear-heavy udder, (b) distended milk cisterns, (c) conical rear teats. (JOHANSSON 1957)

temporary disturbances in development rather than to genetic factors. The defect may occur in one but not the other of a set of identical twins.

It has long been thought that the shape of the udder was in some way related to the shape of the pelvic region (rump). Investigations indicate that there is a slight relationship between the length of the pelvis and the length of the udder and also between the angle of slope of the rump and the slope of the floor of the udder. Cows with a sloping rump seem to have a tendency towards rear-heavy udders. The correlations, however, are slight and of little importance.

In connection with the collection of data on udder proportions in the Swedish dairy breeds, measurements were also taken on the length and diameter of the teats as well as the distances between the teats. On average, the front teats were approximately 1 cm longer than the rear teats, but the diameter was about the same (Fig. 10:5). Both the front and rear teats of the Swedish Friesians (SLB) are, on the average, about 1 cm longer and the diameter 2–3 mm larger than for the Swedish Red and White (SRB) cattle. The variation in both breeds is quite large.

In general, the distance between the front teats is considerably greater than that between the rear teats, which are often so close together as to cause difficulties in milking. A teat index was calculated, showing the distance between the rear teats as a percentage of the distance between the front teats. The average index was 44 for both the SLB and SRB breeds.

In herds milked by machine there often occurs a more or less definite eversion of the mucous membrane of the teat canal (Fig. 10:6). This is probably caused by a too high vacuum and/or by the teat cups being left on too long after the milk has been evacuated. In the more serious cases, the everted membrane becomes inflamed and sores develop. A study showed that pointed teats are predisposed to eversion of the mucous membrane, but that eversion never occurs in teats in which the orifice of the teat canal opens at the bottom of a funnel-shaped depression. It has been suggested that the funnel-shaped teat tip predisposes the udder to infections, in that the milk which remains in the depression serves as a breeding ground for bacteria, which then find their way into the udder. There is no evidence that this is the case. The funnel-shaped teat-opening, in any case, has the advantage that eversion of the mucous membrane does not occur. In a study of 22 pairs of identical twins at Wiad it was shown that the shape of the tip of the teats was very similar within pairs but that the variation between pairs was considerable. This indicates that the shape of the tip of the teats is genetically determined.

Supernumerary teats are much more common in some breeds than in others. More than 90 per cent of the supernumerary teats are located caudally to the four normal teats. The heritability of the number of supernumerary teats in the Swedish data was relatively low, only 0·23. This is in good agreement with the observations on identical twins, where it quite often occurs that one of the twins has one or more supernumeraries, but not the other twin. However, in an investigation with British dairy breeds (WIENER, 1962), the heritability of caudal supernumerary teats was estimated to be 0.63 ± 0.05.

The following figures show the heritabilities obtained from investigations with Swedish dairy breeds for udder indices and teats (within breeds):

	Heritability	Standard deviation
Teat length	0.98 ± 0.20	1·88 cm
Front-to-rear index	0.76 ± 0.12	6·19%
Distance between teats on right and left side	0.50 ± 0.22	2·63 cm
Teat diameter (at the base)	0.38 ± 0.22	0·77 cm
Teat index (teat placement)	0.36 ± 0.22	13·93%
Left-to-right index	0.08 ± 0.08	3·34%

The conclusion may be drawn that selection for shorter or longer teats, as well as for better proportions between the front and rear of the udder, can be expected to be very effective. The heritability figures for the length and breadth of the udder were found to be relatively low, but this may be due mainly to the fact that the error of measurement is very large. This is in the nature of things since there are no strictly definable points for measuring.

Investigations with pigs have shown that the number of teats is to a certain

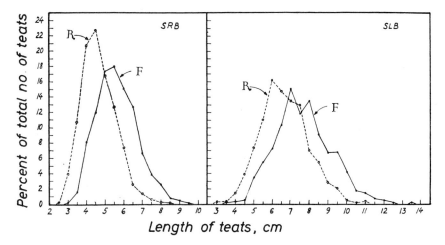

Fig. 10:5. Variation in the length of teats in the Swedish Friesian (SLB) and the Swedish Red and White cattle (SRB).
F = front teats; R = rear teats. (JOHANSSON 1957)

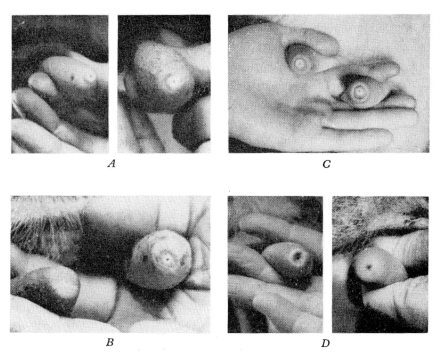

Fig. 10:6. Four different types of teat tips. (*A*) pointed, especially the one on the left with eversion, (*B*) flat, (*C*) saucer-shaped, (*D*) funnel-shaped. (JOHANSSON 1957)

extent heritable. The European wild pig has only 10 teats (5 pairs), whereas the Chinese pig has 14 (7 pairs). Several of the pig breeds in the Western countries have their origins in both of the above-mentioned wild types. On account of the comparatively high fertility of the English Large White breed and the Danish and Swedish Landrace, it is desirable that the sows have 14 functional and symmetrically placed teats. It is not unusual to find asymmetrically placed teats which are associated with small and sometimes non-functional mammary glands. WILLIAMS and WHATLEY (1963), in an investigation with American breeds, estimated the heritability of teat numbers in pigs to be 0.28 ± 0.04. It has been shown also that defects in teat shape, e.g. inverted teats or teats with a constricted base, are heritable and therefore can be eliminated by selecting breeding pigs, both boars and sows, which have normal teats.

THE MILKING RATE

Various devices have been used for measuring the milking rate of cows; a very simple one is shown in Fig. 10:7. The bucket is suspended from a spring balance and the total weight of the milk is read every 15 or 30 seconds throughout the milking. The author has used a similar device with the bucket divided into four sections, so that the milk from each quarter of the udder could be measured after a completed milking. Another type is the 'milkograph', where the weight of the bucket is continuously registered on graph-paper during the milking process. The latest invention in this field is probably the 'electronic four quarter milkograph' (Fig. 10:8) developed by the Alfa-Laval Co., Sweden. The milk flows through separate tubes, one for each quarter, into four containers which are suspended from an electric dynamometer. The load is measured by means of strain gauges which indicate a tension proportional to the load. The tension is converted to electric potential which is measured and recorded with an electronic measuring device. Each weighing can be separately tared, so that the graph shows the net weight of milk. The accuracy of the complete equipment corresponds to a maximum deviation of 0.25 per cent of the actual weight of the milk. This equipment has so far been used only for research purposes.

Fig. 10:9 shows udder evacuation diagrams, made by an 'electronic four quarter milkograph', for two different cows: A with a yield of 13.7 kg milk at a single milking and B with a yield of 10.2 kg. Cow A has a poorly balanced udder, producing only 34 per cent of the total yield in the front quarters. The corresponding figure for cow B is 43.5 per cent but this cow is a very slow milker; the milking time is $9\frac{1}{2}$ minutes. In Fig. 10:10 the weight figures for the four quarters of cows A and B respectively are added and the result is shown as the milk flow rate per 30 seconds, as well as by the cumulated weight of the evacuated milk. For comparison, Fig. 10:12 shows udder evacuation curves obtained by the simple device referred to above (Fig. 10:7). CLAESSON (1963) found that when the rate of milk flow is measured for each quarter separately with the electronic milkograph, the rate is practically constant from the time the flow has started until the quarter is very nearly emptied. The summation curve for all four quarters usually deviates more or less from a straight line, because the milk

Fig. 10:7. Milking machine suspended from a spring balance on a tripod stand. (KEESTRA 1963)

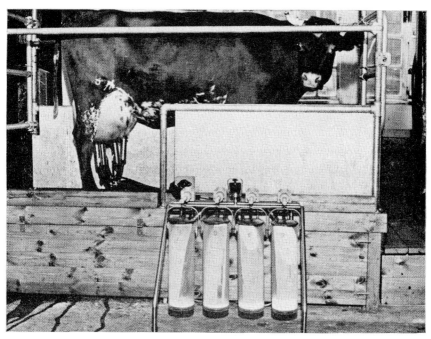

Fig. 10:8. The electronic four quarter milkograph. Alfa-Laval Co., Sweden. Quarter evacuation graphs are shown in Figs. 10:9 and 10:10.

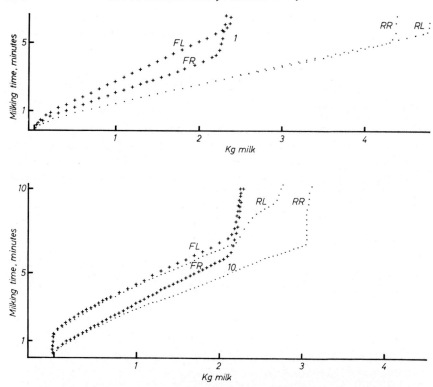

Fig. 10:9. Cumulative milk ejection curves for two cows, A (above) and B (below), obtained by the Alfa-Laval 'electronic four quarter milkograph'. FL = front left; FR = front right; RL = rear left; RR = rear right quarter. (Courtesy of Dr. O CLAESSON)

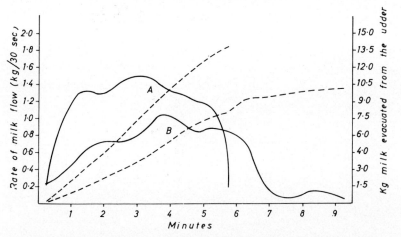

Fig. 10:10. The milk ejection curves for the different quarters, shown in Fig. 10:9, have been added together to show the rate of milk flow from the whole udder of cows A and B expressed as the flow per 30 seconds, as well as the cumulative amount of milk obtained at the various stages of milking. ..... the rate of milk flow (kg/30 sec) ——— the cumulative weight of the evacuated milk at each stage of the milking process.

flow neither starts nor ceases at the same time for all four quarters (Figs. 10:9 and 10). The use of a milkograph with separate automatic registration of the milk flow from the front and from the rear quarters of the udder may be recommended for field tests. Data may then be obtained as to the total yield together with the milking time for the front and rear udder, from which the average rate of flow and the front-to-rear index may be calculated. The graph shows the exact time when the milk flow starts and ceases. With this device the peak flow and the average rate of flow are equally good measures of the rate of milking; but without automatic registration the peak flow is generally a somewhat better measure than the average flow because it is relatively independent of how quickly the flow commences when the teat cups are put on and it is entirely independent of how accurately the time of flow-cessation is read.

All other things being equal, the rate of udder evacuation is determined mainly by the resistance offered by the teat canal to the rate of flow. However, the effectiveness of the 'let down' reflex also plays a part, as well as the length and diameter of the teats. The teat canal can have a different sized lumen in different individuals and the sphincter muscle can also vary in its tension. Damage to the mucous membrane of the teat canal or pronounced eversion of the teat canal can restrict the flow of milk from, originally, normal teats.

BAXTER and co-workers (1950) at the NIRD, Reading, studied the rate of flow in ordinary machine milking and in milking with a cannula inserted into the teat canal so that the resistance was standardised. During milking with the cannula the rate of flow was practically the same for all the cows studied. At increased vacuum there was a greater increase in the rate of flow in milking without the cannula than with the cannula; which indicates that the teat canal opened wider at high than at low vacuum. Other investigations have shown that with the cannula it is possible to milk difficult milkers almost as quickly as easy milkers.

By measuring the diameter of the teat canal with a simple instrument, where no consideration of the tension of the sphincter muscle was possible, a correlation of 0·30 between the diameter of the canal and peak flow was obtained for 68 first-calf heifers. The correlation for cows of varying age was 0·50. The correlation is considerably higher when the cows are at the peak of their lactation (0·59 and 0·67 for first and subsequent lactations respectively) than towards the end of the lactation (0·27 and 0·29 respectively). The size of the teat canal is thus a limiting factor for the rate of flow during milking. (JOHANSSON and MALVEN, 1960.)

The neuro-hormonal udder evacuation ('let down') reflex has been described earlier (p. 28). It is a conditioned reflex, the course of which depends on certain impressions, communicated by feel, sight and hearing, which the cow has learnt to associate with milking. The evacuation reflex functions best if the cowman always follows the same routine at milking and if the cow is in harmony with the environment. Large deviations from the usual routine, as well as disturbing external factors or harsh treatment, mean that the reflex does not function normally and that evacuation of the udder will be more incomplete than usual. Sensitivity varies in different individuals; the nervous cow is always more or less difficult at milking. If a cow becomes used to a vigorous massage of the udder towards the end of milking, it can happen that she does not let down the milk properly until this second reflex stimulation occurs in connection with the massage. The amount of strippings thus becomes unusually large.

In Norwegian, Swedish and Dutch investigations, a negative relationship has been demonstrated between the peak flow and the diameter of the teats as well as the length of the teats. This would seem to be due, at least partly, to the fact that the same teat cups were used for both large and small teats.

Apart from the diameter of the teat canal, the effectiveness of the evacuation reflex and the size of the teats, the rate of flow is influenced primarily by the yield and to some extent by the age of the cow. Such external factors as the vacuum and the pulsation rate of the milking machine have considerable influence, but these factors can easily be standardised and thus eliminated as causes of variation. The vacuum is normally 33–35 cm Hg with a pulsation rate of 40–50 per minute. As already mentioned, the rate of udder evacuation can be considerably increased by increased vacuum, but high vacuum is not recommended because of the risk of teat injury, especially if the machine is not well attended and operates for some time on the empty udder. Normally the time for the suction and pressure phases of the machine are of equal duration, i.e. 1:1. If the suction phase is made three times longer than the pressure phase, a more rapid milking can be achieved. The same thing is possible by increasing the pulsation rate. Just how these two methods affect the health of the udder is not yet clear.

The rate of flow at four milkings, measured at two-month intervals, was determined for each cow in eight progeny groups of Friesian heifers, the first time being about one month after calving. The correlation between the yield of milk at the test milking, and the peak flow was calculated for between and within cows, whereby values of 0·64 and 0·45, respectively, were obtained. In another investigation with 91 cows of varying age, the correlation coefficient between cows was 0·54. In this latter study the udder pressure, which was measured in the teat cistern, was found to be mainly determined by the yield of milk.

On account of the close relationship between the yield of milk and the rate of flow it is usual to correct the peak flow and the average flow for milk yield in order to obtain a better basis for comparison of the rate of milking of different cows and at different milking times. The relationship between the rate of flow and milk yield of a large number of cows is curvilinear. If the cows are divided into groups according to the peak flow, different regressions are obtained for the different groups as shown by Fig. 10:11 (ANDREAE, 1963). Apparently the peak flow decreases with increasing resistance of the teat canal. Correcting the peak or average flow for milk yield will probably favour slow-milking cows. If it were possible to develop a method of measuring the resistance of the teat canal to the flow of milk during milking, a measure of the rate of milking would be obtained that would be independent of both the yield of milk and the method of milking. It has been shown in various investigations that both the peak flow and the average flow during milking vary quite appreciably between different herds because of the differences in the milking routine applied.

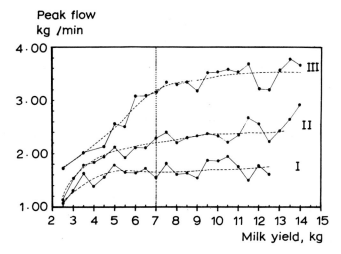

Fig. 10:11. Average peak flow with different milk yields in the period 3rd to 22nd week of lactation.
Three groups of cows classified according to their milking rate at 6–8 kg daily yield.
Group I: 23 cows with less than 2 kg peak flow.
Group II: 26 cows with 2·0–2·5 kg peak flow.
Group III: 22 cows with peak flow higher than 2·5 kg. (ANDREAE 1963)

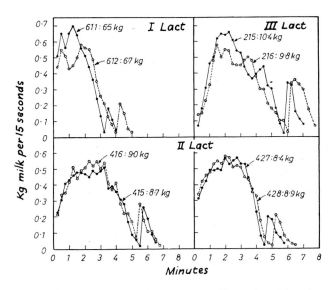

Fig. 10:12. Udder evacuation curves for four pairs of identical twins when both members of a pair have yielded about the same amount of milk per milking. (JOHANSSON 1961)

Tests for rate of milking should preferably be made at approximately the same stage of the first lactation (2–5 months after calving). In later lactations it is more likely that individual quarters or teats have been damaged by infection or injury thus influencing the milking rate and the front-to-rear index.

Studies concerned with the heritability of rate of milking have been conducted partly on twin cows and partly by field trials. Fig. 10:12 shows the evacuation curves for four pairs of identical twins, where both twins of each pair yielded about the same amount of milk. The similarity within pairs is striking in comparison with the differences between pairs. There is good agreement with the results from the twin experiments at ABRO (Edinburgh) and Ruakura. At the former institution the heritability, in the broad sense, of peak flow was estimated to be 0·9. The data were obtained from 25 MZ and 23 DZ twins. At the Ruakura station several series of milking investigations have been carried out. From tests of MZ and DZ twins, the heritability of peak flow was estimated to be 0·7 and in another experiment, in which 138 daughters (half sibs) from 69 sires were tested under uniform conditions, the estimate was 0·76. An interesting observation in the Ruakura split-twin experiment is that animals with low flow rates were affected by poor milking routine to a greater extent than were animals with high flow rates.

The most extensive studies of the rate of udder evacuation under field conditions have been made in Holland by POLITIEK and KEESTRA. In both studies the heritability was estimated from half sib correlations within groups of daughters from A.I. bulls. The milking tests were made on first calvers, as a rule within five months of calving. The following results apply to measurements at one milking per cow (Table 10:2).

The conclusion may be drawn that the heritability of the trait 'rate of milk flow', or rate of udder evacuation, determined at single milkings under field conditions is as high as 0·50–0·60. It should therefore be rather easy to improve this trait, as well as udder proportions, by selection. The most important measure for this purpose would be to consider milk flow rate and udder proportions of the dams when the young bulls are selected for breeding, and to include these traits in the regular progeny-testing of bulls.

DODD and NEAVE (1951) have demonstrated a clear relationship between the rate of milking of cows and their susceptibility to udder infections. Wide teat canals apparently facilitate the entry of infectious organisms into the udder. This

Table 10:2 Heritability estimates for rate of milk flow during milking

	KEESTRA (1963)		POLITIEK (1961): heritability		POLITIEK (1961)	
	Repeatability from one milking to the next	Heritability: 17 A.I. bulls, each with about 25 daughters	24 A.I. bulls, each with about 23 daughters	30 A.I. bulls, each with about 19 daughters	Mean ($\bar{x}$) and standard deviation (s), kg (1058 first calvers)	
					$\bar{x}$	s
Peak flow	0·82	0·65	0·74	0·54	2·04	0·61
Average flow	0·85	0·56	0·81	0·64	1·63	0·53

is probably due to the inability of the sphincter muscle of the canal to close properly when the pressure is released, rather than to the diameter of the canal. If this is the case, then cows with weak sphincter muscles, recognised by leaky teats when the udder is full, should be rejected; and efforts made to produce cows with a good rate of udder evacuation combined with efficient teat sphincters. There is no point in recommending the selection of cows that are hard to milk in order to achieve resistance to udder infections.

THE MILK AND FAT YIELD OF COWS

In discussing the milk production of dairy cows the first question which presents itself is how this is to be measured in terms of the quantitative yield and the period of time during which it is produced.

GAINES (U.S.A.) maintained that the best measure of the 'work' the cow carries out in converting food to milk is the energy content of the total milk yield. Accordingly, he suggested a standardisation of the milk produced on the basis of the calorie content, and he chose milk with a fat content of 4 per cent as the unit. The yield of fat-corrected milk (FCM) is calculated according to the formula $0.4\,M + 15\,F$, where M is the yield of milk and F the yield of fat. This conversion has been used widely, not least in the Scandinavian countries. The yield of butterfat, however, is almost as good a measure of the total calorie production as FCM. This is due, firstly, to the fact that, on the average, approximately half of the calories are associated with the fat and, secondly, that the butterfat content of the milk is rather strongly correlated with the solids-not-fat content ($r = 0.5$). From a genetic point of view, however, it is of more interest to study the variation in the yield of milk and the milk composition separately. According to the analyses which have been carried out on different populations, the correlation between milk and butterfat yield per lactation is very high — about 0.9.

In northern and continental Europe it has been usual to report the milk and butterfat yield of cows per recording year of 365 days. For economic calculations of the profitability of the production this is no doubt convenient, but from a breeding point of view it has several disadvantages. The calvings are spread over all months of the year, and seldom does the lactation period coincide with the recording year. The first recording year will be incomplete for many of the heifers and the second year will include a greater or lesser part of the first lactation. If corrections are to be applied for the age of the cows, the calving season and/or the length of the calving interval, then it is decidedly easier to calculate from the lactation period than from the recording year. The length of the lactation period is highly dependent on how soon the cow becomes pregnant again after calving. It has therefore been internationally agreed that only the yield during the first 305 days after calving be included. Unless some other period is mentioned, the following discussion will be concerned with the yield during this standard period.

Since progeny-testing of bulls should be carried out as early as possible in their lifetime, the use of still shorter periods, e.g. 70 and 180 days, has been tested.

Statistical analyses have shown that satisfactory results can be obtained with such part-lactations. In an American investigation, the heritability of milk yield during the first 100, 200 and 300 days was estimated to be 0·30, 0·36 and 0·42 respectively. In England, ROBERTSON and his co-workers have studied the same problem. The heritability of the production in the first 70 days was 0·36 and for 305 days production, 0·43, the calculations being made on first lactation heifers.

From an economic point of view, it is the average daily yield during the calving interval that is important. Cows that have a high daily yield during the first months of the lactation and then rapidly dry off are undesirable. When recording takes place only once a month, the estimated yield during the first months of the lactation is subject to a high error variation, due to the fact that the first recording day occurs at varying time intervals from the date of calving. Analysis of a large amount of data by VAN VLECK and HENDERSON (U.S.A.) showed that the yield during the first months after calving was relatively variable, compared with the following months, probably because cows are more sensitive to their environment during the early part of the lactation than when the peak of the daily yield has been passed. When the milk yield was recorded once a month, the repeatability of the yield on the recording day for successive lactations was highest for the fifth, sixth and seventh months of lactation. The repeatability of the cumulative yield had almost reached its maximum when it was calculated for the first three months of lactation. The heritability was low for the first month, but reached its maximum for the cumulative yield during the first 3–4 months of lactation; the same was true for the genetic correlation between the part-lactation and the total-lactation yield. The conclusion may be drawn that progeny-testing of bulls could quite well be based on part-lactations of the daughters but that at least three months of the lactation should be included.

In practice, milk and butterfat tests are made at certain intervals and the 'error of measurement' increases with the length of the interval, compared with records made by daily tests. JOHANSSON (1942) found, in an analysis of 100 lactation records, that this error (standard deviation of differences between estimated and actual yields) was 2·7 per cent of the actual 300-day records for milk yield when the tests were made monthly and 5·6 per cent for the bi-monthly tests. The variance within cows of the age-corrected lactation yield of milk or butterfat is about 60 per cent of the total variance within herds, and may be considered as a measure of the error in estimating the production level of individual cows. This 'error variance' is about 4 per cent greater when the tests are made at monthly intervals, and about 11 per cent greater with bi-monthly tests, than with daily milk recording. The 'error variance' of the fat content of the milk increases by about 20 per cent when monthly tests are made and 35 per cent with bi-monthly tests, compared with daily tests. However, with monthly tests the variance within cows is only about 1/3 of the total variance within herds. When it is a question of culling individual cows on the basis of their first lactation yield, the bi-monthly test is distinctly inferior to the monthly test; but in the progeny-testing of bulls the decrease in accuracy of individual records can be counteracted by increasing the size of the daughter groups. The increasing accuracy of milk records achieved with more frequent testing has, of course, to be weighted against the increase in cost.

Some non-genetic causes of variation in yield

The milk and butterfat production of cows per lactation period is influenced by many different factors, some of which are randomly distributed among the different individuals in the population, whereas others are systematically distributed, and so affect certain groups of individuals more than others, e.g. the environmental differences between herds and between different periods of time. Causes of variation such as the age of the cows, the calving interval and length of the dry period are here regarded as environmental, even if the reaction to them can show individual differences. A number of causes of variation, acting singly, have very little effect, but when many factors work in the same direction the result may be considerable.

Age of cows. Generally speaking, the production capacity of cows increases at a declining rate, until the body is fully developed at about 6–8 years of age. After this the capacity decreases at an increasing rate as the ageing of the body proceeds. An approximate picture of this is presented in Fig. 10:13, which is based on data from eight Swedish Friesian herds. Rapidly-growing animals reach their maximum capacity earlier than slower-growing animals, but they also age more rapidly. This appears to be true both of rapid growth rate due to heritable factors and that due to a high level of nutrition during the growing period.

The increase in yield during the first lactation with increasing age at calving is of special interest. Fig. 10:14 is based on the production figures of approximately 47 000 Swedish Red and White heifers with differing ages at first calving. The heifers have been divided into five groups according to the contemporary average yield of the herd. It will be seen that the increase in yield with increasing age at calving (between 23 and 38 months) is appreciably greater in the high-producing than in the low-producing herds. There is a linear relationship between the age at calving (within the interval mentioned) and the milk yield in the first lactation. However, when the average is calculated for all heifers calving within the interval 20–49 months, the relationship is curvilinear. When the age at calving exceeds about three years, there is no further increase in yield for the increased age at calving. Apparently the udder attains full development only after about three pregnancies. Production capacity in the first three lactations depends both on the age of the cow and on the number of calvings.

Length of the calving interval. A short calving interval leads to a lower milk yield in both the current and the succeeding lactation, whereas a long calving interval operates in the opposite direction (Fig. 10:17). The optimum length of the calving interval probably lies between 12 and 14 months since this seems to give the highest average yield over a period of several years. In assessing the individual cow's production during a lactation, consideration should be given to the length of the calving interval. As was mentioned previously, the length of calving interval has a very low heritability, and the effect is therefore evened out when calculations are made on the average yield for several successive lactations of the same cow or a number of daughters of a bull.

Length of dry period. This is rather highly correlated with the length of the calving interval; a short calving interval leaves little room for a long dry period.

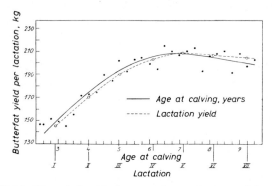

Fig. 10:13. The relationship between the cows' age at calving and their butterfat yield per lactation. Swedish Friesians. (JOHANSSON)

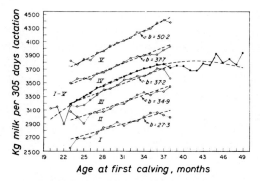

10:14. The relationship between the age of cows at first calving and their milk yield in the lactation which follows.

Within the interval 23–38 months, the cows have been divided into 5 groups according to the mean yield of the herds, and the regression (b) for milk yield on calving age has been given for each group. The average yield of all heifers and for all calving ages is shown by a broken line. (HOFMEYER 1955)

Studies carried out on data from the Swedish dairy breeds have shown that the optimum length of the dry period is 6–7 weeks. A shorter dry period leads to a lower yield in the following lactation. All other things being equal, the length of the dry period is a measure of the cows' persistency for milk production during the lactation. The heritability is about 0·25.

Length of milking interval. This has some influence on the yield of milk. Previously it had been thought that three milkings per day at approximately equal intervals would result in a 10–20 per cent greater yield, than twice-a-day milking, and that a further 5–8 per cent increase could be achieved by milking four times a day. Later investigations, however, partly with identical twins and partly with group experiments, have shown that the figures quoted considerably overestimate the effect of increased number of milkings. With optimum environmental conditions, a greater effect can be expected than when conditions are unfavourable. In twin research at the Wiad station it has been found that with two milkings a day, an extension of the night interval to 15 or 16 hours, and a consequent shortening of the day interval, resulted in only 2–3 per cent lower milk yield during the first 280 days of the lactation than when the intervals between milkings were of equal duration. When milking took place only once a day, there was an average reduction in yield of 50 per cent in the first lactation and 40 per cent in the second lactation. The members of the different twin pairs reacted very differently in their reduction in yield during the first 280 days of the first lactation; the reduction varied from 25 to 85 per cent and was relatively less in high-producing than in low-producing twins.

Season of calving. The conditions prevailing in north-western Europe are such that the season of calving has rather a large effect on the lactation yield, as illustrated in Fig. 10:15, which is based on the same data as Fig. 10:14 (the first lactation of about 47 000 cows of the Swedish Red and White breed). Heifers which calve during the period October to January yield on the average, about 12 per cent more milk than those which calve during May to July. The difference is more pronounced in high- than in low-producing herds.

Management. Such factors as nutrition and husbandry, as well as reproductive disturbances, udder infections, etc., have a great influence on the yield. The question of the relative importance of heredity and environment to the variation of herd averages will be discussed later.

Persistency of yield (shape of the lactation curve)

If the total lactation yield is the same, cows with the most persistent daily production are usually considered to be the most economically advantageous. This is because they can manage with less concentrated food in the winter and are more able to satisfy their nutritional requirements on pasture in the summer than cows which have a high daily yield during the first months of the lactation but then quickly dry off. Nowadays this has not the same importance as previously, because the price difference between roughage and concentrates has altered.

The shape of the lactation curve (persistency of yield) is, unfortunately, difficult to measure satisfactorily, and furthermore, it is influenced by several non-genetic factors, primarily the age of the cows, the level of nutrition and the calving interval. The effect of age and length of the calving interval are illustrated in Figs. 10:16 and 10:17. For practical purposes a simple proportion can be used, e.g. the yield of milk during the second hundred-day period after calving as a percentage of the yield in the first hundred days ($P_{2:1}$). This figure shows how much the yield has declined from the first to the second hundred-day period. In monthly tests, the average yield during the first three recording days after calving and the average yield during the next following three recording days can be calculated and the latter simply expressed as a percentage of the former. The percentage figure obtained ($P_{2:1}$) is not affected by the length of the calving interval. According to JOHANSSON and HANSSON (1940), $P_{2:1}$ has a heritability of about 0·20, ($\bar{x} = 72\cdot5$, $s_x = 12\cdot6$). Those who practise breed selection for the shape of the lactation curve can therefore expect to obtain some effect, even though it will be rather slight.

Several studies have shown that the total yield during the lactation period is determined much more by the maximum yield than by the shape of the lactation curve. In an analysis of data from a Swedish progeny-testing station, about 60 per cent of the variation in yield during 250 days of lactation was due to the highest monthly yield and only 11 per cent to the persistency.

The heritability of the lactation yield (305 *days*)

From the foregoing it will be apparent that milk yield is a highly modifiable trait. Because of the large differences in management, the heritability must be calculated within herds and for relatively short periods of time in order that reliable results can be obtained. It is of interest, however, to study the extent to which the differences in the average production of different herds are genetically determined. This can be done with data from insemination centres, where the same bull has daughters in several herds of varying levels of production. It can also be done with the aid of identical twins, where each twin of a pair is allotted to a different production level. Several such studies have been carried out.

Herd differences and trends within herds

On the basis of data from the Danish bull-testing stations at Själland and Fyn an analysis was made of the relationship between the yield of heifers at the testing stations and the average yield of the herds from which the heifers were delivered. In addition, a study was made of the relationship between the yield of half-sisters of the station heifers in the recorded herds and the contemporary average yield of the herds in question (HOFMEYER, 1955). The results are presented in Fig. 10:18. The yield of the heifers at the testing stations, calculated for 250 days, increased up to a herd average of about 160 kg butterfat, but there was no indication of any further increase with rising herd average. The increase in yield at the stations up to herds with about 160 kg was probably due, largely, to the poor condition of the heifers from the low-yielding herds when they were delivered to the stations rather than to genetic differences in comparison with heifers from the high-yielding herds. The recorded herd half-sisters showed a steady increase in yield with increasing herd average; the regression was 0·714. The results indicate that the genetic differences between the Red Danish herds, which delivered heifers to the testing stations at Själland and Fyn, are very small, in spite of the fact that the differences in yield of these herds are large.

Mention has already been made (p. 149) of a New Zealand experiment, in which 120 pairs of monozygous twins were split at 10 days of age between 20 high- and 20 low-producing herds. Within both groups the twins yielded somewhat less than their contemporaries in the herds. The difference between

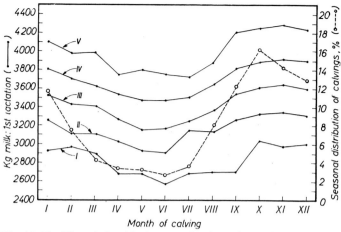

Fig. 10:15. The relationship between the time of year for first calving and the milk yield in the lactation which follows.
The broken line shows the distribution of calvings throughout the year. The figure is based on the same data as Fig. 10:14 with the same method of classification. (HOFMEYER 1955)

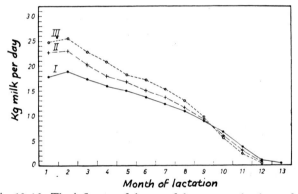

Fig. 10:16. The influence of the age of the cows on the shape of the lactation curve. Mean of approximately 100 Swedish Friesian cows. I first, II second and III third lactation. (JOHANSSON 1961)

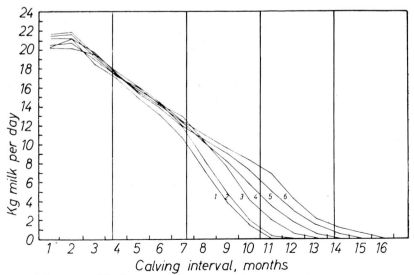

Fig. 10:17. The influence of the length of the calving interval on the shape of the lactation curve.
The curves refer to the following lengths of calving interval: (1) 300–330 days, (2) 330–360 days, (3) 360–390 days, (4) 390–420 days, (5) 420–450 days, (6) 450–480 days. (JOHANSSON and HANSSON 1940)

the twin heifers in the two groups was almost exactly the same as the difference between the two groups of herds in respect of milk yield; but where butterfat content and protein were concerned, the twin groups differed much less than the herds. The conclusion was that there were genetic differences between high- and low-producing herds only in the composition of their milk and not in the yield of milk. Before the second calving, 32 of the MZ twin pairs were moved to the Ruakura experimental station and were kept there during the second lactation

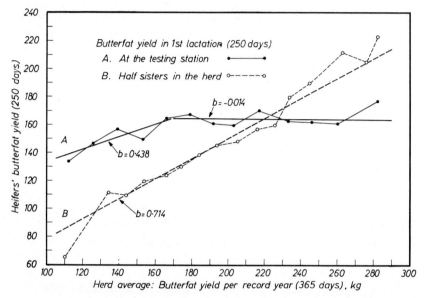

Fig. 10:18. The influence of the herd environment on the butterfat yield in the first lactation (250 days) of Red Danish cows at Själland and Fyn. *A* The yield at the testing stations in relation to the average yield in the herds where the heifers were reared; *B* The corresponding average yield of herd recorded half-sibs to the station heifers. (HOFMEYER 1955)

on a relatively high plane of nutrition. On the average, both twins of the pair yielded the same amount of milk; which showed that the plane of nutrition during the growth period and first lactation did not influence future production capacity. In another series of experiments carried out at the same time, each of the 20 high- and each of the 20 low-producing herds delivered 6 heifer calves to Ruakura, where they were reared and produced under identical conditions. The dams of the heifers had a yield as near as possible to the average of the respective herd. At Ruakura, the average milk yield of the heifers from the two herd groups was practically the same, but the butterfat content differed by nearly half of the corresponding difference between the herd groups. The results of this series of experiments are in complete agreement with the results of the twin experiment. The herd differences in milk yield were completely determined by the environment, whereas approximately 50 per cent of the differences in butterfat and protein content were genetic in origin. Statistical analysis of data from recording societies in England and the U.S.A. indicates that 10-20 per cent of the variation between herds in milk and butterfat yield were genetically determined.

In an A.I. association where the semen of different bulls is largely distributed at random to the herds associated with the bull centre, it can be reckoned that the variation in the average genotype of the herds almost completely disappears during the course of a few cow generations. Of course it is possible to apply different selection intensities among the females in the different herds, but the possibilities of thereby differentiating the herds is relatively small, since the greater proportion of the heifer calves born must be reared for herd replacements. After an insemination centre has been in operation for some ten years, it is to be expected that the associated herds are to a large extent genetically similar, that is, apart from the random variation, which would be most important in the smaller herds. If certain herds are consistently selected to receive the semen from the best of the progeny-tested bulls, then the result can be quite different. When natural service is practised, the herds may be differentiated due to the choice of bulls, though even here the genetic variation is much less than the differences in the herd averages indicate.

Another important fact in this connection is the trend of yields during the course of a number of years. Fig. 18:1 shows the trend of milk yield and butterfat content of all the recorded cows in Sweden from 1910 to 1965. During this period the yield per cow has increased from about 2850 kg to 4870 kg and the butterfat percentage from 3·4 to 4·15. The shortage of feed during the first and second World Wars had a definite effect on milk yield but hardly any on the butterfat content. Particularly since the end of World War II, there has been a considerable increase in yield. In different herds, however, the trend is very different.

The question to what extent the trend in a particular herd is genetic or environmental in origin is of great importance both in heritability studies and in assessing the breeding value of the animals. Attempts have been made to clarify the problem with the aid of advanced methods of statistical analysis, but only with partial success (cf. p. 451). A further development of the use of deep frozen semen, kept in storage for a number of years, would facilitate such investigations. In studying the heritability of milk yield the long term trends are usually eliminated.

The heritability within herds

The heritability of milk and butterfat yield is always calculated 'within herds' for the reasons just mentioned and in a manner described in more detail in Chapter 4, pp. 121-130. The effect of trend within herds is usually eliminated by operating with deviations of individual cows' records from the contemporary herd average. In the past, the heritability estimates have usually been based on regression of daughters on their dams, within groups of daughters from the same sire; but since data from A.I. herds became available, the common procedure is to use the correlation within paternal half sib groups. This has the advantage that the contemporaneity within and between progeny groups can be increased compared with that within and between daughter-dam pairs (cf. Chapter 5, p. 154).

Heritability studies have been carried out on data from recording associations in several countries and from different breeds. In general, practically the same result is arrived at by using butterfat yield as milk yield, when based on comparable records, usually the first 305 days of the lactation. Possibly the heritability figure is a little lower for butterfat yield than for milk yield, but the difference is insignificant and no distinction will therefore be made here. Some investigations seem to indicate that the heritability is higher in high-producing than in low-producing herds, but this does not appear to be general. From European data, and recently also in American investigations, it has been found that the heritability is higher for the first lactation yield than for any of the following lactations. In an analysis of data from Swedish dairy breeds, including 1840 daughter-dam pairs where each cow had completed three lactations, the following heritabilities were estimated by JOHANSSON (1955): First lactation 0.33 ± 0.050, second lactation 0.10 ± 0.047, and third lactation 0.24 ± 0.044. The heritability was estimated to be 0.21 for the average yield in the first three lactations. Similar results have been obtained by DEATON and MCGILLIARD (1964). In this case the first record of the daughter was used as the dependent variable and the various records of the dams were independent variables. Each record was expressed as a deviation from the average of lactations begun in the same calendar year by all other cows in the same herd and breed. The analysis comprised 904 registered Guernsey cows and 1592 Friesians with their corresponding dams. The following correlations were obtained:

Dams' records	Correlation with daughters' first record
Guernsey	
First record	0.149
Second record	0.075
Third record	0.081
First three	0.149 (multiple correlation)
Holstein	
First record	0.256
Second record	0.145
Third record	0.154
First three	0.259 (multiple correlation)

Deaton and McGilliard conclude that 'the first record gives essentially as reliable an estimate of a cow's breeding value, measured by the daughters' performance, as does an appropriately weighted combination of multiple records'.

The low heritability figures obtained for milk and butterfat yield in the second lactation is probably due to the fact that the condition of the cow is more variable at the second than at the first calving. Before the second calving different lengths of the dry period come in as a source of variation.

Some examples of heritability estimates for milk and butterfat yield obtained in various investigations are presented in Table 10:3, together with corresponding figures for butterfat content for comparison.

Table 10:3 Results of heritability studies of milk yield and butterfat content (305 days)

Breed	Yield of milk or butterfat	Butterfat content
Ayrshire, U.S.A. (TYLER and HYATT, 1947)	0·31	0·55
Ayrshire, Scotland (MAHADEVAN, 1951)	0·31*	0·56*
Holstein Friesian, U.S.A. (TABLER and TOUCHBERRY, 1959)	0·27	0·56
Dutch Friesian, Holland (FATTAH-EL-SHIMY, 1957)	0·35*	0·76*
Swedish Friesian (JOHANSSON, 1955)	0·30*	0·59*
Swedish Red and White (JOHANSSON, 1953)	0·39*	0·68*
Six British dairy breeds (RENDEL et al., 1957)	0·43	0·43

* Estimates made only on the yield in the first lactation.

By way of summary, it can be said that the heritability of milk and butterfat yield during the first lactation, estimated from field testing records, is in the vicinity of 0·35. If the average heritability is expressed for a lactation of any order the figure lies between 0·20 and 0·30. The heritability of butterfat content is appreciably higher, on the average, 0·5–0·6. The repeatability coefficient (within herds) for milk yield in the first three or four lactations of the same cow is 0·40–0·50. The repeatability between successive lactations is somewhat higher than, for example, between the first and third or the first and fourth lactations. The fact that the repeatability is only slightly higher than the heritability indicates that, with the breeding methods which are used in practice, the genetic variation between individuals is mainly of an additive nature.

Several heritability investigations have been carried out using data from the Danish bull testing stations. They have resulted in considerably higher values of h^2 for milk and butterfat yield in the first lactation (about 0·6), than when data from field testing records were used. This has aroused a great deal of surprise and different hypotheses have been suggested to explain the differences. On the whole, it is agreed that particular environmental differences have made themselves apparent even between the progeny groups tested at the same time at the same station, so that the heritability estimate has thereby been subjected to a certain amount of inflation. It has been suggested that in order to avoid such environmental differences, the daughters of the bulls under test at the same station ought to be randomly placed in the stalls, instead of in groups and that during the testing, the personnel should be unaware of which bull is sire to the different heifers. That the heritability figures were higher at the testing stations than on the farms is to be expected, partly on account of the greater efficiency of the recording and the better adjustment of the feeding to the individual requirements, and partly on account of the contemporaneity within and between groups at the same station. But one would hardly expect the station estimate to be twice as high. It seems peculiar that the variation in yield between cows within progeny groups is only slightly higher at the stations than when the same bulls are progeny tested in the field while the variation between progeny groups is several times greater. The relatively small variation within groups, however, can be explained simply on the basis of the contemporaneity of the lactations. The 'inflation' which is involved in estimating the heritability of milk yield as a result of experiments with identical twins has been dealt with earlier (pp. 149–156). Finally it must be emphasised here that neither estimates of heritability from identical twin experiments nor those from bull testing stations can be directly applied in practice, where the breeding value must be based on farm records made in different herds and at different times.

In practical breeding, information is required about the probable effect of the selection. This can be obtained taking a sufficiently large random sample of dam-daughter pairs from the population in question and computing the regression of daughters on dams which provides information about the selection effect that can be expected.

MILK COMPOSITION

Up to the end of the 1940s, interest in milk composition was concentrated mainly on the butterfat content, which, by systematic selection, had successfully been increased in several dairy breeds. In the years following, however, the solids-not-fat content of the milk, and especially the protein fraction, has been the focal point. Among the reasons for this is that it has been difficult for butterfat to compete with the very much cheaper margarine. The revaluation of the importance of butterfat and milk protein from a nutritive standpoint has perhaps also played a part. At a number of Dutch dairies the milk is now paid for according to the protein and butterfat content.

Table 10:4 shows the variation of the three main constituents of milk: fat, protein and lactose. It will be seen that fat varies most and lactose least.

Table 10:4 Standard deviation of milk solids according to studies by ROBERTSON *et al.* (*Scotland*), AURIOL (*France*), POLITIEK (*Holland*) *and* VON KROSIGK (*U.S.A.*). *Cf. Tables 10:6 and 10:7*

	Fat %	Solids-not-fat, %	Protein %	Lactose %
Between individual daily samples (within months and herds), ROBERTSON *et al.*				
First month of lactation	0·65	0·42	0·45	0·25
Second to eight month	0·58	0·36	0·24	0·26
Later	0·77	0·50	0·39	0·36
Between lactations averages (according to the four authors above)	0·31	0·24	0·19	0·15

Temporary variations in butterfat content

Butterfat content shows considerable variation from one milking to the next and from day to day, whereas the protein and lactose content vary very little. These temporary variations in butterfat content are due mainly to the fact that the udder is not evacuated to the same extent at each milking. The amount of milk which remains in the udder (the residual milk), varies within wide limits, usually from 10 to 20 per cent of the total milk. The fat, which occurs in the form of small globules with a diameter of approximately $0 \cdot 1 - 15 \mu$, is evacuated with more difficulty than the rest of the milk (the milk serum), and for this reason the fat content of the milk increases during the course of milking, especially in high-producing cows (Fig. 10:19). The first half-litre of milk often contains no more than 1 per cent fat, but the last half-litre may contain 10–15 per cent. Thus, the more completely the udder is evacuated the higher the average fat content of the milk. A large amount of residual milk means that a great deal of the fat remains in the udder and the average fat content of the evacuated milk will be low. This residual milk is not lost, and the greater part can be recovered at the next milking. The average fat content of the milk for several days in succession, or for the whole lactation, is therefore only slightly affected by incomplete emptying of the udder. When high-producing cows are milked twice a day at different intervals, e.g. 9 and 15 hours, the fat content is always much higher after the shorter than after the longer interval. This is due mainly to the more complete evacuation after the shorter interval.

Causes of variation which operate in the long term

The composition of milk from one and the same cow is influenced by a number of different factors.

The stage of lactation. The influence of the stage of lactation is illustrated in Fig. 10:20, which is based on extensive Dutch investigations (POLITIEK, 1957). The colostrum has not been included in the calculations; as is well known, it contains a very high content of protein substances, primarily albumin and globulin. During the first months of lactation there is a reduction in all the milk constituents. The content of fat and protein reach a minimum when the daily yield is at its highest, rising again as the milk yield falls off. The lactose content shows a steady decline throughout the lactation. It is apparent from the diagram that fat and protein content are positively correlated with each other during the course of the lactation; the within lactation correlation has been calculated to be about 0·6.

Age. The age of the cow has some influence on the composition of the milk. Both the content of fat and that of solids-not-fat tend to decrease with increasing age. According to English investigations, the content of fat, protein and lactose, each decrease by about 0·2 percentage units from the first to the ninth lactation.

Nutrition. If a cow is in good condition at calving, then the fat content of the milk during the first months of the lactation will be considerably higher than if the same cow calves in poor condition. The effect on the average fat content for the whole lactation can be of the order of several tenths of one per cent. If the feeding of a lactating cow is drastically reduced, both the amount of milk and the protein content decrease, whereas the fat content increases as long as the milk yield is falling; after which it returns to the normal value.

Feed composition. The composition of the feed has a much greater effect on milk composition than was previously thought. When the amount of bulky feed —primarily hay—or the content of soluble carbohydrates in the ration is reduced, there is a tendency for the fat as well as the protein content to fall. With large alterations in the feed, which have an effect on the rumen flora, the changes in the fat content can be very large. A low protein content in the feed has a depressing effect on the protein content of the milk. Unsaturated fatty acids, such as those found in herring meal and cod liver oil, lead to a reduction in the fat content of the milk, whereas feeding with feeds rich in saturated fatty acids, such as coconut or palm kernel meal, has the effect of increasing the fat content.

High temperature. High temperatures in the environment of the cows have a depressing effect on the fat and lactose content of the milk. Low temperatures have the opposite effect.

Health. The composition of the milk can change radically during conditions of poor health, including mastitis.

Genetic variation

From the foregoing it will be apparent that investigations into the genetically determined variation of milk composition must be made on the average content of fat, protein and lactose per lactation period. These averages should be based on the analysis of samples of a day's milk, taken from each individual monthly or bi-monthly during the course of the lactation. The composition of individual samples is influenced, to a high degree, by temporary external circumstances and the stage of the lactation at which the sample is taken.

Clear differences are found between different breeds, even when they are maintained under similar external conditions; this is illustrated by the averages in Table 10:5.

Table 10:5 Milk composition of American dairy breeds (after ESPE and SMITH, 1952)

	Fat %	Solids-not-fat %	Protein %	Lactose %
Holstein Friesian	3·40	8·86	3·32	4·87
Ayrshire	4·00	8·90	3·53	4·67
Brown Swiss	4·01	9·40	3·61	5·04
Guernsey	4·95	9·54	3·91	4·93
Jersey	5·37	9·54	3·92	4·93

The repeatability and heritability within breeds and herds have been studied by several authors. The following results have been obtained for the repeatability between different lactations of the same cow: fat content 0·58–0·87, solids-not-fat content 0·52–0·76 and protein content 0·41–0·83. A selection of the relatively large amount of data on which estimates of heritability have been carried out is given in Table 10:6. Some figures from Wiad, calculated from twin experiment data, are also included. Robertson's figures for the fat content of milk have been placed in brackets because they are unusually low. The figures from Wiad are not comparable with the others. They have been calculated by comparing the variation within MZ twins with the variation within pairs of contemporary but unrelated cows. In general, it can be said that the heritability

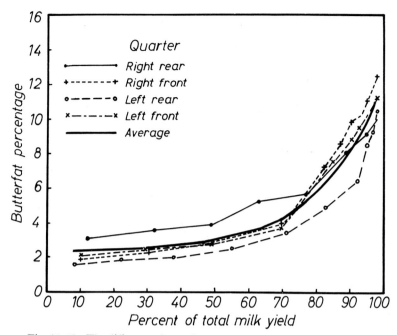

Fig. 10:19. The differences in the fat content of the milk during one milking of a high-yielding cow; each teat separately and the mean for all four teats. (JOHANSSON 1952)

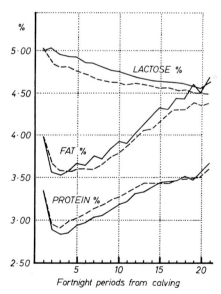

Fig. 10:20. Changes in composition of the milk during lactation.
——— Average for 61 daughters for the A.I. bull Manus.
– – – Average for 111 daughters of the A.I. bull Donald. The test milkings were made at 14-day intervals. (POLITIEK 1957)

Table 10:6 The heritability of the content of different milk constituents

	Fat %	Solids-not-fat %	Protein %	Lactose %
Scottish Ayrshire, ROBERTSON et al., 1956	(0·32)	0·53	0·48	0·36
Dutch Friesian, Vos, 1964 *	0·57	—	0·57	—
Dutch Friesian, Vos, 1964 †	0·77	—	0·71	—
German Black and White, LANKAMP, 1959	0·72	0·83	0·76	—
Five American breeds, VON KROSIGK, 1959	0·52	0·58	0·53	0·07‡
Approximate average	0·60	0·60	0·60	0·35
Twin studies at Wiad, HANSSON, 1956	0·87	—	0·88	0·62

* Paternal half sib correlation.
† Daughter-dam regression.
‡ Lactose + ash, calculated as difference.

for the protein content of milk is as high as that for fat content—about 0·6. The heritability of lactose content is much lower. Milk composition is thus a more highly heritable trait than milk yield.

Genetic differences in protein composition

The protein of the milk is divided into two main components: whey protein and casein. The latter makes up about 80 per cent of the total protein content. By electrophoresis and several other procedures the casein may be divided into subclasses named α-, β- and γ-casein, and the whey protein is similarly divided into β-lactoglobulin, α-lactalbumin, serum albumin and other components. In the last ten years genetic differences have been found in several of these proteins.

ASCHAFFENBURG and DREWRY (Reading, England), who used paper electrophoresis, were able to classify the β-lactoglobulin from individual cows into distinct genetic types. So far three β-lactoglobulin fractions (A, B and C) have been described. The β-lactoglobulin A is the fastest-migrating and C the slowest. Each animal has one or two fractions. The variation is controlled by three co-dominant alleles, Lg^A, Lg^B and Lg^C each giving a single protein fraction. The molecules of the β-lactoglobulin variants are very similar in composition. Each consists of two identical chains of 162 amino acids. Each C-chain differs only in one amino acid residue from the B-chain, whereas the A-chain differs from the B-chain in two places.

Genetic differences have been described in the α-lactalbumin also, but this variation is confined to cattle of Zebu type. All European cattle seem to have only one α-lactoglobulin component. The α-lactoglobulin differences in Zebu are controlled by an autosomal locus α-La with two co-dominant alleles.

The casein has been more difficult to study with electrophoresis, as the various casein components tend to aggregate. However, this technical difficulty is now largely surmounted. Addition of strong urea to the casein prevents aggregation; and subsequent electrophoresis on starch gel (or other suitable supporting media) gives excellent resolution of the casein constituents. ASCHAFFENBURG was able, in 1961, to describe inherited variants of the β-casein. This polymorphism is controlled by three autosomal co-dominant alleles at a locus designated β-Cn. The variants A, B and C occur either singly or in pairs. THOMPSON and co-workers described a similar variation in one of the calcium sensitive α-casein components (α_{S_1}-casein). So far four different alleles have been identified and each give rise to one protein zone in starch gel electrophoresis. The alleles are co-dominant so the heterozygotes have two protein zones (Fig. 10:21). The α_{S_1} and β-caseins seem to be controlled by two closely linked loci.

The amino-acid composition of the α_{S_1}- and β-casein variants was recently worked out in great detail by a team of Dutch and French biochemists. The composition of the two main components is very different but variants of the

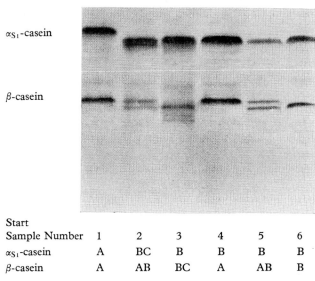

Fig. 10:21. Starch gel electrophoresis of casein reference samples furnished by Dr M. P. Thompson, Philadelphia. Separation and photo by Mr K. Sandberg, Uppsala.

same main component differ only in single amino acids. For instance, α_{S_1} casein B has one glutamic acid residue more and one glycine less than the C variant.

Genetic differences have been shown to occur in the κ-casein. This is a subfraction of the α-casein The κ-casein plays an important role in the clotting of the milk proteins after treatment with rennin. There are at least two genetic variants of κ-casein (A and B) which are determined by two co-dominant alleles. Grosclaude and co-workers (France) have shown that the locus controlling the κ-casein variation is closely linked to the loci for α_{S_1} and β-casein. They investigated progeny from five sires which were heterozygous at the κ-casein locus and the α_{S_1}- or β-casein loci. In all, 109 offspring were studied but no recombinations were found between any of the three loci. The linkage is evidently very close. The results suggest that the variation in the α_{S_1}-, β- and κ-caseins is controlled by a small segment of the same chromosome.

So far, no fewer than 5 genetic systems have been found to influence the composition of the milk proteins. The frequency of the various genes involved differs considerably between breeds. A study of 600 Jersey, Guernsey, Shorthorn and Friesian cattle by ASCHAFFENBURG may serve as an example. The β-casein C did not occur in the Jersey, Friesian and Shorthorn breeds; it was found exclusively in Guernsey. Friesian and Jersey, however, had the A as well as the B variants, while Shorthorn had only A.

ASCHAFFENBURG and DREWRY made a very interesting observation of a relationship between the β-lactoglobulin types and the concentration of β-lactoglobulin and casein. The concentration of each of the β-lactoglobulins varied in proportion to the casein content, but the relation was different for each of the β-lactoglobulin types (Fig. 10:22). At a given content of casein the

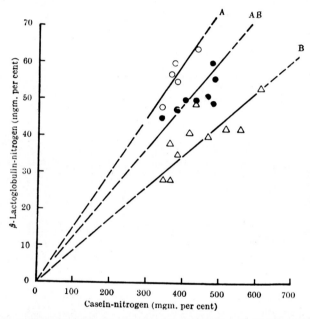

Fig. 10:22. Concentration of the β-lactoglobulins and of casein in the milk of 24 individual cows. (ASCHAFFENBURG and DREWRY 1957)

amount of β-lactoglobulin was considerably higher in type A animals than in type B, while the heterozygotes were intermediate. The results indicate, therefore, that the syntheses of β-lactoglobulin and casein are interdependent and that the relative amounts of β-lactoglobulin synthetised is influenced by the genotype at the β-Lg locus.

Genetic studies of milk protein composition are of fairly recent date. Starch gel electrophoresis of casein gives rise to as many as 20 more or less distinct protein zones. Additional genetic variants will therefore probably be found. The effect of this variation in protein composition on the usability of the milk in the dairy industry is not yet known. However, several such studies are now in progress.

Genetic variation has been detected also in the composition of the milk fat and its content of carotene and vitamin A, as well as in the rennin coagulation time (CLAESSON, 1965).

Correlation between the various milk constituents and between the quantity and composition of the milk

It has been known for a long time that the fat and protein contents of milk are fairly highly correlated, not only 'within cows' in the course of lactation but also, as mentioned previously, in comparisons between different individuals and breeds. Recently, a number of studies have been made to determine the extent of the genetic and environmental causes of the correlation. It is necessary to know the genetic correlations of the various milk constituents in order to assess the possibilities of altering their proportions. Table 10:7 shows some of the correlations which have been estimated in various studies.

Table 10:7 Phenotypic and genetic correlations between different milk constituents (within breeds)

Correlation	Fat/SNF	Fat/Protein	Fat/Lactose	Protein/Lactose
Phenotypic:				
Five American breeds (73 cows) (JOHANSSON and CLAESSON, 1957)	—	0·70	0·04	0·12
Scottish Ayrshire (814 cows) (ROBERTSON and co-workers, 1956)	0·40	0·39	0·14	0·05
Dutch Friesian (23051 cows) (POLITIEK, 1965)	0·70	0·57	0·15	—
Five American breeds (23 herds) (VON KROSIGK, 1959)	0·54	0·62	—	—
Genetic:				
ROBERTSON and co-workers, 1956	0·46	0·48	0·37	0·41
VON KROSIGK, 1959	0·54	0·62	—	—
POLITIEK, 1965	—	0·60	—	—

The table shows that the phenotypic correlations are mainly genetic; the environment seems to play a minor role. If selection is made for a higher protein content of the milk, then not only the protein but also the fat will increase. The correlated selection response can be estimated from the figures in Tables 10:4, 10:6 and 10:7. In the following example the genetic correlation between the two traits is, as usual, denoted by r_A, and the genetic standard deviation by σ_A. The value of σ_A can be written $h\sigma_P$, where σ_P = the phenotypic standard deviation (Table 10:4). The genetic correlation between fat and protein content is assumed to be 0·60.

Estimated increase in protein content when the fat content is increased by 1 per cent by selection

$$= r_A \frac{\sigma_A \text{ (prot.)}}{\sigma_A \text{ (fat)}} = 0.60 \times \frac{0.147}{0.239} = 0.37$$

Estimated increase in fat content when the protein content is increased 1 per cent by selection

$$= r_A \frac{\sigma_A \text{ (fat)}}{\sigma_A \text{ (prot.)}} = 0.60 \times \frac{0.239}{0.147} = 0.975$$

If selection is made for higher protein, without consideration of the fat content, the fat content will increase in any case at about the same rate as the protein content, due to the fact that the genetic variation of fat content is almost twice as large as that of protein content. The proportions of fat and protein can only be altered very slowly by such selection.

HANSSON carried out a study which is of interest in this connection. Using the same twin material (Swedish Red and White) referred to in Table 10:6, he calculated for each cow the ratio between the protein and fat content of the milk and also between the content of lactose and fat. The former ratio was, on the average, 0·84 and the latter, 1·32. The heritability of the protein/fat ratio was estimated to be 0·46 and the lactose/fat ratio 0·84. If the special interest is to increase the protein content without increasing the fat, then it is, without doubt, most rational to base the selection on the protein/fat ratio. The phenotypic standard deviation of this ratio in the data studied by Hansson was 0·037. If it is assumed that the heritability of 0·46 and the variation of 0·037 are representative of the SRB breed and that selection is made only for this trait, then a selection differential as large as the standard deviation can be achieved. This in turn presupposes that the bulls are consistently progeny tested with respect to the protein/fat ratio, and that selection also takes place between these and not only among the females. The gain per generation (about 5 years) that could be expected wouldbe $0.46 \times 0.037 = 0.017$. In order to increase the protein content to the same lev el as the fat content would therefore require $(1 - 0.84)/0.017 = 9.4$ generations, i.e. about 50 years. Even if it is not a completely hopeless task to increase the protein/fat ratio in the dairy breeds by selection, it certainly requires tremendous patience, and involves considerable costs for the production recording which must be carried out. There are, without doubt, much more profitable breeding projects to be engaged in.

A significant negative correlation between the milk yield per lactation and the fat content has been demonstrated when different cows are compared with each other. In a study of data from the SRB, the correlation between cows and within herds was found to be -0.25 and highly significant. Within cows the correlation (0·008) was positive but not significant. This indicates that the genetic correlation is more pronounced than the phenotypic, a fact which has been verified in later investigations. A number of different breeds have been studied and it has been found that the correlation is higher within breeds with a high average fat content, e.g. Jersey and Guernsey, than in breeds with a low average fat content, such as Friesians. The phenotypic correlations have varied between 0·14 and 0·36 and the genotypic, between 0·20 and 0·57. The correlation is more pronounced in young cows than in later lactations. The risk of reducing the fat content by selecting only for milk yield, without consideration of fat content, can be estimated in the usual way. If the genetic correlation between milk yield and fat content is assumed to be -0.40 and the genetic standard deviation is 0.28 percentage units for fat content and 412 kg for milk yield, then the percentage decrease in fat content per kg increase in milk yield is, $-0.40 \times (0.28/412) = -0.00027$ per cent, or 0.27 percentage units per 1000 kg milk. The situation is much more serious, however, if selection is conducted only for higher fat content without regard to milk yield. In this case a reduction in yield of 590 kg milk can be expected for every 1 per cent increase in fat content ($-0.40 \times 412/0.28$). The question of the optimum level of butterfat in the different breeds and consumer requirements is worthy of consideration when the breeding programmes are worked out.

SELECTED REFERENCES

ALFA-LAVAL CO. 1963. *Symposium No. 1 on machine milking* (228 pp.). Tumba, Sweden.

ASCHAFFENBURG, R. 1965. Variants of milk proteins and their pattern of inheritance. *J. Dairy Sci.*, **48**: 128–132.

BARKER, J. S. F. and ROBERTSON, A. 1966. Genetic and phenotypic parameters for the first three lactations in Friesian cows. *Anim. Prod.* **8**: 221–240.

JOHANSSON, I. 1961. *Genetic Aspects of Dairy Cattle Breeding* (259 pp.). Urbana, Ill.

KEESTRA, J. 1963. *Investigations on milking rate in Dutch Friesian cows* (in Dutch) (153 pp.). Diss. Wageningen, Holland.

POLITIEK, R. D. 1966. Probleme der Züchtung auf Milcheiweiss beim Rind. *Internationalen Tierzüchterischen Symposium Oct.* 1965. Berlin.

11

Body Size and Carcass Traits

With the exception of some of the specialised fur animals, all the different species of farm animals are kept, to some extent, for the production of meat; pigs and the specialised breeds of beef cattle and poultry are used exclusively for this purpose. Several breeds of cattle are kept for the dual purpose of producing meat and milk, while sheep are able to produce both meat and wool. The production of meat from horses is of minor importance. Because of the economic importance of meat production, it is natural that a great deal of research has been carried out into the genetic and environmental factors which influence body size, growth rate and carcass quality of farm animals.

BIRTH WEIGHT

Intra-uterine influence. All those factors which contribute to the nourishment of the foetus in the uterus have an effect on the birth weight. In the presence of large number of foetuses there will be less nourishment available for each individual foetus; and, quite naturally, increased litter size in normally polytocous animals as well as multiple births in normally monotocous animals will reduce the birth weight of individual offspring. This aspect was discussed more fully in Chapter 5 (p. 143).

Nutrition of the mother. The nutrition of the mother, quite naturally, influences the birth weight of the young. The relationship between birth weight and the plane of nutrition during gestation has been thoroughly studied in sheep. In an investigation by WALLACE, pregnant ewes were kept on high, medium (no increase in body weight) and low planes of nutrition during the final six weeks of pregnancy. The birth weight of twin lambs born in the medium and low groups was, on the average, 71 and 52 per cent, respectively, of the weight of twin lambs in the high plane group. Inadequate nutrition first manifests itself in the weight of the mother, since her body serves as a reservoir from which nourishment is transferred to the growing foetus.

Seldom has the female reached mature weight by the time the first offspring is produced. Cows reach mature size only after the birth of the fourth or fifth calf. Sheep and pigs are also immature at the time of primiparity. It is to be expected, therefore, that the *age of the mother* influences the birth weight of the offspring, so that full-grown mothers produce heavier progeny than younger mothers. VENGE in a study of the Swedish Red and White cattle found that the birth weight of calves from first calvers was about 33 kg, whereas the weight of calves from cows which had completed three lactations was about 36 kg. A similar result has been noted in the Red Danish dairy cattle, though in this case the birth weight was somewhat higher. There is also a close relationship between the age of ewes and the birth weight of lambs.

The influence of sex on birth weight. Males are generally heavier than females at birth. In the Swedish Landrace, male pigs weigh about 50 g more than female pigs at birth and it can be reckoned that ram lambs weigh about 5 per cent more than ewe lambs of the same breed. In the investigations, referred to above, on the birth weight of calves in the Red Danish and Swedish Red and White breeds, it was shown that bull calves outweighed heifer calves by about 2·4 kg in the former, and by about 1·5 kg in the latter breed.

The influence of inheritance. Weight at birth and maturity differs greatly between breeds. According to American investigations the weight of a four-year-old Jersey cow is approximately 70 per cent of the weight of a Friesian cow of the same age. The difference in birth weight is even greater; Jersey calves are only about 60 per cent of the weight of Friesian calves. Similar differences have been shown between the birth weight also of certain breeds of pigs and sheep.

The breed differences indicate that inheritance influences the birth weight. If breeds of different birth weight are crossed with each other, the birth weight of the offspring is usually somewhere between the averages of the two parent breeds, though the dam has a stronger influence than the sire on the birth weight of the offspring. This maternal effect was clearly demonstrated in the crossing between the Shire horse and the Shetland pony discussed on p. 66.

At an experiment station in Montana (U.S.A.), KOCH and CLARK (1955) carried out a large-scale investigation into the influence of inheritance on various traits of the Hereford breed. Birth weight was included in the investigation which comprised 4553 calves from 137 bulls. The birth weight figures were corrected for the age of the dams and for the sex of the calves. The heritability, estimated from the intra-class correlation between paternal half sibs was found to be 0·35. The intra-class correlation for maternal half sibs was about three times as great as that for paternal half sibs; which shows that, apart from inheritance, the intra-uterine environment of the calves is of great importance for the birth weight.

PRE-WEANING GROWTH

In the majority of farm animals the growth of the young after birth is mainly dependent on the milk production of the mother as well as her general mothering ability. If the milk yield is insufficient, the growth of the young will be correspondingly retarded. The importance of the ewes' milk supply to the growth of lambs has been demonstrated by F. N. BONSMA (1939) working with South African breeds of sheep. The correlation between the growth of the lambs and the milk yield of the ewes in the first 77 days after birth was 0·81. The relationship was especially strong during the first 14 days. Those factors which favourably affect the milk yield, e.g. increase in the age of the ewes up to 5 or 6 years of age, have a positive effect on the growth rate of the lambs.

The number of young also has an effect on the growth rate. HAMMOND found that single born Suffolk lambs increased by about 2·4 kg per week in the first month, but that twins and triplets started with a growth rate of 1·5 kg per week and were 3–4 months old before maximum growth rate was reached. The weight difference between single lambs and multiple-birth lambs becomes less with increasing age.

The litter size has a very definite effect on the growth rate of individual pigs. This is illustrated in Table 11:1, which shows the relationship between litter size and growth rate of piglets to 3 and 8 weeks of age. The data refer to some 5,000 piglets of two Norwegian breeds.

Table 11:1 *The relationship between litter size and average weight of pigs* (after BERGE, 1949)

Number of pigs born alive per litter	Birth weight kg	3 weeks of age weight kg	3 weeks of age survivors %	8 weeks of age weight kg	8 weeks of age survivors %
1–5	1·42	6·01	85·5	18·0	82·0
6–10	1·27	5·11	84·7	15·1	81·0
11–15	1·18	4·72	82·4	13·8	77·9
16–21	1·08	4·49	71·2	13·3	70·5

The *heritability* of weaning weight is consistently lower than that of birth weight. This is rather natural, since growth rate is highly dependent on the mothering ability of the dams. In Koch and Clark's investigation with Herefords it was found, for example, that the heritability for weaning weight was 0·24 compared with 0·35 for birth weight.

POST-WEANING GROWTH RATE

After weaning, the growth rate is dependent upon the ability of the animal to ingest, digest and convert bulky and concentrated food of various kinds into animal tissue. The efficiency of this conversion process exhibits considerable variation between individuals. The factors which influence growth rate after weaning have been the object of a great deal of investigation, particularly with pigs.

Growth rate and feed conversion in pigs

In 1907 special stations were established in Denmark for recording the growth rate, feed conversion and carcass quality of pig progeny of selected breeding animals. Two female pigs and two castrated males from the same litter were selected and sent to the testing stations at 8–10 weeks of age. Up to 1950 the test pigs were group-fed, but since then they have been individually fed. As far as possible all the pigs receive the same quality of feed and are slaughtered at about 90 kg liveweight. Similar testing stations have been established in several other countries, but generally, as for example in Sweden, group feeding is practised.

Some results from the Danish pig progeny testing stations are summarised in Table 11:2.

Table 11:2 Changes in the feed conversion and growth rate of pigs at the Danish pig progeny-testing stations during the first fifty years of testing

Year	Number of pigs tested	Average daily gain, g.	Number of Scandinavian feed units/kg gain
1908–1912	2,318	560	3·74
1912–1917	3,780	568	3·76
1919–1925	4,584	577	3·61
1925–1930	10,336	637	3·40
1930–1935	13,943	633	3·35
1935–1940	14,876	640	3·26
1940–1945	10,580	644	3·29
1945–1950	12,772	656	3·21
New testing stations with individual feeding			
1950–1955	16,443	672	3·04
1955–1960	18,488	683	2·97
1960–1964	20,356	684	2·94

It will be seen from the table that from the commencement of testing the growth rate has improved by about 22 per cent and that in the same period the feed consumed per kg growth showed a corresponding decline. The rapid improvement shown in the beginning of the 1950s would appear to be due mainly to the better environment of the stations with individual feeding facilities. At the same time as the growth rate has been increased, a reduction of the fat content of the carcass has been achieved and consequently a reduction in the calorie content. The backfat thickness was reduced from 3·92 cm in 1908 to 2·61 cm in 1964 (cf. p. 320). As the energy content of the body tissues has been reduced, the energy utilisation has not been improved to such an extent as the reduction in number of feed units per kg growth would indicate. These changes in quality will be further discussed on p. 321.

Sex has a very strong influence on the growth rate of pigs. With group feeding, castrated males grow significantly faster than gilts; but with individual feeding the situation is reversed. The faster growth rate of castrated boars with group feeding is probably due to the fact that they are more aggressive at the feed trough.

The influence of heredity on the daily growth rate of pigs from 20 to 90 kg liveweight at the pig testing stations has been the subject of several investigations. JOHANSSON and KORKMAN (1950) estimated the heritability of growth rate in Swedish breeds of pigs to be 0·26 and similar results have been obtained for other breeds. The heritability of this trait has also been studied with data from the Danish pig progeny testing stations with group and individual feeding. JONSSON found that with group feeding the heritability was about 0·2 but with individual feeding it was about three times as high. The heritability was estimated from the intra-class correlation of paternal half sibs of the same sex. The variation between groups of half sibs with the same sire was about the same in both cases, but the variation within litters was considerably less with individual feeding. This reduction of the within litter variation resulted in a corresponding increase of the intra-class correlation and consequently also increased the heritability (cf. p. 126). This increase in heritability probably will not lead to a corresponding increase on selection progress. The small variation within litters is mainly a result of the absence of competition between litter mates and this will not have the same consequences as an elimination of random environmental variation. This point will be discussed further in Chapter 16.

Feed utilisation, i.e. the number of feed units required for each kg increase in weight, has a heritability of about the same order as growth rate; which is natural, since the amount of feed is adjusted according to the growth. Thus, pigs which grow fastest also have the best utilisation of feed. Both the phenotypic and genetic correlations between these two traits lie between -0.8 and -0.9.

The effect of inbreeding and crossing on the production traits of farm animals will be dealt with more fully in Chapter 14. It is pertinent, however, to mention here the effect on the growth rate, viability and feed utilisation of pigs. Extensive investigations in the U.S.A. and in England show that inbreeding has a very strong negative effect on these traits. Crossing inbred lines results in a pronounced improvement in litter size and growth rate in comparison with inbred lines. This applies especially when the females are themselves products of line crosses. However, when the crosses are compared with non-inbred control groups, the absolute improvement is somewhat doubtful.

Breed crosses, on the other hand, have been tried with success in many countries. A crossbreeding experiment with Swedish Landrace and Yorkshire has recently been terminated at the Wiad station. Details of the plan and results of the experiment will be given in Chapter 14. Reciprocal crosses were made between the two breeds. Comparisons were then made between contemporary litters from full sisters, where the one litter was produced by mating with a boar of the same breed and the other litter was crossbred. In this experiment the crosses were consistently better than the purebreds in respect of litter size and total weight of the litter, as well as the growth rate and individual weight of the pigs.

It is of special interest to note that the crossbred litters were more uniform; there was less variation with regard to number of pigs per litter, growth rate and age at slaughter than there was with the purebred pigs. Uniformity is a trait of great practical importance, and it is further evidence for the advantages of crossbreeding.

The growth rate of cattle and sheep

Sex has a strong influence on the growth rate of the young also in cattle and sheep; the male grows faster than the female. In a Swedish investigation with monozygous cattle twins, BRÄNNÄNG (1966) studied the effect of castration on the growth rate of bull calves up to 25 months of age. One bull of each pair was castrated, the other serving as a control. Castration was carried out at one month of age with five pairs; and in two other groups, consisting of four and five pairs, at six and twelve months respectively. Castration led to a reduction in the growth rate, which, when averaged over the whole period from one to twenty-five months, was about 40 g per day. The time of castration appeared to have no significant effect.

In the investigation by KOCH and CLARK into the variation of various production traits in Hereford cattle, the influence of heredity on the weight of the calves at one year of age and on their growth rate was studied. The heritability of weight at one year was about 0·5 and of daily growth rate during the first year about 0·4. Other investigations have given similar results. The relationship between body-weight and growth at different ages is given in Table 11:3, where the

Table 11:3 *Genetic and phenotypic correlations between weight and rate of growth of cattle at different ages. The genetic correlations are given above the diagonal* (KOCH and CLARKE, 1955)

	Birth weight	Weaning weight	1 year weight	Growth rate birth–weaning	Growth rate weaning–1 year
Birth weight		0·63	0·40	0·46	0·06
Weaning weight	0·39		0·54	0·98	−0·03
1 year weight	0·34	0·47		0·51	0·83
Growth rate birth–weaning	0·21	0·98	0·44		−0·05
Growth rate weaning–1 year	0·04	−0·33	0·67	0·36	

genetic correlations are given above the diagonal and the phenotypic correlations below the diagonal. The results indicate that to some extent it is the same genes that influence the weight at all three ages. Selection for high birth weight should therefore automatically lead to higher weight at weaning and at one year old.

In Germany, LANGLET and GRAVERT (1963) have carried out progeny testing of bulls with respect to carcass characteristics. Various systems of rearing have been employed. One system was intensive rearing in stalls, where bull calves had an average daily gain of 1100 g. They were slaughtered for baby beef at 10 months of age. The difference between progeny groups was greatest with this high intensity rearing. The heritability of daily gain was estimated to 0·42.

SKJERVOLD (1958) has studied the various factors which influence the growth rate of three Norwegian cattle breeds. The experiment included 21 progeny groups each of 8 bull calves from A.I. dairy bulls. Each progeny group was divided into two equal subgroups of which one received more intensive feeding than the other. The calves started the experiment when they were one month old and were slaughtered when they were 18 months. During the first stall feeding period there was no difference in the daily gain either between breeds or between progeny groups within breeds. During the grazing period and the second stall feeding period, however, the differences were statistically significant. The heritability of daily gain in these two periods was estimated to be 0·55 and 0·65 respectively. The genetic correlations between the daily gains in the different periods varied from 0·7 to 0·9, the relationship being strongest between the growth rates in the two stall feeding periods.

In Sweden, A.I. dairy bulls have been progeny tested on the basis of the growth rate of their calves in the field. LILJEDAHL and LINDHÉ (1964) summarised the results of 57 progeny groups of Swedish Friesians. Each progeny group consisted of 12 bull calves placed on different farms at about 10 days of age. The experiment was divided into four blocks and in each block all the farms received at least one calf from each A.I. bull, so that on the average all progeny groups in the same block were exposed to the same environmental conditions. By allowing a number of bulls to have progeny groups in several blocks the results from all progeny groups could be made comparable after correcting for environmental differences between the blocks. The difference in average daily gain from 10 days to 20 months of age was very large between the various progeny groups. In one experimental block consisting of 17 progeny groups, the best group had an average daily gain of 345 g carcass weight and the lowest group a gain of 291 g carcass weight per day. The heritability of growth rate was estimated to 0·29.

Studies of the daily gain of sheep from weaning to one year of age indicate that it has a relatively high heritability of 0·5–0·6.

To a considerable extent the production of meat from cattle is the result of crosses between beef and dairy breeds. A large-scale crossbreeding experiment is now nearly complete in Sweden; in the experiment, heifers of the Swedish Red and White (SRB), numerically the largest breed in the country, have been crossed with bulls of the Swedish Friesian, Red Danish, Aberdeen Angus, Hereford and Charollais breeds. In order to obtain a satisfactory comparison between breeds, use was made of a relatively large number of randomly selected young bulls from each breed, e.g. 41 Friesians and 31 Aberdeen Angus, but only 12 Charollais since these were the only bulls of this breed available in the country. So far 909 calves have started the experiment. The calves have been reared on large farms which took 10 calves of the SRB and each cross. In this way it was possible to compare the purebreds with all the crosses on the same farm. There was a large variation between the various breed crosses, as far as growth rate was concerned. When the growth rate of the purebred SRB bulls is given the value 100, the crosses with the different breeds had the following comparative values: Aberdeen Angus 101, Red Danish and Hereford 109, Charollais 110 and Friesian 112. It is interesting to note that the cross with the Friesian breed gave calves with the highest rate of growth. It is obviously an advantage to be able to cross the SRB with another dairy breed since the heifer calves can be used for milk production if so desired. Another advantage with the Friesian cross was that the frequency of calving difficulties and stillborn calves was considerably less than in crosses with the Charollais and also less than in the purebred Friesians.

Changes in body proportions during growth

As an animal increases in weight and size there is a change in its body proportions. These and other questions were studied by HAMMOND and his associates. Different tissues and parts of the body reach the maximum rate of growth at different stages. Organs which are of decisive importance for the normal function of life such as the brain, the nervous system and digestive organs are well developed at birth. After birth, priority is given first to the development of bone tissue, next to muscle growth, and finally to fat storage. The newborn contains only a very small amount of fat.

As far as bone is concerned, growth follows a special pattern. The head attains maximum rate of growth shortly before and after birth. The metatarsal bones also show relatively early development. From these two regions growth gradients proceed backwards and upwards respectively so that maximum growth in the various parts of the trunk takes place later than in the head and the lower parts of the limbs. At the same time as the bones are growing in length they are also growing in thickness. An example of these changes is provided by an investigation carried out by Hammond into the changes which take place in the hind legs of Suffolk ewes from birth to four years. The weight changes are summarised in the table below (after HAMMOND, 1960).

Bone	Weight changes in bones of the hind leg of Suffolk ewes, as percentages of the weight of the cannon bone.		
	Birth	5 months	4 years
Cannon	100	100	100
Tibia	197	245	272
Femur	217	285	324
Pelvis	142	430	569

There is a definite growth gradient from the lower part of the hind leg towards the pelvis. At birth the weight of the pelvis was not quite 1·5 times the weight of the cannon bone but at four years of age it was nearly 5·7 times heavier. When the muscle and fat tissues are taken into consideration the differences in proportion are even more pronounced.

HAMMOND showed that in sheep the hind quarters increased proportionally more than the fore quarters with increasing age. This does not seem to be the case in cattle. BRÄNNÄNG (1960) studied the carcass proportions in approximately 300 animals of three Swedish dairy breeds. In bulls there was a marked decrease in the percentage of hind quarters from 3 months to 2 years of age. There was a similar tendency in steers and heifers, though much less pronounced than in the bulls.

The effect of plane of nutrition on body proportions

The growth of young animals consists mainly of an increase in bone and muscle tissue. The young animal stores only small amounts of fat even when on a high plane of nutrition. A high plane of nutrition, on the other hand, accelerates development, so that sexual maturity and full-grown size is reached earlier. The effect of the intensity of feeding on growth rate has been studied by HANSSON (1953) and co-workers at the Agricultural College of Sweden. In one of these experiments 16 pairs of MZ twins of the Swedish Red and White breed were divided into four equal groups. One twin of each pair received standard feeding while the other four twins in each group were given 60, 80, 120 and 140 per cent of the standard feeding up to 25 months of age. The feed was nutritionally balanced for all the groups. The animals on the 140 per cent level were fed so intensively that they were unable to consume all the feed offered. Their actual consumption was consequently only 124 per cent of the standard control. At 25 months of age the standard-fed animals weighed, on the average, 432 kg, whereas the groups on 60 and 140 (i.e. 124) per cent levels weighed 329 and 462 kg respectively. During the first half-year there was an almost linear relationship between the daily growth rate and the plane of nutrition; this continued also in the second half-year period, the high-plane animals showing the highest growth rate. At one year of age, however, the high-level animals had used up the greater part of their growth capacity so that in the second year the standard-fed animals showed the greatest rate of growth. The animals fed at a low level did not at any time in the first two years attain the same rate of growth as the standard-fed animals.

In Fig. 11:1 are shown representative twin pairs from each of the four experimental groups at ages of 22-24 months. In the low-plane animals the head and fore-quarters are comparatively better developed than the rump and depth of the barrel. They have not, at two years of age, reached the proportions typical of heifers of the SRB breed. The high-level animals, on the other hand have already attained a good depth of barrel and well developed rump.

After 25 months of age the experimental twins were fed according to the same standard as their control twin sisters. The differences between the twins became progressively less, and the heifers reared on the low level eventually reached almost the same size and body proportions as their respective twin sisters. It should perhaps be pointed out again that the composition of the feed to all groups was properly balanced and the experiment was concerned only with quantitative over- or under-feeding. It is true, of course, that a simultaneous deficiency not only of calories but also of essential vitamins and minerals can easily result in permanent retardation of development.

In an American feeding experiment with MZ twins from beef breeds, WINCHESTER and ELLIS (1956) allocated one twin of each of 10 pairs to a low level of feeding for 3-4 months of the first year; the other twin was fed normally. The feeding level of some of the experimental twins was so low that some of them actually lost weight. At the end of the experimental period all the twins were fed generously and the previously under-fed animals grew as fast as, or faster than, their normally fed partners. The animals were killed when they were judged to be mature for slaughter. The normally fed animals were, on an average, somewhat younger at slaughter. The size and body proportions of twins belonging to the same pair were very similar, and the authors stated that these similarities of body proportions at the time of slaughter was the most striking result of the whole experiment.

It is thus clear that a low level of feeding results in an extended period of growth; from which it follows that if the animals are slaughtered at a given age before growth is complete, then the muscle development of the low level animals will be less than that of the well fed animals. It is mainly the relatively later developing hind-quarters and loin region, economically the most valuable parts of the carcass, which show lack of muscle development with low-level feeding. In the Swedish experiments just referred to, there was a very large difference in growth rate between the twin pairs. Differences of the same type also occur

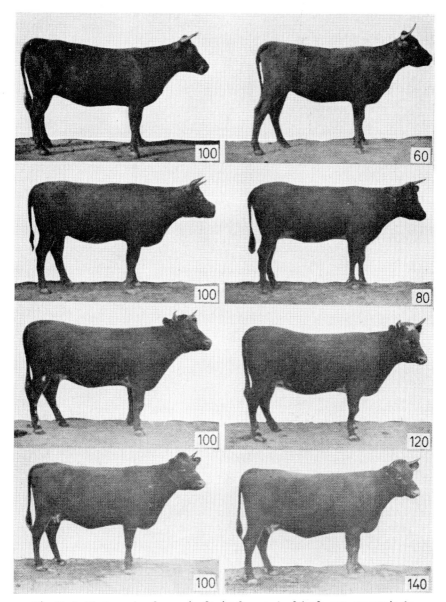

Fig. 11:1. Photographs of one pair of twins from each of the four treatments in the rearing intensity experiment at Wiåd, Sweden.
The standard-fed control twin of each pair is on the left and the figures under each twin indicate the level of feeding. See text. (HANSSON *et al.* 1953)

between breeds. In unimproved breeds of farm animals, the growth rate extends over a relatively long period. Heredity can therefore have an elevating or depressing effect on growth rate in the same way as level of feeding. This situation is illustrated in Fig. 11:2.

It is important that the time of slaughter is adjusted to the time of optimum body proportions with regard to carcass requirements. Certain breeds pass through the period of maximum bone and muscle growth relatively quickly; after which the animal begins to lay down larger amounts of fat. In the Middle White breed of pigs suitable proportions of lean to fat are reached at about 40–50

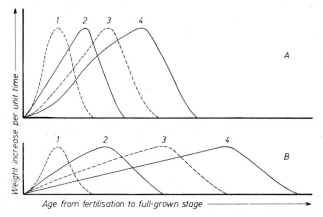

Fig. 11:2. Curves showing the relative growth rate of different parts of the body and tissues and how these are influenced by the plane of nutrition and heredity. *A* an early-maturing breed or high plane of nutrition, *B* a later-maturing breed or low plane of nutrition. The respective curves show (1) parts with early maximum growth, e.g. head and cannon bone, (2) humerus and neck, (3) muscles and femur, (4) fat, loin and rump. (PALSSON 1955)

kg liveweight, whereas in the Large White or Landrace breeds the storage of depot fat first becomes important at a weight of 80–90 kg. Middle White pigs must therefore be slaughtered very much earlier than pigs of these latter two breeds. Similar breed differences occur in sheep and cattle.

THE INFLUENCE OF HEREDITY ON BODY SIZE

The investigations into the growth of MZ cattle twins, referred to earlier, indicate that heredity has a strong effect on the size of the mature animal. In cattle and sheep the heritability of mature body-weight has been estimated to be between 0·4 and 0·5. In cattle, the influence of heredity has also been studied with respect to various body measurements. A large-scale investigation of this kind has been carried out at Iowa State University, U.S.A., where for several decades the height of the withers, body length, chest depth, heart girth, paunch girth and the liveweight of all the animals in the University herd of Holstein-Friesians have been measured. LUSH, TOUCHBERRY and associates have estimated the heritability of these traits from the correlations between dams and daughters. They also calculated the genetic correlations between the various measurements.

The influence of age on the heritability of various body measurements of cattle (after BLACKMORE, McGILLIARD and LUSH, 1958 and TOUCHBERRY, 1951) is illustrated in the following table.

Trait	Heritability at the age when the measurements were taken			
	6 months	1 year	2 years	3 years
Withers height	0·34	0·44	0·86	0·73
Chest depth	0·24	0·34	0·79	0·80
Body length	0·17	0·19	0·63	0·58
Heart girth	0·18	0·28	0·55	0·61
Paunch girth	0·25	0·29	0·41	0·26
Body weight	0·14	0·21	0·53	0·37

The heritability at 3 years of age is based on 187 dam-daughter pairs. The other coefficients are computed from extended data from the same herd including 334 dam-daughter pairs. The genetic component of the variation is much greater at 2 and 3 years of age than at the lower age for all traits, except paunch girth. This measurement is influenced not only by the muscle development but to a very high degree by pregnancy and the rumen content at the time of measurement. These causes of variation are especially important in older animals, and the heritability is therefore relatively low.

The genetic correlation between certain body measurements is quite high but between others it is of low magnitude. This is illustrated by Table 11:4 which

Table 11:4 Genetic and phenotypic correlations between various measurements of body size in cattle (after TOUCHBERRY, 1951). *Genetic correlations above the diagonal.*

	Withers height	Chest depth	Body length	Heart girth	Paunch girth	Weight
Withers height		0·81	0·80	0·65	0·31	0·70
Chest depth	0·74		0·76	0·84	0·51	0·72
Body length	0·67	0·71		0·56	0·18	0·83
Heart girth	0·63	0·81	0·58		0·61	0·81
Paunch girth	0·27	0·43	0·40	0·79		0·69
Weight	0·53	0·67	0·70	0·88	0·84	

shows above the diagonal the genetic correlation between various traits calculated on a dam-daughter basis (cf. p. 132), and below the diagonal the corresponding phenotypic correlations between traits in the same individual. All the traits were measured at three years of age. The genetic correlations have obviously large standard errors. However, it seems to be that typical skeletal characteristics, such as height at withers and body length, show a high genetic correlation, while the genetic relationship is much weaker between skeletal traits and those traits which are influenced by the muscle development and fat deposition, e.g. heart and paunch girth. It seems reasonable to assume that certain genes have a stimulating effect on skeletal growth in general and that these genes thereby influence not only the height of withers and depth of chest but also body length and weight. Similarly it is probable that heart girth and paunch girth are influenced by genes which have a general effect on muscle development.

WRIGHT (1932) classified the genes which influence body size into three categories: (1) those which have a general effect on growth and therefore influence all body measurement and the weight; (2) those which influence only a group of traits, e.g. the size of the skeleton but not the muscle development; and (3) genes which influence only a single trait. These relationships are illustrated schematically in Fig. 11:3.

TOUCHBERRY assumed that the gene systems specific for skeleton growth influenced withers height, body length, chest depth and body weight, but not heart and paunch girth. He further assumed that the group of genes specific

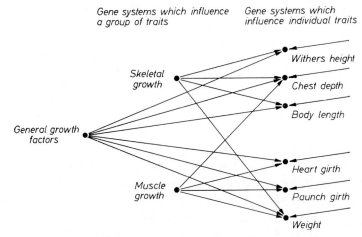

Fig. 11:3. Schematic diagram illustrating how body size is influenced by general growth factors and by group and trait specific gene systems. (TOUCHBERRY 1951)

to muscle development were involved in the control of chest depth, heart and paunch girth as well as weight. With these assumptions it was possible to estimate the influence of the general growth factors as well as the group specific and trait specific gene systems on the genetic variation of various traits. Between 50 and 70 per cent of the total genetic variation in all the traits, except paunch girth, was estimated to depend on the gene systems, which influenced the growth in general. There was also a tendency for the group specific gene systems to be more important than those which influenced only a single character.

LUSH and his co-workers found further evidence in support of the hypothesis of gene systems specific for the skeleton and muscle development. They determined the genetic correlation between the above-mentioned body measurements and the weight at different ages. It was found that with increasing age there was an increase in the genetic correlations between traits belonging to the same group, e.g. between body length and height of withers, but a decrease in the correlation between traits from different groups, e.g. between height of withers and paunch girth. On the other hand, the phenotypic correlations between traits in the same individual were consistently higher at 6 months and 1 year than at 2–3 years of age. The reduction as age increased was greatest for those traits belonging to different groups, i.e. the skeletal and muscle growth groups. The changes in the genetic and phenotypic correlations with increasing age indicate that those gene systems which have a general effect on growth have a relatively strong influence at early ages and that group-specific, and most likely the trait-specific, gene systems become more important with increasing age up to maturity.

The data from milk recording societies and the heifer testing stations have been used to study the influence of heredity on the various body measurements of Red Danish dairy cattle. The heart girth of heifers is measured about two months after first calving in connection with the normal milk recording, and reports from the Danish heifer testing stations contain data about the weight and certain body measurements of all animals in the month of March (approximately five months after calving). MASON and his associates (1957) found that the heritability for heart girth was about 0·4 in both sets of data. The heritability of withers height was a little higher (0·51), and for body-weight it was estimated to be 0·37. UDRIS (1961) carried out an analysis of other data from the Danish heifer testing stations, which included, in addition to the withers height and heart girth, measurements of the width of the hips and thurl. These two latter traits were found to have heritabilities of 0·54 and 0·47 respectively.

Both the above investigations resulted in lower coefficients of heritability than those obtained for corresponding traits in the Iowa analyses referred to previously. The main reason for this is probably that there was considerable variation in the age of weighing in the Danish data, whereas this variation was very small in the American investigation. Both MASON and UDRIS computed the phenotypic and genetic relationships between the various body measurements. The genetic correlation for heart girth and withers height was about 0·4 which was considerably less than in the American data (0·65). The highest genetic correlation obtained by Udris was between thurl width and heart girth (0·67) and the lowest was between width of hips and heart girth (0·32).

A summary of the influence of heredity on birth weight, growth rate, body weight and other body measurements is shown in Table 11:5. Traits which are economically important, such as growth rate, feed conversion, and weight at maturity, all have medium heritabilities so the opportunities for improvement by selection seem to be fairly good.

Table 11:5 The heritability of birth weight, growth rate, mature weight and certain body measurements of cattle, sheep and pigs

Trait	Cattle	Sheep	Pigs	
Birth weight	0·4	0·3		
Weaning weight	*0·2	0·3		
Daily gain from weaning to slaughter	*0·4 †0·3–0·6		‡0·3	§0·6
Feed conversion	*0·4		‡0·2	§0·5
Mature or nearly mature weight	0·4	0·4		
Wither height	0·5–0·8			
Heart girth	0·4–0·6			

* Beef cattle.   † Dairy cattle.   ‡ Group feeding.   § Individual feeding.

MEAT QUALITY

Quality demand and methods of assessment

The value of a carcass is determined by many factors. Among them is the killing-out percentage, i.e. the weight of the carcass as a percentage of the live-weight; the conformation of the carcass and the relative proportions of meat and bone; the amount and distribution of the fat and finally the colour and structure of the lean and fat (Fig. 11:4). All these factors are influenced by the age and stage of development of the animal. It is necessary, therefore, in studying the influence of heredity on quality characteristics, to base the investigation on animals that have reached the same stage of development. In Scandinavia and a

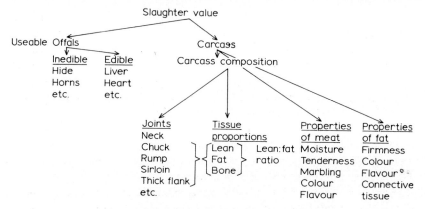

Fig. 11:4. Factors which determine slaughter value.

number of other countries pig breeding is based on the results of the testing stations mentioned earlier. From the information recorded at these stations it has been possible to obtain satisfactory estimates of the influence of heredity on the various quality characteristics. The situation with regard to sheep and cattle is not quite so clear though several investigations have been carried out in recent years.

Certain quality traits can be determined by actual measurement, e.g. the back-fat thickness of pigs; other traits must be judged entirely by subjective methods of scoring, e.g. the distribution of subcutaneous fat on beef cattle carcasses. Research is in progress to develop methods by which quality traits, now assessed subjectively, may be objectively measured, e.g. meat tenderness. From a breeding point of view it would be valuable if the quality traits could be measured on the live animal.

Carcass conformation and leanness. The relative proportions of the head, feet and other less valuable parts should be as small as possible, whilst the hind quarters should be well developed and well fleshed. The percentage of bone should, of course be small. These characteristics cannot be measured without an expensive dissection of the different tissues. The alternatives are subjective scoring or measurement of other traits which have a high correlation with the trait to be assessed.

In Danish and Swedish pig progeny testing, leanness is determined by measurements made on photographs of the cross section exposed after cutting the side at the level of the last rib and at right angles to the line of the back (Fig. 11:5). The area of the *m. longissimus dorsi*, or 'eye' muscle, is then measured by means of a planimeter and the abdominal muscles, collectively known as the 'streak', are measured as shown in the figure. The area of the 'eye' muscle should be as large as possible and the muscles in the streak should extend right up to the 'eye' muscle, as otherwise the value of the streak is much reduced.

In cattle the possibilities of indirectly measuring bone percentage and leanness have been investigated. According to SKJERVOLD, (1958) the weight of the cannon bone (*metatarsal*) in relation to the carcass weight is a reasonably

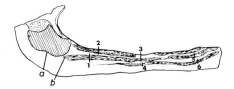

Fig. 11:5. Cross-section of a bacon side behind the last rib. The 'eye muscle' area has been specially outlined. The backfat thickness is measured at *a* and *b* and the thickness of the streaky muscles is measured at points 1–6. (HELLBERG 1961)

good measure of the bone percentage of the carcass. The fullness of the muscles in the rump and thighs may, according to the same investigation, be estimated with a relatively high degree of accuracy from an index obtained by multiplying the length of the thigh by its circumference measured at a certain distance from the pubic bone. The partial correlation between this index and the weight of lean in the hind quarters was 0·7, when the carcass weight was held constant.

Several American studies with beef cattle indicate that it is possible to obtain a reasonably good estimate of the leanness on the live animal by dividing the heart girth by the height of withers. The higher this index, the higher will be the killing out percentage and amount of lean at a constant liveweight.

Colour of lean. The colour of the lean is due mainly to the myoglobin in the muscles. The lean of young animals is pale, whereas in older animals, especially old horses, it is dark coloured. Pigmeat is light red to pink, and it is often evaluated according to a scale of points. The colour of pigmeat has become increasingly important in recent years on account of the spread of a condition called 'muscle degeneration' or 'pale, soft, exudative porcine musculature'. The muscles of affected pigs show, after slaughter, a pale, greyish-pink colour and their juice-holding capacity is greatly diminished. In severe cases the meat is unsaleable.

Fat content, distribution and colour. There are two types of body fat, depot and intra-muscular. With regard to depot fat, only the subcutaneous fat can affect the carcass quality, since the fat around the kidneys, intestines and other viscera are part of the offal. The carcass should be covered by a thin and even layer of subcutaneous fat, which prevents the meat from drying out and thereby pro-

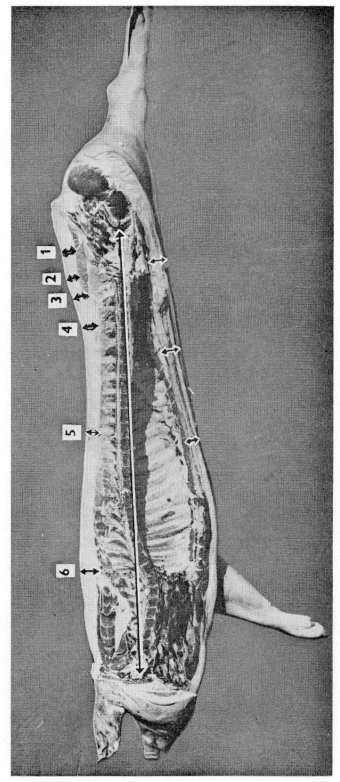

Fig. 11:6. A half pig carcass showing the points for measurements of body length and the thickness of back fat and belly.

motes the desirable formation of lactic acid in the muscles. Lean which is covered by a protective layer of fat becomes tender and juicy; but the fat layer should not be too thick. The requirements for fat thickness vary from country to country. In Sweden, for example, for pigs to be classified in the highest grade the back-fat thickness must not exceed 25 mm at the thinnest part of the back

Fig. 11:7. Beef hindquarters. The one on the left has an insufficient covering of subcutaneous fat whereas the one on the right has the right amount of finish with an even distribution of fat over the surface of the carcass. (SoL-Film)

(Fig. 11:6). Neither cattle nor pigs can qualify for the higher grades if the subcutaneous fat is too thin. Such carcasses are used mainly for manufacturing purposes. Fig. 11:7 shows examples of beef carcasses with different degrees of fat coverage.

Since the back-fat thickness of pigs is a decisive factor in assessing the quality of the carcass, attempts have been made to measure the fat thickness directly on the living animals. A number of different methods have been used, the most promising of which would appear to be ultrasonic measuring. When ultrasonic waves are propagated at right angles to the skin of the animal some of the waves are reflected back by the interface between the skin and the fat, others by the interface between the inner and outer fat layers, and still others from between the interface of the fat and muscle. The time that elapses between the signal and the echo is proportional to the thickness of the fat; and with the aid of special electronic equipment the thickness of fat can be read directly. The accuracy of these measurements approaches that of measurements taken on the carcass. Similar measurements have been performed on cattle, to study muscle development.

The *intra-muscular fat* is presumed to have a positive effect on meat quality since it is deposited between the muscle fibres and bundles. The effect of this when cooking is to make the meat juicy and tender. Because it resembles marble in appearance, this characteristic of meat is often referred to as 'marbling'.

The *colour of the fat* is of particular importance, since the consumer demands an almost colourless fat. Pigs normally have colourless fat; the fat of cattle, sheep and horses, on the other hand, is very often more or less yellow-coloured. The intensity of the yellow depends on age, feeding and the individuality of the animal; and it is due to carotene, the precursor of vitamin A. Carotene is found in large quantities in plants and some of it is stored in the fat. From a nutritive point of view, yellow fat can hardly be regarded as a disadvantage, but the consumer believes that the yellow colour of the fat indicates that the meat is from an old animal. There is undoubtedly some truth in this belief since the older animals generally have a yellower fat than younger animals due to the fact that when the depot fat is metabolised there is less breaking down of the carotene. Nevertheless, deep yellow colour is not always a sure sign that the meat is from an older animal, since older animals can have pale yellow fat and younger animals can have deeper coloured fat. That inheritance plays a part in fat colour is supported by the fact that certain breeds of cattle, e.g. Jersey, have a yellower fat than other breeds, e.g. the Friesians. In Icelandic sheep predisposition to deep yellow fat has been shown to be due to a recessive gene.

Tenderness and texture of meat. Organoleptic characteristics of meat, such as the aroma and flavour, cannot be measured objectively; they must be evaluated subjectively by special taste panels after the meat has been prepared according to standard procedures. Relationships have been found between the aroma and colour and also between colour and flavour; the darker the meat, the stronger the flavour. Males have darker meat than females and older animals darker meat than young animals.

The tenderness of meat is evaluated by organoleptic methods, i.e. taste panels, and also by physical measurements of the resistance to shearing and crushing, etc. Other methods of assessing tenderness are based on measurement of closely related characteristics, such as the amount of connective tissue and the thickness of the muscle fibres. Organoleptic studies are generally performed by a panel of specially trained tasters who score the tenderness according to a scale of points from 1 to 10, where the highest score is allotted to the tenderest meat. Physical devices for measuring tenderness have been constructed on the basis of several different principles. The so-called 'denture tenderometer' is mainly a mechanical reproduction of the human chewing action. The most widely used apparatus for assaying meat tenderness is the Warner-Bratzler shear, which measures the force required to shear a sample cut of the meat. The correlation between shear resistance and tenderness as assessed by organoleptic methods is usually about -0.8.

Muscles which are in constant motion and are subjected to the most stress and strain become tougher and harder than muscles which perform very little work, e.g. the fillet muscle (*m. psoas*). There is a strong relationship between the degree of tenderness and the type and extent of connective tissue which surrounds the muscle fibres and bundles of fibres. The superficial rib muscle consists of well-defined muscle fibre bundles surrounded by well developed connnective tissue. The meat from the rib muscles is tough and requires a prolonged cooking. In the fillet muscle hardly any segmentation of the muscle by connective tissue can be distinguished, and the meat is consequently tender and easily digested.

Carcass quality in pigs

A prerequisite for a good yield of bacon and chops is that the slaughter pig has a long body. The length is highly dependent on the number of thoracic and lumbar vertebrae, which in domesticated pigs varies between 20 and 23. In a study of Norwegian pigs BERGE found that pigs with a total of 20 thoracic and lumbar vertebrae had a length of 77·9 cm from the atlas, or first cervical vertebra, to the sacrum and 88·1 cm to the anterior tip of the symphysis pubis. When the number of vertebrae increased to 23 the corresponding measurements were 86·0 and 94·3 cm, respectively; a very substantial increase. The total length increased by 1·84 cm per vertebra. Simultaneously with the increase in number of vertebrae there was an increase in the number of ribs. The fact that the number of vertebrae is under strict genetic control is illustrated by the results from the following matings (after BERGE, 1949).

Number of thoracic and lumbar vertebrae		Number of offspring
Parents	Offspring	
20 × 20	20·32	87
21 × 21	21·03	355
22 × 22	21·83	1860
23 × 23	22·56	33

Since the number of vertebrae is genetically influenced, the body length is affected to a large degree by inheritance. The heritability of carcass length of pigs slaughtered at 90 kg liveweight has been estimated in several countries. There is close agreement between the results which indicate a heritability of about 0·5. Back-fat thickness and belly thickness also show a high heritability of between 0·5 and 0·7.

Systematic selection results, therefore, in quite rapid improvement of body length as well as back-fat and belly thickness. This is clearly illustrated by the results from the Danish testing stations. During the period from 1930 to 1964 the average carcass length increased from 89·4 cm to 96·3 cm, an increase of

8 per cent. In the same period the back-fat thickness decreased from 3·94 cm to 2·61 cm, a reduction of no less than 34 per cent (Fig. 11:8). The reduction in average back-fat thickness was also accompanied by a reduction in the standard deviation from 0·44 to 0·30.

The back-fat thickness of carcasses from pigs reared at the Scandinavian pig testing stations is measured at the shoulder, mid-back and over the 'rump' muscle (*m. gluteus medius*) (Fig. 11:6). The aim is to achieve a carcass with a uniform distribution of fat along the back. One way of measuring the uniformity

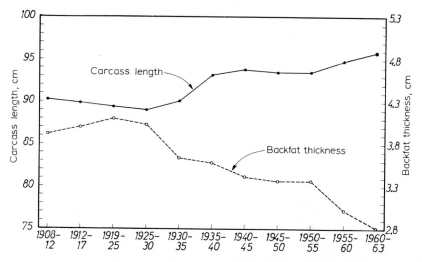

Fig. 11:8. Changes in body length, backfat and belly thickness of Danish Landrace pigs in the period from 1908 to 1963. The pigs were slaughtered at 90 kg liveweight. (Data from Danish pig progeny testing)

is to calculate the standard deviation of the back-fat measurements; the less the standard deviation the more uniform the distribution of fat. OSTERHOFF (1956) found that the heritability of this index of uniformity was of the order of 0·2 for pigs slaughtered at 90 kg liveweight.

The extent to which the relative proportions of the various joints are genetically determined, e.g. the weight of the ham as a percentage of the weight of the whole carcass, has also been studied; Canadian investigations showed that the heritability of weight of ham and shoulder was between 0·4 and 0·5.

A number of quality characteristics which are assessed by subjective scoring have also been the object of heritability studies. The shape of the ham and firmness of fat have heritabilities of about 0·4, whereas the genetic influence on the variation in points for leanness appear to be even greater.

JONSSON (1965) estimated the heritability of the area of the *longissimus dorsi* and the total lean and fat at the cut immediately after the last rib. The heritability of muscle area and the area of the *longissimus dorsi* was about 0·3, whereas the heritability of the fat area was about 0·5. The investigation was based on data obtained from 2709 barrows and 2714 gilts tested at the Danish progeny testing stations between 1958 and 1962. The colour of the lean showed much larger variation in the gilts than in the castrates, the respective heritabilities being estimated to 0·44 and 0·16.

Inbreeding and crossing appear not to affect the various quality traits as long as the inbreeding is not too intensive. In the crossing experiment at the Agricultural College of Sweden, referred to earlier, some twenty measurements and quality characteristics were studied. No definite positive or negative effect of the crossing could be demonstrated; which indicates that the positive changes achieved for growth rate and vigour were, at least, not counteracted by negative alterations in quality traits.

Many of the production characteristics of pigs are correlated with each other when measured at a constant market weight. A long body is generally associated with a thinner back-fat and less fullness in the hams than a shorter, more compact body. Table 11:6 summarises the results of investigations into the genetic and phenotypic correlations between various production characteristics of pigs from Danish and Swedish progeny testing stations.

Table 11:6 Genetic and phenotypic correlations between various production traits in pigs. The phenotypic correlations refer to two traits in the same individual and are given below the diagonal. The genetic correlations are given above the diagonal (after JOHANSSON and KORKMAN, 1950; FREDEEN and JONSSON, 1957 and JONSSON, 1958)

		L	Ba	Be	G	Fe	H	F	M
Body length	L	–	−0·4	+0·2	+0·1	−0·01	−0·1	−0·2	
Back-fat thickness	Ba	−0·3	–	−0·5	−0·2	+0·2		−0·2	−0·9
Belly thickness	Be	−0·1	+0·2	–					
Daily gain	G	0	−0·1		–	−0·9			
Feed units/kg gain	Fe	0	+0·2		−0·8	–			
Points for ham	H	−0·3					–	+0·3	
Points for fat firmness	F	−0·2	+0·4				+0·2	–	
Points for leanness	M		−0·8						–

When the same trait has been studied in several independent investigations, the results have shown close agreement. The genetic correlations are of special interest from a breeding point of view, since they indicate the possibilities for simultaneous improvement of two or more traits. The relationships appear to be such that selection for increased body-length will automatically bring about a reduction in back-fat thickness, an increase in daily gain, and a reduction in the number of feed units required for each kg liveweight gain; in other words, a simultaneous improvement of all four traits. On the other hand, this selection may be expected to result in some reduction in the points for shape of ham and firmness of fat. It can also be seen from the Table that an increase in back-fat thickness also increases the number of feed units per kg growth which must be expected, since fat is richer in energy than lean. The genetic correlation between these traits is thus positive. It is surprising, however, that the genetic correlation between back-fat thickness and belly thickness indicates that an increase in the former brings about a decrease in the thickness of the belly.

Carcass quality in cattle

A number of investigations into the influence of inheritance on various quality traits have been carried out with American breeds of beef cattle. The heritability of dressing percentage appears to be quite high (about 0·6) and the same applies to the area of the *longissimus dorsi*, a trait which is a good indicator of the leanness of the carcass. The carcass quality, expressed as the total score allotted for several characteristics, has a medium heritability of 0·3–0·4, but the results from different investigations vary widely, primarily due to the fact that in several cases the results are based on too small samples.

Quality characteristics were also studied in the investigations of SKJERVOLD in Norway and LILJEDAHL and LINDHÉ in Sweden, referred to earlier, in which they studied the progeny testing of A.I. dairy bulls for beef production. There were highly significant differences between the progeny groups with regard to dressing percentage in the Swedish data. The means for the progeny groups varied from 49·1 to 52·6 per cent and the heritability of dressing percentage was estimated to 0·21. One half of the carcass from each animal was cut into joints. The standard procedure in Sweden is to divide the sides into fore and hind quarters by a cut between the 10th and 11th ribs. The hind-quarter is, of course, the most valuable. The relative weight of the hind-quarters varied widely between the progeny groups and the heritability was estimated to 0·31. The area of the *longissimus dorsi* between the 10th and 11th ribs had a heritability of 0·20 and the same value was obtained for the average thickness of the subcutaneous fat and also for the percentage sirloin plus rumpsteak. Similar results were obtained in the Norwegian investigation, which also included boning of the carcasses. This latter study indicated that there were further genetic differences for several important traits, e.g. bone percentage.

Information about the influence of inheritance on the more intangible quality traits such as tenderness, flavour and juiciness is extremely limited. A number of investigations indicate that there are breed differences in respect of tenderness. KINCAID (1962) reported recently the results of a large cross breeding experiment which included some 2700 cattle in the southern part of the U.S.A. Brahman and Brahman × European breeds of cattle had distinctly tougher meat than the British breeds and Charollais. Other investigations have also pointed to the fact that Brahman and other Zebu breeds have a tough meat though the variation within breeds is quite large. Attempts have been made to estimate the heritability of tenderness but there has been little agreement between the results.

The effect of sex on carcass quality

The male animal has different, and with respect to carcass quality, inferior body proportions to those of the female. The shoulders and neck are more strongly developed, whilst the hind-quarters are relatively less developed than those of females. In addition, the muscle fibres are darker and coarser and the meat has a stronger flavour. The sex difference in flavour first becomes apparent after sexual maturity. The flavour of boar meat is so strong that all male pigs reared for slaughter are castrated.

A comparison of the carcass quality of group fed castrated males and gilts from the same litter shows that the latter have a consistently better quality. The length of gilt carcasses, slaughtered at 90 kg liveweight, are 5–10 cm longer and the back-fat is 2–3 mm thinner than castrated males. The points for leanness and shape of the hams is also better for gilts.

In the rearing of beef cattle it is often a question whether the bulls should be castrated or not. Castration results in two main changes in the steers, a relatively higher deposition of fat and a different conformation to that of bulls. The work of BRÄNNÄNG (1966) with identical twins, briefly referred to previously (p. 304), has contributed to the clarification of the effects of castration on carcass quality. All the animals were slaughtered at 25 months of age. At that age the liveweight of the bulls was 8 per cent higher than that of the steers, though they had received the same amount of feed. In spite of the fact that the bulls had a relatively heavier hide and head, the dressing percentage was significantly higher than in steers; which was due to the deposition of mesenteric fat in the latter. The carcasses of the steers were also much fatter. The bull carcasses contained 10 per cent separable tallow and 70 per cent lean whereas the steer carcasses had 16 per cent tallow and 64 per cent lean. The percentages of bone, ligaments and tendons were approximately the same. Furthermore the fat content of the tallow and lean was much higher in the steers, which means that their tissue had a higher energy content and consequently required more feed units per kg liveweight gain. Approximately 11·6 feed units per kg growth (carcass weight basis) were required for the steers compared to 10·5 for the bulls.

In so far as carcass conformation was concerned, the steers had a better distribution between the fore- and hind-quarters. The thighs and rump of the steers were relatively, but not absolutely, larger than in the bulls; but no differences in the points for leanness or other quality traits could be found between the experimental groups. The somewhat better carcass conformation of the steers did not economically counteract their lower dressing percentages and higher feed consumption per kg liveweight gain than those of the bulls.

Similar results were obtained by WITT and ANDREAE (1965) in a castration experiment with 10 monozygous twin pairs slaughtered at 18 months of age. One member of each pair was castrated when 6 months old. The average live

Table 11:7 The heritability of various carcass quality traits. In the case of pigs all assessments are made after slaughter at 90 kg liveweight

CATTLE		PIGS	
Trait	Heritability	Trait	Heritability
Dressing percentage	0·6	Body length	0·5
Points for carcass quality	0·3	Back-fat thickness	0·5
Cross section of 'eye muscle'	0·3	Belly thickness	0·5
Bone percentage	0·5	Weight of ham	0·5
		Points for ham conformation	0·6
		Points for fat firmness	0·4
		Points for leanness	0·7

weight at time of slaughter was 469 kg for the bulls and 419 kg (i.e. 10·7 per cent less) for the steers. The body proportions at 18 months were quite different. The withers height of the steers was 1·5 cm higher, while the width of chest, hips and pelvis was significantly greater in the bulls. The cost of feeding and management was approximately the same for bulls and steers, but the total income per animal was no less than 152 D marks (approx. £14) more for the bulls.

In summary, it can be said that the carcass quality of cattle and pigs shows quite high heritability (Table 11:7), and it may be assumed that the situation is very much the same in sheep.

CONCLUDING REMARKS

It is clear that the more important biological factors concerned with economical production of beef and pigmeat, feed conversion, growth rate and carcass quality, can all be improved by systematic selection. This is illustrated by the results already achieved in pigmeat production, even if all the improvement cannot be attributed to genetic causes. The rapid advances in pig breeding have been made possible by the introduction of testing stations, where a more objective evaluation of the genetic potential of breeding animals for production traits can be carried out.

Improvement of the growth rate and meat quality of cattle and sheep also requires some form of performance and/or progeny testing of males which are to be used extensively, e.g. A.I. bulls. This introduces the problem of improving the milk and beef production simultaneously, the possibilities for which are dependent on the genetic correlation between these traits. So far, only a very few investigations have been carried out which throw light on this important but complex problem.

MASON and co-workers (1957) and others have studied the relationship between milk yield and height of withers, as well as between milk yield and body weight in March, of heifers at Danish bull testing stations. When the age at first calving was held constant the correlation between height of withers and milk yield was found to be 0·36 and between body-weight in March and milk yield −0·07. However, the latter correlation cannot be taken as an indication of a negative correlation between body-weight and milk yield. According to a study by JOHANSSON (1954), the growth of high yielding heifers between calving and March was insignificant and in several cases there was even a loss in weight. The phenotypic correlation between the weight immediately after calving and milk yield was however positive, being 0·22.

A positive genetic correlation between body-weight and milk yield in the first lactation was obtained by MILLER and McGILLIARD (1959), in a study of 6000 cows from the Friesian, Guernsey and Jersey breeds. The weight was determined by a tape-chest girth measurement taken within 30 days of calving. After eliminating the influence of age it was found that the relationship between body-weight and milk yield was curvilinear. There was an increase in yield with weight up to a certain weight, 900–1000 lb for Friesian heifers, after which there was no further increase. Genetic correlation between body-weight and milk yield within herds was estimated to be 0·3.

The investigation by BLACKMORE and co-workers, referred to earlier, into the genetic relationships between various body measurements of American Holstein cattle also included a study of the effect of body size on the yield of fat corrected milk of 334 dam-daughter pairs. The body measurements were taken at 6, 12 and 24 months of age. The genetic correlations varied considerably according to the way in which they were calculated and were not statistically significant. There was, however, at all ages a positive genetic relationship ($r = 0\cdot2-0\cdot3$) between milk yield and withers height which is in agreement with the results obtained by Mason. There was a negative genetic relationship between milk yield and chest girth but there was no genetic correlation between milk yield and the weight at the various ages.

These results indicate that selection for increased milk yield should result in a simultaneous alteration of the body proportions with an increase in withers height and a decrease in chest girth. The investigations referred to earlier showed that there was a positive relationship between carcass quality and the ratio of chest girth to withers height. Selection for increased milk yield may therefore be expected to lead to a reduction in carcass quality. To some extent this is supported by the results of an earlier investigation by COOK into the relationship between the carcass quality of 83 dairy Shorthorn steers and the yield of FCM of their dams. The following results were obtained.

Milk to muscle/ratio	−0·07
Milk to percentage fat in carcass	−0·17
Milk to carcass quality score	−0·20

The sons of cows with a high milk yield had, on the average, leaner carcasses than sons of cows with a lower milk yield. The muscle to bone ratio was influenced only very slightly by the milk yield of the dams but the changes in the subjective

Table 11:8 *Correlation between milk yield and daily gain of half-sib groups in progeny tests* (after JOHANSSON, 1964)

Breed	Sex of animals raised for slaughter	Weight at slaughter	No. of progeny groups	No. per group	Correlation between half sib groups	Author
Dairy Shorthorn	Steers	450 kg.	43	6–8	0·27	Mason (1962)
Red Poll	Steers	450 kg.	49	8–10	0·05	Mason (1962)
British Friesian	Steers	450 kg.	33	8–10	0·01	Mason (1962)
Red Danish	Bulls	220 kg.	31	14	0·03*	Mason (1962)
German Friesian	Bulls	350 kg.	25	10–12	0·22	Langlet (1963)
Spotted Mountain	Bulls	500 kg.	9	10–12	0·13	Bogner and Burgkart (1961)
Israeli Friesian	Bulls	400 kg.	15	15–60	0·06	Bar-Anan and Levi (1963)

* Within stations.

carcass quality scores were relatively large. It must be remembered, however, that in this case the data were very limited.

The results of BLACKMORE and co-workers' study of the relationship between the body-weight at different ages during rearing and the milk yield of the mature animals suggest that growth rate and milk yield are to a large extent genetically independent. In several other investigations, however, a positive relationship has been found between growth rate and milk yield. LANGLET (1963) studied the progeny test results of German Friesian bulls, which were tested on the milk yield of their daughters and on the carcass quality of their sons. About 10 bull calves were included in each progeny group and their growth rate, feed conversion and carcass quality were evaluated at special testing stations. The daughter groups were large and distributed over several different herds. The correlation between the deviation of the daughter groups from the contemporary herd average milk yield and the growth rate of their paternal half brothers was estimated to 0·22.

The progeny testing of bulls on the basis of the meat production of their sons and milk yield of their daughters has also been carried out in the Scandinavian countries, England and Israel. In Denmark the testing takes place at special testing stations, whereas in England and Israel it is done in the field. The results of these progeny tests are summarised in Table 11:8. The correlation between the rate of gain of the sons and the milk yield of the daughters are small and not significant. However, with the exception of the Red Poll group all the correlations point in the same direction, i.e. the two traits have a slight positive correlation. In Langlet's investigation data was collected on the milk yield of the dams whose bull calves were put on test. The correlation between the growth rate on a carcass basis (carcass weight divided by age in days) of 231 bull calves and the milk yield of the dams (305-day lactation) was statistically significant and of the order of 0·13.

In summary, it appears from the investigation so far carried out that there is a positive relationship between growth rate and milk production. On the other hand, selection for high milk yield may be associated with changes in body proportions so that the withers height increases and the chest girth decreases. Until more research results are available it is impossible to assess the magnitude and economic importance of these correlated changes in body proportions. Should a reduction in chest girth be due mainly to a reduction in fat deposition, the change need not be regarded as a problem; the consumers' demand is for leaner meat, and an adequate layer of subcutaneous fat can be obtained by suitable feeding during a short period before slaughter. However the question is quite different if selection for high milk yield should also lead to a deterioration in fullness of the muscles. Further research is necessary in order to clarify these important problems.

SELECTED REFERENCES

AMERICAN MEAT INSTITUTE FOUNDATION. 1960. *The Science of Meat, and Meat Products* (438 pp.). W. H. Freeman and Co. London.

BLACKMORE, D. W., MCGILLIARD, L. D. and LUSH, J. L. 1958. Genetic relations between body measurements at three ages in Holsteins. *J. Dairy Sci.*, **41**: 1045–1049.

CRAFT, W. A. 1958. Fifty years of progress in Swine breeding. *J. Anim. Sci.*, **17**: 960–980.

GRAVERT, H. O. 1962, 1963. Untersuchungen über die Erblichkeit von Fleischeigenschaften beim Rind. *Z. Tierzücht. ZüchtBiol.*, **78**: 43–74, 139–178.

HANSSON, A. and CLAESSON, O. 1960. Report on the research with monozygotic cattle twins. *Publs. Eur. Ass. Anim. Prod.*, **9**: 23–75.

JOHANSSON, I. 1964. The relationship between body size, conformation and milk yield in dairy cattle. *Anim. Breed. Abstr.*, **32**: 421–435.

JONSSON, P. 1965. *Analyse af egenskaper hos svin av den danske landrace (Analysis of Characters in the Danish Landrace Pig)* (490 pp.). Landhusholdningsselskapets forlag, Köpenhamn. English summary (34 pp.).

TERRILL, C. E. 1958. Fifty years of progress in sheep breeding. *J. Anim. Sci.*, **17**: 944–959.

WARWICK, E. J. 1958. Fifty years of progress in breeding beef cattle. *J. Anim. Sci.*, **17**: 922–943.

WENIGER, J. H., STEINHAUF, D. and PAHL, G. H. M. 1964–5. Metoden zur Bestimmung der Fleischbeschaffenkeit. 1–3. *Fleischwirtschaft*, **44**:45.

WITT, H. M. and ANDREAE, U. 1965. Kastrationseffekte bei männlichen eineiigen Rinderzwillingen. *Z. Tierzücht ZüchtBiol.*, **81**: 1–45.

12 Wool Production and Fur Quality

Sheep wool has a unique position among mammalian hair because of its durability, spinning properties and ability to absorb dyes and moisture. These properties are determined by the specific structure of the wool fibre. Like other mammalian hair, wool fibre is surrounded by a protective layer, the *cuticle*, which consists of scale-like keratinised cells. The main body of the fibre is made up of the *cortex*, which consists of long fusiform cells tightly packed and lying parallel to the axis of the fibre. The strength and durability of the wool fibre is due to the effective packing of the cortical cells (Figs. 12:1 and 12:2). In some coarse wools there is a *medulla* within the cortex. The medulla consists of relatively large rhombic or cube-shaped cells with air spaces between them. The finer wool fibres are completely without a medulla. The occurence of the medulla decreases the strength of the fibre, and the greater the proportion of cortex to medulla the stronger the fibre.

TYPES OF WOOL HAIR

Three main types of hair can be distinguished in sheep. They may all occur in the same individual:

(a) *Short*, thick *guard hairs* filled with medulla. They are found on the limbs and head of the majority of sheep breeds. The fleece of the wild Mufflon sheep consists mostly of these short guard hairs. They lack the ability to absorb dyes and cannot be used in textiles.

(b) *Long guard hairs*, over 3 cm in length. There is often a medulla but not to the same extent as in group (a). This hair type occurs in the north European short-tailed and in the English long-woolled breeds of sheep, and the Karakul, but is mixed with fine-wool fibres in the undercoat. Brittle guard hairs with air spaces in the medulla and dull, lifeless appearance is usually called *kemp*. The occurrence of too much kemp is regarded as a serious fault in the wool. The same is true of the *mane hairs*, which are straight and stiff like the mane in horses.

(c) *True wool fibre*. The fineness can vary considerably from the coarse and sometimes medullated Cheviot wool (mean dia. 35 μ) to the very fine wool of the Merino sheep (dia. about 16 μ). The undercoats of the sheep breeds with mixed wool are usually very fine and therefore referred to as down hairs.

The wool fibres are formed in hair follicles which appear as invaginations of the epidermis into the dermis (Figs. 12:1 and 12:3). The fibre grows from a papilla at the bottom of the hair follicle. A number of sebaceous and sweat glands open into the upper part of the fibre canal. The primary follicles are formed about 60

TYPES OF WOOL HAIR

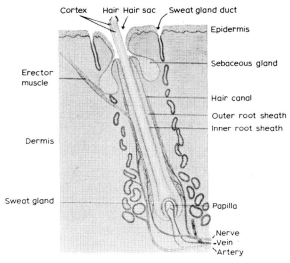

Fig. 12:1. Hair follicle with hair. (From DOEHNER 1958. *Handbuch der Tierzüchtung*, I. Verlag Paul Parey, Hamburg)

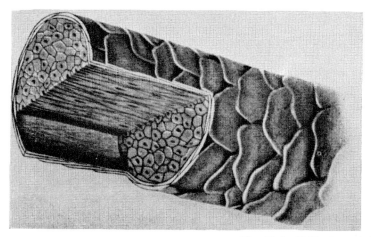

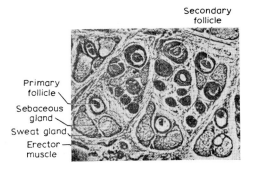

Fig. 12:2. Half schematic illustration of a wool fibre. (REUMUTH 1954. *Handbuch der Schafzucht und Schafhaltung*, IV. Verlag Paul Parey, Hamburg)

Fig. 12:3. A horizontal cross section of skin from a Romney lamb, which shows a group of three primary follicles with secondary follicles as well as sebaceous and sweat glands (magnification *ca* × 100. Courtesy of Dr H. B. CARTER, ABRO, Edinburgh.)

days after fertilisation, and in conjunction with these a number of secondary follicles develop. The time for the formation of the secondary follicles, and their number, appears to be very important for the fineness of the wool fibres. The secondaries develop very soon after the primaries in the Merino breed, but in Karakul sheep development of the secondaries is delayed until about 6 weeks after the formation of the primaries. The different follicle types appear to compete for the substances from which the fibre is built. In the Karakul, the primaries succeed in taking the lead and the animal therefore produces coarse guard hair. In the Merino the many secondary follicles can keep pace with the primaries with the result that all the wool fibres are fine, with about the same diameter. With rare exceptions, the Merinos completely lack guard hairs in the fleece.

The wool fibres lie arranged in tufts of about 30–40 hairs held together by the wool wax or grease and special binding hairs to form *staples*.

The wool grease, which is formed in the sebaceous and sweat glands, is of great importance for the flexibility and durability of the wool. It consists of cholesterins mixed with fatty acid esters, free fatty-acids and various salts. Lanolin is manufactured from wool grease, and is used in cosmetics and the production of salves and skin creams. The amount of wool grease varies widely between different breeds; together with dirt and other impurities which are fastened in the fleece, the wool grease may constitute as much as 50 per cent of the total weight of the unwashed fleece.

WOOL QUALITY

In assessing the quality of wool, attention is paid to the quality of the individual fibres and staples as well as the character of the fleece as a whole. The wool fibres are assessed on a number of characteristics, such as length, fineness, uniformity and waviness (crimps) together with strength and elasticity.

The *length* of a wool fibre is defined as its length after one year's growth. The most usual measure is the *staple length*, which is the length of the wool in the crimped condition. The absolute length is also measured, i.e., when the fibres are in the stretched condition. In the Lincoln breed absolute fibre lengths of up to 450 mm have been recorded.

The *fineness* of the fibres is a deciding factor for the manufacturing use of wool. The finer the wool, the longer the yarn that can be obtained from a unit weight of the wool. In the spinning trade the measure of fineness is the number of skeins of yarn of a given length which can be spun from 1 lb (0·45 kg) of washed wool. These are called 'spinning counts' and should not be confused with the 'quality numbers' which are also concerned with fineness but are allotted to the unwashed wool.

In scientific studies the diameter of the fibres is used as a measure of the fineness. The diameter can be determined in several ways, for instance, from photographs of the cross section of fibres under the microscope and with the aid of the projection microscope. Some examples of the shape and size of cross sections of wool fibres from Swedish sheep breeds are shown in Fig. 12:4.

Fibres, from the Swedish Landrace, of the carpet-wool type show a wide variation in diameter, and the guard hairs from some animals can be highly medullated.

Obviously, in order to ascertain the suitability of the wool for the various processes, it is necessary to measure *uniformity* and the *frequency of medullated hairs*. The uniformity can be measured in several ways, such as calculating the coefficient of variation of the diameter of the wool fibres.

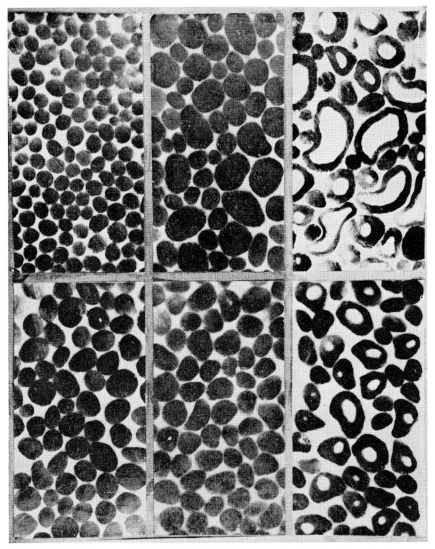

Fig. 12:4. The appearance of the cross section areas of typical wool samples from Swedish Landrace sheep (*top*) and British breeds (*bottom*).
Top left: fine-wool type; *Middle*: good; *Right*: medullated, coarse carpet wool.
Bottom left: Oxford Down; *Middle*: good; *Right*: coarse, poor quality Cheviot wool.
(Courtesy of DR S. SKÅRMAN, the Agricultural College of Sweden)

Sheep wool is characterised by the crimp and spiral twist of the fibres. Earlier, much attention was paid to the crimp, as it was believed that there was a strong relationship between the number of crimps per unit length and the fineness, i.e. the finer the wool, the more crimps. It has been shown, however, that the crimp is a rather poor measure of fineness, even though it does so happen that the fine wool breeds, such as the Merino, have considerably more crimps per cm than breeds with coarser wool, e.g. the Cheviot or Swedish Landrace.

The *strength* or *soundness* of the wool fibre is measured by determining the tension required to break the fibres. With regard to the individual hairs thicker fibres are strongest, but per unit of weight or in relation to the cross section area, the thinner fibres are much stronger. Another trait of importance is the *elasticity*. Good wool should be elastic; otherwise the material produced from it easily gives rise to 'knees' and 'elbows'.

In addition to the characteristics of the individual fibres, the shape of the wool staples are assessed, as well as the fleece as a whole. A cylindrical staple indicates that the wool fibres are of even length and about the same fineness. Staples of a wool with a high content of guard hairs are conical in shape, and the fleece has an open appearance. In Fig. 12:5 the appearance of wool staples from the three main types within the Swedish Landrace are illustrated: the carpet, the fur, and the fine wool types. In the first two types mentioned the staples are pronouncedly conical. Three typical wool samples from British breeds are also shown in the figure.

In wool-producing breeds a high density, i.e. large number of fibres per mm^2 skin area is highly desirable. Merino sheep have about 60 hair follicles per mm^2, whereas the Karakul sheep have only about half as many. The fleece should also have a good coverage of the body. Further, colour type, curliness and lustre are important characters in breeds used for the production of pelts for the fur trade. High lustre is also a requirement of wool which is used for special knitting and weaving handicrafts.

FACTORS WHICH INFLUENCE THE YIELD AND QUALITY OF WOOL

There is a close relationship between the amount of wool produced per individual and the quality. The factors which influence the one characteristic very often also affect the other. Unfortunately, an increase in the amount of wool is usually accompanied by a fall in the quality.

The yield of wool

The unwashed fleece weight is generally used as a measure of the yield of wool. It is true that it is the yield of clean wool that is decisive from an economic point of view; but the phenotypic correlation between the yield of unwashed and clean wool, from sheep of the same breed kept under the same environmental conditions, is so high (approximately 0·9) that the former measure can be said to be satisfactory. MORLEY (1951) has found that the genetic correlation between the yield of unwashed and clean wool in the Australian Merino is about 0·7.

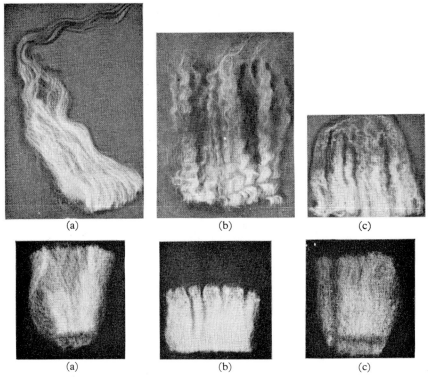

Fig. 12:5. Samples of different types of wool.
Top: Swedish Landrace (a) carpet-wool (b) fur-wool (c) fine-wool *Below*: Wool from three British breeds: (a) Oxford Down (b) Shropshire (c) Cheviot.

Fig. 12:6. Merino ewe with numerous skin folds.

Non-genetic factors. Age has some effect on the fleece weight; with increasing age, up to a certain limit, there is an increase in yield. According to a number of concurring investigations the effect of age is due to (*a*) the diameter of the fibres increasing with age, (*b*) the staple length decreasing with age, (*c*) pregnancy decreasing both diameter and length. During the first years the diameter, and therefore the yield, increases. After that, (*b*) and (*c*) take the upper hand and some decline can be observed.

Body size also influences yield. In a study of Swedish sheep breeds, JOHANSSON and BERG (1939), obtained a correlation of 0·36 between the liveweight of the ewes and their annual wool production from three to five years of age. Similar results have been obtained in several other sheep breeds. This relationship is quite natural, since the wool-producing area is larger in large sheep than in smaller individuals of the same breed. In the Merino breed there are animals with large and numerous folds of the skin (Fig. 12:6). Since these animals have a larger skin area than those with a tighter skin, it was believed earlier that the skin folds had a positive influence on wool yield. More recent studies have shown that this is not the case. Only the amount of wool wax increases. Selection is therefore being made against skin folds as they give rise to uneven wool quality, present difficulties at shearing and provide brooding grounds for parasitic insects.

Naturally, the nutritional *condition* of the sheep has a great influence on the fleece weight. Pregnancy and suckling lower the amount of wool produced. JOHANSSON and BERG, in a study of four breeds of Swedish sheep, obtained the following averages, expressed relative to the wool production of a ewe with a single lamb as 100: ewe with twins 97·3, single pregnancy without lactation 103·8, empty 112. In a number of other breeds, e.g. the German meat-type Merino, the reduction in the amount of wool produced with twin births was much greater than in the Swedish breeds. The differences are most probably due to the feeding conditions. Direct under-nourishment has a very strong negative effect on the yield of wool. The animal produces so-called 'hunger-wool', which is recognised by approximately normal length but decreased diameter, and brittle fibres.

Genetic factors. Investigations into the influence of heredity on fleece weight have been carried out mainly on the fine-wool breeds of the Merino type and also on some of the long-wool breeds of British origin. The coefficient of repeatability for the wool production of ewes from year to year is about 0·6 in sheep of the Merino type. Studies by TERRILL and HAZEL (1943, 1947), MORLEY (1955) and others indicate that the heritability of fleece weight in the fine wool breeds is about 0·4 when based on the correlation between the dams and their progeny or between paternal half sibs. The heritability appears to be somewhat higher for the amount of clean wool. According to concurrent reports from New Zealand and Canada the heritability for wool production in the long-wool Romney is considerably lower, about 0·15. In the British Welsh Mountain sheep which have wool of intermediate length the heritability has been estimated to 0·7. The results indicate that heredity has, in general, a relatively strong influence on wool yield but that this influence may be less in some breeds than in others. The reasons for these breed differences are unknown.

In Australia, the Merino data have been used to study the influence of heredity on the density of the fleece and the number of follicles per unit skin area. The heritability of both these traits was found to be rather high (0·5–0·8).

The wool covering on different parts of the body varies widely between different breeds of sheep. The Cheviot and Swedish Landrace have no wool on the head, whereas with the Shropshire only the tip of the nose is free from wool. TERRILL and HAZEL studied the heritability of scores for face covering in several breeds and crosses. In the Rambouillet the heritability was 0·56, and about the same result was obtained for three new American breeds formed from crosses between long wool breeds and sheep of the Merino type.

Wool quality traits

The influence of non-genetic factors. The majority of factors which affect fleece weight have an effect also on the quality. As already pointed out, under the *influence of increasing age* the fibre diameter increases and the length declines, i.e. the quality is lowered. This is illustrated by an investigation carried out by BERGE (1942) with the Norwegian Cheviot breed. The growth in length per day during the first year was, on the average, 0·52 mm, but only 0·44 mm in the third year. At the same time the thickness of the fibres increased from 32·8 μ to 34·9 μ.

Nutrition has a very big influence on the wool quality. A high level of feeding gives thicker and longer fibres. Periods of high level and adequate feeding interrupted by periods of low level and deficient feeding give rise to an uneven wool with low durability. An investigation with Swiss Landrace compared the wool produced during the summer pasture grazing with the winter period indoors. The diameter of the summer wool was 2·4 μ larger and had greater durability and elasticity.

SKÅRMAN and NÖMMERA (1954) found similar differences in a study of 95 Swedish Cheviot rams. The diameter of the fibre in the wool produced during the winter months was 7·8 μ less than the autumn wool. A significant correlation was found, however, between the fibre diameter of the wool of rams in the spring and autumn, calculated within years and flocks.

The effect of sex on wool quality is that rams have a coarser wool than ewes. The difference in the Swedish breeds was found to be very small (1–2 μ). In a study of the American Rambouillet, it was found that at weaning ram lambs had about 4·4 mm longer wool staples than ewe lambs.

The influence of heredity on wool quality. The large breed differences with regard to wool length, fineness and medulla content indicate that these traits are genetically determined. HARDY (1950) found the following values for the daily growth in length of the wool fibres in six different breeds:

	mm		mm
Rambouillet	0·20–0·33	Corriedale	0·38–0·48
Merino	0·26–0·32	Oxford Down	0·49–0·52
Shropshire	0·32–0·46	Romney	0·56–0·57

The daily growth in length was thus only half as great in the fine wool breeds, Rambouillet and Merino, as in the Romney sheep. The length of the wool is a typical quantitative trait where breed crosses lead to an intermediate result. HAZEL and TERRILL (1943) studied a large number of dam-daughter pairs and groups of paternal half sibs of the Rambouillet breed in the U.S.A. and found that the heritability of staple length at weaning, as well as at one year of age, was about 0·4. Similar results have been obtained in the Merino and Romney breeds.

On the basis of wool characteristics, the Swedish Landrace is divided into carpet-, fur- and fine-wool types. Matings between these types occur quite often, and studies by SKÅRMAN of the lambs and one-year-olds from these crosses clearly show that the wool types are influenced by several gene pairs. The types are not constant in matings *inter se*.

The heritability of *medullated guard hairs* in the lamb fleece or the adult sheep has been thoroughly investigated in several breeds. Normally Romney lambs have no great amounts of guard hairs, but some lambs have a high frequency of long, coarse medullated guard hairs, and this gives them a 'hairy' appearance. DRY (1955) and others are of the opinion that the hairiness is determined by an incompletely dominant gene N, which, apart from giving rise to a large number of coarse guard hairs, also stimulates the growth of horns. The homozygote NN is horned in both sexes. The gene also reduces the crimpiness of the wool. The adult animals have a 'carpet-like' wool type with a high frequency of kemp, which is shed at certain periods. Other investigations with the Romney breed indicate that hairiness can also be caused by homozygosity for a recessive gene. The inheritance is probably not so simple as was first thought, though clearly the type is under strong genetic control.

BRYANT found correlations of 0·25 and 0·28 between sire-progeny and dam-progeny respectively as to the occurrence of kemp in the fleeces of the Scottish Blackface breed. The heritability was thus of the order of 0·5, so that selection against this trait should give good results.

The number of crimps per unit staple length appears to be under relatively

Table 12:1 The heritability of wool yield and quality

Trait	Breed	Heritability
Yield, unwashed wool	Merino (MORLEY, 1955)	0·40
	Rambouillet (TERRIL and HAZEL, 1943)	0·28
	Romney (MCMAHON, 1943)	0·10–0·15
	Welsh Mountain (DONEY, 1956)	0·70
Yield, clean wool	Merino (MORLEY, 1955)	0·47
	Rambouillet (TERRIL and HAZEL, 1943)	0·38
Wool cover on head	Rambouillet (TERRIL and HAZEL, 1943)	0·56
Staple length	Merino (MORLEY, 1955)	0·52
	Rambouillet (TERRIL and HAZEL, 1943)	0·36
No. crimps per unit length	Merino (MORLEY, 1955)	0·47
Quality number	Romney (RAE, 1948; 1950)	0·27; 0·41
Occurrence of kemp	Scottish Blackface (BRYANT, 1936)	0·53

strong genetic control. In the Australian Merino the heritability has been estimated to about 0·4.

There are very few direct investigations into the influence of heredity on the fibre diameter. SHELTON and co-workers estimated the heritability to 0·57, based on a small amount of data from the Rambouillet breed. RAE and others calculated the heritability of 'quality numbers' of fleeces of the Romney breed as being between 0·3 and 0·4.

The majority of traits which influence the yield and quality of the wool have, as shown above, a relatively high heritability. A summary of the influence of heredity on wool traits is given in Table 12:1.

Genetic relationship between wool traits

From the heritability coefficients shown in Table 12:1 it will be clear that selection for the majority of the individual traits which determine the fleece weight and quality of the wool should result in rapid improvement. This is especially true of quality traits. Experience in practical sheep breeding has shown, however, that selection for a single trait, e.g. fleece weight, leads to a deterioration in quality, and vice versa. These practical observations have been verified in scientific investigations.

RAE made estimates of the genetic correlations between the various wool traits in the Romney breed. The fleece weight was correlated positively with the staple length (0·25), and also with the occurrence and extent of medullated guard hairs (0·28); but it was negatively correlated (−0·47) with quality numbers. The staple length was positively correlated with the occurrence of medullated coarse guard hairs (0·41) and highly negatively correlated with the quality number (−0·73). Similar results have been obtained by MORLEY in studies with the Australian Merino. There was a positive genetic correlation of 0·39 between the clean fleece weight and staple length. The latter trait was negatively genetically correlated with the number of crimps, points for skin folds and body weight.

It would appear, therefore, to be very difficult to combine in the same breed high fleece weight and high quality. Breeds which produce wool for top quality cloth (in the clothing industry), e.g. Merino and Rambouillet, will apparently in the future continue to yield only a relatively small amount of wool, even if some improvement can be achieved by selection. If the fine wool breeds are to remain competitive, the lower fleece weight must be compensated for by a higher price of the finished products.

PELT CHARACTERISTICS

The lamb pelts from certain sheep breeds with mixed wool are used in the fur trade. The most important of these is the Karakul which originates from the steppe regions of the Caspian Sea. The breed is now widely distributed in many countries. The indigenous Gotland sheep and the Icelandic sheep together with some other indigenous northern European breeds are also used for the production of these pelts. Colour, curliness and lustre are the characteristics which determine the value of the pelts in the fur trade.

The Karakul lamb at birth has a very characteristic curl in its wool (Fig. 12:7). The fleece consists wholly of lustrous guard hairs which form round, or oval, very tight, spirally twisted curls. At birth the curls lie very close to the skin but after a few days the spirals begin to straighten out as the hair grows. When the lamb is a month old one can see the inside of the spirals, and in the adult sheep the hair is practically straight. *Age* has thus a pronounced effect on the quality of the Karakul. The best pelts are obtained from the 'unborn', so called 'breitschwanz' or broad-tail lambs, the mothers are slaughtered about 3 weeks before the normal time for lambing. The Persian pelts (Persian

Fig. 12:7. Karakul lamb, one day old

lamb) are obtained from lambs killed a few days after birth. The more inferior pelt quality, 'astrakan', is obtained from a somewhat older lamb with longer wool and more open curls.

In the investigation by SKÅRMAN, referred to earlier (p. 172), into the factors which influence fleece colour in lambs of the Gotland sheep, the influence of age on the formation of curls and fleece quality was also studied. At two months of age only about 25 per cent of the lambs had typical curls, but at five months about 75 per cent of the lambs were curly. In the Gotland sheep therefore, age has an opposite effect on the formation of curls to that in the Karakul. The curl is, however, not of the same type in the two breeds; in the Karakul the curl is much tighter.

Colour is very important for the price of the pelts. At the auctions conducted by the Swedish fur breeders association in 1965 an average of 55 kronor was paid for light grey lamb of the Gotland breed pelts, whereas the medium grey and dark grey commanded an average price of about 100 kronor. Black and white pelts of the Gotland breed are useless in the fur industry.

The influence of heredity on the quality characteristics of the pelts is still incompletely investigated. As far as the indigenous Gotland sheep are concerned, studies by SKÅRMAN indicate that colour, curliness and lustre are quantitative traits influenced by a large number of genes.

The pelt quality in crosses between the Karakul and other sheep breeds has been studied by workers at several places, e.g. ADAMETZ (1917). The typical curl of the Karakul has a tendency to be dominant. The quality of the curls in the crosses was inferior, however, to that in the purebred Karakul. Some sheep breeds, e.g. the E. Friesian milk sheep, are more suitable for crossing with the Karakul than other breeds. The Merino cross is characterised by a very poor pelt quality. At Beltsville YAO and co-workers (1953) made a study of the influence of heredity on the variation of various lamb pelt quality traits. The investigation was based on data from 728 Karakul lambs, or crossbred lambs with predominantly Karakul ancestry, together with their 207 dams and 22 sires. The heritability was estimated from the half sib correlation and the regression of the progeny on the parents. The following traits were assessed on a score system with a scale of 10 points: type of curl, tightness and size of curl, texture, lustre and hair length. The results of the heritability estimates varied within different groups of data (pure Karakul and different breed crosses). The size of the curls and tightness, as well as the lustre, showed relatively high heritability (about 0.4). The heritability of curl type was 0.25, whereas for hair length and texture the figures were only 0.05–0.10. It would appear that selection for some of the important traits would quickly lead to an improvement in quality.

SELECTED REFERENCES

CARTER, H. B. 1959. Wolleistung, Woll- und Pelzqualität. *Handbuch der Tierzüchtung*, Vol. III, Chapter 11. Edited by Hammond, Johansson and Haring. Verlag Paul Parey.
RAE, A. L. 1956. The genetics of the sheep. *Adv. Genet.*, **8**: 190–253.
TERRILL, C. E. 1958. Fifty years of progress in sheep breeding. *J. Anim. Sci.*, **17**: 944–959.

13 Production Characters in Poultry

In Europe and North America the bulk of eggs consumed are produced by chickens, though some duck eggs are also used, mainly by the confectionery trade. Most of the poultry meat eaten is also derived from chickens and only about 15 per cent is derived from turkeys, ducks, and geese. Market forecasts do not suggest much change in the relative consumption of products from the different species though there may be some small proportionate increase in the amount of turkey meat. The following discussion will be largely confined to chickens with some mention of meat production from turkeys.

Whereas it was customary in the early decades of this century to produce eggs and meat from the same breeds of chicken, it has been found that greater performance and efficiency can be achieved by the breeding of separate varieties for the production of the two commodities. Not only has specialisation occurred but a change in the type of genetic material used for production as well. Self-contained breeds, designated as such by morphological features such as feather colour and comb type, have been replaced by 'hybrids'. The latter are two, three and four-way strain crosses which are designated by their performance characters. The strains involved are derived from several of the original breeds and also from composite populations for which no breed title would be appropriate. This type of selection and breeding system will be discussed later (Chapters 14 and 17).

Apart from the separation of egg and meat production chickens for efficiency reasons, there is a sound genetic reason for this. The fewer characters for which the breeder has to select, the greater the chance of success and the faster the rate of progress. Having separated the two selection aims, i.e. eggs or meat, it is reasonable to investigate the relative economic importance of the characters in the two types of chicken. A discussion of egg production characters will be followed by one on meat production characters.

EGG PRODUCTION CHARACTERS

Several analyses of economic data have been made in an attempt to identify the characters of importance for egg production. It is apparent that egg numbers alone can account for up to 90 per cent of the variation in economic return. Other characters which have some, if minor effect, are egg weight, body weight, feed consumption, viability and egg quality. To these traits must be added the fertility and hatchability of eggs. It is logical to consider these in the same order as the life sequence of the chicken.

Fertility and hatchability

Whilst there is a considerable amount of evidence on the effects of nutrition, management systems and egg storage treatment on fertility, there have been few genetic investigations. Genetic analyses which have been reported indicate, as might be expected, that the heritability of fertility as a compound character is low, probably of the order of 0·05 and that improvement by selection would be difficult. Not only are the whole of the reproductive physiology systems of both the male and the female domestic fowls involved in this character (normally traits are affected by the physiology of one individual), but it should be realised that fertility problems are also closely related to social and mating behaviour patterns and that data collected from birds which are flock mated may be very different from data collected on pair mated or artificially inseminated birds.

It is obvious that any single gene having a large detrimental effect on fertility could be relatively easily eliminated from a population and, unless there is some opposing advantageous pleiotropic effect, the gene is likely to be eliminated by natural selection alone. Apparently, in the pleiotropic category is the dominant autosomal gene for rose comb found in the White Wyandotte breed. Whilst dominant for comb morphology, the gene is simultaneously recessive for reduced fertility. It has been conclusively shown that the gene's effect in the homozygous rose-combed male is to reduce the sperm's fertilising ability but not to reduce the viability of embryos derived from matings involving such males. It is extremely difficult to distinguish between low fertilisation and a high incidence of early embryonic death, so that fertility and hatchability of fertile eggs are frequently considered together as hatchability of eggs set. There are several known majo gene effects which cause embryo abnormalities and death, and these can be easily eliminated providing there are no heterozygote advantage effects which tend to maintain stable polymorphisms.

Research by MORTON *et al.* (1965) into relationships between blood groups, transferrins and albumens in adults, and mortality of progeny during incubation and in the first two days post hatching, indicates extremely complex genetic relationships. It is not surprising that reported estimates of heritability of hatchability of fertile eggs and of all eggs set have been about 0·10–0·15. It must be stressed that hatchability is not an easy trait to measure, partly because of the considerable variation between eggs in the time required in the incubator for the chicks to hatch. Some data analysed for genetic parameters has been based on the number of chicks hatching by a particular time after setting in the incubator. Studies indicate that heritability decreases the longer the time over which eggs are allowed to hatch and that the heritability of hatching time (the time between setting in the incubator and the emergence of the chick from the egg which does not include any pre-incubation storage time at temperatures of approximately 12°C post laying) is quite high, being about 0·50 for half sib estimates. Selection for hatching time in each direction has been successful and resulted in changes in egg size. There is a positive genetic correlation between egg weight and hatching time and a negative correlation between egg weight and hatchability of fertile eggs.

Egg production

The number of eggs produced in a certain time interval, e.g. a year, is generally used as a measure of the production capacity of a hen under specified environmental conditions, but sometimes the total weight of eggs is used. However, since egg number and the average weight of eggs are negatively correlated it is preferable to regard them as separate traits.

The average egg production per hen housed in Great Britain has increased during the last twenty-five years from about 165 to 235 eggs per year. The main part of this increase is probably due to an improvement in feeding and management, but genetic improvement of the stock has also contributed. Professor HUTT at Cornell University, U.S.A. has compared a strain of White Leghorns with unselected Jungle Fowls under the same environmental conditions. The mean egg production of the survivors to 500 days of age was 181 for the White Leghorns but only 62 eggs for the Jungle Fowl. Of course this is an extreme case, but in Germany HARING and GRUHN compared 'hybrid chicks' from certain line crosses with purebred White Leghorns and with two strains of the New Hampshire breed, one light and one heavy type. The average egg production per hen housed to 500 days of age was 248, 201, 177 and 203 respectively. These examples indicate the importance of breeding for production traits in poultry.

It would be interesting to know to what extent the average egg production per hen can be raised by further genetic and environmental improvement. Obviously no answer can be given. However, the limit may be determined by the fact that the average interval between the laying of two eggs in the same series is about 26 hours, with a variation from 24 to 30 hours, and that birds with two fully functional ovaries and oviducts have not been reported. Hens that produce 365 or more eggs per year will probably continue to be rare for some time; but it is not necessarily the upper limit of their ability.

Hens generally reach their highest egg yield in the first laying year. The egg production in the second year is, on an average, only 70–80 per cent of the production in the first year, and the third year's production is 70–80 per cent of the second year's production. Hens are kept for egg production usually for one laying year only, or at the most two years. However, some birds which have been handicapped in the first year by unfavourable environmental influences may in the second year show a yield as high as, or even higher than, in the first year, provided that they are able to recuperate from the first year's damage.

Recording the egg production of individual hens is not new and, according to the literature, the first efficient trap-nest was constructed and used in Austria in 1879. About the turn of the century, trap-nesting as a method of recording the egg yield of individual hens was used at several experimental stations in the U.S.A. and spread rapidly to other parts of the new and the old world. In recent years with the increase in the use of laying cages and the adoption of satisfactory artificial insemination techniques, the labour involved in recording egg production has been considerably reduced.

There are three commonly accepted ways of expressing the egg production of a group of birds: survivor production, hen day production and hen housed production. *Survivor production* is the number of eggs per bird for those birds which survive the recording period. *Hen day production* is total number of eggs laid by the flock in a given period divided by the sum of the number of hens alive on each day of the period (hen days). This measure, like the previous one, does not include any measure of viability. The third measure, *hen housed production*, is a combination of egg production and viability, being the total number of eggs laid divided by the number of birds alive at the start of the recording period. All three measures can be combined with sexual maturity if needed.

The common practice has been to record the egg production for each bird during a period of one year, starting on a certain date, e.g. the first of October or

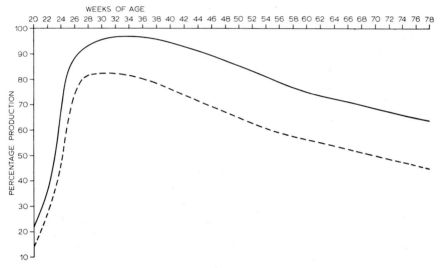

Fig. 13:1. Percentage hen-housed egg production curves for a flock of modern 'hybrids'.
The upper line shows the egg production the birds are bred to achieve when all environmental factors are favourable. The lower line is a warning line; if production falls below this line, management is at fault.

on the date when the first egg is laid. Later, the American geneticist F. B. HUTT suggested measuring egg production up to 500 days of age of the bird; this method has several advantages and is now used at many American and European testing stations. The effect of age at maturity is automatically included, and the records of groups with different hatching dates can be directly compared; which is not the case when the recording period starts at the same date for all birds and groups. A typical egg production curve is shown in Fig. 13:1.

Usually, the egg production was recorded for each hen for each day of the week, with the possible exception of Saturday and Sunday. The rising cost of labour has stimulated the use of simplified and cheaper methods, and several investigations have been made to elucidate this problem. The main results of two such investigations are presented in Table 13:1. The errors of the records,

Table 13:1 Reliability of trap-nest records when the number of recording days is reduced

Trap-nesting days per week	FABER, 1960 Error: No. of Eggs	% of Y	r_{xy}	WHEAT, 1956 Error: No. of Eggs	r_{xy}
6	—	—	—	3·3	0·99
5	6·0	3·0	0·98	5·1	0·99
4	7·4	3·7	0·97	7·0	0·98
3	9·0	4·5	0·96	9·3	0·96
2	12·9	6·4	0·90	12·8	0·93
1	17·3*	8·6	0·82	19·8	0·85

* The hens were trap-nested 5 days a month.

when trap-nesting is carried out only on certain weekdays, are stated as the standard deviations of $(X_n - Y)$ where Y = the actual egg production in the first laying year as verified by daily trap-nesting, and X_n = the estimated production when the hens are trap-nested n days per week; r_{xy} = the correlations between the X_n and Y records. There is very good agreement between the results of the two investigations. Obviously, the reason for the rise in error with decreasing number of trap-nesting days is that individual hens have a variable length of pause from laying, and when they are trap-nested at longer intervals the chance that laying and trap-nesting do not correspond with each other increases. The relative error, when trap-nesting is carried out four days a week, is approximately the same as that of milk records based on monthly tests. If reliable records are desired for egg production of individual hens, trap-nesting should be carried out at least three days per week, but when records are kept only for family or group comparisons one or two days per week should be satisfactory.

At the present time, selection is based mainly on records for the first part of the first laying year. It has been shown that the heritability of part-year records is almost as high as that for yearly, or 500-day, records. LERNER and co-workers estimate that the genetic correlation between the egg production from the commencement of laying to the end of November and in the whole year is 0·74, and that between production to the end of January and in the whole first year is 0·85. Similar results have been obtained by other workers. One reason that many breeders use part-period records is that the generation interval can be reduced and thereby the yearly gain by selection increased.

This practice, however, may have some undesired effects. Whilst the early part-year record has a high positive genetic correlation with the whole-year record, in some strains it has a negative genetic correlation with production in the later part-year record. Selection on early part-records in these strains may have the effect of altering the two part-records in opposite directions so that the total egg production remains unchanged. There is evidence that this has already happened in some populations selected on early part-year record. The solution appears to be, either to select on a longer record and so increase the generation interval, or to select on a part-period record taken for three or four months in the middle of the production period.

In the early part of this century several attempts were made to explain the individual variations in egg yield as due to mono- or dihybrid segregation of genes. When it became clear that all such attempts were in vain, the research workers of the twenties distinguished between five components of egg production and tried to give the variation in these components a Mendelian interpretation. Although even this hypothesis was an over-simplification of a complicated phenomenon, the five 'components' are interesting enough to be mentioned: (1) *Age at sexual maturity*, i.e. age at first egg. Under otherwise equal conditions, early-maturing birds lay more eggs than late-maturing ones. The heritability of this trait is estimated to be $0.15-0.30$. (2) *Laying intensity*. This is expressed as percentage hen-day production in a certain period, e.g. October to January. (3) *Broodiness*. The manifestation of broodiness depends on the hormone prolactin, and is highly heritable. Therefore, it is easy to reduce the frequency of broody hens to a rather low level by selection; but so far it has not been possible to eliminate the trait entirely. (4) *Winterpause*, i.e. the laying pause exceeds 7 days without the manifestation of broodiness. It is difficult, however, to distinguish between normal pauses and a 'winterpause' as different traits. With modern feeding and management, there should be no winterpause. (5) *Laying persistency* towards the end of the laying year. Normally hens start to moult when they are 16–20 months old; some moult slowly and others rapidly, some continue laying while moulting, but most stop laying. Hens that start moulting late in the season are on an average better producers than those that start early, and these differences are partly genetically determined. It is possible, however, to postpone moulting by a gradual increase in the number of daylight hours. Late moulting means higher persistency. When the total egg production is measured from the commencement of laying, persistency has a greater influence on the records obtained than when the yield is measured from hatching to 500 days of age; early maturity is then more important. Hutt has shown, however, that hens with high egg yield to 500 days of age are usually persistent layers. There is also a positive correlation between the appearance of winterpause and early maturity.

On the basis of hen-day production within strains, the genetic correlation between total production (for one laying year or up to 500 days of age) and both the early-part record (first four months of lay) and late-part record (last four months of lay) is positive, being about $0.6-0.8$; but the genetic correlation between the two part records seems to vary, between strains, from $+0.6$ to -0.7.

The heritability of part-records (three or four months from first egg) of hen-day egg production has been estimated to be $0.25-0.30$; and that for the production for a complete laying year, or to 500 days of age, is very similar ($0.25-0.35$). The heritability of hen-housed production to 500 days is much lower, being only $0.05-0.10$.

The effects of environment on egg production are considerable and complex, and these largely come under the headings of nutrition, light and physical en-

Egg-size

Egg-size, which is nearly always measured as weight, is at its lowest at the commencement of the laying period, but it increases steadily for about seven or eight months, after which there is only slight increase, if any at all, for the rest of the laying period. A typical curve of egg-weight change in the first laying year is

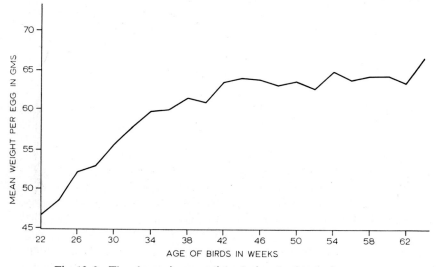

Fig. 13:2. The change in egg weight during the first laying year.

shown in Fig. 13:2. In general, the first egg laid in a series, following a pause, is somewhat larger than the subsequent eggs. When a hen has reached her maximum egg-size in the first laying year, approximately the same size is maintained in the second year; but later egg-size decreases with increasing age. Normal egg-size varies considerably between breeds, strains and inbred lines. Under standard environmental conditions, a range of 40–60 g per egg can be expected, but the rate of increase during lay is very similar irrespective of the egg-size when laying commences. This increase is approximately 6 g per dozen eggs per week.

For selling purposes, nearly all countries in Europe and North America grade eggs in some way by weight. The two most common practices are either to classify eggs into arbitrarily designated weight-grades and sell them by number at different prices for different weight-grades, or similarly to assign eggs to weight-grades and sell by weight at a price which is differential depending on the weight-grade of egg making up the weight sold. Either of these practices results in a reducing price advantage for increasing egg-size as the basic size goes up. At the top of the scale, egg-size increases above the weight of the top grade are of no economic advantage. For selection purposes, economic return declines with increasing egg-size and there is an upper threshold above which increasing egg-size is not economically worthwhile. Also, as market requirements usually involve a proportion of eggs falling into each of the weight grades, there is obviously an intermediate optimum egg-size, which depends on the economic differentials between the weight-grades and the average price per egg. The selection aims for egg-size con therefore be stated only in respect of specific market requirements.

In order to select for egg-size it is very costly to measure and record the weight of all eggs from a bird throughout a laying period. Several analyses have been published, which indicate that both mean egg-weight and market value of all eggs for a full laying period have a correlation of 0·8–0·85 with the mean weight of three or more days' eggs collected at any time between 32 and 55 weeks of age of the bird. Such a small sample is estimated to have a heritability of about 80 per cent of the heritability of the mean egg-weight of all eggs, which from half sib estimates within breeds and strains is of the order of 0·5 with a range of 0·2 to 0·7. Several selection experiments for both large and small egg-size have been successfully conducted, usually with positive correlated changes in body-weight. Inbreeding has generally caused some reduction in egg-size though most of the reduction can usually be explained as a correlated response to the reduction in body-weight which is also a usual consequence of inbreeding.

Environmental factors, including nutrition, light pattern, temperature and ventilation, particularly during rearing, can influence early, and hence mature, egg-size to a considerable extent. However, the extent of the direct effect of environment on egg-size as compared with the indirect effects through body-size and sexual maturity are not known. The interrelationships of these three traits are complex. Within breeds, there are positive genetic and environmental correlations between any two of the three traits, with the exception of the environmental correlation between body-weight and sexual maturity and possibly that between sexual maturity and egg-size—which are negative. Since the genetic correlation between egg-size and egg-number is negative but the environmental correlation is positive, combined selection for these traits is not easy.

Viability

Viability implies resistance to any of a very large number of causes of death. Among such causes are both 'genetic defects' and infectious disease. It is reasonable to assume that 'genetic defects' have a wide range of severity, which relates to the ease with which they can be detected. 'Genetic defects' in the present context, can be divided into those that cause anatomical abnormality which generally reduces bird performance and those that cause fairly major metabolic disturbance necessitating major changes in the diet, or other aspects of the environment, if the bird is to survive or perform satisfactorily. The breeder is generally not concerned about the grosser genetic defects, since these are relatively easily eliminated from the strains he uses. More interest and effort is applied to selecting strains which are resistant to infectious disease. HUTT (1958) has reviewed research on both 'genetic defects' and genetic resistance to infectious disease. He produces evidence to show that selection for resistance to disease can be effective; but he also shows conclusively that resistance to one pathogenic organism does not necessarily or usually imply resistance to another pathogen. Therefore, it would be necessary to select strains specifically for resistance to all diseases they are likely to encounter. This would be an enormous task, made more difficult by the problems of maintaining regular and standard exposure to the disease and of avoiding the transfer of passive immunity so that exposure shall have an equal chance of affecting all individuals. Also, because of the relatively short generation interval and large population size of pathogenic organisms, they are likely to mutate as populations much faster than their hosts. Hosts resistant to the original pathogen might not be resistant to the mutated pathogen. The task of selecting resistant hosts would be unending, and would probably never keep pace with the mutating pathogens. It is now generally agreed that, where alternative effective vaccination, or elimination by slaughter programmes, are available, selection for resistance to specific disease is not worthwhile.

The last conclusion does not preclude the possibility of selection for non-specific resistance to infectious disease. Such resistance might include such physiological aspects as general antibody response mechanisms, and tissue and cell wall permeability. It has not so far been shown that such a form of resistance exists in chickens; but many selection programmes are designed on the basis that it does. Consistent differences in viability can be found between strains and crosses, even when comparisons are made on several farms and hence under several different conditions of exposure to disease. The results of such a comparison are found in Table 13:2. Such viability differences can obviously be made the aim of selection and are also presumably non-specific disease resistance differences.

Table 13:2 Differences in viability during lay (20–64 weeks of age) between three strain crosses concurrently tested on eleven farms. There were approximately 80 birds per strain cross on each farm at the start of the test. The data are expressed as percentage mortality

Farm	Strain cross		
	A	B	C
1	8·4	17·4	14·2
2	8·1	21·7	8·5
3	7·3	20·0	9·1
4	8·9	24·2	13·2
5	20·4	40·6	15·0
6	2·9	20·0	4·2
7	6·1	9·2	6·3
8	3·1	9·4	2·4
9	12·2	15·3	13·2
10	10·9	24·8	31·9
11	2·0	14·6	3·8
Average	8·2	19·7	11·1

Viability is an all-or-none trait, and hence presents some special problems for biometricians and breeders. It is obvious that, for quantitative analysis, an individual can have one of only two values; and for this reason the trait is generally measured on family groups as the ratio of birds dying to those hatched. Comparisons and analyses of such data are difficult because of differences in family group size. A useful treatment of this problem in poultry has been published by ROBERTSON and LERNER (1949). Estimates of heritability of viability from such data have yielded values of 0·05–0·1 for both half and full sib estimates (cf. p. 127).

As was mentioned above, it is usual (for selection purposes) to include viability with egg production in a composite trait, called hen-housed production. This may not be the most efficient method of selecting for these two traits. Recent evidence of the genetic correlations between hen-day production and viability shows that it is about +0·4, and that more progress would be achieved by the use of a conventional selection index of the two traits.

Body-weight and feed consumption

The importance of body-weight in laying chickens is connected more with its relation to feed consumption, sexual maturity and egg-weight than with its direct effect on economic return. In countries with a well developed broiler industry the value of the carcass of the hen at the end of the first laying year rarely exceeds twice the cost of the day-old chick and generally has a much lower value. In some instances disposing of the carcasses has become a problem, so that the smaller the bird the better. The relationships with other traits, however, cause considerable problems in deciding the weight of the bird to be selected. Genetically and environmentally, within strains the relationship between body-weight and feed consumption, sexual maturity and egg-weight are all positive; but between body-weight and egg production the genetic correlation is negative and the environmental one positive (Table 13:3). Therefore, what is desirable

Table 13:3 Genetic (above diagonal) and environmental (below diagonal) correlations between components of egg production (where known)

	Age at first egg	Hen-day production	Egg-weight	Mature body-weight	Viability	Specific gravity	Albumen height
Age at first egg		−0·4	0·4	0·4	0·7	−0·1	−0·2
Hen-day production	−0·3		−0·5	−0·1	0·4	0·1	−0·1
Egg-weight	0·2	0·2		0·25	−0·2	0	0·5
Mature body-weight	−0·2	−0·1	0·3		0	0	−0·3
Viability						0	−0·2
Specific gravity							0·3

on economic grounds is antagonistic to the direction of selection for body-weight. On the one hand, reduced food consumption and increased egg production will be produced by a lower body-weight; on the other, earlier maturity, greater egg-weight and higher carcass value will be derived from a larger body-weight. A selection index is, therefore, probably the best solution.

The heritability of adult body-weight at a certain age is fairly high, being between 0·2 and 0·6 for paternal half sib and offspring-parent estimates respectively. It is known, however, that there are genetic differences in the shape of the growth curve, so that selection based on a single measurement of body-weight could have a differential effect on the whole growth pattern, depending upon the age at which the measurement is made. For instance, selection for increased weight at point of lay might result in a bird which grew rapidly up to that age but put on little weight thereafter, whereas similar selection for body-weight at the end of the laying year might result in a bird which grew more rapidly throughout its life. The former result may be desirable for a laying bird, since rapid growth to point of lay will advance maturity and increase egg-size; whilst slow later growth keeps feed consumption to a minimum. Still another alternative is to select on the basis of body-weight at both the beginning and the end of the laying year. This has two complications. Firstly female body-weights taken between 17 and 22 weeks of age are probably an unreliable indication of differences in mature body-weight, because the birds' reproductive organs are developing rapidly at this time and much of the variation in body-weight is merely variation in maturity of sexual organs. Secondly selection based on body-weight at the end of the laying year leads to a longer generation interval. The usual practice is to select on the basis of a single body-weight measurement taken between 22 and 32 weeks of age.

Feed consumption is closely correlated to body-weight and, though it has been shown that there are genetic differences in feed consumption after body-weight differences have been accounted for, little if any research has been published on selection for feed consumption other than as a correlated response to body-weight. The main problem involved in directly selecting for feed consumption is the difficulty and cost of accurately recording feed intake. This problem is discussed in more detail later in Chapter 16.

Egg quality characters

These characters have become of increasing importance in recent years, as marketing outlets have placed greater emphasis on the sale of quality products. In a review of the genetic basis of egg quality by BAKER (1960) the following traits as well as egg-size were considered: egg-shape, shell quality and colour, albumen quality, blood and meat spots, and yolk quality. The economic importance varies with the method of market payment in different countries. Egg-shape, shell quality and blood and meat spots are probably of major importance, since they have a direct effect on the return obtainable for an egg. In some areas, particularly the United Kingdom, parts of Northern Europe and the north-eastern United States, shell colour is also economically important, since higher prices can be obtained for brown eggs than for white.

Egg-shape. This trait is of particular importance to packers and transporters because of its influence on the proportion of broken and cracked eggs. A typical range in shapes would be from almost a perfect sphere to regular ovoid and pear shaped. An objective assessment which has been used to measure egg-shape experimentally is given by the formula (Breadth/Length) × 100.

An index of 74 is considered optimal and a variation between 72 and 76 as satisfactory. It can be seen that the narrower or longer the egg the lower the index. Though the index is somewhat limited in that only two variables of shape are included, these two are largely responsible for determining the suitability of shape for the type of containers at present used for moving eggs.

Seasonal variations in shape have been reported, as has variation between eggs within a clutch from the same hen. Reports on the heritability are few but vary widely in their estimates. Some workers have found heritabilities of 0·10–0·20 whilst others using similar or different methods of analysis have reported heritabilities of 0·60.

Evidence of sex-linked effects on egg-shape have been published. Information concerning the correlations between egg-shape and other production traits is limited. The estimates reported are a positive genetic correlation with egg-size ($+0.4$) and a negative environmental correlation (-0.3). All other estimates with other traits have proved to be very small and therefore are considered to be of little consequence.

Shell quality. This trait is important because it affects market value through resistance to cracking and because it affects hatchability. Though it has not been proven, it seems likely that the optimum shell quality for the two purposes may not be the same. The structure of the egg shell has been described by ROMANOFF and ROMANOFF (1949) and the factors which contribute to shell quality and strength are shell thickness, porosity of the shell, membrane thickness, the mineral content of the shell and the protein matrix. There is considerable difficulty in defining, and hence in measuring, shell strength. The final objective is to produce eggs which do not crack under field conditions. Since normally only between 2 and 7 per cent of eggs handled in packing stations are found to be cracked when passed over a candling lamp, it is extremely difficult to obtain correlations other than zero or very small correlations, between strength measurments, shell characters and the proportion of eggs cracked under commercial conditions. There is a high negative correlation between shell thickness and the proportion of cracks; but the mean shell thickness and the value of the regression vary between strains, indicating that other shell characters are responsible for shell strength. Many research workers have designed methods for measuring shell strength and then related the measurements to shell characters. Only in few cases, however, has a relationship between the strength measurement and the proportion of cracks under field conditions been obtained. The various strength measurements reviewed by TYLER (1961), which include impact, crushing and piercing tests, are therefore open to considerable question as to their efficacy in achieving the objective of reducing the proportion of cracked eggs.

The measurement of the thickness of the shell without membranes (removed by boiling the shell in sodium hydroxide solution) is easy to carry out with micrometer callipers but is a rather slow process. There is considerable variation in shell thickness at different places on the same egg and for this reason it is necessary to take the average of between four and six readings taken along the axis of the longitudinally bisected egg. The percentage of shell in the egg, which is highly correlated with shell thickness, can be estimated from the specific gravity of the egg. The latter can be measured by a simple method without destroying the egg. Sodium chloride solutions of different concentration are prepared (e.g. specific gravities of 1·06, 1·065, 1·07, etc. up to 1·10) and it is determined in which one of these solutions the egg just keeps floating. A satisfactory specific gravity reading for shell quality is considered to be not lower than 1·07 or 1·08. This method has been criticised because variation in atmospheric conditions, particularly temperature, can appreciably alter the readings obtained and because transfer of eggs from one solution to another very quickly alters the concentration of the original solutions. Many of these drawbacks can be overcome by using a hydrometer graduated for specific gravity and a single container of water. The porosity of the shell and membranes combined can be determined from the evaporation loss under constant temperature and humidity conditions. Variations in the texture of the outer layer of the shell and shell mottling are commonly found, and attempts have been made to assign objective measurements to these characters. Considerable unevenness of the shell surface may be caused by injuries in the oviduct, in some cases resulting from outbreaks of infectious bronchitis.

Heritability estimates for shell quality traits are not numerous but indicate a value of 0·3 (full sib estimate) for shell thickness and between 0·3 and 0·6 (sire and dam component combined from sib analysis) for specific gravity. Selection experiments designed to change shell thickness have proved successful in both upward and downward directions. Estimates of phenotypic and genetic correlations between shell thickness or specific gravity and other production traits are negative for egg production and egg-weight, and positive for albumen quality.

Shell colour. Shell colour of eggs depends on the pigment oöporphyrin, which is deposited mainly on the outer surface of the shell. The pigment is related to haemoglobin of the blood. Even white eggs contain small amounts of oöporphyrin. In the early part of the laying year more pigment is deposited on the egg shell than towards the end of the year.

Until recently, measurement of shell colour has depended on the subjective comparison of eggs with a standard colour range of eggs which could be numbered for subsequent quantitative analysis of the data. However, at least two instruments which estimate reflectivity of the shell are now available for measuring shell colour. Several hundred eggs an hour can be handled by these instruments.

Shell colour varies considerably between breeds. Some breeds, such as the Barnevelder, Maran and Orpington, produce eggs with a deep brown shell colour, whereas the eggs of the Rhode Island Red and Sussex are less pigmented and more reddish brown, and those of the White Leghorn are largely pure white. Shell colour is considered to be a quantitative character and heritability estimates are variable, ranging from 0·3 to over 0·9 for half sib analyses. There is consistently a higher dam than sire estimate, and though this has been interpreted as evidence of dominance effects, the interpretation may be wrong. This is suggested by the fact that in crosses between strains with brown and strains with white shelled eggs the F_1 is usually intermediate. Phenotypic and genetic correlations, estimated from very limited data from a selection experiment,

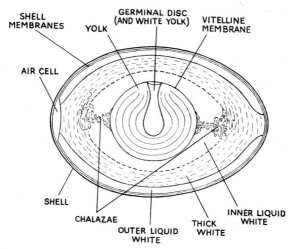

Fig. 13:3. If a cross section is made through an egg at the time of lay its internal arrangement resembles this diagrammatic illustration.

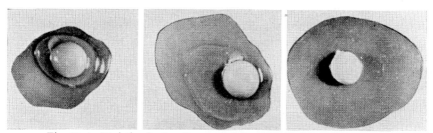

Fig. 13:4. Variation in the albumen height of eggs when broken out on a horizontal glass plate. *Left*: egg with high albumen quality; *Middle*: average; and *Right*: low albumen quality. (OLSSON 1962)

indicate that increase in egg numbers was associated with a dilution in shell colour, whilst selection for egg weight maintained shell colour almost constant. Other published evidence shows that the percentage of coloured meat spots increases with darker shell colour.

Some South American breeds produce eggs with blue or green shell colour and the blue colour depends on an autosomal dominant gene. The green colour is assumed to be due to a mixture of brown and blue pigment.

Albumen quality. The interior structure of an egg is shown in cross section in Fig. 13:3.

The albumen in the egg can be divided physically into four layers. Surrounding the yolk is a layer of thick albumen, which extends in a thread to both poles of the egg and is called the chalazae and chalaziferous layer. Next is the layer of inner thin albumen, followed by a deep layer of thick albumen. Finally there is a further layer of outer thin albumen which is found just under the inner shell membrane. The consumer considers a desirable egg to be one which when broken out stands firm with a relatively high proportion of thick albumen. An egg which has a spreading 'watery' white is not considered desirable. It may be due to the strain of bird, to certain respiratory diseases, to seasonal effects, particularly the temperature to which the birds are subjected, and to poor egg storage conditions. The range of albumen quality found in eggs is shown in Fig. 13:4.

Albumen quality has been measured in several ways: for instance, the viscosity has been measured as well as the weight of albumen, an index of the height and width of thick albumen and the area of thick albumen. The commonest measurement is the height (in millimetres) of the albumen at the periphery of the yolk. This can be measured by breaking the egg onto a flat surface and measuring the height with a tripod micrometer. This albumen height is often expressed as Haugh units, which is a simple correction of height for the effect of egg-size. It has been found that the regression of albumen height on egg-size varies considerably between breeds and strains, and therefore the regression estimate used to calculate Haugh units may be misleading for comparisons of albumen quality between breeds or strains. All estimates of albumen quality should be made at a standard time, optimally one day after the egg was laid, so that comparisons are not confused by differential storage-time effects.

The heritability estimates of various albumen quality estimates are variable, and depend on the method of analysis and strain or breed. The estimates range from $0 \cdot 1$ up to $0 \cdot 7$, depending on strain; and success has been reported in selecting for strains with differing amounts of thick albumen. Correlations with other production characters indicate both positive and negative phenotypic and genetic relationships with egg production, and consistently positive correlations with egg-weight.

Blood and meat spots. Approximately 1–2 per cent of all eggs candled are found to contain blood or meat spots, which usually appear in the border between the yolk and the albumen. Until recently it was considered that meat spots were degenerate stages of blood spots, which arise as intrafollicular haemorrhages. It is now known that blood and meat spots are of separate origin, and that meat spots originate from lesions in the mucosa of the oviduct. Meat spots show fluorescence in ultra-violet light but blood spots do not. The size of the inclusions and the colour (red to white) of the meat spots vary. After a relatively heavy haemorrhage a large part of the albumen may be stained by blood. For experimental purposes, the number of inclusions are counted in eggs which have been broken out and the inclusions can also be scored for size by visual assessment. Blood spots are more frequent than meat spots, and eggs containing either type are removed commercially by candling. The frequency of blood and meat spots is considered to be genetically independent. Selection for high and low incidence of blood spots has proved successful, but their incidence could not be eliminated in the low line. The realised heritability was 0·5, but other estimates derived by component analysis of data from unselected material have been much lower—of the order of 0·1–0·25. Because of the generally very low incidence of spots it is to be expected that the estimate of heritability will be partly frequency-dependent. On an average, the frequency is considerably lower in Leghorns than in the heavier breeds.

Yolk characters. The colour of the yolk appears to have some consumer appeal, deep yellow yolks being preferred. Since the yolk colour can be almost completely determined by certain harmless additives in the ration, this trait is of little direct interest to the geneticist. An estimate of heritability of colour under a standard nutritional regime is 0·15. Probably of greater importance is yolk-weight and its proportion of the total egg-weight. The heritability of yolk-weight has been found to be as low as 0–0·1 (full sib analysis) and as high as 0·4 (half sib analysis). Genetic and phenotypic correlations between yolk-weight and egg-weight have been estimated as positive and high (0·82 and 0·58 respectively) but these are part-whole correlations; more important are the high negative genetic ($-1·0$) and phenotypic ($-0·4$) correlations found by JAFFE (1964) between yolk-weight/egg-weight ratio and egg-weight, suggesting that selection for increased egg-weight would result in a lower proportion of yolk in the egg.

Miscellaneous characters. In recent years a number of characters have been mentioned as of importance to egg production and as possible criteria for selection. These characters include bird temperament, the consistency of the droppings, and feet condition (for instance selection against curled toes). Methods of measuring these characters are still being sought; so parameters such as heritability have not yet been estimated. The economic value of some of these characters is open to doubt and it is obvious that, if they are included in indices for selection, the rate of progress for the economically much more important characters will be reduced.

A summary of the h^2 estimates for egg production traits is presented in Table 13:4.

Table 13:4 Approximate range of heritability estimates for egg production traits

Hen-day egg production in the first laying year (No. of eggs):	
Part time record, 3 or 4 months' lay	0·25 – 0·30
One year from date of first egg, or to 500 days of age	0·25 – 0·35
Hen housed egg production one year from date of first egg, or to 500 days	0·05 – 0·10
Age at sexual maturity (age at first egg)	0·15 – 0·30
Egg-size (representative samples of eggs from each hen)	0·40 – 0·50
Egg-shape	0·25 – 0·50
Shell colour (within a population with brown eggs)	0·30 – 0·90
Thickness of shell	0·25 – 0·60
Colour of yolk (on standardised ration)	0·10 – 0·40
Firmness of albumen	0·10 – 0·70
Frequency of blood and meat spots	0·10 – 0·50
Fertility	0·00 – 0·05
Hatchability	0·10 – 0·15
Viability	0·1 – 0·15

MEAT PRODUCTION TRAITS

Until the 1930s, chicken meat for human consumption was largely derived from old hens which had completed their laying life or from surplus cockerels from egg-producing or dual-purpose breeds, such as the New Hampshire and Sussex which were specially grown for the meat trade. From the 1930s in America and much later in the United Kingdom, special strains and 'hybrids' were developed for meat and broiler production, other than those developed for egg production. Broiler chickens were developed first as pure breeds and more recently as two and three way cross 'hybrids'. The relatively high efficiency of these 'hybrids' specifically for meat production has severely depressed the economic value of the carcass of the laying hen to the point where it is frequently less than the cost of the day-old pullet.

The characters of major importance in the development of broiler chickens are growth rate, feed efficiency, mortality, feathering, skin colour and carcass quality. To these traits, which refer to the performance of the broiler chick itself, should be added the reproductive performance, particularly hatchable egg production of the female parent. The probable negative genetic correlation between egg production and growth rate, which has already been mentioned, makes the task of simultaneously improving these two traits in one strain extremely formidable. The more effective method of developing female parent strains (sometimes a two way cross) specifically for hatchable egg production, using an index of egg production and progeny growth rate, and male parent strains specifically for growth rate and carcass quality of the progeny, has been widely adopted. In a situation in which the genetic variance is largely additive the latter method may have no advantage; but because of the hybrid vigour frequently demonstrable for egg production, it is proving more successful than selection for all traits within a single population.

Growth rate

The growth rate of typical broiler 'hybrid' chicks is shown in Fig. 13:5. The heritability of growth rate to a given age is very similar to the heritability of weight at that age. For the range eight to twelve weeks, which is the age during which the majority of broiler chickens are slaughtered, the heritability of both characters has frequently been estimated to be 0·4 to 0·8 and there is considerable evidence of sex linked effects. Recently, however, in some strains which have been heavily selected for growth rate over many generations, heritability has been found to be very low, occasionally almost zero. The problem of how to improve further such plateaued populations is not easy to solve, but has been successfully overcome in some instances by amalgamating several strains and selecting in the composite strain so produced.

As selection for growth rate has proved successful, it has been found practicable successively to reduce the age at slaughter from about twelve weeks down to nine weeks or less. The market requirements are for a range of weights of processed animals and these are usually provided in a crop of broilers by the differential weight between the two sexes and the usual 11 or 12 per cent coefficient of variation around the mean slaughter weight. In cases where fairly widely divergent weights are required, however, the best procedure may not be to grow the same hybrid to different ages and weights. There is evidence that there are genetic differences in growth rate between hybrids, such that it depends on the proposed age and weight of slaughter which is the optimum hybrid to grow. Changes in the weight ranking of different hybrids may be critical in the age range eight to twelve weeks. Under these circumstances, it is preferable to decide the required weight for slaughter and select for the shortest time to reach

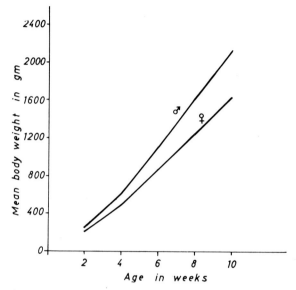

Fig. 13:5. Growth of male and female chicks on a lightcycle of one hour light:two hour dark. (From CHERRY and WARWICK, 1962, Br. Poult. Sci., 3:31–40)

that weight, rather than to select for increased weight at a particular age. Another problem which arises is that the weight differential between sexes increases as the mean weight of both sexes increases. There is evidence that the genetic correlation for weight between the two sexes is not equal to 1·0, so that it may be possible to select for a decreased differential weight between sexes at the same time as for increased weight of both sexes. This would necessitate the use of a selection index involving the genetic parameters for growth in the two sexes separately and the genetic correlation between them.

There is a high phenotypic correlation between the weight of day-old chicks and the weight of the eggs from which they hatch. An effect of egg-weight on chick-weight has been found up to ten weeks of age; but the proportion of variation accounted for by this and other maternal effects is usually very small. Because of the relationship between egg-weight and chick-weight it is hardly surprising that negative phenotypic and environmental correlations between broiler-weight at ten weeks and age of female parent at sexual maturity have been reported. However, the genetic correlation is positive.

Feed efficiency

Most of the variation in feed consumption and feed conversion between strains or breeds can be accounted for by differences in body-weight. While genetic differences other than those attributable to weight can be demonstrated, it is considered not worth selecting for feed conversion efficiency, except through growth rate, because of the difficulty and high cost of accurately recording feed consumption. The heritability of feed conversion, uncorrected for body-weight, has been reported to be as high as 0·6.

Viability

Chick survival is an important trait in the economics of broiler production; but, as previously stated, the heritability is extremely low (probably 0·05 or less). Mortality in chicks within the first few weeks of life is very dependent on conditions during incubation and during the period of transit from incubator to brooding facilities. With the widespread vaccination of female parent birds, the immunological status of the progeny will be such that chick mortality in the first few weeks of life will be generally very low.

Carcass quality

Carcass quality is extremely difficult to define objectively, but three traits which have a close connection are worthy of note. Firstly, the eviscerated weight can easily be measured and is closely related both phenotypically and genetically to liveweight. The heritability of eviscerated weight is as high as for liveweight. Secondly, the shape of the breast muscles over the keel bone has been considered in several investigations into assessing carcass quality. The measurement most commonly used has been breast angle; but other combinations of breast width, cross sectional breast area and body depth have also been used. The breast muscles should be wide and deep, so that there is no projection of the keel bone above the surrounding tissues. The heritability of breast angle at eight weeks of age has been estimated at 0·4, and the genetic and phenotypic correlations with live weight are about 0·3 and 0·4 respectively.

Thirdly, breast blisters or keel bursae are commonly found in broilers, and their causes are both genetic and managemental. They are the largest single cause of downgrading of carcasses in processing plants. The incidence of breast blisters in a strain increases with increase in weight, and the correlation between incidence and the weight of individual birds in the flock is high. The incidence is higher in males than females, even after correction for weight differences. The heritability of breast blisters has been estimated at 0·20 by half sib analysis. Any management factor which predisposes to poor bone formation, such as low calcium in the diet or a shortage of vitamin D, also predisposes to a high incidence of breast blisters, because the birds tend to perch on the ground more frequently and so cause irritation of the tissues over the keel bone.

Feather and skin characters

Two aspects of feathering are important to the production of meat chickens. The first essential is that the chicks should feather rapidly. Slow feathering, with resultant comparatively bare areas of skin, leads to increased numbers of breast blisters and scarring of the skin. This in turn means a higher proportion of downgraded and poor quality carcasses in the processing plant. Secondly, white feathers are preferred to coloured ones because of the appearance of the carcass after plucking. Remains of dark-coloured pin-feathers left on the carcass detract from its value. The usual method of avoiding these detrimental features is to produce hybrids from parents, one of which at least is homozygous for the dominant white gene; this inhibits black pigment formation. Unfortunately, strains of birds carrying the recessive white gene are not entirely devoid of pigment, the pinfeathers may be more or less pigmented and pigmented spots often appear in the skin on the breast of the chicks. The pigment deposition increases gradually for the first few weeks after hatching and, though it decreases later, it may still be evident when the broilers are processed. Generally, the pigment deposition is more marked in female than in male chicks. These problems do not arise with fast feathering dominant white chicks. Some research has suggested that the dominant white feather gene is either linked with genes for reduced growth rate or has an adverse pleiotropic effect on growth rate. More recent evidence has cast doubt on the suggestion so that no final conclusion can be reached at present.

Most of the breeds from which broiler strains have been developed were yellow skinned, and for the majority of markets for which broilers are produced this character is satisfactory. In the United Kingdom, however, where the consumer has generally associated yellow skin with the carcass of an old laying hen, the preference is for a white skinned broiler. Strains of broilers carrying the white skin gene have now been developed by intercrossing the yellow skinned broiler strains with breeds that have white skin, such as the Light Sussex, and by subsequent selection for growth rate and white skin in several generations after the F_1. Such localised consumer market requirements can be a considerable brake on progress in such more important economic characters as growth rate.

A further skin colour defect to be avoided is that of black or blue melanin pigmentation, generally on the shanks. In past years such pigmentation, common in broiler strains, was much disliked by processors and packers and more recently has been strongly selected against by breeders. Few broiler strains now have any melanin pigmentation of the skin.

OTHER POULTRY

Turkeys, ducks and, particularly in Canada, geese are important sources of meat. The amount of genetic research on characters of economic significance in ducks and geese has so far been small, but considerable effort has gone into research on turkeys. Breeds such as the Broad Breasted Bronze and the Beltsville Small White (selected for high reproductive ability and small size, quick maturing, compact, meaty body) have been widely used and selected to produce several strains for large-scale meat production. The characters of importance are similar to those required for broiler production, namely high growth rate and good carcass quality. Because a turkey produces far fewer eggs than a farmyard hen, the cost of the day-old poult is of greater economic significance relative to other traits than that of the chicken. In the turkey greater emphasis is placed on egg production and hatchability in relation to growth and carcass characters than in chickens for broiler production. The heritabilities of characters in turkeys are similar to the estimates of similar characters in the chicken. For a summary of heritabilities see ALTMAN and DITTMER (1962).

SELECTED REFERENCES

ALTMAN, P. L. and DITTMER, D. S. 1962. *Growth, including Reproduction and Morphological Development.* Federation of American Societies for Experimental Biology, Washington 14, D.C.

BAKER, C. M. ANN, 1961. The genetic basis of egg quality. *Br. Poult. Sci.*, **1**: 1–16.

FABER, H. VON. 1960. Methoden zur Verkürzung der Fallnestkontrolle unter Berücksichtigung der Familienauslese. *Arch. Geflügelk.*, **24**: 35–37.

HUTT, F. B. 1958. *Genetic Resistance to Disease in Domestic Animals.* Cornell University Press, Ithaca, New York; and Constable, London.

JAFFE, W. P. 1964. The relationship between egg weight and yolk weight. *Br. Poult. Sci.*, **5**: 295–298.

MORRIS, T. R. 1963. The effect of changing daylengths on the reproductive responses of the pullet. *Proc. 12th World's Poultry Congress, Sydney, Symposia*: 115–124.

MORTON, J. R., GILMOUR, D. G., McDERMID, E. M. and OGDEN, A. L. 1965. Association of blood-group and protein polymorphisms with embryonic mortality in the chicken. *Genetics*, **51**: 97–107.

ROMANOFF, A. L. and ROMANOFF, A. J. 1949. *The Avian Egg.* John Wiley & Sons, Inc., New York; and Chapman and Hall, London.

TYLER, C. 1961. Shell strength: its measurement and its relationship to other factors. *Br. Poult. Sci.*, **2**: 3–19.

14 Breeding Methods

This chapter and those following will deal with the practical application of the research which has been referred to in the preceding chapters. Let us first summarise some of the basic principles which were discussed in detail in Chapter 4.

In animal breeding one works with populations of varying size, such as breeds, subgroups within a breed, herds or family lines. The characteristics of the population are determined by the individuals which make up the population, while the traits of each individual are determined by its own genetic constitution and the environment. The population undergoes continuous renewal as individuals cease to exist and are replaced by new ones. If the population were extremely large, the genetic constitution would be unchanged from generation to generation with random mating and no selection. It is seldom that such a system of multiplication, called *panmixia*, exists but it can be used to advantage as a basis for comparison to study other systems in the way that sea level is used as the base for measuring heights.

Deviations from panmixia can be of two types:

1. *The zygote frequency* is altered, without necessarily involving a simultaneous change in the gene frequency. The frequency of zygotes can be changed by the application of a particular *mating system* which differs from random mating; the animals can be mated together according to their genetic likeness (relationship) or according to their phenotypic resemblance. This is what is meant when one speaks about *breeding methods*.

2. *The gene frequency* is altered by natural or artificial selection, which also brings about a change in the zygotic frequency (p. 101). *Natural selection* is based entirely on the viability of the individual and the individual's capacity for reproduction in the prevailing environment. *Artificial selection*, on the other hand, is applied by man himself and can be directed at any trait, even though the changes may lead to an impairing of the animal's adaptation to the natural environment. These questions will be dealt with under the heading 'methods of selection' (Chapter 16).

Animal breeding means a planned genetic alteration of the population so that the animals can better fulfil the demands dictated by production requirements. In breeding work, one must, as a rule, combine a suitable method of breeding with effective selection. It must also be realised that natural selection is always at work, to a greater or lesser degree, especially when inbreeding is applied.

SPECIES, BREEDS, STRAINS, LINES AND FAMILIES

At this point it will be well to define the systematic concepts which will be used in the following.

The animal kingdom is divided into *species* on the basis of both morphological and physiological characters. Most important, however, is the reproductive discontinuity, i.e. two species do not interbreed, or do not produce fertile progeny when mated together. This discontinuity can have different causes and vary in the degree of manifestation. Horses and asses can be mated with each other and produce full viable offspring, but they are sterile. There are some cases reported in the literature of the female progeny (mules) from a mating between a mare and a jackass having been fertile, but these are rarities. These two species are thus clearly separated from each other. The Tibetan yak and ordinary domestic cattle can be mated together and produce progeny; the female progeny are fertile but the males are sterile. Only after two back-crossings to domestic cattle are fully fertile male progeny produced, and these will then carry one-eighth of the yak's and seven-eighths of the cattle's heredity. There is no reproductive discontinuity between Zebu cattle and the European cattle breeds, both the male and female progeny from crosses between them being fully fertile. By means of such crosses it has been possible to produce new breeds, e.g. the American Santa Gertrudis breed. Due to the morphological differences between the Zebu and European cattle they can be regarded as subspecies of one and the same species. From the foregoing it will be apparent that the dividing line between species is, in many cases, fluid and not always as sharp as one might be led to believe.

Within a species, different *breeds* can be distinguished. From a biological point of view, a breed can be defined as a population of animals which differ from those in other populations within the same species in respect of definite genetically determined traits. The traits which characterise a breed can be qualitative —e.g. hair type, hair colour, horns—or quantitative such as size, body type, milk yield or fat content of the milk. It is possible to produce a breed, which is homozygous for one or several qualitative traits, but as a rule this does not distinguish it from the other breeds, since a number of breeds can have the same external characteristics. Apart from this, these qualitative breed characteristics are of little or no importance, at least in the larger farm animals. Production traits, which decide the economic importance of the breed, show continuous variation; and no clear dividing line can be drawn between breeds, even when the breed averages differ rather widely from each other. As mentioned earlier, these traits are under the influence of a large number of genes, and a considerable degree of heterozygosity exists within all breeds with regard to all quantitative traits. Consequently, biologically speaking, there are no pure breeds of farm animals, and it would not appear to be possible to produce one.

As used in practice the term *pure breed* refers to animals which are registered, or eligible for registration, in the breed herd book. These 'purebred' animals constitute a selected group, intended for use in breeding. The requirements for an animal's acceptance in the herd book vary with time and place. The breed concept in practical breeding is conventional rather than biologically determined.

The division into breeds is justified, however, since populations which constitute the breeds have been specialised for different purposes and for differing local conditions.

Although a misnomer, breeding within the same breed has, since olden times, been called *pure breeding* and matings between animals of different breeds has been called *cross breeding*. The term *crossing* will be used here in a wider sense, for example, even when mating takes place between strains or lines within the same breed. Crosses between different breeds will be referred to as *breed crossing*. Very often a breed can be divided into different *strains*, which from a breeding

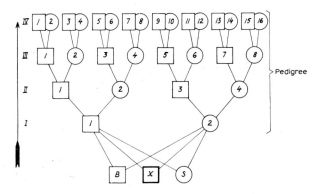

Fig. 14:1. Pedigree of a group of sibs *B*, *X* and *S*. The different ancestral generations have been numbered backwards in time from the parental generation. Consanguineous mating reduces the number of ancestors in the pedigree.

point of view are more or less isolated from each other due to geographic conditions or, when in some respects, the aim of breeding is different. This kind of breeding will be called '*strain breeding*'.

A *line* is meant to indicate a collection of animals, which, as a result of inbreeding, are more closely related to each other than to the other individuals in the strain or breed. If an inbreeding coefficient of at least 0·375, corresponding to two generations of full sib mating, is reached, then such a line is called an *inbred line*. *Family* will be used to indicate a collection of full sibs or half sibs. Consequently, all individuals within a family are equally closely related to each other. The old custom of referring to male and female lines, e.g. boar lines and cow families, will not be followed here. These terms are meaningless, unless at the same time it is stated how many generations back the ancestral sire or dam, which gave its name to the line or family, is to be found. An individual inherits half its genes from the sire and half from the dam, and it is therefore no more justified to point out a particular male ancestor than to point out a female the same distance from the individual whose pedigree is being examined (Fig. 14:1).

A REVIEW OF DIFFERENT MATING SYSTEMS

The aim of a particular mating system can be either to increase, or decrease, the homozygosity of the progeny, compared with the parents, or in some cases, to maintain the degree of homozygosity unchanged. This constitutes the basis of the following classification.

A. Random mating

This is not always the same thing as unplanned mating, since the object can be simply to keep the genetic constitution of the population unchanged, for example in experimental studies, where a *control population* is required for comparing and measuring the effect of other mating systems.

B. Breeding for increased homozygosity

Very often, but not always, this means breeding within a breed.

1. *Inbreeding* can be defined as mating between animals which are more closely related to each other than the average relationship between all individuals in the population, as for example a breed. Breeding within small populations leads automatically to a certain amount of inbreeding. This was discussed in more detail in Chapter 4 (p. 104). The method of measuring relationship and degree of inbreeding by calculating *the coefficient of relationship*, R, and the *coefficient of inbreeding*, F, was also dealt with in the same chapter, pp. 106–110. From a practical point of view, it can be said that mating between animals that are less closely related than cousins has so little effect on the degree of inbreeding that it may be disregarded. Matings between half-cousins leads to only a very slight increase in the coefficient of inbreeding.

Inbreeding can be either occasional or consistently carried out for several generations. In the latter case, three main types of inbreeding systems can be distinguished:

(*a*) *Close inbreeding*, such as mating between sibs or between parents and progeny (incest), in order to achieve inbred lines with a relatively high degree of homozygosity. One of the methods most often used is full sib mating for many generations. The same effect can be achieved by consistently backcrossing the progeny to the younger of the parents (p. 110). How quickly these inbreeding systems lead to homozygosity, when no selection takes place, is shown in Table 4:5, p. 110 and Fig. 14:4. It can be estimated that after some ten generations the number of heterozygous loci in individuals in the inbred lines has decreased to about 12 per cent of the number at the beginning of the inbreeding. The genetic differences between the lines becomes more and more pronounced until almost complete homozygosity is reached within the lines (cf. p. 108). Half sib mating is very much slower in reaching homozygosity, but it is also less risky. By the application of less consistent and milder inbreeding the same results may be achieved, though it takes a longer time to do so (cf. p. 384).

(*b*) *Strain formation*, a considerably milder form of inbreeding. In the long run it leads to increased homozygosity within the strains.

(*c*) *Linebreeding*, inbreeding within an ancestral line with the object of increasing a particular male or female ancestor's proportion of the genetic constitution of the progeny. The most intensive form of linebreeding is backcrossing to the same parent for several generations in succession (Table 4:5, p. 110). Usually, a much milder form of linebreeding is employed, such as mating a female with a grandsire, or uncle who will carry half of the grandsires genes.

2. *Mating based on phenotypic resemblance.* This system of mating in its purest form, pays no regard to degree of relationship, the mating combinations being selected only according to the external resemblance of the animals; the mating is of 'like with like' as it was called in olden times, e.g. large bulls with large cows and small bulls with small cows. In inbreeding, identical genes are brought together regardless of their effect; but when mating is on the basis of external resemblance, genes which have a similar effect are brought together, regardless of whether they have a common origin or not. Consequently, this system of breeding has very much less influence on the degree of homozygosity than inbreeding has. The method of mating 'like with like' can effectively contribute to the differentiation of the breeding material when the heritability is high for the traits in question, but the effectiveness declines with lower heritabilities. It is also possible to mate 'unlike with unlike' and produce an intermediate form, but this of course leads to an increase in heterozygosity. In practical animal breeding, attention is paid both to relationship and to external resemblance in the selection of mating combinations.

C. *Breeding for increased heterozygosity*

Outbreeding is the opposite of inbreeding; the relationship of the animals which are mated together is less close than the average relationship within the population. Mating between inbred lines or strains within the same breed is thus a form of outbreeding. In what follows no distinction will be made between such outbreeding and crosses between different breeds. The following types of crossbreeding can be distinguished:

1. *Single two-way crosses.* Two different populations (inbred lines, strains or breeds) can be crossed with each other to produce an F_1 generation which is used only for production purposes and not for breeding. As will be discussed later, the F_1 individuals usually exhibit hybrid vigour, especially when inbred lines are crossed with each other. However, the method has the disadvantage that a relatively large number of the parental types has to be maintained in order to continue the crossing. When two inbred lines of the same breed are crossed, the progeny are said to be *in-crossbred*.

2. *Three-way crosses.* The first generation crossbred females are crossed with males of a third line, strain or breed, thus utilising the hybrid vigour of the dams; this is especially important in the crossing of inbred lines.

3. *Four-way or double crosses.* Populations A and B are first crossed with each other, and so are C and D, to obtain the F_1 generations F_{AB} and F_{CD}. These are then crossed together to give the 'double hybrids', $F_{(AB.CD)}$. This method is used extensively in poultry breeding for crosses between inbred lines which have low viability. It is necessary to maintain only a relatively small number of animals of the lines A, B, C and D, which generally show lowered viability (cf.

p. 372). For the production of commercial chicks crosses are made between the 'hybrids' F_{AB} and F_{CD}, which usually have normal viability.

4. *Back-Cross.* Usually, the F_1 females are backcrossed to males of one of the parental populations for example, by the combination $F_{(AB.A)}$ for the production of commercial animals. This method is advantageous especially when the F_1 females, on account of their hybrid vigour, are better mothers than females from either of the parent populations (maternal influence). This is very much the case in pig-breeding; but for economic reasons planned rotational crosses are to be preferred.

5. *Rotational crossbreeding.* Crossbred females are mated with males from either of the parent populations with the proviso that the matings are alternated

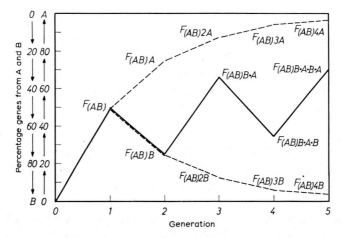

Fig. 14:2. Schematic illustration of the genetic effect of continuous rotational backcrossing to the parental breeds.
The two broken lines from the first generation of crossing (F_{AB}) represent repeated backcrossing to A or B and the solid line rotational backcrossing to A and B. The scale on the left of the diagram indicates the proportion of genes from the parental breeds present in the progeny. (Modified after LAUPREACHT 1958)

(rotated) for each new generation. In American pig-breeding this method has achieved widespread use in crosses between different breeds. The following types of rotational crossbreeding have been tried:

(a) *Criss-crossing.* Breeds A and B are crossed to produce an F_1 generation, then F_{AB} sows are back-crossed to boars from breed B, the $F_{(AB.B)}$ sows are then mated back to boars from breed A and so on (Fig. 14:2). The method has advantage over the two-way cross, that one can continue to use the crossbreed sows for breeding and it is necessary to purchase only 'purebred' boars.

(b) *Three-way rotation.* F_{AB} females are mated with boars from a third breed C. Boars from breed A are used on the next generation of females and boars from breed B on the following generation of females, and so on.

(c) *Four-way rotation.* Males from a fourth breed D are used on females of the combination $F_{(AB,C)}$. Thereafter males from breeds A, B, C and D are used in succession for each new generation.

After rotational crossing has been applied for a number of generations, a situation of equilibrium is reached with respect to the proportions of the different breeds in the genetic constitution of the crossbreds. In two-way crosses, equilibrium means that 67 per cent of the genes of the progeny come from the breed to which back-crossing last took place, and 33 per cent from the other breed. In three-way crosses, 57 per cent of the genes come from the breed of the last used male, 29 per cent from the previous breed and 14 per cent from the third breed, which will be used for the next back-crossing. After equilibrium is reached, the highest proportion of genes in the genetic make-up of the progeny, which come from the males of the breed last used, can be calculated from the formula $(50 \times 2^n)/(2^n - 1)$, where n denotes the number of breeds used in the rotation crossing. If two breeds are used, the figure will be 66·7 per cent; if three breeds, 57·1 per cent; and if four breeds, 53·3 per cent. Increasing the number of breeds further in rotational crossbreeding has little effect on the degree of heterozygosity. Three breed crosses have given good results in pig and have also been used to advantage in cattle and poultry breeding. The same rotation method can, of course, be used for crosses between inbred lines or 'strains' within breeds.

6. *Topcrossing.* Topcrossing is a term used for the mating of inbred males to females of non-inbred populations. It is questionable whether the greater homozygosity of these males for production traits can compensate for their lower fertility and vigour.

7. *Crossing with recurrent reciprocal selection.* In each of two breeds or strains, progeny-testing is carried out by crossing with the other breed or strain (reciprocal crossing). Those animals which produce the best progeny from such crosses are then used for multiplying their own breed or strain. The object is to change both populations gradually, so that they give better results in crosses with each other (cf. p. 448).

The following two methods of crossbreeding are concerned only with increasing the heterozygosity of the population in the initial stages; with continued crossbreeding the heterozygosity will generally begin to decline again.

8. *Grading-up.* Backcrossing to the same breed takes place generation after generation. The object is to change a mixed population to a 'pure breed'. With each new generation, the proportion of genes from the original mixed population decreases to half the proportion present in the previous generation, so that after four generations it has decreased to 6·25 and after five generations to 3·125 per cent (Fig. 14:2). This method of breeding has been widely used in many countries where it was desired to change cattle of mixed ancestry to a particular recognised breed type, e.g. Ayrshire or Friesian. After four or five generations of successive crossing to herdbook registered bulls the resulting progeny are accepted as 'pure bred'.

9. *Crossing for the production of a new breed.* The great majority of our present-day breeds of farm animals have been founded by crossing different breed types with each other in an attempt to combine their desirable traits in the new breed.

During the last decade several new breeds of beef cattle, sheep and pigs have been produced in the U.S.A. by such combination breeding. The consolidation of a new breed, following the foundation crosses, generally demands a certain amount of inbreeding combined with intensive selection.

INBREEDING AND CROSSING IN THE LIGHT OF EXPERIMENTS

Ideas about inbreeding and crossing, and their advantages and disadvantages, have varied from time to time. Mention has already been made of the famous English animal breeders, Robert Bakewell, the Colling brothers and others. During the latter half of the eighteenth century and beginning of the nineteenth, they practised very intensive inbreeding with cattle and sheep, and to some extent even with pigs, and were able to achieve very good results. The application of this method, however, is not completely without penalties, since it often leads to lowered fertility and reduced vigour, so called *inbreeding depression*.

A number of experimental studies on the effect of inbreeding had already been carried out during the nineteenth century, but it was only after the breakthrough of Mendelism at the turn of the century that the work really began. The American inbreeding work with maize and Sewall Wright's inbreeding experiments with guinea-pigs contributed immensely to unravelling the mystery of the detrimental effects of inbreeding.

When maize is self-fertilised, there is considerable inbreeding depression during the first generations. This gradually declines and after 8-10 generations reaches an inbreeding minimum, after which there is very little further depression. With inbreeding, the original heterozygous material is split up into a number of lines which differ more and more from each other until a constancy is reached within the lines; for all practical purposes these represent 'pure lines'. The depression is apparent within all the lines, but the degree of intensity varies. Many lines die out because of their lack of ability to reproduce.

When the inbred lines of maize are crossed with each other there is a marked improvement in the F_1 generation, and some crosses will give higher yields than conventionally selected seed maize give. In 1912 one of the American pioneers in maize genetics, G. H. SHULL, coined the term *heterosis* to describe the hybrid vigour obtained from crossing. The production of 'hybrid maize' is based on this hybrid vigour. In comparative field trials it gave 25-30 per cent higher yield than conventionally selected seed maize. The seed used in practice has usually been obtained by crossing two different F_1 populations, thus obtaining a 'double hybrid' (cf. p. 365). The F_1 generations may be more uniform, but the inbred lines are poor seed producers and therefore the double-cross method has been found more economical. Only very few of the inbred lines give the desired yield increase in crosses, and so it is important to test thoroughly which combinations give the best results.

Heterosis is said to occur when the average of the generation produced by crossing is higher for a particular trait, or several traits, than the average of the parental populations. The percentage of superiority is used as a measure of the degree of heterosis. It has been suggested, however, that, for practical purposes, it would be better to measure the heterosis effect according to the superiority over the best of the parent populations. If it is not possible to produce something better than this, then there is no purpose in cross breeding.

Wright carried out consistent full sib mating of guinea-pigs for more than thirty generations. The inbred lines reached a relatively high constancy in respect of coat colour and a number of morphological traits, but the fertility and other viability traits showed considerable variation. During the first twelve generations the inbreeding depression was very pronounced and several lines died out. A number of morphological malformations and peculiarities appeared, but the most serious inbreeding depression was in fertility and viability. When after twenty generations the inbred lines were crossed with each other, a definite hybrid vigour was obtained in the F_1. Resistance to tuberculosis increased by 20 per cent, body size by 12 per cent, average litter size at birth by 10 per cent and growth rate from birth to weaning by 11 per cent. At the same time the frequency of stillborn young decreased by 7 per cent, mortality between birth and weaning by 11 per cent, and the interval between litters by 30 per cent.

The presentation of these extremely interesting results raised the question of whether something similar could be achieved in farm-animal breeding. Large-scale inbreeding experiments with poultry and pigs, and to a less extent with sheep and cattle, were commenced at the American research stations to elucidate this question. After the end of the World War II, inbreeding and crossing experiments were carried out in Great Britain also. A short summary of the results of these experiments is given below.

Poultry

In 1932 an experiment was begun at Iowa State College, in which inbreeding was conducted with 23 lines of White Leghorn. For control purposes, 3 lines of non-inbred stock were maintained. The degree of inbreeding varied between lines but, as a rule, full sib and half sib mating was applied. By 1946 an inbreeding coefficient of 0·85 had been reached in some cases. Selection was made for egg hatchability and viability of the chicks, but in spite of this there was a deterioration in these traits as well as in body size and egg production. Fig. 14:3 shows the average decline of egg production in relation to the degree of inbreeding. The first year of laying was divided into three phases: $P1$ = from date of first egg to 31st December; $P2$ = 1st January to 30th April; $P3$ = 1st May to 364 days from laying of first egg. PT = the egg production during the whole of the first year. The number of eggs was calculated as a percentage of the number of feeding days of all birds for each period. This figure, which is a measure of the laying intensity, decreased by 0·43 per cent units for each 1 per cent increase in the degree of inbreeding. This means that with an inbreeding degree of 0·5, corresponding to three generations of full sib mating, the laying intensity decreased to 38 per cent, compared with 60 per cent at the commencement of inbreeding. Crosses between inbred lines resulted in an increase in laying intensity of F_1 females in the first year to 62·9 per cent, i.e. slightly higher than for the non-inbreed stock.

Another inbreeding experiment has been in progress at the University of Minnesota since 1937. The aim of this experiment was to study inbreeding as a tool in breeding work with poultry. Initially full sib mating was used, but after some years this was abandoned and since then the degree of relationship in mating has varied according to the reaction within the different lines. Continuous selection has been made for egg production, body size and viability. By 1950 the experiment included the production records of 9000 laying hens and information on about 25 000 chicks. In several of the lines, inbreeding coefficients of

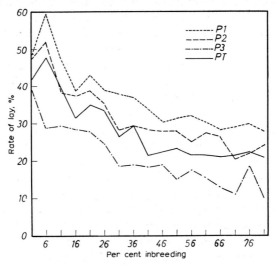

Fig. 14:3. The influence of degree of inbreeding on the laying intensity of hens. $P1$ = the laying intensity to 1st January, $P2$ = 1st January to 30th April, $P3$ = 1st May to end of recording year, PT = the whole year. (STEPHENSON et al. 1935).

0·6 to 0·7 had been reached. It was calculated that, in spite of the selection which had been applied, for each 10 per cent increase in inbreeding production fell, on the average, by 9·3 fewer eggs per hen per year. At the same time the hatchability decreased by 4·4 per cent and the age at laying of the first egg increased by 6 days. The effect on body size was insignificant and the egg-weight remained virtually unchanged. In almost all the lines, a number of defects such as crooked breast bones, mis-shaped beaks and cleft palate, segregated out. Several of the inbred lines were tested in crosses with each other; and in a number of combinations consistently good results were obtained. The research workers suggest that lines should be tested in crosses when an inbreeding coefficient of 0·375 to 0·50 has been attained.

At the University of California, Berkeley, eight lines of White Leghorn were started in 1944. All the lines were produced from 'production flocks', which served as controls during the course of the experiment. Within each line selection took place for only one trait, either egg production per hen or the average weight of the eggs. Within two lines, selection was for high, and within another two lines for low, egg production. The same thing was done for egg-weight. Several other traits were also recorded, in order to study how they were influenced by inbreeding and selection. Egg-weight gave definite results both in the positive and negative directions and did not appear to be affected by the inbreeding. Egg production, on the other hand, decreased during the first three generations of inbreeding in all the lines, even those selected for high production. During the fourth and following generations, the effect of selection for high production began to show results. When an inbreeding coefficient of about 0·5 had been reached, the different lines were crossed with each other. Crosses between lines that had been selected in opposite directions gave generally better results than crosses between lines that had been selected in the same direction. The genetic variation within the lines, in respect of egg production and body-weight, appeared not to decrease with advancing inbreeding. It was thought that probably the homozygosity of the lines proceeded more slowly than the inbreeding coefficients indicated, due to the fact that the genetic constitution was better balanced with a higher than with a lower degree of heterozygosity.

Several inbreeding experiments have been planned specifically to answer the question of what can be gained by crossing inbred lines to produce 'hybrid poultry' similarly to the production of hybrid maize. At Beltsville, U.S.A., comparisons have been made between the result of crossing inbred lines ($F = 0·6 - 0·8$) of Rhode Island Red and White Leghorn with populations of the 'pure' breeds bred in the conventional manner. The highest egg production was obtained when inbred Leghorn males were mated with inbred Rhode Island Red females. In the first laying year the crosses gave an average of 245 and the pure breeds 215 eggs, a superiority for the former of 14 per cent. A certain increase in production would most probably have been obtained by crossing the breeds without inbreeding, but hardly as much as that just mentioned.

One problem which has received some attention has been the relationship between incrossbred and inbred line performance. The general conclusion was that, though the relationship was not high, few useful incrossbreds would be missed if the worst 25 per cent of inbred lines were discarded before being crosstested.

At Reaseheath in England, five inbred lines of White Leghorn were tested in crosses. For comparison a commercial flock of Leghorn hens was maintained at the research station, and the females were mated to purchased non-inbred males. With regard to viability and egg production, several of the line crosses were superior to the control birds but in body-weight and egg-weight they were somewhat inferior. Inbred lines gave results in crosses which were not related to the production of the lines themselves.

In the U.S.A. several very large private concerns have for many years produced inbred lines, and by crossing them have produced 'hybrid chicks', which are sold at a higher price than normal commercially produced chicks. Unfortunately, it is usually impossible to get a definite idea of how strongly the lines are inbred and how they are combined in crossing. One of these firms states in an advertising brochure that it conducts full sib mating for four generations and thereafter tests the lines in crosses with each other, sometimes by first crossing to a line which has previously shown good general combining ability. Blood grouping is also carried out in an effort to identify which lines can be expected to give the most valuable combinations (cf. p. 209). The day-old chicks which are sold to the egg producers are, as a rule, 'double hybrids' obtained by combining two F_1 generations from single crosses.

Commercial types of 'hybrid chicks' have been the object of comprehensive testing in the official 'random sample tests' (cf. p. 472). They have, on the whole, given good results. The superiority is much less than it is with hybrid maize, but even an increase in egg production of a few per cent and an equally large reduction in chick mortality and mortality in the first laying year are of considerable economic importance.

At the end of the 1940s a regional poultry breeding laboratory was organised at Purdue University in order to co-ordinate the breeding research in poultry in the Central and Northern States. The inbred lines and crosses produced by the various State experimental stations are tested at Purdue under standardised environmental conditions. In this way it should be possible to throw light on the question of the economic results of producing 'hybrid chicks'. The cost of producing inbred lines and their testing in crosses with each other is of course very high. The question is whether it is possible to achieve approximately the same results with other and less costly methods. Large-scale research is conducted also in Great Britain.

It has been shown that, even with the crossing of strains (within the same breed) that for many generations have been bred separately, a considerable increase in viability and laying capacity is obtained in the F_1 generation. Strain breeding leads eventually to a genetic differentiation of the material, even if this is not so pronounced as with intensive inbreeding. As an example, mention can be made of the work at Cornell University, where two strains of Leghorn, which had been commercially bred for thirteen years, were crossed. The inbreeding coefficients of the two strains were 0·13 and 0·08, respectively. The production of the F_1 generation exceeded the average production of both strains by 25 eggs per hen up to 500 days of age; and the hatchability of the eggs increased by 5 per cent.

Even breed crosses, without previous inbreeding, have given good results in many cases. At Iowa State University a comparison was made between strain crosses within breeds and breed crosses. The investigation was carried on for three years, with commercially bred strains of New Hampshire, Rhode Island Red, Plymouth Rock and Australorp. The hatching percentage was highest and chick mortality lowest with the breed crosses. Egg production per hen was 12 per cent higher for the breed crosses, and the strain crosses were 10 per cent higher than the average of the pure-bred groups. Danish trials have shown that in crosses between different laying breeds, the superiority of the crossbreds over the 'pure' breeds is more pronounced under unfavourable than under favourable environmental conditions. This is most likely due to the fact the crossbred birds are genetically better buffered against external stresses. Rotational crosses with

Pigs

Inbreeding experiments with pigs were begun at several American research stations in the 1920s. In 1936 the U.S. Department of Agriculture organised the Regional Swine Breeding Laboratory at Iowa State University, in order to co-ordinate the work of the State experimental stations. By 1956 the co-ordinated experiments of ten States had produced and tested more than 100 inbred lines from different breeds; the inbreeding coefficients within these lines was between 0·3 and 0·6. It was soon found that the inbreeding depression resulting from consequent full sib mating was considerable, even though the different lines varied greatly in this respect. Professor Winters (Minnesota), in particular, used therefore a flexible system, in which the degree of inbreeding was continuously adjusted according to the animals' reaction, i.e. the extent of the inbreeding depression. In this way there were greater possibilities of counteracting depression within the lines by selection. On the other hand, there is the risk that the reduction in heterozygosity will be much less than the estimated inbreeding coefficient would indicate. Other research workers have suggested that one should first try full sib mating with a large number of lines for a couple of generations, and then select hard among the lines; after this, less intensive inbreeding can be conducted with the most vigorous lines.

Reduced litter size at birth, increased mortality of the young pigs and decreased growth rate have been the characteristic features of most inbreeding experiments. The following data from the Regional Swine Breeding Laboratory may be quoted as an example (Table 14:1). It is interesting to note that with regard to litter size and viability of the pigs the inbreeding of the dam has contributed to the depression effect to about the same extent as the inbreeding of the pigs themselves.

Table 14:1 Change in performance for each 10 per cent increase in the inbreeding coefficient (DICKERSON et al. 1954)

	Inbreeding of the Litter	Dam of the litter	Combined effect
No. of pigs at birth	−0·20	−0·17	−0·38
No. of pigs at 21 days	−0·35	−0·31	−0·65
No. of pigs at 56 days	−0·38	−0·25	−0·63
No. of pigs at 154 days	−0·44	−0·28	−0·72
Av. weight at birth	0·02	−0·06	−0·03
Av. weight at 21 days	0·08	−0·11	−0·03
Av. weight at 56 days	0·03	0·06	0·08
Av. weight at 154 days	−3·44	−0·13	−3·57

An inbreeding experiment was started in 1950 at the Animal Breeding Research Organisation (ABRO), Edinburgh, with 88 full sib groups of the Large White and Wessex breeds. The plan was to carry out three generations of full sib matings before line-cross tests were made. Selection was applied within and between lines for litter size as well as for pig weight at weaning and at 6 months of age. Inbreeding depression was manifested already in the first generation when the pigs, but not the dams, were inbred. This depression was markedly increased in the second generation, where the decrease in individual weight at 180 days of age amounted to 14·6 kg on an average. At this stage no less than 68 lines were lost because of deterioration in fertility. The carcass quality of the inbred pigs was about equal to that of the outbreds. A pronounced increase in the frequency of congenital abnormalities was another feature of the inbred lines, as is shown by the following figures from DONALD (1955), based on 319 inbred and 169 outbred litters with 3006 and 1693 piglets respectively.

Abnormality	Inbred %	Outbred %
Kinky tail	4·6	2·1
Skeletal deformities	1·1	0·4
Hernia	1·5	0·4
Other types	0·8	0·9

Only 8 of the lines were eliminated because of segregation of defects. Kinky tail and hernia do not show simple recessive inheritance; when two 'kinky tail lines' were crossed, only a few piglets with this defect appeared in the litters, so produced, probably because of the heterosis of the crossbreds.

The depression accompanying inbreeding is mainly a result of the decline in general vigour and fertility, and is only to a minor extent caused by the increased segregation of congenital defects. In the U.S.A. it has not been practicable to carry the inbreeding farther than two or three generations before the crosses are made, in some cases only one generation. As a rule the line crosses have been superior to the inbreds in litter size and viability of the piglets, but on an average they have not been superior to the foundation stock. Three-line crosses have produced better results than two-line crosses, probably because the dams have been crossbred in the former but inbred in the latter case. CRAFT (1958) published the following averages from the Regional Swine Breeding Laboratory breeding project (Table 14:2).

Table 14:2 Average results of inbreeding and crossbreeding swine at ten co-operating experiment stations in central states of the U.S.A. (CRAFT, 1958)

	Numbers of litters	No. of pigs per litter at birth	Per cent pigs weaned
Within breeds			
No inbreeding	1564	8·89	67·9
Inbred (Av. 30%)	9424	7·07	61·9
Two-line crosses	1572	7·78	71·3
Three-line crosses	1181	8·30	71·8
Breed crosses			
Sows inbred	1323	8·28	74·6
Sows crossbred	2190	9·51	77·2
New breeds or lines from cross foundation	3180	8·95	71·7

It is uncertain to what extent the averages of the different groups are really comparable; but supposing that they are, the best results have been obtained from breed crosses when the dams have been crossbred. This is in agreement with the conclusion reached by Winters (Minnesota) and with results obtained at the Beltville Centre, that the results of line crosses improve with increasing genetic diversity between the lines. At the Missouri Station it was found that line-cross dams reached puberty 30 days earlier than the inbred sows, that on average they ovulated more eggs and produced 1·85 more piglets per litter. Furthermore, sterility and reduced fertility were more frequent in the inbred lines than in the line-crosses.

On the whole, the development of inbred lines and their use for line crosses have given less promising results when applied to swine than in poultry breeding. The cost of producing highly inbred lines and their testing in various cross combinations is very high, and at present it seems uncertain whether this method of swine improvement will yield profitable results. Numerous experiments have also been made on crossing breeds without previous inbreeding. FREDEEN (1957), who has reviewed these experiments up to 1956, points out that in most cases adequate comparison with the parental breeds has been lacking but in spite of this the results tend to show that an increase in the number of viable pigs per litter may be expected in breed crosses and that crossbred dams nurse their piglets better than purebred sows.

One method used in studying the effect of breed crosses has been to make double matings, i.e. the sow is mated to a boar of her own breed and then immediately in the same heat period re-mated to a boar of another breed, thus providing the purebred and crossbred pigs with the same maternal environment.

LUSH et al. (1939) double mated 36 Duroc and Poland China sows and found that the crossbred pigs excelled their purebred litter mates by 15·4 per cent in survival from birth to weaning, and were 2·5 per cent heavier at birth and 10·7 per cent heavier at 8 weeks of age. Winters (1935) was the first to advocate criss-crossing and rotational crossing, basing his recommendations on actual breeding experiments. In one of his early experiments he made crosses between Duroc, Poland China and Chester White. When evaluation was based on four traits (number of pigs at birth and weaning, daily gain after weaning, and feed conversion) the superiority of the crossbred pigs over the purebreds was as follows: first cross (F_1) 6·3 per cent, back-cross 7·5 per cent, and three-way cross 11·7 per cent.

SMITH and KING (1964) made a statistical study of 34 800 litters on 800 farms in Great Britain, mainly representing the Large White, Landrace and their crosses. Generally, there was a lower mortality in crossbred litters; they had 2 per cent more pigs at birth and 5 per cent more pigs at weaning than the purebred litters. The total litter weight at weaning was 10 per cent higher for the crossbreds. The litters from crossbred dams excelled the purebreds in number of pigs per litter by about 5 per cent at birth and 8 per cent at weaning, thus showing the importance of heterosis in the dams. All comparisons were made within herds.

At the Wiad Research Station in Sweden a crossbreeding experiment has been carried out between Large White and Landrace pigs comprising a total of 619 litters from 187 sows and 144 boars. The experiment started with 12 pairs of full sisters and the same number of unrelated boars of each breed, representing different herds and ancestry in various parts of the country. In each full sib pair one female was mated to a boar of her own breed and her sister was mated to a boar of the other breed. For production of the second litter from these sows the breed of the boars was changed; usually each boar was allowed to serve only four sows, i.e. two of each breed in the first generation and thereafter two purebred and two crossbred sows; after this new boars were procured. As far as possible, each sow produced four litters; the third litter was sired by a boar of the same breed as the one that sired the second litter, and for the fourth litter a boar of the other breed was used. Farrowing was concentrated in spring and autumn, and therefore sows which produced four litters had one purebred and one crossbred litter in the spring season and one purebred and one crossbred litter in the autumn. Replacement sows were obtained as a rule by selecting two females from the third litter of the sows, selection being based on their previous records. Purebred litters were produced throughout the experiment for controls; F_1 sows were backcrossed to boars of the dam's breed, after which criss-crossing was practised.

Table 14:3 Deviation of crossbred litters from the purebreds at different ages (SKÅRMAN, 1965)

	Purebred litters, av.	Deviation from purebreds		
		F_1	First backcross	Second backcross
Litter size				
At birth	10·44	+0·37	+0·68	+0·82
At 3 weeks	8·47	+0·33	+0·69	+0·42
At 8 weeks	8·20	+0·51	+0·59	+0·47
At 20 weeks	8·12	+0·63	+0·57	+0·55
Early postnatal mortality	10·58	−6·67	−2·79	−1·14
Litter weight, kg				
At birth	14·60	+0·96	+0·08	+0·68
At 3 weeks	46·25	+4·00	+5·17	+2·28
At 8 weeks	111·16	+16·33	+8·88	+12·11
At 20 weeks	377·16	+43·74	+29·38	+41·22

Table 14:3 presents a summary of the results obtained with regard to size and weight of litters at four stages of development. The figures show an increase in the litter size at birth of 3·5 per cent in the first crossbred generation and 6·5 per cent in the backcross generation; at three weeks of age the increases were 3·7 per cent and 8·1 per cent respectively. The corresponding increases in total litter weight at 20 weeks of age were 11·6 per cent and 8·0 per cent respectively. The results are in agreement with previous finding, in that F_1 mothers produce larger litters and nurse them better than purebred mothers; but the growth rate of the young is higher for the first generation crossbreds than for backcrosses as soon as

the litter is independent of the mother for its feed supply. Why the second backcross generation has been equal, and in some respects superior, to the first backcross generation is difficult to explain.

Other findings in this crossbreeding experiment may be summarised as follows. The frequency of stillbirths was 4·17 per cent for the purebreds and 2·98 per cent for the crossbred generations, corresponding figures for congenital defects were 3·20 per cent and 1·93 per cent respectively. The variation in individual pig-weight within litters was lower for the crossbred than for the purebred pigs. The frequency of reproductive failures was highest for the purebred sows. With regard to carcass quality, there was no significant difference between purebred and crossbred pigs; and the differences between breeds, Large White and Landrace, were also very small. Where such differences exist, as for example, in percentage meat in the side and size and shape of ham, the crossbreds were usually intermediate to the two pure breeds.

It may be concluded that crossbreeding generally yields somewhat better results than pure breeding with regard to litter size at birth as well as viability and growth rate of the pigs, and that the variation in weight for age within the litters decreases. Certain breed crosses, however, show greater heterosis than others, probably because of differences in genetic diversity and degree of heterozygosity. Rotational crossbreeding with three breeds may be expected to yield better results than criss-crossing with only two breeds, since a higher degree of heterozygosity will be obtained in the crossbred pigs.

According to Hazel (1963) nearly all slaughter pigs in the U.S.A. are products of crossbreeding, and crossbred sows are used to a large extent in commercial herds. Smith and King (1964) state that in Great Britain over 60 per cent of the litters produced are crossbred.

In the U.S.A. a number of firms produce 'hybrid boars' for breeding purposes. As a rule, no information is forthcoming about how these boars are produced, but advice is given to farmers as to how the boars are to be used. The method seems to be this: if a farmer is sold an F_1 boar from a cross between breeds A and B, then after this boar has been used the farmer is recommended to use a boar from a cross between C and D and the next time a boar from another cross —in other words, rotational crossing with hybrid boars. The breeders appear to have found that this method gives good results.

Sheep and goats

Inbreeding experiments, including crosses between inbred lines, with sheep and goats have also been carried out at American research stations, though on a small scale. At the University of Illinois a herd of French Alpine goats were intensively inbred without any pronounced inbreeding depression. Crosses between different sheep breeds have been practised on a large scale in Great Britain for many years with very good results. Ewes of the Scottish Blackface, for example, have been mated to Border Leicester rams and 'fat lambs' produced by mating the F_1 ewes with Suffolk rams. For similar purposes crosses are made in the U.S.A. between sires of the mutton breeds, e.g. Hampshire or Suffolk, and wool-type ewes from the 'Western Ranges'. According to Winters, rotational crossing with sheep has about the same advantages as in pig breeding.

Cattle

Several of the American experimental stations have conducted inbreeding experiments with cattle. The aim has been to try a combination of relatively mild inbreeding and selection as a method of breeding in dairy cows rather than to produce inbred lines for crossing with each other.

At Beltsville an inbreeding experiment was in progress from 1913 to 1943. The foundation females were cows of mixed ancestry, which were mated to a proven Holstein bull. The aim of this experiment was to study the possibilities of building up a high-producing herd by using a proven bull, and his sons, on cows of average production. Daughters, and in some cases even granddaughters, were back-crossed to the same bull, so that the progeny obtained carried 75 and 87·5 per cent, respectively, of the genic constitution of their paternal ancestor.

This intensive linebreeding was accompanied by rather pronounced inbreeding depression symptoms. The number of services required per conception increased from 2·0 for non-inbred cows to 3·6 when the inbreeding coefficient of the calves reached 0·5. The mortality of inbred calves less than one month old was 15 per cent. When the same bull was mated with unrelated cows, there were no deaths among the 43 calves born; but in the 'purebred' Holstein herd served at the same time, 29 calves died out of the 300 born (9·7 per cent). The effect of inbreeding on body-weight and production is illustrated by the relative figures shown in Table 14:4.

Table 14:4 *The effect of inbreeding on body-weight and milk yield, relative figures (after* SWETT *et al.* 1949)

Degree of inbreeding	Birth wt. of calves	Weight of heifers at 18 months	Cows which completed at least one lactation			
			No.	Weight of Pituitary	Udder	Yield of butterfat
Non-inbred	100	100	22	100	100	100
F: 0·10 – 0·29	100·1	92·5	15	96·7	95·1	101·5
0·30 – 0·49	95·9	89·6	27	92·2	83·8	93·0
0·50 – 0·69	83·7	87·3	7	80·2	54·1	82·6

The inbred animals were smaller at birth and grew more slowly than the non-inbreds; it was also observed that they were listless and had coarse coats, which is an indication of lowered condition. A number of endocrine glands were smaller, as shown in Table 14:4. The relative figures for the weight of the udder applies to full udders of lactating cows which were slaughtered. Milk and butterfat yield declined with increased inbreeding; the butterfat content of the milk decreased during the first generations of inbreeding, but the reason for this would appear to be that the original cow population had a relatively high butterfat content (4·43 per cent) and that mating took place with Holstein bulls.

At Davis in California, a Jersey and a Holstein herd were inbred, and after 25 years the inbreeding coefficient in the Jersey herd was on the average 0·15, for particular individuals over 0·40. Several recessive defects segregated out; calf mortality increased and the body size of the animals decreased. As in a number of other inbreeding experiments, the depression in body size was especially apparent in the early stages of development; the inbred animals grew more slowly than the non-inbred but eventually attained about the same size. For each 1 per cent increase in inbreeding the weight of the females was lower, at birth by 0·28, at 6 months by 0·47 and at $4\frac{1}{2}$ years of age by 0·10 per cent. In the Holstein herd also the inbreeding resulted in increased calf mortality and reduced growth rate. A significant negative regression of milk yield on degree of inbreeding was demonstrated for 164 Holstein cows, the daughters of 22 bulls. Milk yield fell by 94 kg per lactation for each 1 per cent increase in inbreeding.

Since 1930 a closed herd of Holstein cattle, consisting of about 70 cows, has been maintained without the introduction of fresh stock at Iowa State University. In 1955, the average inbreeding coefficient for the whole herd was 0·074 with the highest value of 0·34; yield data were available on 534 cows, the daughters of 69 bulls. The decline in production in the first lactation was 24 kg for each 1 per cent increase in the degree of inbreeding, after eliminating the differences between the different progeny groups. The decline was somewhat less in later lactations, probably because the inbred animals developed more slowly. This investigation also showed that the inbreeding had a more depressive effect on the growth rate of the young animals than on their mature body size. The butterfat content of the milk was not affected by the inbreeding.

An interesting experiment was started in 1948 at the Wisconsin Experiment Station. Results have been reported by MI, CHAPMAN and TYLER (1965). The foundation animals consisted of 6 proven sires of the Holstein breed, together with one or two outbred sons and 20 outbred daughters of each sire. Each of these groups was designated as a sire line. Matings were planned to produce inbred daughters from each sire line as well as outbred daughters, the latter by crossing each sire line with two other lines. Inbreeding and line crossing was continued by the same mating system in later generations. Outbred heifers born in the herd were mated to randomly selected A.I. bulls for the production of a control group. There were, thus, non-inbred progeny from non-inbred dams (0–0), inbred progeny from non-inbred foundation dams (I–0), inbred progeny from inbred dams (I–I), and 2-line cross progeny (2Lx), 3-line cross progeny (3Lx), etc. Detailed observations were made on growth, reproduction and production of all the experimental animals. When the report was published it gave 760 production records from 310 cows sired by 31 bulls of the 6 sire lines, in addition to 174 outbreds which had completed their first lactation. The difference between the inbreds and outbreds was used as a measure of inbreeding depression or advantage of heterozygosity. Considering the actual yield during the first lactation, the inbreds produced 591 kg milk and 15 kg butterfat less than the outbreds, and, although they were 24 days older on an average, they weighed 27 kg less at the first calving. The average inbreeding coefficient for all the inbreds was 25·5 per cent. The linear regression of yield on per cent inbreeding was −23·5 kg milk and −0·63 kg butterfat but there was wide variation between the lines in this respect. The 2-line cross animals produced, on the average, 826 kg more milk and 33 kg more butterfat than the inbreds; they weighed 44 kg more, and were 3 weeks younger at their first calving. The change in production with age did not differ significantly between the groups. The frequency of calf infections and reproductive disturbances was higher for the inbreds than for the other two groups. The following concluding statement by the authors may be quoted: "Unpredictability of inbreeding in stock of different genetic origins, tendency to lower reproductive and productive efficiency, and lack of uniformity in performance of inbred individuals discourage the development of inbred lines as a general means of improving dairy cattle. Development and maintenance of inbred lines would be too costly, and crossing of such lines may not provide individuals clearly superior to individuals resulting from outcrossing."

Several statistical analyses have been carried out on data from practical breeding in order to show how temporary consanguineous matings influence the body size and milk yield of the progeny. An investigation was carried out by TYLER et al. (1949) with three herds of American Holstein-Friesians. The inbreeding coefficients of a total of 117 females were between 0·06 and 0·37. It was found that, within progeny groups, the milk yield decreased by 33 kg for each 1 per cent increase in inbreeding. The girth of the animals also tended to be less, especially at 18 months of age, but the height of the withers was not affected. ROBERTSON (1954) compared the first lactation yields of 82 inbred British Friesian heifers, all resulting from sire × daughters matings, with the yield of contemporary, non-inbred half-sisters in the same herds. The reduction in yield per cow as a result of inbreeding was 13·3 kg for each 1 per cent increase in inbreeding.

A similar investigation was carried out by HANSSON et al. (1961) on the daughters of A.I. bulls in the Swedish Red and White (SRB) and Swedish Friesian (SLB) cattle. The data included 12 897 SRB and 10 927 SLB cows. The frequency of close consanguineous matings was low; only 1·1 per cent of the SRB and 3·0 per cent of the SLB cows had inbreeding coefficients higher than 0·10. The regression, within herds and sires, of the daughter's milk yield in the first lactation (305 days) on degree of inbreeding (F in per cent) was $-14\cdot3$ in the SRB and $-10\cdot1$ kg in the SLB. This indicates that, on the average, sire-daughter mating resulted in a decrease of 358 kg in milk yield in the former and 258 kg in the latter breed. The figures are in good agreement with Robertson's results based on British data.

In summary, it can be said that inbreeding with dairy cattle, according to all the investigations to date, results in slower growth rate, reduced milk production and lowered fertility.

Crossing between different dairy breeds has been tried both in the U.S.A. and Europe. The research has not always been planned so as to give an objective answer to the question of the eventual advantage of crossing over pure breeding. In the first place, those animals which are used for crossing should be random samples of their respective breeds; secondly, the purebred and crossbred animals should be compared under the same external conditions. A crossbreeding experiment begun at Beltsville in 1939 and now completed, was criticised as not fulfilling either of these two conditions; some progeny-tested bulls were used for the crossing, and no control groups of the original breeds were maintained. Four different breeds were used, Holstein-Friesian, Jersey, Guernsey and Red Dane. The average production in the first lactation is shown in Table 14:5.

Table 14:5 Yield in the first lactation of cows in the crossbreeding experiment at Beltsville (FOHRMAN et al. 1954)

	No. of cows	Age at calving Yr:Mo.	Average for lactation		
			Milk kg	Butterfat %	kg
Females used for crossing	55	2:6	4781	4·55	206
Two-way cross, F_1	55	2:2	5915	4·53	266
Three-way cross	58	2:2	6061	4·44	276

Although no general conclusions can be drawn from these figures, it may be said that the crosses in the way they were made, have given good results. A crossbreeding experiment between British Friesian, Ayrshire and Jersey cattle is in progress at the Animal Breeding Research Organisation, Edinburgh. The experiment was begun in 1947 with 40 heifers from each breed, purchased from a number of different herds. Use has been made of young A.I. bulls, which represent a cross section of the males in the breeds. By reciprocal crosses and simultaneous pure breeding, six groups of crosses and three 'purebred' groups are obtained and compared with each other. So far, only preliminary reports have been published, but the indications are that the average production of the crosses is very similar to the averages of the parent breeds, with the exception of butterfat yield, which is somewhat higher in the Jersey crosses. Furthermore, the fertility of the cows and the viability of the calves has been higher in the crossbred than in the purebred animals. A similarly planned crossbreeding experiment with dairy breeds is also in progress at Beltsville, U.S.A.

At three American experimental stations (Illinois, Indiana and Ohio) crossbreeding experiments are being carried out, in which only the females are random samples of their respective breeds and only two bulls are used per generation from each breed. Each bull produces progeny with cows of its own breed and with cows of another breed. At the Illinois experimental station, crossbreeding between Holstein and Guernsey has been conducted for a number of years. The crossbred calves have shown lower mortality, higher weights at 18 to 25 months of age, and better condition than purebred calves. The height at the withers and other measurements of skeleton size in the mature animals have been about the same as the average of the parental breeds. The production results of the crossbred heifers in the first lactation showed that milk yield was 4·5 and butterfat yield 14·6 per cent higher than the average production of the contemporary purebred heifers. Formerly, crossing between different beef breeds was practised in Great Britain in order to increase growth rate and meat quality of the progeny, but in the last two decades or so, crosses between bulls of the beef and cows of the dairy breeds have attracted more attention. According to the Report of the

Production Division of the Milk Marketing Board for 1964–1965 beef bulls accounted for 33 per cent of the total inseminations of dairy cows. When different cross combinations have been compared with steers of the dairy breeds, it has been found that killing-out percentage and meat quality is improved by crossing. Similar comparative experiments have also been carried out in the Scandinavian countries, but at the moment selection from within the dairy breeds for better meat production plays a greater role than crossing with bulls of the beef breeds.

To summarise, the investigations referred to have shown that intensive inbreeding leads to considerable deterioration of those traits which are decisive for the general viability and fertility of the animals. Birth weight and growth rate also tend to decline. The occurrence of genetically determined anatomical and physiological defects increases, but they are of minor importance compared with the decline in general viability. A number of quantitative traits, such as butterfat content of the milk and egg size in poultry, are very little affected by inbreeding; the depression is dependent on the genetic constitution of the inbreeding material and varies, therefore, from case to case.

By close consanguineous breeding for several generations it has been possible to produce inbred lines of poultry, the inbreeding coefficients of which have been calculated to be $0.5–0.8$. 'Double hybrids' have been produced by crossing these inbred lines and as far as viability and egg production are concerned, they are to some extent superior to the pure breeds. Breeding work with pigs has not met with much success in this respect, since the lowered viability of the inbred animals has made it impossible to inbreed with such intensity or for so many generations as with poultry. In line crosses with pigs the gain in vitality has been less than with poultry; rotational crossing between different pig breeds, on the other hand, has given good results. Breed crosses have also been tested in breeding sheep and cattle, and in general some increase in vitality has been noted.

THEORETICAL EXPLANATIONS OF INBREEDING DEPRESSION AND HYBRID VIGOUR

The maize geneticists first believed that the increase in vigour, which always accompanied crosses between two inbred lines, was due to the heterozygosity itself; the gametes which united with each other at fertilisation had different genic constitutions and this was supposed to stimulate the processes of life. This *stimulation theory* soon gave way, however, to the *dominance theory*, according to which, the inbreeding depression was due to the segregation of recessive genes which in the homozygotes reduced viability. In crosses between highly homozygous inbred lines the dominant and recessive alleles were brought together and the latter could not then exert their effect. EAST and JONES (1919) stated that homozygosity for all advantageous genes constituted the most favourable conditions for growth and reproduction. On the contrary, no one has succeeded in producing, even in maize, a homozygous line which exhibits the same viability as the heterozygous crossbreds. It was then thought that this might be due to linkage between favourable and unfavourable genes, whereby the union of entirely favourable genes was prevented; but this explanation is hardly sufficient.

Inbreeding depression and its mirror image 'hybrid vigour' (heterosis) probably have several causes. It is certain that the inbreeding tends to increase the homozygosity of the offspring, with the following consequences:

1. Segregation of recessive genes which often have unfavourable effects in the homozygous condition.

2. Reduced frequency of heterozygous gene pairs. In so far as favourable interaction of alleles at the same locus (overdominance) occur, reduced heterozygosity is associated with lowered viability.

3. Alteration in the gene combinations. This in turn alters the interaction between genes at different loci (epistatic effect).

All three of these changes probably contribute to the decline in fertility and general vigour which invariably follows intense inbreeding when it is practised for several generations. Our breeding populations of farm animals are highly heterozygous and, as such, are balanced for a harmonious interaction between the majority of genotypes and their environment. Increased homozygosity disturbs this equilibrium. On the other hand, one may expect that in many cases a favourable effect will result from further increasing the heterozygosity by crosses between populations which have been bred as separate units for many generations.

The experiments which have been performed with laboratory animals and the different kinds of farm animals have clearly shown that the products of crossing inbred lines, or strains, are better buffered against unfavourable environmental conditions than the individuals within the lines or strains.

Since crosses are less sensitive to the environment, there is less variation of vital traits in their F_1 generation, than in more or less homozygous lines or strains. This has been shown to be true with a number of breed crosses. Inbreeding does not succeed in increasing the uniformity of farm animal breeds in respect of such vital traits as fertility, viability and growth rate. This can be achieved for a number of qualitative traits, however, and also for quantitative traits determined mainly by additive inheritance and only slightly modifiable, e.g. egg-size in poultry and butterfat content of milk in cattle.

Inbreeding depression and hybrid vigour have, in the main, the same basic causes. The heterozygosity and gene interaction which was lost by inbreeding can be restored by suitable crosses. Even Charles Darwin (1868) pointed out that inbreeding operates gradually but with accumulated effect, whereas crossing usually exhibits full effect in the first generation. This still retains its validity. Fig. 14:4 shows how quickly consistent inbreeding, according to a particular system, leads to homozygosity. Theoretically, full sib mating for twenty generations in succession should mean that the different lines which are formed can be regarded as 'pure lines'. In practice, because natural selection favours the heterozygotes, homozygosity proceeds much more slowly, especially with the milder forms of inbreeding.

A scheme of the different mating combinations with regard to the relationship between the mates is given in Fig. 14:5. It is apparent from Fig. 14:4 that continuous half-cousin mating can increase the homozygosity by only about 4 per cent. On the other hand, ordinary cousin mating should eventually lead to complete homozygosity, at least theoretically. Accordingly, in the scheme shown in Fig. 14:5 a latitude of 9 per cent deviation from zero has been allowed, corresponding to the mean of full and half-cousin coefficients of relationship. Outside these limits, matings are designated as either inbreeding or outbreeding. It should be stated that crossing between highly inbred lines of the same breed

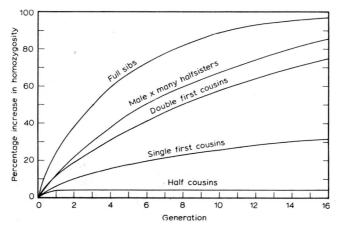

Fig. 14:4. The increase in homozygosity with continuous inbreeding according to a given system of mating when neither natural nor artificial selection is operating. The zero point is the homozygosity at the commencement of inbreeding, regardless of whether this is high or low. Theoretically all the inbreeding systems of at least single first cousin mating should, sooner or later, lead to complete homozygosity. Half-cousin mating results in only a slight increase in the degree of homozygosity (4%) no matter how long the mating system is conducted. (WRIGHT 1921)

can result in a greater increase in heterozygosity than breed crossing, without previous inbreeding.

It seems probable that there is an optimum level of heterozygosity for our populations of farm animals; this level no doubt varies with different environmental conditions. For those breeds of farm animals which have been the object of selection and a certain amount of inbreeding for a long time, it can be assumed that the optimum level of heterozygosity is higher than can be attained by 'random mating' within the populations. It is here that crossing with other populations would be an advantage. Just how genetically different the populations to be crossed with each other ought to be if the best result is to be achieved can be decided only by crossbreeding experiments under the environmental conditions in which the offspring will be used.

For the commercial animal breeder, an increase in the degree of heterozygosity of the herd by crosses between strains or breeds involves much less risk than inbreeding. One must unconditionally accept that inbreeding, on the average, gives worse results than 'random mating' or outbreeding. Inbreeding should be used only by an experienced and well informed breeder, working according to a well thought out plan and with a particular object in view. In certain cases, it can be advantageous to mate a first class male, proved by progeny-testing, with his own daughters; but it must be remembered that intensive inbreeding for several generations always results in a reduction in the fertility and viability of the progeny, and the economic losses may considerably outweigh the eventual gain. To produce inbred lines and then test these in crosses with each other is

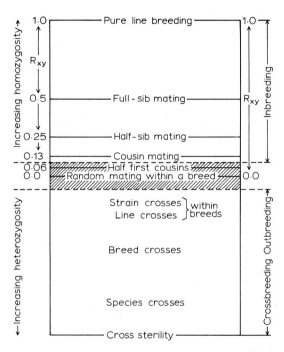

Fig. 14:5. Schematic illustration of different mating combinations with respect to the degree of relationship between the individuals mated with each other.

beyond the resources of the individual farmer. This work must therefore be delegated to national research stations or to large private enterprises. As far as the larger farm animals are concerned (sheep and goats, pigs, cattle and horses), there is still no evidence that any economic advantage is to be gained in this direction, at least in relation to the cost involved.

SELECTED REFERENCES

BOWMAN, J. C. 1959. Selection for heterosis. *Anim. Br. Abstr.*, **27**: 261–273.
FREDEEN, H. T. 1956. Inbreeding and swine improvements. *Ibid.*, **24**: 317–326.
———, 1957. Crossbreeding and swine production. *Ibid.*, **25**: 339–347.
GOWEN, J. W. (editor). 1952. *Heterosis* (552 pp.). Iowa State University Press, Ames, Iowa.

LOOSLY, J. K. (editor). 1958. A review of 50 years of progress in animal science, 1908–1958. *J. Anim. Sci.*, **17**: 909–1135.

MI, M. P., CHAPMAN, A. B. and TYLER, W. J. 1965. Effect of mating system on production traits in cattle. *J. Dairy Sci.*, **48**: 77–84.

15 Estimation of Breeding Values

A distinction is often made between breeding and commercial herds as well as between pedigree and non-pedigree livestock. Regardless of this distinction it is desirable for all herds to improve the economically important traits of the animals, irrespective of whether or not they are registered in a herdbook. It is always important to estimate the breeding value of the animals which are expected to reproduce their kind. The following basic requirements would apply to any breeding animal:

1. Normal reproductive capacity.
2. Records of the animal itself and/or its nearest relatives for traits which are of interest from a breeding point of view.

The results of breeding depend to a large extent on the possibilities of assessing the breeding value of the animals and how these possibilities are utilised. When selecting animals for breeding the aim is partly to prevent the spreading of undesirable genes in the population, e.g. genes for anatomical or physiological defects, and partly to increase the frequency of genes responsible for desirable production traits. We shall discuss first how carriers of undesirable genes, which are often recessive to their 'normal' alleles, can be detected and eliminated.

TESTING SUSPECTED CARRIERS FOR UNFAVOURABLE GENES

Unfavourable recessive genes exist in all species and breeds of farm animals. The fact that more hereditary defects have been reported in some breeds than in others is probably due to disparities in the accuracy of observation rather than to actual differences in the frequency of defective genes. Natural selection tends to reduce the frequency of these genes but new ones arise by mutations so that every population will carry some unfavourable recessives. The number may, however, be reduced to a minimum by planned selection measures.

In large breeding populations the average frequency of non-viable segregants will always be low, but in certain lines or strains a temporary accumulation may take place and cause considerable loss to the breeders. In the early stages of the artificial insemination era in cattle breeding, attention was drawn to spreading unfavourable recessive genes from carrier bulls to a large number of offspring. In the individual case this danger exists but, considering the population as a whole, artificial insemination is no more risky than natural breeding. There is no reason to assume that carriers of lethal or semi-lethal genes should be more preferred in A.I. than in natural breeding, and therefore the average risk would be the same in both cases under otherwise comparable conditions. In fact the possibilities of reducing the frequency of these genes are increased by A.I. Private breeders often try to conceal the fact that their bulls or boars have thrown abnormal offspring, but the member of an A.I. association usually reports it because he does not want it repeated. Owing to their greater use, the carriers are usually discovered earlier in artificial than in natural breeding.

Dominant genes with lethal effect in the heterozygous condition eliminate themselves as soon as they become effective, and semi-lethal dominant genes cannot persist for more than a few generations in the breeding populations. Recessive genes may, however, be multiplied and passed through several generations before their presence is known. The efficiency of mass selection against these genes decreases with decreasing frequency of heterozygotes. For example, when the frequency of male and female carriers is 20 per cent, the segregation of homozygous recessives will be $0 \cdot 2^2 \times 0 \cdot 25 = 0 \cdot 01$, or 5 per cent of the total number of heterozygotes, but when the frequency of carriers has decreased to 5 per cent, the segregation of recessives will be only $0 \cdot 05^2 \times 0 \cdot 25 = 0 \cdot 000625$, or $1 \cdot 25$ per cent of the heterozygotes. In order to accelerate the process of elimination it is necessary to find ways and means of identifying the heterozygotes and prevent them from breeding. In some cases lethal and semi-lethal genes are incompletely recessive to their normal alleles; for example, achondroplasia in Dexter cattle and the creeper fowl, and porphyria in pigs; the heterozygotes can then be identified without any special tests. The gene for 'snorter' dwarfs in beef cattle shows some effect in the heterozygotes, although not enough for their identification in the majority of cases. In the fowl, a recessive gene prevents the homozygous hens from transferring the riboflavin provided in the diet to their eggs, with the result that all embryos in the deficient eggs die during incubation. Buss et al. (1959) found that the average level of riboflavin in the blood and eggs of heterozygous hens was only about 60 per cent of that in non-carriers. Owing to the variation in riboflavin level, however, only some of the carrier hens could be identified by this phenotypic test. In man, a large number of hereditary metabolic defects are known, where the heterozygotes can be discovered by chemical tests. In farm animals, only a few inherited metabolic defects are known at present, but this situation will probably change with the intensification of research in this field.

When the carriers cannot be distinguished by their phenotype, *breeding tests* may be applied. It is important that these tests are efficiently planned and carried out as early as possible in the life of the individual. Usually the tests can be limited to the males, where it is easier and more reliable owing to the larger number of offspring. If the male carriers are eliminated, there will be no segregation of homozygotes. Obviously, a female which has produced defective homozygotes should be eliminated from further breeding.

The number of offspring needed to test a suspected carrier for a certain recessive gene with complete penetrance in the homozygote may be estimated for any desired level of significance. Let us assume that a bull can be tested by matings to known carriers of the gene in question, and that we wish to reduce to 0·05, or 0·01 the probability (P) that he passes the test without throwing any defective calves if he is a carrier. For each single offspring from the mating Aa × Aa the probability of being the dominant type (AA or Aa) is 0·75 and for n offspring the probability is 0.75^n. The numbers n corresponding to the probabilities mentioned may be calculated from the formulas $0.75^n = 0.05$ and $0.75^n = 0.01$, i.e. 10·4 and 16·0 respectively. This type of test would be particularly suitable in swine-breeding, where many young are produced by each fertile mating. Two litters from such matings should be sufficient for the test. With decreasing penetrance of the gene an increasing number of offspring would be needed to obtain the same accuracy of the test. For example, if the penetrance was 50 per cent the expected segregation of defectives would be 12·5 instead of 25 per cent, and the formula $0.875^n = 0.01$ would give 34·5 as the value of n.

The general formula for calculating n, the number of offspring needed to reduce P to a certain level, is $s^n = P$, where $s =$ the probability for non-segregation with regard to each offspring.

When the homozygotes are viable and fertile, the most efficient test is to mate the male to females which manifest the trait. The probability for segregation of aa from the mating Aa × aa is 0·5 when the penetrance of aa is complete. The probability of non-segregation in n such matings is 0.5^n. When $P = 0.05$, $n = 4.3$ and when $P = 0.01$, $n = 6.6$. Obviously, this test can be used only for relatively harmless defects. Sometimes, however, the penetrance may be incomplete, or the inheritance perhaps still more complicated. In such cases, matings of suspected males to defective females may be possible and advisable. *Atresia ani* in pigs (cf. p. 227) may be mentioned as an example. It is questionable whether the inheritance of the defect can be explained according to the monohybrid scheme, but if that is the case, the penetrance of the gene in a double dose is somewhere between 40 and 50 per cent, according to the results of mating defective boars and sows to one another. If a normal boar in matings with defective sows produces 40–50 per cent defective progeny it would indicate that the boar is a homozygous carrier (aa); if the boar is heterozygous (Aa) 20–25 per cent defective progeny would be expected.

Table 15:1 *Number of fertile matings needed to test bulls and boars for recessive genes. $q =$ the frequency of heterozygous females in the population*

	1. The male is mated to his own daughters, or to daughters of known heterozygotes*	2. Artificial insemination on a random sample of females from the population		
$q =$	0·5	0·2	0·1	0·04
A. Bulls: one calf from each cow ($m = 1$)				
$P = 0.05$	23	58	118	298
$P = 0.01$	35	90	182	459
B. Boars: 10 pigs per sow ($m = 10$)**				
$P = 0.05$	5	14	30	78
$P = 0.01$	8	22	46	120

* The frequency of the recessive gene in the population is usually very low and has little influence on the calculations; therefore it is disregarded.
** The figures for boars would apply also to cockerels producing 10 chicks per hen.

Another possibility of testing a male for recessive genes is to use him on a group, or population, of females containing a certain fraction of carriers (Aa). If this fraction is denoted by q and the fraction of non-carriers is denoted by p, and $q+p = 1$, the general formula for estimating the number of fertile matings (one litter from each female) needed for the test would be $(p+qs^m)^n = P$, where s = the probability that an individual offspring of a carrier female is not aa, m = number of young per litter, and n = number of females. The frequency of a recessive lethal gene in the female population is equal to half the frequency of heterozygous carriers of the same gene. For monotocous animals $m = 1$; $s = 0.75$ when the penetrance of aa is complete. Table 15:1 shows the number of females needed to reduce the value of P to 0.05 and 0.01 respectively.

Table 15:1 is divided into two vertical sections. Section 1 applies when the male is mated to his own daughters, or to daughters of known heterozygotes. In Section 2 it is assumed that the male is mated to a random sample of the population and it would apply especially when artificial insemination is practised. The two sections will be discussed separately.

1. In many cases, bulls and boars have been tested by mating them to a certain number of their own daughters. Section 1 of the table shows the number of daughters needed for such tests. If the male is a carrier for one or several recessive genes, 50 per cent of the daughters will be carriers of the same genes ($p = q = 0.5$). This testing method has the advantage that it applies to all recessive genes carried by the male and with clearly manifested effect in the homozygotes, irrespective of whether the gene is known to occur in the population or not. The method has at least two disadvantages, however. One is that the male reaches a fairly advanced age before the test is completed, and the other is that the sire × daughter matings involve a rather high degree of inbreeding ($F_x = 0.25$) which, on an average, will cause a decrease in fertility, viability, growth rate and in cows also in milk yield. This testing method should therefore only be used when other and less costly methods are not available.

When the male is mated to daughters of males or females which are known to be heterozygous for a certain recessive gene, the test applies only to this gene. The accuracy of the test will be the same as in sire × daughter matings. On the other hand, inbreeding can be avoided and the test can be completed while the bull is still young. The difficulty is to obtain a sufficiently large number of daughters of known carriers.

2. With artificial insemination, when the semen from young bulls or boars is distributed at random to the females of the population, an 'automatic' and early test for recessive genes is accomplished, and this makes the planned tests superfluous. Section 2 of Table 15:1 shows the number of fertile inseminations needed by semen from bulls or boars in order to decrease the probability that a male carrier is not revealed as such to 0.05 and 0.01, respectively. When 10 per cent of the cows in a population are carriers of a certain recessive gene, a carrier bull must produce a normal calf with each of 182 cows in order to reduce the probability that he is not detected to 0.01. If the frequency of heterozygotes is only 4 per cent, 459 calves must be produced for the same reduction. This is not

difficult to accomplish with A.I.; the test is obtained as a by-product of the regular progeny test for milk and butterfat yield. The 'automatic' test is possible also in pig-breeding. That the number of pigs per litter varies is of much less importance for the accuracy of the test than that the minimum number of sows is reached.

An argument against these 'automatic' tests is that if a lethal or semi-lethal gene does not occur in the population, or occurs with very low frequency, a male may distribute this gene to a large number of offspring without being identified as a carrier. Completely recessive lethal genes do no harm, however, as long as they exist only in the heterozygotes. As soon as they reach so high a frequency that segregation is likely to occur, the carriers will be detected. When the carrier males are eliminated from breeding, the segregation of the defect ceases.

A prerequisite for the efficiency of the 'automatic' tests is a careful registration of congenital defects in the new-born animals, and also that all such defects are reported to the insemination centre. This registration should be fairly easy to arrange when the inseminators visit the herds regularly in order to fulfil their other duties. When types of defects with unknown genetic background appear, test matings should be arranged in order to throw light on the mode of inheritance. Such registration of defects would also facilitate investigations into their frequency in different breeds and under various environmental conditions. The biochemical defects are usually more difficult to diagnose than the morphological malformations, and more research is needed to increase the knowledge about these defects. Practising veterinarians should be on the look out for metabolic disturbances which may be suspected to have a genetic basis.

The economic importance of lethal and semi-lethal genes in farm animals has sometimes been exaggerated. They are no serious threat to the animal industry as a whole, but they are important enough to deserve consideration in breeding schemes. The accumulation of subvital genes, with less spectacular effect in individual cases, may cause much greater losses. Quite often, genes which have a favourable effect on one character, for example growth rate or body conformation, may have an unfavourable effect on the general fitness of the animals, for example in respect of fertility, viability, or resistance to diseases. Therefore, the progeny test of males should also take into consideration the fitness of the offspring under the prevailing environmental conditions.

BREEDING VALUE FOR QUANTITATIVE TRAITS

Quantitative traits in which the variation can be studied by visual assessment of the animals and by performance tests are, generally, very modifiable. The only thing which can be directly determined under the prevailing environmental conditions is the *phenotypic value* of the traits. When applied, for example, to the milk production of cows, the phenotypic value gives a very uncertain assessment of the individual animal's genotype. Furthermore, this trait is sex-limited; it is not manifested in the male. Nevertheless, it is necessary to obtain an estimate of the genes that the bulls can be expected to pass on to their offspring, i.e. their *breeding value*. This can be done with the aid of certain general methods which have been developed by animal geneticists during the past few decades. The accuracy of the estimation depends largely on the data which are available. It is also possible to assess the accuracy of the estimate, when this is carried out in a particular way. The breeding value can be estimated for a particular trait, or for several traits in combination. Where individual traits are concerned, the breeding value can be estimated on the basis of one criterion or several criteria, such as the pedigree and collateral relatives. We shall first discuss the estimation of the breeding value for a particular quantitative trait and afterwards take up the question of estimation for a combination of several traits.

Definition of the breeding value concept

The breeding value of an individual is determined by the genes he transfers to his offspring. Each offspring receives a random sample half of the sire's and dam's genes. The assessment of breeding value is based on an estimate of the average effect of those genes which the individual concerned passes on to its offspring in random matings within the population. In this connection no consideration is taken of the animal's age, as this is only of quantitative importance, since a younger animal has the prospects of leaving more progeny in the future than an older animal.

The difference between *general breeding value* and *special breeding value* has already been pointed out (p. 116). The general breeding value is determined by the additive (average) effect of the genes in the combinations in which they can appear within the population concerned. The special value, on the other hand, depends on the deviations from the genes' additive effect caused by dominance and epistasis. When, for example, certain males are mated not at random with the females of the population but with specially selected groups, the differences between their progeny groups can depend, to some extent, on how the genic constitution of the males matches with the genes from the respective groups of females (specific combining ability). This is especially apparent when inbred lines or strains are mated with each other and also in breed crosses, usually referred to as 'nicking'. In statistical studies of data from cattle breeding it has not been possible to demonstrate the occurrence of significant 'specific combining ability' with respect to the milk production of cows in matings within the same breed (SEATH and LUSH, 1940, HANSSON et al., 1961). 'Nicking' is probably of much less importance in matings within breeds and strains than the practical breeder imagines.

Interaction effects can occur between heredity and environment; which means that the magnitude of differences between individual breeding values or even their ranking may be different in different environments (Fig. 15:1). Genotypes suitable for intensive animal production are perhaps less suitable under primitive environmental conditions and vice versa. Only the general breeding value, which depends on the genes' additive effect under fairly similar conditions, can be estimated with the general methods. The special breeding value, in particular crosses or with highly specialised environmental conditions, must be experimentally tested from case to case. On the other hand, it may be assumed that in matings within a particular breed, in its natural environment, the deviations from the general breeding value as a result of non-additive inheritance or interactions between the genotype and the environment, are relatively small and therefore of minor importance. In progeny tests of males it may be desirable to split the progeny groups into several environmental levels, corresponding to the environmental variation in the field. The results would then be applicable to the 'average environment' in a certain area. The following discussion is limited to the general breeding value of the animals.

The phenotype of the individual, or more precisely, our measure of it, will be denoted by P; the estimated general breeding value by $\hat{A}$ and the true breeding

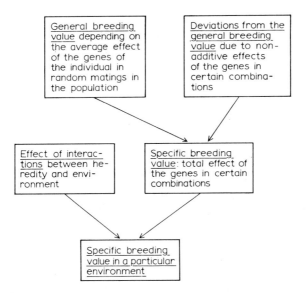

Fig. 15:1. Scheme of the definitions relating to breeding value.

value by A. The average phenotypic value of the population ($\bar{P}$) indicates its average breeding value under the prevailing environmental conditions ($\bar{P} = \bar{A}$). The breeding value is a property of the individual itself and the population from which its mates are drawn. 'One cannot speak of an individual's breeding value without specifying the population in which it is to be mated' (FALCONER, 1960). The population average ($\bar{P}$) will therefore be used as a base in all comparisons between individuals and groups. What is needed, is to estimate how much the individual or group deviates from $\bar{P}$ in a positive or negative direction. It is often preferred to express this deviation as a relative value (per cent of $\bar{P}$) instead of in absolute values. The relative breeding value shows perhaps more clearly how much better or worse than the population average the individual may be as a breeding animal.

Various possibilities of assessing the breeding value of an individual

An assessment of the individual's breeding value can be based on the phenotypic value of (1) ancestors, (2) the individual itself, (3) collateral relatives and (4) the progeny.

These criteria are often available in the order in which they are given here, though the information on collateral relatives is, in many cases available earlier than the phenotypic value of the individual itself.

Our discussion of the problem of estimating breeding value is based on the following assumptions:

1. Inheritance of the traits is additive and the interaction, or correlation, between heredity and environment has no significant effect.

2. There is no inbreeding. A low level of inbreeding has only a slight effect on the result of the estimation; but, if the inbreeding is relatively high, it is necessary to calculate the actual coefficients of relationship. If inbreeding has been carried out continuously for several generations, so that the homozygosity has increased appreciably, then an assessment of the general breeding value will be rather dubious for the reasons which should be apparent from the previous chapter.

3. The phenotypic values, used in the assessment, are representative. This means that either all available values are included or a random sample of these. If, for example, cows are to be assessed on the basis of their own milk yield, this can be done either according to the yield in the first lactation or perhaps on the average of the first three lactations, but there should be no selection of the available data. It has been suggested in the past that only the highest of the records in two or three successive lactations should be used. Such a procedure favours those cows which have had the largest number of lactations, since the possibilities of selection increases with the number of lactations. The same principle applies when a male, e.g. a bull, is assessed on collateral relatives or on progeny. All collaterals and offspring with available data must be included, or a random sample taken.

4. All environmental influences are randomly distributed within the population in which the comparisons are being made, or corrections are applied for the effect of systematically operating factors. It is this condition that is most difficult to fulfil. In pigs, systematic differences between contemporary litters occur due to the different suckling capacities of the dams and their general mothering ability (maternal influence). In poultry, environmental differences occur between groups of chicks hatched at different times, etc. All these differences can be satisfactorily eliminated. The environmental differences are sometimes 'deliberately produced'—as when presumptive dams of bulls are fed and managed better than their stable mates in order to boost their production figures—thus producing a correlation between heredity and environment. In this way the dams of the bulls appear better than they actually are. Errors of this sort have a misleading effect in selection. For the present, it will be assumed that all environmental influences are randomly distributed. The correction of systematic errors will be dealt with later.

The heritability of a particular trait has previously been defined as the fraction of the total phenotypic variance which is additively genetic; h^2 shows the regression of the breeding value on the phenotype. The deviations of the individuals' probable breeding value ($\hat{A}_X$) from the average for the population can be estimated by $h^2(X-\bar{P})$. If required, the population average can be added, thus $\hat{A}_X = h^2(X-\bar{P})+\bar{P}$.

The method of path coefficients can be very useful in demonstrating the value of the various criteria which contribute to the breeding value of an individual (cf. p. 95). A diagram of this type is shown in Fig. 15:2. The breeding value is denoted by A and the phenotype by P, as in the above. The individual to be

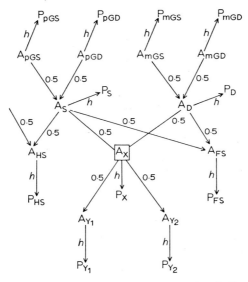

Fig. 15:2. Diagram of different criteria in estimating the breeding value of an individual (Ax).
The individual's phenotypic value = Px, S = sire, D = dam, pGD = paternal granddam, pGS = paternal grandsire, mGD = maternal granddam, mGS = maternal grandsire, HS = half sib, FS = full sib, Y = progeny and h = the correlation between breeding value and phenotypic value.

assessed is shown by X and its breeding value by A_X; the sire is indicated by S, the dam by D, a half-sister by HS, a full sister by FS and the daughters by Y_1 and Y_2. The correlation between the breeding value of parent and daughter (or son), i.e. the genetic correlation (= the coefficient of relationship), is 0·5 if there has been no inbreeding. The correlation between breeding value and phenotype of an individual has been denoted as $\sqrt{h^2}$. To obtain the correlation between the individual's breeding value (A_X) and the phenotype of a particular relative of X it is necessary to multiply all the intervening paths; for the dam it will be $0·5h$, for granddam or half-sister $0·25h$, for a full sister $(0·25+0·25)h$ and for a daughter $0·5h$. The regression of A_X on the phenotype of a relative, e.g. the dam, is obtained by multiplying the correlation coefficient by the ratio of the standard deviation of A_X and the phenotypic standard deviation of the dam, thus, $\sigma_{A_X}/\sigma_{P_D}$. It has already been shown (p. 118) that this ratio is equal to h and consequently the regression of A_X on P_D is equal to $0·5h^2$. The corresponding regression of

A_X on granddam's phenotype (P_{GD}) is $0.25h^2$ and so on. On the average the same result can be expected from selecting a young bull on the basis of *one* randomly selected full-sister's yield, as on the dam's or *one* daughter's yield.

When the individual has many sibs, or many daughters, its breeding value for traits with low heritability can be estimated with more accuracy by using the average of the family group than from the individual's own phenotype. The correlation between the breeding value of the individual (A_X) and the phenotypic mean of n half-sisters, full sisters or daughters is obtained by multiplying the correlation between A_X and the phenotypic value for one member of the group by $\sqrt{n/\{1+(n-1)r_g h^2\}}$ where r_g is the genetic correlation (coefficient of relationship) between family members, and $r_g h^2$ is the phenotypic correlation in a randomised environment. The regression of $\hat{A}_X$ on the mean of n family members is obtained, as before, by multiplying the correlation coefficient by the ratio of the standard deviations. This ratio is $\sqrt{n/\{1+(n-1)r_g h^2\}}$. The regression coefficient, b, is thus $nr_g h^2/\{1+(n-1)r_g h^2\}$ (cf. Table 15:2).

Table 15:2 Correlation between the breeding value of an individual, e.g. a bull, and a certain criterion (I), such as the phenotype of half sibs, full sibs or progeny. The individual itself is not included in the sib groups

Information on phenotypes	Correlation: R_{AI}	Regression: $b_{A/I}$
1. One half-sister's	$0.25h$	$0.25h^2$
2. Average of n half-sister's	$0.25h\sqrt{\dfrac{n}{1+(n-1)0.25h^2}}$	$\dfrac{0.25nh^2}{1+(n-1)0.25h^2}$
3. One full-sister's	$0.5h$	$0.5h^2$
4. Average of n full-sisters'	$0.5h\sqrt{\dfrac{n}{1+(n-1)0.5h^2}}$	$\dfrac{0.5nh^2}{1+(n-1)0.5h^2}$
5. One daughter's or son's	$0.5h$	$0.5h^2$
6. Average of n daughters' (or sons') when they are from different dams	$0.5h\sqrt{\dfrac{n}{1+(n-1)0.25h^2}}$	$\dfrac{0.5nh^2}{1+(n-1)0.25h^2}$

The maximum value of the correlation between A_X and the mean phenotype of a large number of half sibs is 0.5, for full sibs 0.71, and for progeny from different dams 1.00. The maximum value of the regression of the individual's breeding value on the mean phenotype for both full and half sibs is 1.00, whereas the corresponding value for the regression of A_X on the mean phenotype of the progeny is 2.00. When an animal is mated to a random sample of the population, its breeding value deviates from the population mean by twice the deviation of the progeny from the same mean.

Those correlations and regressions which are of interest in estimating the breeding value of the individual on the basis of sibs and progeny are presented in Table 15:2. The stated values of r_g assume that there has been no inbreeding.

The estimation of the breeding value on the basis of the individual's own progeny by one single female gives the same result as estimating on the basis of an equally large group of full sibs. This gives information about the results of only one mating combination; other combinations can give deviating results. The estimation of the breeding value on the basis of a sib group is especially valuable when the trait is not manifested in the individual itself, e.g. in the assessment of young bulls for milk production and generally for traits with low heritability. The most accurate estimate is nevertheless on the basis of the progeny, when the progeny are out of different dams and their number is relatively large.

Information from several sources may be combined in an *index* (I) on which the selection is based. The combination is so arranged that the maximum correlation is obtained between the breeding value and the index; this correlation is denoted by R_{AI} since it concerns a 'multiple correlation' between the breeding value and all the components of the index (cf. p. 95). The relationship between the R_{AI} and the selection effect will be discussed in the following paragraph.

It has been shown previously (p. 133) that the selection response (Re) per generation can be estimated by multiplying the heritability with the selection differential, S, i.e. $Re = h^2 S$. In order to have a better basis of comparison with regard to the selection differential for different traits or criteria, it is often expressed in 'standard deviation units' and denoted by i (selection intensity) $= S/\sigma_P$. Since $h\sigma_P = \sigma_A$ and R_{AI} is the correlation, h, between the breeding value and the index used, then $Re = R_{AI} i \sigma_A$. From this it is apparent that the selection response is determined partly by the correlation between the breeding value and the index (R_{AI}), partly the selection intensity i, and partly the variation of the breeding values within the population σ_A or, using an expression from LUSH (1961), by (accuracy in estimating the breeding value) × (selection intensity) × (variation in A). If any one of these factors is reduced to zero, the selection effect will be nullified. With the selection intensity unchanged and the standard deviation σ_A given, the selection effect is proportional to R_{AI}. The generation interval is assumed to be unchanged.

The assumption that environmental influences are randomly distributed within the population is seldom strictly true. Systematic environmental differences are generally found between herds, as well as between years and seasons within herds. In addition, the age of the animals has some effect on the manifestation of their traits.

Attempts have been made to correct the phenotypic values for a number of 'non-genetic' factors, for example the age of the animals when the traits are measured. Applied to individuals, these correction factors are very uncertain and if corrections are made for several causes of variation at the same time, the result can be most unreliable since, among other things, many causes of variation are correlated among themselves. In view of this, the assessment of the animals should be based, as far as possible, on comparable phenotypic values, for instance, the milk yield during the first lactation when this begins at a particular time of the year, or the weight of a sow's first litter. When it can be applied, this method is definitely to be preferred to the use of corrections.

As far as herd differences and trends within herds are concerned, the individuals can be assessed either on their deviation from the contemporary herd mean (or from the mean of contemporary herd mates), *or* an attempt can be made to estimate that part of the variation between herds and periods which is genetically determined. The applications will be discussed together with the various methods of assessment.

Individual performance

Selection which is based solely on the individual's own phenotypic value is called *mass selection*. This sort of selection gives good results when the heritability is high, but the effect decreases with falling heritabilities. One possibility of increasing the accuracy with low heritabilities is offered when the trait is manifested and can be measured several times on the same individual, e.g. repeated lactations of a cow, or several litters from a sow. As mentioned earlier (p. 120) the heritability for the mean of n successive production results of the same individual is $(h_n^2) = nh^2/\{1+(n-1)r\}$, where r is the coefficient of repeatability. The heritability is assumed here to be equally large for each single result. However, the heritability of milk yield in cows, for example, is so much lower in the second than in the first lactation, that the cow's heredity for milk production is more accurately assessed on the basis of the first lactation than on the mean of the first two or three lactations (cf. Chapters 10 and 17).

Even if the heritability is the same for each record included in the mean, the gain in using the mean is less than h_n^2 indicates. This is due to the following reasons, which will be discussed in more detail in the next chapter (p. 442):

(1) The selection differential is reduced since the variation decreases in calculating the mean;

(2) The generation interval increases with every new record which is included before selection takes place;

(3) If continuous selection is carried out every time a new record is available, as is usually done, e.g. after completion of first, second, third, etc. lactation of dairy cows, individuals with several records constitute a highly selected group, within which further selection will be comparatively ineffective.

The estimation of the breeding value of a cow on the basis of her own performance can be done in the following way:

1. The cow is assessed only with respect to her deviation from the contemporary herd mates, without considering herd differences. If the contemporary herd mates, or the herd as a whole, have a phenotypic mean $\bar{H}$, then the breeding value can be calculated according to

$$\hat{A}_X = h^2(P_X - \bar{H});$$

2. If the heritability h_H^2 for the herd differences within the population is known, an addition can be made for the genetically determined part of the particular herd's deviation from the population mean, and the estimation of the individual's breeding value then becomes,

$$\hat{A}_X = h^2(P_X - \bar{H}) + h_H^2(\bar{H} - \bar{P}).$$

This assumes that all the phenotypic values, including the population mean, apply to animals of the same age. The same principles can be applied for other production traits and classes of livestock.

The pedigree (performance of ancestors)

In general, the estimation of the breeding value of an individual for quantitative traits is relatively low in accuracy when based on the phenotypic values of ancestors. This is due, in the first place, to the incomplete heritability as well as the different possibilities of combinations resulting from Mendelian segregation in traits with polygenic inheritance. Furthermore, the records of the ancestors may be made several years ago under environmental conditions rather different from those existing today. The value of the pedigree as a guide in choosing breeding animals was very much exaggerated by breeders in the past, though of course the pedigree estimate has its importance, since in many cases there is no other alternative. If the inheritance were wholly additive and $h^2 = 1$, then statistically, each parent would 'determine' 25 per cent of the variation in the progeny and the other 50 per cent would have to be attributed to Mendelian segregation. Since the traits are modifiable, however, this 'contribution' from each parent reduces to $0.25h^2$ and if h^2 is only 0.1, then $0.25h^2$ is only 0.025. The regression of the individual's breeding value on the phenotypic value of the dam (or sire) is $0.5h^2$, and for each ancestor further back this reduces by half; the regression on the granddam is thus $0.25h^2$, on the greatgranddam $0.125h^2$, and so on. This applies with reference to a single ancestor, on the sire or dam side, when intervening ancestors are not considered.

Since the individual obtains half its genes from the sire and half from the dam, the sire and dam side of the pedigree should be accorded equal weight. In many cases, however, there is much more accurate information about the ancestor's breeding value on one side than on the other, and in such cases the sire and dam must have different weights, as when the sire of a young bull is progeny-tested on a large number of daughters, but the dam has only one record of production. The breeding value of traits manifested in both sexes can generally be estimated *both* on the basis of the individual itself *and* on the ancestors. The question then is, how much weight to give to the various criteria in order that the calculated index, I, will have a maximum correlation with the individual's breeding value. This question is rather complicated, since in many cases its solution involves calculating the partial regression of A_X on each criterion used, e.g. the cow's own milk yield, the dam's, the maternal granddam's yield and the average daughter's yield of the sire and maternal grandsire. The problem has been dealt with by several authors, e.g. LE ROY (1958); schemes for calculating the weighting factors have been published by SKJERVOLD and ØDEGARD (1959). The weights, k, are 'standard partial regression coefficients' which are identical with Wright's 'path coefficients' from the ancestor's breeding value to the individual's phenotype. These regression coefficients refer to variables expressed in standard deviation units rather than in units of actual measurements.

Assuming that the sire is progeny tested the heritability (h_1^2) of the sire's 'phenotypic' deviation from the population mean $2(\bar{Y} - \bar{H})$ is obtained from

Table 15:3 *Weight factors k for the deviations of the phenotypic value of the individual itself and those of ancestors from the population mean, P, and the multiple correlation R_{AI}, between the breeding value of the individual, A, and the calculated index, I*

Information	Weights	R_{AI}
(1) Formulae		
X	$k_X = h^2$	$\sqrt{h^2}$
$S+D$	$k_1 = 0 \cdot 5h_1^2;\ k_2 = 0 \cdot 5h_2^2$	$\sqrt{0 \cdot 5(k_1+k_2)}$
$X+S+D$	$k_X = \dfrac{h_X^2(4-h_1^2-h_2^2)}{4-h_X^2(h_1^2+h_2^2)};\ k_1 = \dfrac{2h_1^2(1-h_X^2)}{4-h_X^2(h_1^2+h_2^2)};$ $k_2 = \dfrac{2h_2^2(1-h_X^2)}{4-h_X^2(h_1^2+h_2^2)}$	$\sqrt{k_X+0 \cdot 5(k_1+k_2)}$
$D+mGS$	$k_2 = \dfrac{h_2^2(4-h_6^2)}{2(4-h_2^2xh_6^2)};\ k_6 = \dfrac{2h_6^2(1-h_2^2)}{2(4-h_2^2xh_6^2)}$	$\sqrt{0 \cdot 5k_2+0 \cdot 25k_6}$
$D+mGS+mGD$	$k_2 = \dfrac{h_2^2(4-h_5^2-h_6^2)}{2\{4-h_2^2(h_5^2+h_6^2)\}};\ k_5 = \dfrac{2h_5^2(1-h_2^2)}{2\{4-h_2^2(h_5^2+h_6^2)\}};$ $k_6 = \dfrac{2h_6^2(1-h_2^2)}{2\{4-h_2^2(h_5^2+h_6^2)\}}$	$\sqrt{0 \cdot 5k_2+0 \cdot 25(k_5+k_6)}$
4GP	$k_3 = 0 \cdot 25h_3^2;\ k_4 = 0 \cdot 25h_4^2;\ k_5 = 0 \cdot 25h_5^2;$ $k_6 = 0 \cdot 25h_6^2$	$\sqrt{0 \cdot 25(k_3+k_4+k_5+k_6)}$

(2) Examples	Individual	Parents		Grandparents				R_{AI}
	k_X	k_1	k_2	k_3	k_4	k_5	k_6	
$h^2 = 0 \cdot 3$ for each individual:								
X	0·30							0·55
$X+D$	0·28		0·11					0·58
$X+S+D$	0·27	0·11	0·11					0·61
$X+S+D+4GP$	0·26	0·10	0·10	0·04	0·04	0·04	0·05	0·63
$S+D$		0·15	0·15					0·39
$S+D+4GP$		0·13	0·13	0·06	0·06	0·06	0·06	0·44
4PP				0·08	0·08	0·08	0·08	0·27
$h_X^2 = 0 \cdot 3;$ for P and GP: $h^2 = 0 \cdot 6$								
$X+D$	0·27		0·22					0·61
$X+S+D$	0·23	0·23	0·23					0·68
$X+S+D+4GP$	0·22	0·20	0·20	0·06	0·06	0·06	0·06	0·69
$S+D$		0·30	0·30					0·55
$S+D+4GP$		0·26	0·26	0·07	0·07	0·07	0·07	0·57
4GP				0·15	0·15	0·15	0·15	0·39
$h^2 = 0 \cdot 6$ for each individual								
X	0·60							0·77
$X+S+D$	0·51	0·15	0·15					0·81

Symbols		h^2	Weight	
Individual (X)	$= x$	h_X^2	k_X	$S+D = P$
Sire (S)	$= 1$	h_1^2	k_1	$pGS+pGD+mGS+mGD = 4GP$
Dam (D)	$= 2$	h_2^2	k_2	
Paternal grandsire (pGS)	$= 3$	h_3^2	k_3	
Paternal granddam (pGD)	$= 4$	h_4^2	k_4	
Maternal grandsire (mGS)	$= 5$	h_5^2	k_5	
Maternal granddam (mGD)	$= 6$	h_6^2	k_6	

the formula $0{\cdot}25nh^2/\{1+(n-1)0{\cdot}25h^2\}$ = the regression (b) of future daughters on those tested (cf. p. 407). This applies also to the grandsires (h_3^2 and h_5^2, in Table 15:3).

In the first part of Table 15:3 some examples are given of formulae for calculation of the weights for combining different criteria into an index, I, for the estimation of the breeding value of an individual, A_X, and also the correlation between the breeding value and the index, R_{AI}. The symbols used are defined at the foot of the table. Examination of the formulae shows that they can be simplified considerably if the same heritability is applicable for each one of the ancestors as for the individual itself. However, this is not always the case. Assume, for example, that the breeding value of a young bull is to be estimated on the basis of the progeny-test of the sire on 20 daughters and on the dam's own yield, when the heritability for the first lactation is 0·3. Even if the dam has produced for several lactations, the h^2 should not be given a higher value than 0·3, for reasons which will be discussed later (p. 443). For the sire, h^2 can be calculated to be $\dfrac{20 \times 0{\cdot}25 \times 0{\cdot}3}{1+(20-1)0{\cdot}25 \times 0{\cdot}3} = 0{\cdot}62$ (cf. p. 407). The weights in this case will be $0{\cdot}5 \times 0{\cdot}62$ for the sire and $0{\cdot}5 \times 0{\cdot}3$ for the dam, which means that, when the only available information is the estimated breeding value of the parents, then the average of these two values is the best estimate of the breeding value of the individual. In the example cited, $R_{AI} = \sqrt{0{\cdot}5(0{\cdot}31+0{\cdot}15)} = 0{\cdot}48$. The estimation of the dam's breeding value was uncertain ($h^2 = 0{\cdot}3$), however, and it may therefore be desirable to improve this estimate by considering also the maternal grandparents, especially the maternal grandsire if he is progeny-tested on a large number of daughters. If the maternal grandsire was progeny-tested on 50 daughters $h_5^2 = 0{\cdot}81$ for him; for the granddam $h_6^2 = 0{\cdot}30$. On the dam's side of the pedigree then the weights are $k_2 = 0{\cdot}118$, $k_5 = 0{\cdot}155$ and $k_6 = 0{\cdot}057$, which corresponds to a correlation of 0·34 between the individual's breeding value and the criteria from the dam, maternal grandsire and granddam, instead of 0·27 when only information on the dam was available. If a heritability of 0·6 for both dam and maternal granddam could have been reckoned with, then the weights would have been $k_2 = 0{\cdot}242$, $k_5 = 0{\cdot}116$ and $k_6 = 0{\cdot}077$, and the correlation would have been 0·41. The more uncertain the breeding value estimate of the dam on the basis of her own phenotypic merit, the more valuable the information on her parents; conversely, the more accurately the breeding value of the dam can be estimated on the basis of her own phenotype, the less there is to be gained from considering even maternal grandparents. The same argument can naturally be used for the sire and its parents; the formula on line 5 in Table 15:3 can then be used by substituting the appropriate values of h^2 for the respective ancestors. If the sire is progeny-tested on a large number of daughters, it is meaningless to consider even his parents. When the progeny test is based on 50 daughters, $h^2 = 0{\cdot}81$, then the correlation between the son's breeding value and the estimated breeding value of the sire will be $0{\cdot}5 \times 0{\cdot}90 = 0{\cdot}45$. If the paternal grandsire is included and has been progeny-tested on 50 daughters ($h^2 = 0{\cdot}81$), then R_{AI} increases to only 0·46 (cf. p. 402).

When criteria are available on the individual's own phenotype, there is less need for considering the ancestors, and this is especially so with increasing heritability. The formula on line 3 in Table 15:3 shows the weights to be applied when the assessment is based on the individual itself and both the parents $(X+S+D)$. If the heritability 0·3 can be applied to each, then R_{AI} increases from 0·55 to 0·61 when the assessment is done with $X+S+D$ instead of only X, and if $h^2 = 0.6$, R_{AI} increases from 0·77 to 0·81 by including $S+D$.

Fig. 15:3 shows the accuracy, R_{AI}, that can be achieved in estimating the breeding value of an individual from different criteria and with varying heritabilities for the trait being considered. With increasing heritability, the assessment of both ancestors and progeny becomes less important in estimating the breeding value if information is available on the individual's own phenotype. When it can be practised, mass selection gives the best results at the higher heritabilities. As the heritability falls, the importance of progeny-testing increases. This is especially true with sex-limited traits, where the breeding value of the individual can be assessed on the sire's and maternal grandsire's progeny, which is equivalent to estimates on the basis of its own and the dam's paternal half sibs. The diagram has general validity, though it is drawn specially with reference to the character of milk yield in dairy cattle.

The breeding index for an individual can be calculated in the following way:

$$I = k_X(X) + k_1(1) + k_2(2) + k_3(3) + k_4(4) + k_5(5) + k_6(6), \text{ and}$$

$$R_{AI} = \sqrt{k_X + 0.5(k_1+k_2) + 0.25(k_3+k_4+k_5+k_6)}.$$

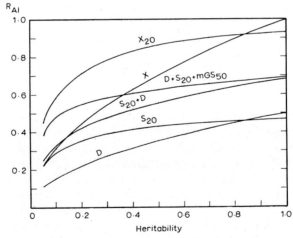

Fig. 15:3. The accuracy (R_{AI}) of various estimates of an animal's breeding value with regard to a certain trait at different levels of heritability.
X = the individual's own phenotype
X_{20} = progeny test on 20 daughters (or sons)
D = the dam's phenotype
S_{20} = the sire's progeny test on 20 daughters (or sons)
$S_{20}+D$ = the sire's progeny test on 20 daughters and the dam's phenotype
$D+S_{20}+mGS_{50}$ = the dam's phenotype + sire's and maternal grandsire's progeny tests on 20 and 50 daughters respectively.

The symbols (X), (1), (2) etc. indicate the respective deviations from the contemporary herd (or population) men. The k-factors are defined in the text and in Table 15:3.

ROBERTSON (1959) has devised a new method for the weighting of information from various sources (own phenotype, pedigree and progeny) in the estimation of breeding values. The concept 'standard progeny record' (s.p.r.) is introduced, and, as applied to dairy cattle, is defined as the first single lactation record of a daughter. The weighting factor for each piece of information is the equivalent number of standard progeny records. The number, n, of s.p.r. equivalent to the individual's own record can be obtained from the formula $h^2 = 0.25nh^2/\{1+(n-1)0.25h^2\}$ according to which $n = (4-h^2)/(1-h^2)$; when $h^2 = 0.3$, $n = 5.3$. When the progeny test of a sire has $h_1^2 = 0.8$, the equivalent number of the son's daughters, m, is $0.5(0.8) = 0.25mh^2/\{1+(m-1)0.25h^2\}$; i.e. when $h^2 = 0.3$, $m = 3.1$. The sire can never provide more evidence about his son's breeding value than $(4-h^2)/3h^2$ of the son's own daughters. When $h^2 = 0.3$ this maximum is reached at 4·1 daughters of the son. Table 15:4 shows equivalent s.p.r. at eight heritability levels.

Table 15:4 Equivalent number of standard progeny records at various levels of heritability. X = own record; D = dam's record; and S = progeny test of sire on 50 daughters.

	Standard progeny records		
h^2	X	D	S
0·05	4·1	1·0	9·1
0·10	4·3	1·0	6·3
0·20	4·6	1·0	4·2
0·30	5·3	1·0	3·1
0·40	6·0	1·0	2·4
0·50	7·0	1·0	2·0
0·60	8·5	1·0	1·7
0·90	31·0	1·0	1·0

The figures in the table show how the importance of pedigree information decreases with rising level of h^2 when information is available on the individual itself.

Before applying the weighting factors all records are expressed as deviations from the contemporary herd (or population) mean and converted to the scale of the animal's own performance; this is done by doubling the deviations when they are transferred from one generation to another. The proper weights can be taken from the table; when the individual itself is progeny-tested, the weight is given by the number of daughters. After weighting each piece of information, the average final deviation is calculated and regressed towards the population mean by the factor $0.25nh^2/\{1+(n-1)0.25h^2\}$.

Collateral relatives

The average phenotype of full and half sibs can give valuable information about the individual's breeding value, especially when the heritability of the trait is low

404 ESTIMATION OF BREEDING VALUES

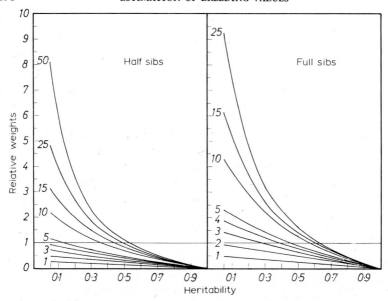

Fig. 15:4. Diagram to show the relative weights to be applied to the phenotypic mean of a sib group (half sibs and full sibs respectively) when the individual's own phenotypic value is given the weight $b_1 = 1$. The number in the sib groups is given on the left of each curve. The individual is not included in this number.

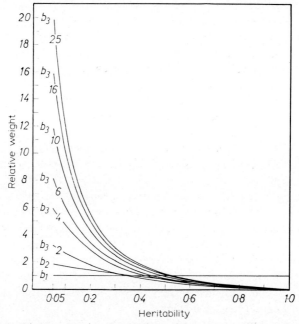

Fig. 15:5. Diagram showing the relative weights to be applied to the phenotypic means of full sib groups in which the individual is included (b_2) together with that applying to the mean of all full sib groups with the same sire as the individual (half sibs with each other) (b_3).

The weight for the individual's phenotypic mean is $b_1 = 1$. The number in each full sib group is 4. The number of full sib groups in the calculation of b_3 is given for each curve. Both Figs. 15:4 and 15:5 are based on the assumption that environmental influences are randomly distributed.

and the sib groups are large. Taking the sire's and maternal grandsire's progeny test into consideration means, of course—as was pointed out previously—an estimate of the breeding value of the individual and its dam on the basis of their paternal half sibs. The application in the estimation of a young bull's breeding value has already been mentioned but will be discussed more fully later (p. 461).

When assessing an individual's breeding value in pig and poultry breeding there is the possibility of combining information about its own phenotype with information from full and half sibs. It is the best pigs from the best litters and the best cocks and pullets from the best sib groups that are selected for breeding. The relative weight which should be given to the individual itself and the mean of full or half sib groups is shown in Fig. 15:4. From the diagram it will be seen that a full sib group ought to be given more weight than a half sib group with the same number of individuals in the group.

When a boar has sired several litters from different sows, or a cock several full sib groups out of different hens, then, in assessing their sons, information may be obtained about (1) the son's own performance, (2) the full sib group to which he belongs and (3) the groups which are half sib to him. Fig. 15:5 shows the relative weights which ought to be accorded each of these three criteria, at differing heritabilities and when four individuals are included in each full sib group but with differing numbers of sib groups from the same sire. The individual itself has been included in the full sib group to which it belongs, and this full sib group is included in the average of all half sib groups.

Consideration of the mean of full and half sib groups is especially important for traits which are manifested in an all-or-none manner but from a genetic point of view are quantitative, e.g. disease resistance. For example, disturbances in dairy cows, such as mastitis or cystic ovaries, may or may not occur, although it can be assumed that the genetic variation of the resistance is continuous. If, under otherwise similar conditions, the frequency of these disturbances is appreciably higher among the daughters of a particular bull, then it may be assumed that he has passed on to his progeny genes for low resistance. If, for example in pig breeding, it is desired to carry out selection for increased resistance to a certain disease, e.g. atrophic rhinitis, then not only should the females which show the disease be excluded from breeding but so also should all sib groups in which the frequency of the disease, under similar conditions, is conspicuously high. This type of selection could be used to advantage in poultry breeding with the aim of genetically improving the resistance to such diseases as leukosis, certain avitaminoses, inflammations of the oviduct, and others. Only healthy birds from full sib groups in which the disease does not occur or has a very low frequency should be selected for breeding.

One source of error, which one must be on the look out for in all family selection, occurs because sibs, especially full sibs, usually live under more similar environmental conditions than unrelated individuals. Differences can occur, for example, between litters from different sows as a consequence of maternal influence and between litters from the same sow due to different sizes of the litters, time of farrowing and so on.

Progeny-test

The performance and quality of the progeny give the final answer about the result of using an animal for breeding. When a male is mated to a random sample of females from his own population and there is no systematic environmental differences between the progeny groups, the correlation between his general breeding value (A_X) and the average phenotypic value of the progeny increases towards unity with increasing number of offspring, rapidly when the heritability is high and more slowly when it is low (cf. Fig. 15:6). The relative value of the progeny test, however, as compared with individual performance, decreases with increasing heritability.

Progeny-testing has found large practical application in dairy cattle. Initially, research work was directed at developing methods of investigating the bull's influence on the daughter's milk yield and butterfat content. This was later followed by studies of the bull's influence on other traits of the progeny, such as rate of milking, growth rate and carcass characteristics.

In progeny-testing bulls, usually only one daughter or son is included from each female to which the bulls were mated. With boars and cocks the situation is different, in that usually several full sib groups are included, which are half sibs of each other. The basic principles are nevertheless the same, no matter which trait or class of animal is considered. We begin with a relatively detailed account of the progeny-testing of bulls for the inheritance of milk production, and then discuss briefly its application to other traits and classes of animals.

Progeny testing of bulls

Milk and butterfat yield. The sources of error which must be taken into consideration in progeny-testing can be summarised as follows:

A Randomly distributed sources of variation (which can be reduced by increasing the number of daughters per bull):

1. the genetic variation within sib groups as a result of Mendelian segregation;
2. randomly distributed environmental factors and errors of measurement in milk recording.

B Systematic errors (which, in the main at least, can be eliminated by corrections or by using only comparable data):

1. the effect of age on yield capacity of cows.
2. the possibility that dams constitute a selected group genetically different from the herd mean.

C Systematically distributed environmental differences between progeny groups, in that they have been kept in different herds, or at different periods of time, or both. (These errors are not reduced by increasing the number of offspring but must be eliminated by other means.)

As has already been explained, environmental differences between herds and time periods may be eliminated by paying attention only to the difference in yield of the daughters ($\bar{Y}$) and the herd or the contemporary herd mates of the same age ($\bar{H}$). If only the first lactation of the daughters is considered, there are the advantages that the material is less selected, and that the assessment can be made earlier than if several lactations are included. Another advantage is that the heritability of the first lactation record is relatively high.

If the dams are unselected they can be ignored in the progeny-test calculations, and the breeding value of the bulls can then be estimated as:

$$\hat{A}_X = 2b(\bar{Y} - \bar{H}); \quad b = \frac{(0\cdot 25h^2 n)}{1 + (n-1)0\cdot 25h^2}.$$

If the dams, D, are selected, such that they have a production considerably higher or lower than the mean for the herd, a correction can be made as follows: $\hat{A}_X = 2b[(\bar{Y} - \bar{H}_y) - 0\cdot 5h^2(\bar{D} - \bar{H}_D)]$; $\bar{H}_y$ = herd or stable mates average contemporary with the daughter's lactation, and $\bar{H}_D$ = the corresponding contemporary average of the dams. With artificial insemination, it is usually unnecessary to make any correction for the dam's influence, if the bull's semen is distributed at random within and between herds and if the number of daughters is relatively large.

The regression, b, of a later group of daughters on a similar number of earlier daughters is calculated according to the formula shown above. It may be regarded as the repeatability of the 'daughter level' ($\bar{Y} - \bar{H}$) which is equivalent to the heritability of the bull's 'phenotypic value' $2(\bar{Y} - \bar{H})$ (cf. p. 410). If systematic errors occur, the accuracy of the estimate is reduced. Assume, for example, that only the mean of the daughters is considered, but not the deviation from the herd mean, and that each bull is tested in only one herd. Then assume that the bulls are later used on many other herds, for example by artificial insemination. An environmentally determined correlation, c^2, will then exist between individuals within those groups of daughters on which the bull was tested, but not between those daughters and the future daughters of the same bulls. The repeatability of the daughter average then becomes

$$b = \frac{n(0\cdot 25h^2)}{1 + (n-1)(0\cdot 25h^2 + c^2)}.$$

The repeatability is shown by the diagram in Fig. 15:6 for four different conditions,—when $c^2 = 0$ and $h^2 = 0\cdot 3$ and $0\cdot 6$; and when $c^2 = 0\cdot 0625$ and $0\cdot 125$ where $h^2 = 0\cdot 3$ in both cases. The diagram illustrates first, that increasing the number of daughters is very important when the heritability of the trait is low. It is also apparent from the diagram that the repeatability of the test reaches a maximum with a relatively small number of daughters when systematic environmental differences occur between the daughter groups, and also that this maximum becomes lower the greater the part played by these systematic differences. American investigations have shown that this is actually the case in practice. The artificial insemination association in the State of New York previously purchased bulls which were progeny-tested on the basis of the daughters' yield in individual herds. Later the same bulls were progeny-tested within the A.I. association (cf. p. 411). The daughters' mean yield in individual herds did not show any correlation with the A.I. progeny test (i.e. $b = 0$); but, when the natural service test was calculated as the daughters' average yield minus the average yield of the stable mates ($\bar{Y} - \bar{H}$), the repeatability was 0·2. However, when both the first and second tests were conducted within the association, the repeatability was 0·63 when the number of daughters was 20, and 0·81 when 50 daughters were included in the test. The result shows that progeny-testing based only on the mean of daughters from a single herd without regard to the herd mean, is valueless. Even when the differences between herd means are eliminated, the results are more accurate when the testing takes place

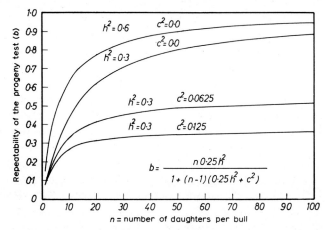

Fig. 15:6. The repeatability (b) of the progeny test of bulls with different heritabilities and varying number of daughters when there are systematic environmental differences between the groups ($c^2 > 0$), and when there are no such differences ($c^2 = 0$).

within an A.I. association than within individual herds, because the number of daughters is greater and the daughters are spread over a large number of herds. In the Swedish Red and White breed a correlation of 0·343 was obtained for the progeny test of 147 sire-son pairs with A.I.; but the corresponding correlation for 659 sire-son pairs, tested in individual herds, was found to be only 0·150. The progeny-testing method is described on p. 411. Similar results have been obtained by the Milk Marketing Board in England.

Where individual herds with natural mating practice are concerned, there are undoubtedly heritable differences between them. If consideration is to be given to these differences in the progeny-testing of bulls, an estimate must be made of the heritability of herd differences (h^2_H). According to the studies which have been carried out so far, h^2_H for milk yield in the first lactation may be about 0·1. The estimate of the bull's breeding value in relation to the population mean can then be obtained as follows:

$$\hat{A}_X = 2b[(\bar{Y} - \bar{H}_y) - 0.5h^2(\bar{D} - \bar{H}_D)] + h^2_H(\bar{H} - \bar{P}).$$

For bulls in service in A.I. associations, there would appear to be very little to be gained by considering either the dams or the heritability of the herd means;

consequently the breeding value of the bulls can be estimated according to $\hat{A}_x = 2b(\bar{Y} - \bar{H}_y)$.

Practical application of bull progeny-tests. Cow-testing associations were already organised in Denmark in 1895, and in 1902 the progeny-testing of bulls began on the basis of the results recorded. Until quite recently, these progeny-tests have been founded on dam-daughter comparisons without special reference to the herd means. In more recent years, however, a new principle has been applied. The daughters from the bull to be tested are divided into two groups according to the herd mean, with 180 kg butterfat as the dividing line between the groups. The average calving age of the daughter groups is computed together with their average yield (milk, fat percentage and butterfat) for each group during 130 and 305 days of the first lactation. In addition, the contemporary herd mean (kg butterfat) is given, and the daughter's yield is expressed as a percentage of the herd mean.

During World War II difficulties were experienced in Denmark, as in other countries, in obtaining comparable dam-daughter data for the progeny test of bulls; so special bull-testing stations were set up; these began operating in 1945. The procedure for the testing stations is that 18–20 daughters from each bull that is to be tested are transferred to the testing station some time before the first calving, and there feed consumption and production are carefully recorded. Studies on the rate of milking are also carried out. Admission takes place about 1st September and calving is reckoned to take place between 1st October and 15th November, when the heifers are between 27 and 33 months old. On the average, recording at the stations includes 304 lactation days per heifer for each group though there is an individual variation from about 250 days to one year. At each station there are, as a rule, at least two, and at some as many as nine, progeny groups on test simultaneously. Since 1960, about 100 bulls a year have been tested at some 30 testing stations. The results are published annually, giving information for each individual cow as to feed consumption, milk yield, fat content, milking rate, body weight and measurements, etc. Statistical analyses of the testing results have been carried out by several research workers. These analyses tend to show that neither the environmental differences between years and stations have been completely eliminated nor the environmental differences between contemporary groups at the same station. Heifers sent to the stations have not been selected on the basis of the production of their dams; but certain environmental differences between groups are probably due to the fact that all test heifers from each bull have been placed together as a group at the same station, and there milked and cared for by the same herdsman (TOUCHBERRY, ROTTENSTEN and ANDERSEN, 1960).

The variance within progeny groups has been surprisingly low, compared with the variance between groups; this may be due mainly to the fact that the members of the groups are strictly contemporary with regard to age and season of calving, and consequently the genetic differences between groups are manifested more clearly than under field conditions (cf. p. 154). TOUCHBERRY *et al.* (1960) found that the genetic correlation between station tests and field tests of the same bulls was only 0·68 for milk yield, whereas the corresponding correlation between independent field tests was 0·94. They suggested that "there is either a large interaction between sires and level of management, or that the between-sire components from test station data are inflated with environmental differences". The second explanation was favoured. In view of the results from the Ruakura split-twin-pair experiments and other twin studies (p. 151), it seems likely that both factors have contributed to the difference between the correlation coefficients. In the field tests, the daughters of each bull were distributed over many herds on different levels of management; which would seem to be an advantage.

The greatest drawback of the station method is that the tests are too expensive to allow testing enough bulls for an efficient selection after completion of the tests. With artificial insemination and widespread use of milk recording, it is possible to organise efficient progeny-testing in the field on a much larger scale and at a lower cost than can ever be done at special testing stations. The greatest advantage of the testing stations is that the recording can be extended to other important characters than milk yield and butterfat percentage—for example, milking rate and udder proportions, the protein content of the milk, reproductive disturbances, etc.

The station method has been tried on a small scale in Great Britain, Germany and Sweden.

After World War II the most significant change in the methods of bull progeny-testing on field records has been the discarding of dam-daughter comparisons in favour of comparing daughters with herd-mates. The new methods have been developed together with the spread of artificial insemination, in connection with which they are particularly well suited.

The change appears to have been first made in New Zealand. Significant contributions to the methodology of progeny testing have been made by J. L. LUSH and C. R. HENDERSON in the U.S.A. and by ALAN ROBERTSON in Great Britain. There are some differences between countries in the testing procedure, but the basic principles are the same.

In New Zealand at least 10 daughters are required for an official progeny test; the test is repeated annually until there are 20 daughters; any subsequent testing is at the owner's request. Cows aged 5–9 years are considered as mature, and records for younger cows are corrected to mature level; separate age correction factors are used for different agricultural areas. Since practically all cows calve in the spring, season of calving does not need to be considered. Records of all daughters of a bull are used for the progeny test, except when the lactation has been obviously abnormal or made after the cow was 10 years old. Each lactation record of a daughter is compared with the average age-corrected records of all cows in the same herd and the same year, excluding those which are daughters of the bull. The sire's daughter-level is estimated as $\hat{Y} = \bar{Y} - b(\bar{H}_m - \bar{P})$ where $\bar{Y}$ = the mean of all available daughter lactations, $\bar{H}_m$ = the stable mates' average, $\bar{P}$ = breed average within the region, and b = the within-sire regression of daughters on herd level. According to statistical analysis of field data, $b = 0.9$; which means that the heritability of herd differences would be 0·1 and the estimate of the daughter's level could be just as well written $\hat{Y} = (\bar{Y} - \bar{H}_m) + 0\cdot 1(\bar{H}_m - \bar{P}) + \bar{P}$ (cf. p. 408). The difference between the estimated daughter level and the breed average is called 'the difference from expectancy', and this difference is regressed towards the breed average by the familiar factor, $0\cdot 25nh^2/\{1 + (n-1)0\cdot 25h^2\}$; the product shows the 'rating' of the bull.

In the State of New York (U.S.A.) more sophisticated methods have been developed for estimating the daughter's average and the average of contemporary herd-mates. All cows aged between 23 months and 14 years are included in the progeny tests, and their records are corrected to mature yield at 6–7 years of age. If the cows have been milked oftener than twice daily, their records are corrected also for frequency of milkings. Short lactations are corrected to the standard length of 305 days, provided they are not caused by normal drying off. The daughter average is calculated as a weighted mean of the individual daughter averages; the weighting factors take account of repeatability and number of lactations in each daughter's average. The stable mates' average includes all cows in the herd except those which belong to the progeny group in question; the records of the stable mates are treated and averaged in the same way as those of the sire's daughters. The breed average is calculated for three successive years. All records are classified into one of three seasons of calving and all comparisons between daughters, stable mates and breed are made within years and seasons. The daughter-level for a sire is then estimated as $\hat{Y} = \bar{Y} - 0\cdot 9 (\bar{H}_m - \bar{P})$, the same symbols being used as before, although the method of calculating $\bar{Y}, \bar{H}_m$ and $\bar{P}$ is different. The individual $\hat{Y}$'s are then combined into an average for all bulls in the A.I. stud, and the merit of each sire is estimated as the deviation of its $\hat{Y}$ from the stud average. The calculation of this average is based on weights which take into account the variation among sires and the number of daughters by each sire. The results of the progeny tests are reported thrice yearly.

In *Sweden* all dairy bulls are progeny tested as soon as they have 10 daughters with first lactation records; the test is then repeated each time the number of daughters has been doubled. When a daughter is culled or otherwise eliminated before she has completed a 305-day record, her yield is corrected to the 305-day level; the minimum requirement is 46 days lactation. These daughters are included in the progeny group as a precaution against selection effects. Otherwise no corrections are applied to the individual records, but the average for the progeny group is corrected to a calving age of 28 months (Swedish Red and White and Swedish Friesian). Only records from calvings between 20 and 36 months of age are considered. For each daughter the basis of comparison is the yield of all contemporary cows in the same herd where the daughter made her first record. The average contemporary herd yield is corrected to first lactation yield by using the regression (b) of all first calvers within the same breed on their contemporary herd average. The average contemporary yield of first calvers $(\bar{H}_1) = \bar{P}_1 + b(\bar{H}_t - \bar{P}_t)$, where $\bar{P}_1$ = breed average for first calvers, $\bar{H}_t$ = the actual herd average for all contemporary cows irrespective of age, and $\bar{P}_t$ = the breed average for all tested cows in the same period of time. In 1966,

$\bar{P}_1 = 4453$ and $\bar{P}_t = 5512$ kg milk for the Swedish Friesian breed; for the period 1963–65, $b = 0.78$. The difference between the age-corrected daughter average ($\bar{Y}_1$) and the estimated yield of contemporary first calvers in the same herds ($\bar{H}_1$) is expressed as a percentage of $\bar{H}_1$ and this difference is regressed towards the population mean in the usual way, depending on the number of daughters in the progeny group. After adding 100, a relative figure (called the F-value) for the estimated merit of the progeny group is obtained. The F-value refers to fat-corrected milk (4 per cent fat content). The average fat content of the daughters and of the herd contemporaries is calculated without any corrections. For the progeny test of young A.I. bulls their semen is used for a number of cows which is large enough to obtain about 75 milk-recorded daughters from each bull.

In Great Britain the progeny tests are based on complete first lactation records of 200 days or more (up to 305 days) of the daughters and their contemporary stable-mates. The bull's daughters are compared with daughters of other bulls having their first lactation in the same herd and the same year, thus avoiding age-corrections and special measures for the elimination of herd effects. For each herd the difference is calculated between the average yield of the bull's daughters, $\bar{Y}$, and their contemporaries, $\bar{H}$, and each difference is weighted according to the harmonic mean of the number of daughters, n_1, and their contemporaries, n_2, in the same herd: the weighting factor $W = (n_1 \times n_2)/(n_1 + n_2)$. The weighted differences are added and their sum is divided by the sum of the weights: $\Sigma W(\bar{Y} - \bar{H})/\Sigma W =$ the contemporary comparison (CC). The number of 'effective daughters' $= \Sigma W$.

The published results contain the following information for each bull which is tested on at least 20 'effective daughters': the actual number of daughters and the number of 'effective daughters', average for the age at calving of the daughters, milk and fat yield and butterfat percentage in the first lactation, and the weighted difference in milk yield, CC, between the daughters and their contemporaries, i.e. the 'daughter-level' without corrections. The number of daughters which have completed a second and third lactation and the corresponding yield average are also stated. In addition, the 'relative breeding value', RBV, of the bull is calculated, based on the contemporary comparison and the estimated regression, b, of future daughters on those included in the progeny-test.

$$RBV = \frac{[2b(CC) + \bar{H}]100}{\bar{H}},$$ where $\bar{H}$ is the average yield of the herds in which the test is made. Another RBV has also been calculated and refers to the average yield of all tested cows of the same breed (or region) taking into account the genetic differences between the herds:

$$RBV(\text{breed}) = \frac{[2b(CC) + 0.2(\bar{H} - \bar{P}) + \bar{P}] \times 100}{\bar{P}},$$ where $\bar{P}$ is the breed and $\bar{H}$ the herd average and 0.2 is the heritability of herd differences. When A.I. has been practised for several cow generations, however, the genetic differences between herds are probably so small that they can be disregarded.

In calculating the value of b, the heritability of the first lactation yield is assumed to be 0·3. The importance of using the regression factor b decreases with an increasing number of daughters but seems to be rather important when W is less than 50.

This review of progeny-testing methods in Denmark, New Zealand, New York State, Sweden and Great Britain shows that the published results generally refer to the 'daughter level', which deviates from the population mean only half as much as an estimate of the breeding value.

The progeny-tests should be completed as early as possible in the life of the bulls; which is one valid reason for testing on the first lactation yield only. Preliminary tests should preferably be made on part-lactation records, e.g. the first four months after the first calving. There are also other reasons for considering only the first lactation: (1) The heritability of the first record is considerably higher than that of the second and somewhat higher than for any of the following records; and (2) selection will have very little influence on the daughter averages when only first lactation, complete or corrected incomplete, records are used. Nothing would then be gained by correcting the first record to mature yield, but corrections to a standard age at first calving may be advisable.

From these points of view the British method of progeny-testing would seem to be superior to the other methods, but two disadvantages may be pointed out. One is that it cannot be applied in small herds, because no contemporaries may be present, and the other is that the season of calving is not considered. Calving in the same year is no guarantee for contemporaneity. In this respect the New York State method has an advantage.

Milk composition and milking traits. The fat and protein content of milk are less modifiable by the herd environment (feeding and management) than the yield of milk and butterfat. It is therefore less important to eliminate or reduce herd differences when progeny-testing for fat and protein content. The average fat content of the milk of a bull's daughters gives more reliable information about his breeding value than does the average milk yield. However, when all the daughters have been in the same herd, their mean should be compared with the contemporary herd mean. There are two reasons for this: (1) It is likely that the cows are selected for the trait in question, especially in bull breeding herds; (2) the fat content of the milk is positively correlated with the level of nutrition. When, as with A.I., the daughters of each bull are distributed over a number of herds, such sources of error in the calculation of the means are mainly eliminated.

During recent years bulls have also been progeny-tested for udder shape and milking characteristics of the daughters. As was shown earlier, the proportion between front and rear udder, measured by the amount of milk from the front glands as a percentage of the total milk, is relatively highly heritable. Both this trait and the rate of milking (peak flow or average flow) are determined for each daughter at, for example, two milkings during the third to fifth month of lactation. The most extensive application of the recording of udder proportions and milking rate has so far been in Holland. In measuring udder proportions there is no need to reckon with any environmental differences between herds, as there is in measuring rate of milking, which is highly sensitive to the milking routine in the herds and to the yield of milk. This must be taken into consideration in progeny-testing for this trait.

Growth rate and carcass quality. Between the two World Wars, work was started in Germany and Holland to improve the carcass characteristics of Friesian cattle for meat production. Selection was made for relatively short-legged animals, with pronounced arching of the ribs, wide rumps and well developed muscles (cf. Table 18:1, p. 476). During the past two decades a similar endeavour has been made in Denmark and Sweden.

As has already been mentioned (p. 305), studies have been commenced on the influence of the bulls on the growth rate and carcass traits of their male offspring. In Denmark, Norway and Germany this has been done by collecting a certain number of bull calves from a half sib group at a testing station where they are reared. The feed conversion and growth rate are recorded up to slaughter, after which various measurements are made on the carcass to determine the quality. The number of recorded offspring per sire has varied from 8 in Norway to 12 in Germany. A valuable study of testing methods was carried out by SKJERVOLD (1958) but continued research in this field is necessary. We shall return to the question of progeny-testing versus individual performance testing for carcass quality traits in Chapter 17 (p. 464).

Progeny-testing of boars

In Swedish pig progeny-testing, which operates mainly along the lines of the Danish system, two females and two castrated males from each litter to be tested are delivered to the testing station when the average weight is between 15 and 18 kg. The test pigs are selected impartially by the testing authority on the basis of the weights at 3 and 7 weeks of age. At the testing stations the feed conversion is recorded on a group basis and the growth rate individually. The pigs are slaughtered at 88–90 kg and a number of measurements are taken on the carcasses, which are also scored subjectively for certain quality characteristics (cf. p. 319).

In order that this pig-testing shall provide a satisfactory basis for progeny-testing of the boars, it is essential that the pigs sent to the stations be representative of their litter, and further that the dams of these litters be representative of the female breeding material in the population or at least in the herd.

When a boar is progeny-tested on the basis of a number of test litters, the following two factors must be considered, in addition to what has been said about bull progeny-testing.

1. The pigs from the same litter are full sibs ($r_g = 0.5$, when there is no inbreeding), but pigs from different dams and the same sire are only half sibs ($r_g = 0.25$). Therefore, attention must be paid not only to the number of pigs which have been tested but also to the number of litters from different dams.

2. Those pigs which belong to the same litter (test group) are more influenced by the similarity of their environment during foetal development, suckling and at the testing station than are pigs from different litters. In an analysis of data

from the Swedish pig progeny-testing, the correlation was calculated between full sibs of the same sex when these full sibs were born in the same litter, and also when they were born in different litters. The following results were obtained:

Correlation between full sibs of the same sex belonging to

	Same litter	Different litters
Carcass length	0·47	0·32
Backfat thickness	0·36	0·25
Belly thickness	0·36	0·20
Daily rate of gain per pig	0·21	0·13
Age at slaughter	0·48	0·29

The correlations are consistently larger for the pigs born in the same litter than for those born in different litters. If the sows are to be progeny-tested the test litters should be out of different boars.

The repeatability of the progeny-test, b, can be calculated as follows (cf. p. 407):

$$b = \frac{mn0\cdot25h^2}{1 + 0\cdot25h^2 n(m+1) - 0\cdot5h^2 + c^2(n-1)}, \quad \ldots [1]$$

where m is the number of tested groups (litters from different sows), and n is the number of tested pigs per group.

If c^2 is sufficiently small to be ignored, the formula simplifies to:

$$b = \frac{mn0\cdot25h^2}{1 + 0\cdot25h^2 n(m+1) - 0\cdot5h^2}. \quad \ldots [2]$$

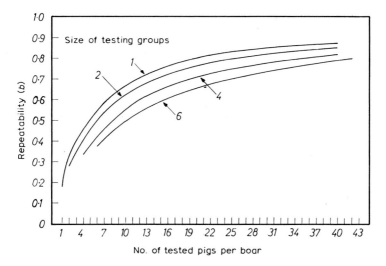

Fig. 15:7. The effect of the size of the test litter group on the repeatability of the progeny test when a given number of pigs are tested per boar. The heritability of the trait is assumed to be 0·6 and $c^2 = 0$.

Fig. 15:7 shows how the accuracy of the progeny-test increases with an increasing number of tested groups per boar and with differing numbers of pigs per group, when the c^2 term is ignored and the heritability of the trait is 0·6. If only one pig from each litter is selected at random, the accuracy of the result is, on the average, about the same with 7 pigs (litters) as when three groups of 4 pigs from each litter are tested, i.e. 12 pigs. The greater the environmental differences between the different litters, the larger will be the c^2 factor and the more important it is to test many litters from different sows; as a rule it is then necessary to reduce the size of the test groups. There is less accuracy lost in increasing the size of the test groups at low than at high heritabilities.

A valid question from a practical point of view is whether, for the same number of pigs per boar, the groups should be larger or smaller. If the pigs are individually fed, as is the case in the new Danish testing stations, then one might as well test one or two pigs instead of four per litter. It has been argued that the fewer the number of pigs taken from the litter for testing, the less representative they will be, in that the possibilities of selection increase. It should be possible to avoid this by selecting the pigs at random whose weight are within the limits $\bar{x} \pm 2\sigma$, where $\bar{x}$ is the average weight of the pigs in the litter and σ is the standard deviation in litters of the same size. The average weight of the male and the female pigs should be calculated separately. As already mentioned, the analysis of data from the Danish pig progeny-testing showed a much higher heritability for growth rate of the pigs in the new testing stations with individual feeding than for pigs which were group fed under the old system. This observation is so interesting from a methodological point of view that some illustrative figures will be given (Table 15:5).

Table 15:5 *Means and standard deviations for daily gain, carcass length and backfat thickness for hogs and gilts with individual and group feeding. $\bar{x}$ = mean and σ_i = SD within litters and sex* (JONSSON, 1959)

	Daily gain, g		Carcass length cm		Backfat thickness mm	
	$\bar{x}$	σ_i	$\bar{x}$	σ_i	$\bar{x}$	σ_i
Group feeding:						
Barrows	685·2	53·4	92·9	1·66	35·2	3·04
Gilts	670·4	53·8	93·5	1·61	32·4	2·27
Heritability (within sexes)	0·24		0·42		0·56	
Individual feeding:						
Barrows	668·1	24·1	93·2	1·67	34·7	2·42
Gilts	678·8	25·4	93·8	1·61	32·2	2·23
Heritability (within sexes)	0·63		0·48		0·50	

The figures show that the gilts on group feeding had a lower growth rate and on individual feeding a higher growth rate than the barrows, probably because with group feeding they were more timid in the contest at the feeding trough. The variation in growth rate within groups and sexes decreased by more than

50 per cent in the change-over from group to individual feeding. Besides giving each individual an equal chance to eat, the improved housing conditions at the new testing stations has probably had some effect in this respect. The estimates of the heritability of growth rate (within sexes) were 0·24 for group feeding and 0·63 for individual feeding. This difference was entirely due to the reduction of variance within tested groups, since the variance between sires was practically the same under both systems of feeding. Therefore, it seems unlikely that the efficiency of selection among the tested boars would increase in proportion to the heritabilities mentioned. Furthermore, it is open to question whether the results obtained with individual feeding are applicable to ordinary practical conditions where the pigs are kept in relatively large groups. This question could be answered by testing the same boars under both group and individual feeding conditions.

Under both systems, the progeny-testing of boars for growth rate and feed conversion is actually less accurate than calculated according to formula [2] due to systematic environmental differences between groups (the c^2 factor) acting before (maternal influence) as well as after commencement of the station test. Testing the boars on more litters from different sows, preferably from different herds, would help considerably to decrease the systematic errors, even if the total number of tested pigs is the same, e.g. when 2 pigs from each of 10 litters are tested instead of 4 pigs from each of 5 litters (cf. Fig. 15:7). The best policy would probably be to test only 2 pigs from each litter and to utilise the results not only for the progeny test of sires but also for selection among the full sibs of the tested barrows (cf. Chapter 17, p. 467). Environmental correlations between litter mates will be less pronounced for traits which show high heritability with both systems of feeding, e.g. carcass length and backfat thickness.

The foregoing discussion of progeny-testing of boars applies also to cocks. A standard progeny-test for a cockerel would involve mating him with 10 to 15 non-related females, constituting a random sample of the strain. Progeny from such matings would be hatched over a two- or three-week period to produce 20 to 40 pullets for testing. Selection based on half sib family groups of this size would lead to more efficient selection gains than mass selection for characters with a heritability of approximately 15 per cent or less (cf. LERNER 1958). Obviously, in some cases, such as for sex-limited traits like egg production, mass selection is not feasible and progeny-testing is essential.

In order to facilitate an early selection between males, the first progeny-test comparisons must be based on part records of the daughters' first laying year, say, up to 1st January. Since the hatching dates of the pullets influence their part-time egg record, comparisons between males should be made on pullets hatched within a fairly short time of each other. Alternatively, the part-time egg records can be corrected for hatch effects by taking each pullet record as a deviation from the mean of the records of all pullets of the same hatch. A second selection between males can be made when the daughters have completed their first laying year; and at this time selection can be based on both egg production and pullet mortality.

Some breeders have suggested that it is unnecessary to trap the individual pullets and that progeny group records from the different males are all that is required. By using the second method a considerable amount of labour can be saved but the method also has some disadvantages. When progeny groups are kept and fed separately, there may be environmental differences between them that are difficult to estimate and may obscure the real genetic differences between males. Further, records are not obtained for the individual pullets or for the full sib groups. Individual trap-nesting has several advantages; but the question is, does the extra information and accuracy obtained justify the increased cost? In both the United Kingdom and the United States there is considerable use of battery cages, including single bird cages. With this type of housing, recording is much easier than on deep litter with trap-nests. If it is necessary to breed from the pullets, this is accomplished by means of artificial insemination.

In the improvement of broiler chickens, progeny-testing is necessary only for such characters as carcass quality, egg production and chick mortality, all of which have low heritabilities. The heritability of growth rate is often so high that individual selection (in both sexes) would ensure satisfactory selection progress, at least in the early stages of improvement. After several generations of selection, when genetic variation has been reduced, it may be necessary to introduce some form of progeny-testing or family selection.

It should be mentioned that, in order to avoid the necessity of keeping birds for more than 18–20 months of age, many breeders are now using some form of family selection based on full or half sib testing instead of progeny-testing males. This means that instead of using selected progeny-tested males for the reproduction of strains, males that are full or half sibs to the pullets on test are used. This method avoids overlapping generations and, what is more important, reduces the generation interval quite considerably.

Diallel mating

The maternal influence on the result of progeny-testing can be eliminated by diallel mating, which is relatively easy with pigs and poultry but almost impossible with cattle because of their slow rate of multiplication. If, for example, 6 sows are available at the same time and in the same environment for the testing of two different boars, then the sows are divided at random into two groups of 3 sows. One group is mated with boar A and the other group with boar B. For the next litter the boars are exchanged between the two groups. In this way each boar produces a litter with each of the 6 sows and the two groups of litters are then compared with each other. The difference between the breeding value of the two boars is estimated as twice the difference between the mean of their progeny. It is also possible to test three or four boars, so-called *polyallel* mating; but if each boar is to produce an equal number of litters with all the sows, the test takes a very long time. The time can be shortened by a system of 'rotational mating', which is planned so that the boars can be compared with each other in pairs. If any of the links in the chain fail, however, not all the comparisons can be made.

This mating system always has the disadvantage that the comparisons are limited to the male animals being tested, and the progeny groups cannot easily be compared with the population mean. This is most probably one of the reasons why this form of progeny-testing has not achieved much practical application in animal breeding.

Selection indices for several quantitative traits

In practice, selection can seldom be limited to a single trait, and several traits have to be considered simultaneously. The old method of evaluation of the animals by scoring with different maximum points for different traits and then adding the points allotted, was an attempt to achieve a balance between the traits. The weight given to individual traits should express their comparative economic importance. This must be regarded as an important step in striving for rationalisation in animal production.

HAZEL (1943) and LUSH have developed the principles for estimation of the individual's 'total breeding value' (net merit) for two or more traits. The principle is the same as in the estimation of the individual's breeding value on the basis of, for example, its own phenotype together with the phenotypes of the paternal and maternal ancestors. The weights used are standardised partial regression coefficients for the different traits included in the selection index. Through these coefficients, the multiple correlation, R_{AI}, between 'total breeding value' (net merit) and the selection index I reaches its maximum value. The idea behind the selection index is that the superiority of a particular trait balances out the deficiencies of one or more other traits, so that the economic result of the breeding work shall be the best possible. A weighting factor k is computed for each trait, and the index for the individual's net merit, expressed as the deviation from the population mean, is obtained by summing the products of the weight factor for the trait in question with its deviation X from the corresponding herd or population mean: $I = k_1(X_1) + k_2(X_2) \ldots + k_n(X_n)$; (cf. p. 400).

For calculation of the weighting factors reference may be made to Hazel's publication. The following information is needed for constructing the most efficient selection index for two or more traits:

1. The heritability of the various traits.
2. Their relative importance (economic weight).
3. The phenotypic and genetic correlations between the traits.

Calculation of the economic weighting factors can, in many cases, be very difficult. It is necessary to find out how a certain change in the different traits influences the economic result of the production under the prevailing market conditions. The economic weighting factors are not constant for any length of time but are likely to undergo changes caused by market trends and also by the effects of differential selection pressure on the different traits which enter the index. In dairy cattle breeding, for example, the relative importance of meat production has increased over the past two or three decades compared with milk production, and there is no longer the same incentive to increase the butterfat content of the milk as there was earlier. In the improvement of a population of farm animals it is impossible to take into consideration temporary fluctuations in the economic importance of the traits; but changes in the trend of a long time nature must receive attention.

Knowledge of the phenotypic, and especially the genetic, correlations between various traits is unfortunately very limited. The correlations may have different value in different breeds and also under different environmental conditions. For example, several investigations indicate that the negative correlation between milk yield and butterfat content is more pronounced in dairy breeds which produce milk high in fat than in breeds producing milk low in fat. Furthermore, the estimates of genetic correlations which have been made have usually large sampling errors due to small number of animals, and in some cases there are probably also systematic errors involved. More investigations on suitable data are needed in this field.

Two examples, however, may be presented from analyses of the construction and application of selection indices for several traits which show more or less pronounced phenotypic and genetic correlation *inter se*.

TABLER and TOUCHBERRY (1955) analysed data from the American Cattle Club comprising 2810 dam-daughter pairs from 414 Jersey herds. The daughters were the progeny of 756 sires with at least two dam-daughter pairs from each sire. The first single lactation available was used, after age correction, and for each cow, milk yield (M), fat yield (F), fat percentage (f) and type classification (T) were considered. The cows were divided into 5 classes, according to type. The heritability estimates from these data were: for milk yield 0·25, fat yield 0·20, fat percentage 0·56, and type rating 0·25. The phenotypic correlation between milk and fat yield was found to be 0·88 and between milk yield and fat percentage −0·36; corresponding genetic correlations were 0·72 and −0·50 respectively. The correlations of type rating with milk yield, fat yield and fat percentage were small and insignificant. Five different selection indices were constructed and their effect when applied to a population of Jersey cows was estimated Selection was assumed to take place only among the females, 20 per cent of the animals being culled on their index value.

In calculating the 'milk equivalent', one kg fat was considered to have the same economic value as 17·6 kg milk with a fat content of 5·3 per cent. Fat percentage as such was given the value zero. The economic value of type varies between individual herds; one extreme is zero and the other extreme may be that one standard deviation of milk should be given equal weight as one type-grade. Five indices were calculated of which estimated accuracy varied from 0·448 for I_1, where only the fat yield was considered, and 0·558 for I_2 which took into account both milk and fat yield. However, the 'milk equivalent' of the total selection response was highest when only the milk yield was considered. When type was included in the index and considered equally important as milk and fat yield (per standard deviation unit), the expected improvement in type was very slight but there was a reduction of 15 per cent in the genetic change of the 'milk equivalent'. Apparently, to select for type in a dairy breed can be profitable only when type is very well paid for on the market, thus compensating for the reduction in progress with regard to milk production.

Hogsett and Nordskog (1958) studied the application of selection indices to data from 15 lines of poultry at Iowa State University. Records were available on the March egg-weight, body-weight and the winter egg production (rate of lay from December to February inclusive) for 1938 daughters from 566 dams and 85 sires. Table 15:6 shows the economic weights, heritabilities, phenotypic and

Table 15:6 *Selection index for poultry* (Hogsett and Nordskog, 1958)

Trait	Relative economic weight	Heritability	Correlation between traits	
			Phenotypic	Genetic
Egg-weight (E)	$+1\cdot14$/g	0·40	$E-W$ $+0\cdot35$	$+0\ 50$
Body-weight (W)	$-2\cdot48$/lb	0·50	$E-R$ $+0\cdot05$	$-0\cdot05$
Laying rate (R)	$+1\cdot00\%$	0·10	$W-R$ $-0\cdot05$	$-0\cdot50$

Selection indices	Restriction	Relative net efficiency, %
$I_1 = 1\cdot27E - 37\cdot74W + R$	None	100
$I_2 = 4\cdot29E - 41\cdot32W + R$	$(Re)E = 0$	92
$I_3 = 2\cdot61E - 1\cdot94W + R$	$(Re)W = 0$	54
$I_4 = 0\cdot66E + 4\cdot14W + R$	$(Re)E = 0; (Re)W = 0$	50
Selection on R alone		65

genetic correlations used in the calculation of four different indices for the overall merit of the three traits mentioned. The table also shows the estimated relative efficiency of the indices. The weighting factors for egg-weight (E) and body-weight (W) are expressed as multiples of the weighting factor for rate of lay (R in percentage units). The relative economic weights for egg-weight and rate of lay were based on the monetary value of the increase in egg-yield during the first laying year, whereas the corresponding factor for body-weight took into account the increase in maintenance requirement of the birds with increasing weight and therefore it became negative. According to index I_1 the genetic change in the traits associated with one standard deviation unit of selection in the index was calculated to be $-0\cdot70$ grams E, $-0\cdot26$ lb· W and $2\cdot22$ per cent R. Because body-weight has a negative weighting and is highly heritable, the use of index I_1 causes a relatively large genetic change in body-weight in the downward direction which ultimately would lead to undesirable consequences. Neither could a decrease in egg-weight as much as $0\cdot7$ g per generation be tolerated for long, assuming that selection would be equally effective over several generations. One alternative is to construct an index designed to maximise the genetic-economic value but restricted such that there would be no genetic change (Re = 0) in one or two of the traits concerned   The authors found that placing such restriction on egg-weight (Index I_2) would cause a reduction in net efficiency of only 8 per cent compared with index I_1 but placing the same restriction on body-weight (Index I_3) reduced the net efficiency by 46 per cent. Selection on laying rate (R) alone would reduce the net efficiency by 35 per cent. The example shows that the problem of constructing selection indices may be rather complicated in some cases.

The weighting factors, k, of the indices referred to in these two examples include, for each trait, the contribution of the estimated breeding value (with proper adjustment for the correlation between traits) *and* the assigned economic value of one unit change in the trait to the 'net merit' of the individual. Several authors have pointed out, however, that it may be practical to keep these two things separate, by first calculating a partial index for each trait ($I_1, I_2 \ldots I_n$) and then multiplying each partial index by the corresponding economic weight ($a_1, a_2 \ldots a_n$). The total selection index would then be $I_T = a_1 I_1 + a_2 I_2 \ldots + a_n I_n$. The advantages with this method are that the effect of heritability and inter-trait correlations show up clearly for each trait, apart from their economic importance, and that the economic weight factors can easily be changed when needed without repeating the laborious calculations in obtaining the partial I's which have a biological foundation (KARAM et al., 1953).

When the traits are not correlated with each other, or when the phenotypic and genetic correlations are so weak as to be of no account, the calculation of a selection index is considerably simplified, since the weight factors become simply ah^2, where a denotes the economic weight of the trait under consideration, and h^2 is the heritability, hence $I = a_1 h_1^2(X_1) + a_2 h_2^2(X_2) \ldots + a_n h_n^2(X_n)$.

When an index is computed according to this principle, it will be found that there is no point in including traits with very low heritability or very slight economic importance, since they have very little influence on the size of the index. Nor is it necessary to include all those traits to which attention is paid in breeding. For example, in the selection of young bulls for progeny-testing, the first requirement is their breeding ability in respect of capacity to serve and semen production. If these requirements are not fulfilled the bull would be rejected, regardless of other merits. The selection index is most applicable to such quantitative traits as milk yield and butterfat content, growth rate and muscle development. Furthermore, it can be used for combining information on a certain trait in the individual itself and close relatives as discussed earlier in this chapter.

Practical breeders have, no doubt, always tried to consider the different traits according to their alleged importance, and they have also considered the ancestors and the family of the animals selected for breeding. The problems which occur in this connection have been elucidated by the theoretical analyses referred to above. It may be premature to recommend general use of the more complicated selection indices in practical selection procedures, especially when reliable estimates of the genetic correlations between the various traits are not available; but the basic principles evolved from research along these lines will certainly be a good guide in any appraisal of the breeding worth of the animals.

SELECTED REFERENCES

FEWSON, H. 1963. Untersuchungen über die Effektivität verschiedener Selektionsmassnahmen unter besonderer Berücksichtigung der züchterischen Verbesserung der Milchmenge und des Fettgehaltes beim Rind. Z. Tierzücht. ZüchtBiol., **71**: 101–161.

HAZEL, L. N. 1943. The genetic basis for constructing selection indices. *Genetics*, **28**: 476–490.
LE ROY, H. L. 1958. Die Abstammungsbewertung. *Z. Tierzücht. ZüchtBiol.*, **71**: 328–378
——, 1963. Merkmalskorrelation. *Ibid.* **79**: 217–236.
PIRCHNER, F. 1964. *Populationsgenetik in der Tierzücht.* (210 pp.). Hamburg.
SKJERVOLD, H. 1964. Testing and selection methods in pig breeding. *Inst. Animal Breeding and Genetics*, (*Inst. Husdyravl.*) Agr. Coll. Norway, *Medd.* **165**: 16 pp.
——, and ÖDEGÅRD, A. K. 1959. Estimation of breeding value on the basis of individual's own phenotype and ancestors' merits. *Acta Agric. scand.*, **9**: 341–354.
SEARL, S. R. 1964. Review of sire-proving methods in New Zealand, Great Britain and New York State. *J. Dairy Sci.*, **47**: 402–413.

16 Selection Response and Selection Methods

Extensive and long-term selection experiments have been carried out with various species of animals in order to determine the effects of selection on quantitative traits. The most important experiments, from a theoretical point of view, have been performed with the fruit fly (*Drosophila melanogaster*) and with mice, both of which reproduce very rapidly. The generation interval of the fruit fly is about two weeks and that of mice 8–12 weeks. Selection experiments with farm animals have been carried out mostly with pigs and fowls. Obviously, the selection results obtained with fruit flies and mice cannot be expected to find direct application to the larger farm animals; but certain general principles have been elucidated by these experiments.

These general principles will be discussed first in the light of some selection experiments designed to change the body size of the fruit fly, mice and fowls. Next, selection experiments concerned with production traits in farm animals will be dealt with, and finally the selection methods which can be used in practical animal breeding will be discussed. First, however, some terms must be defined.

After continuous selection in a particular direction the selection response sooner or later diminishes and finally ceases. It is said that the *selection limit* has been reached; but since this limit is seldom definite the term *selection plateau* is often used. If, after selecting in a particular direction, selection is then made in the opposite direction, i.e. back towards the original value, this is referred to as *reverse selection*. If selection ceases and the line is allowed to reproduce by random mating the term *selection pause* or *suspended selection* is applied.

CHANGES IN THE THORAX LENGTH OF FRUIT FLIES

F. W. ROBERTSON and REEVE (1952) studied the changes in the length of the thorax of fruitflies following selection for increased and decreased length. They used mass selection, i.e. selection was based on the individual's own phenotype. Out of 100 flies of each sex, the 20 largest and the 20 smallest were selected for further breeding. An unselected control line was maintained at the same time. The change in the thorax length in the selected lines was measured as the deviation from the contemporary mean value of the control line. The experiment was repeated with three unrelated strains with very similar results. The detailed results from one of the strains is shown graphically in Fig. 16:1.

The selection had an immediate effect in both the positive and the negative direction. The total alteration was greatest, however, in selection for decreased length, i.e. the selection effect was *asymmetric*. When reverse selection was applied after five generations of selection the lines rapidly returned to the original level. After about ten generations of selection in the lines selected upwards and 15–20 generations in those selected downwards, all systematic changes ceased; the selection limit had been reached. When the lines were allowed to reproduce without selection there was some regression towards the original level in the lines selected for longer thorax. There was no such regression in the lines selected for shorter thorax. Selection in the opposite direction, after the selection limit had been reached, resulted in a rapid return to the control-line

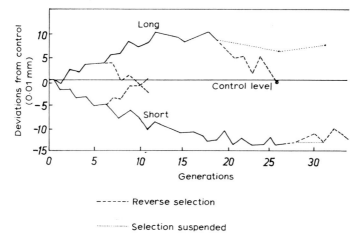

Fig. 16:1. The result of selection for long and short thorax in the fruitfly.
(F. W. ROBERTSON 1955)

value of those lines which had been selected for longer thorax, but this did not happen in the lines selected for shorter thorax.

The reason why the selection ceased to have an effect was therefore different in the up and the down lines. In the latter, the genes for shorter thorax had probably been fixed in the homozygous condition, whereas the longer thorax lines must have been still quite heterozygous.

In Chapter 4 it was stated that the relationship between the response to selection (Re) and the selection differential (S) was a measure of the heritability; thus $h^2 = Re/S$. Since the response to selection was greatest in the downward direction, it might be expected that the heritability would actually be higher in this direction than in the selection for increased length; but this was not the case. The difference in selection response in the first 10 generations was due instead to the fact that the variation was larger in the downward selected lines, and this in turn resulted in a higher selection differential. The heritability of thorax length, calculated from the results of selection in the first ten generations, was estimated to be about 0·3 for both lines. In the lines selected for increased length the variation was almost constant during the course of the experiment. There was, on the other hand, an increased variability in the lines selected for decreased length.

SELECTION EXPERIMENTS WITH MICE

FALCONER has conducted extensive selection experiments with mice. In one of the experiments (FALCONER, 1955) selection was carried out for high and for low weight at 6 weeks of age. As in the experiments with the fruit flies, the selection had an immediate effect both in the positive and negative directions. The response to selection, however, was very different in the two lines; and the asymmetry remained even after taking into account the differences in the selection differentials. Falconer related the selection response to the pooled selection differentials in the different generations. The regression of selection response on the selection differential was much higher in the line selected for low weight than in the high-weight line. The effective heritability in the former line was estimated at 0·52 and in the latter at 0·18. Selection for low body-weight was therefore nearly three times as effective as selection for high weight.

The weight of the mice at 6 weeks was determined by the weight at 3 weeks (weaning) and the growth rate after weaning. The asymmetry between the two lines of selection was caused entirely by the different effect of selection on weaning weight, which decreased appreciably in the line selected for low 6-week weight but remained nearly constant in the line selected for high weight at 6 weeks. Selection for low body-weight therefore brought about simultaneous changes in the mothering ability. There was no direct selection for mothering ability since the selection was made within litters. The changes in mothering ability, therefore, afford an example of a *correlated selection response*.

SELECTION FOR SINGLE TRAITS IN THE FOWL

For nineteen years LERNER (1958) and his co-workers at the University of California selected for increased shank length in the mature fowl. In addition to studying the immediate effect of the selection on shank length, observations were also made on the correlated changes in vigour and body proportions. The males used for breeding were selected on the basis of the shank length of full and half-sisters, whereas the females were selected mainly on their own phenotype. The selection, or S-line, was compared with the control line (P) which was maintained without selection. After 12 generations another line (SS), was formed from the S-line. There was no further selection in the SS-line. The changes in shank length in the three lines are shown in Fig. 16:2. The selection resulted in a rapid increase in shank length in the first 7 generations but from the 7th to the 18th generation there was no consistent progress. There was, however, an overall increase up to the 16th generation; thereafter there was a sharp decrease. From the commencement of the SS-line there was a rapid reduction in shank length.

There was a high degree of inbreeding in the S- and SS-lines, which increased from 0·1 at the start of the experiment to nearly 0·6 in the 18th generation. The degree of inbreeding in the control line was held at a low level.

By the time the 12th generation was reached the hatchability of eggs in the S-line was only 57 per cent of that in the control line and there was a continued decrease with selection. When the SS-line was started, there was an immediate increase in hatchability even though the degree of inbreeding continued to increase. The selection for increased shank length had also a detrimental effect on the vigour of the birds. When selection ceased and the shank length decreased, there was an improvement in vigour.

The increase in shank length resulted in other pronounced changes in body conformation, the length of the spinal column increased and there was an alteration in the posture of the birds (Fig. 16:3).

It is obvious that the selection for increased shank length caused a disturbance of the harmonic gene balance that existed in the population before the selection started. The slower progress after the 7th generation was partly explained by the

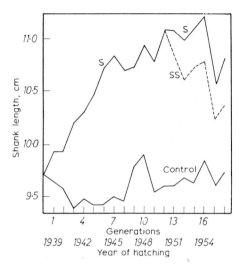

Fig. 16:2. (Left) The result of 18 years of selection for increased shank length in the fowl. (See text.) (LERNER 1958)

Fig. 16:3. (Below) Representative females from Lerner's selection experiment for shank length in White Leghorn.
Left: Hen from the control line (no selection). *Right*: Hen from the selection line (12 generations of selection for long shank). (From LERNER 1954: *Genetic Homeostasis*. Oliver and Boyd, Edinburgh)

reduced viability of birds with long shanks. During the first few generations the same number of sexually mature progeny were obtained from the birds selected for breeding, irrespective of their shank length. The *expected* and *effective selection differentials* were therefore approximately the same. After 11-15 years of selection, the effective selection differential was only about 60 per cent of the expected. Since mortality increased, a larger proportion of the line had to be used for breeding in order to maintain the lines constant in number. For these two reasons the effective selection differential decreased from 0·38 cm in the 2nd to 5th generations to 0·20 in the 11th to 15th generations. It was possible to explain about two-thirds of the decline in progress per year after the 7th generation by the change in the effective selection differential.

Reproductive fitness is a composite trait determined in the fowl mainly by egg production, fertility, hatchability and adult viability. NORDSKOG and GIESBRECHT (1964) at Iowa State University studied the influence of selection for specific metric traits on reproductive fitness. Five lines from a common White Leghorn base population have been selected for six generations. Line A, which consisted of 16 sires and approximately 200 dams per generation, was selected for high egg production, lines B and C for high and low body-weight respectively, while lines D and E were selected for high and low egg-weight. Each of these latter lines was made up of 8 sires and approximately 100 dams. The selection in line A did not produce any consistent change in egg production nor in any of the other fitness components. Body-weight increased by 54 per cent in line B, while fitness declined steadily by 47 per cent. In line C, body-weight decrease by 26 per cent and fitness by 16 per cent. Egg-size increased by 15 per cent in line D while fitness decreased by 39 per cent. Finally, in line E egg-size decreased by 14 per cent and fitness by 16 per cent. The selection for upward or downward changes in body-weight and egg-size was thus very effective. The changes, however, were accompanied by substantial declines in reproductive fitness. The decline in fitness was more severe with upward than downward selection.

Egg-size and body-weight were also genetically correlated so that a change in one direction for one of the traits was accompanied by a change in the same direction in the other character. From the correlated response during four generations of selection the genetic correlation between body-weight and egg-size was estimated at 0·56.

The decrease in fitness may be either a direct result of the phenotypic deviations in the metric trait as such, or it may be due to disturbances in the genetic balance. Nordskog and Giesbrecht favoured the latter explanation. In fact it seems quite natural that strong selection would tend to decrease general fitness in much the same way as inbreeding. Strong selection for a single trait, e.g. body-weight, will make the animals more homozygous for all loci influencing body-weight. At the same time adjacent loci on the chromosomes involved will also tend to become homozygous and the balance of the genotype will be disturbed, with a consequent decline in fitness.

Reproductive disturbances are a great problem in several species of farm animals, particularly the American turkey breeds. In a recent paper, Nordskog interpreted the low reproductive fitness in the turkey as due to decreased genetic buffering, the decrease resulting from intense selection for body conformation and growth rate. The decline in reproductive fitness started in the early 1940s concurrently with a very rapid improvement in the production traits, and the subsequent intense selection for these has been accompanied by a further decrease in fertility, hatchability and egg production so that a critical level is now being approached. Nordskog summarised these changes in a diagram (Fig. 16:4) which probably has much relevance to all species under intense

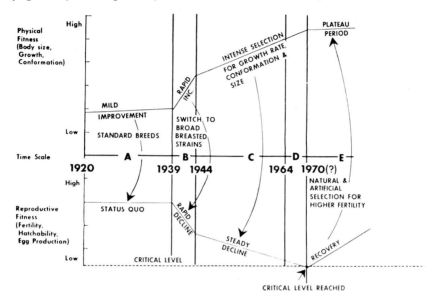

Fig. 16:4. A schematic picture of how past and future breeding improvement in physical fitness (body size, growth and conformation) may be related to reproductive fitness (fertility, hatchability and egg production). (NORDSKOG 1964)

selection. It was forecast that a critical level of reproduction would be reached by 1970, when attention would have to be focused on the reproductive traits, the increase of growth rate and conformation being subsequently temporarily halted for a time.

THE SELECTION PLATEAU

The experiments referred to in *Drosophila*, mice and poultry have all shown that progress under intense selection for one single trait—though very effective in the beginning—will ultimately come to a halt. The causes of this selection plateau vary. The genetic balance may be upset by continuous selection with a resulting decline in viability and reproductive fitness and an equilibrium between natural and artificial selection will develop. However the total variation at the plateau does not necessarily decrease. Part of this variation is no doubt genetic, as is evidenced by the fact that reversed selection at the plateau may bring the line back towards the origin.

Populations at a selection plateau have been studied particularly in poultry. DICKERSON (1965) reported on a closed commercial Leghorn flock which had been selected for increased egg production for more than twenty years without any consistent positive result. A repeat mating procedure was developed, by which birds from different generations could be compared at the same age within the same year. The procedure also made it possible to compare groups of chickens with the same average genotype in successive years so that the environmental trend could be determined. This procedure showed that the genetic merit did not increase as a result of the continuous selection in the flock at the plateau. However the progeny from unselected parents did not lay as many eggs as progeny from selected parents. The heritability for the individual traits determining egg production was fairly high. The same results were obtained three years in succession; so it is unlikely that the differences were due to experimental error. Continuous selection was needed in order to maintain the genetic merit. Suspended selection led to a drop in production.

Dickerson attributed to gene interaction the positive regression of progeny on parent. Under random mating the favourable epistatic combinations were assumed to be broken up, and as a result egg production declined.

Additional evidence for the importance of gene interaction in strains approaching a plateau was obtained by Dickerson in an experiment in which the egg production of three-way crosses was compared with that in the constituent two-way crosses. The following comparisons were made between crosses involving four strains

$$A\male \times (B.C.)\female \text{ vs. } (A\male \times B\female) \text{ and } (A\male \times C\female)$$
$$A\male \times (D.B.)\female \text{ vs. } (A\male \times D\female) \text{ and } (A\male \times B\female)$$
$$A\male \times (D.C.)\female \text{ vs. } (A\male \times D\female) \text{ and } (A\male \times C\female)$$

The single cross and pure strain females used for the matings were produced concurrently from randomly chosen parents of the three female strains. The total egg production was about 4 per cent poorer in the three-way than in the two-way crosses. Dickerson concluded, therefore, that the genotypes of strain B, C and D were each organised to complement A in a somewhat different manner, and that the gametes from BC, DB and DC female parents included cross-overs and recombinations of chromosomes from different strains which did not complement A gametes, as well as those produced by the B, C and D parents.

The basic quantitative selection theory rests on the assumption that the response in each generation builds upon the gains in the preceding generation. When much of the genetic variance is due to gene interaction, however, this assumption is not true. GRIFFING (1962) has demonstrated that on relaxation from selection the cumulative epistatic response to selection disappears, leaving as a residual the genetic gains predicted on the basis of independent non-interacting loci. The rate of disappearance of these epistatic contributions is a function of the intensity of the linkage between the loci involved.

CONCLUSIONS FROM THE SELECTION EXPERIMENTS WITH DROSOPHILA, MICE AND FOWLS

1. Traits which show additive genetic variation can be altered rapidly by selection.

2. The selection response is greatest in the first generations, declines with continued selection, and finally ceases. The limit for further selection response has then been reached.

3. When selection is made in opposite directions, the difference between the means of the two lines at the selection limits considerably exceeds the standard deviation of the original population. In the experiments with *Drosophila* and mice, the difference between the means of the two lines selected in opposite directions was ten times greater than the phenotypic and twenty times greater then the genetic standard deviations of the original populations.

4. There is often a difference in the response per generation between lines selected in opposite directions and also in the accumulated selection response at the selection limits.

The asymmetry between upward and downward selected lines can often be attributed to differences in the effective selection differential due to differences in viability and fertility of the lines. Furthermore, the variation, and therefore the possibilities for selection, may be larger in one line than the other. In long-term selection experiments asymmetry is to be expected even after adjusting the selection response for the effective selection differential. This is because the selected lines are often very small populations and consequently particular rare genes may be present in one line but not in the other. Other genes may have very different frequencies in the lines. These genetic differences need not necessarily influence the mean value of the trait at the start of the selection, but may have considerable importance later. Selection which favours heterozygotes and maternal influence can also cause asymmetry in the response to selection.

The reasons why the response to selection declines and finally ceases after a number of generations have already been discussed. To some extent, the reasons are the same as for asymmetry in the response to selection in different directions. Part of the remaining genetic variation may be additive, but the selection response is then retarded because of a decline in the viability and fertility of individuals with extreme phenotypic values. Further, it is often found that the residual genetic variation is non-additive. It was stated earlier that the selection limits were by no means definite. If the lines are allowed to undergo a period of random mating, new gene combinations may be created by crossing-over with the result that selection can be effective again. Another possible way to break the deadlock is to cross two lines with each other.

Before proceeding to discuss selection in farm animals it is necessary to point out a fundamental difference between selection in practice and selection in laboratory experiments. Such experiments are usually carried out with closed lines and, apart from mutations, the responses to selection are determined by the genetic material present in the lines at the beginning. In farm animal breeding, on the other hand, new genetic material may be introduced by matings with unrelated animals. The long-term possibilities of changing the population by selection are therefore better than indicated by the selection experiments with laboratory animals.

SELECTION RESPONSE ON DIFFERENT PLANES OF NUTRITION

In practice, breeding males are often selected on the basis of results obtained in other, usually better, environmental conditions than those in which their progeny will produce. The effect of this on the response to selection has been illustrated by experiments with mice by FALCONER (1960) and also in experiments with pigs and fowls.

Falconer studied the response to selection of the growth of mice from 3 to 6 weeks of age when on high or low planes of nutrition. The original population was split into two groups of which one was fed generously and the other on a low plane of nutrition from 3 to 6 weeks of age. Each group was divided into two lines; in one line selection was made for high, and in the other for low, rate of growth. From 6 weeks of age all the mice kept for breeding were fed on a high plane. All selection was made from the first litter of the dams. The progeny from the second litter were used to study correlated selection results when the progeny were fed on the opposite plane of nutrition to that on which their parents had been selected. The lines on the high plane were denoted by H+ or H− depending on whether selection was made for high or low growth rate. In a similar way the symbols L+ and L− were used for the lines on the low plane of nutrition. The offspring in the second litter of H-line dams were thus fed on a low plane, whereas the second litter offspring from the L-line dams were fed on a high plane. Growth on the two planes of nutrition was regarded as two separate traits.

The results of the experiment are summarised in Fig. 16:5. The solid lines show the response to direct selection, e.g. the effect of selection for high growth rate on a low plane of nutrition (L+). The dotted lines show the indirect or correlated selection response on the other plane of nutrition, e.g. growth changes on the high plane of nutrition of animals from the L+ lines.

In the first generations direct selection was more effective for increased growth on the high plane. The difference between the two lines gradually diminished, and in the long term the increase in weight on the high plane was the same irrespective of whether the selection was based on the growth of animals on the high or low plane of nutrition. Direct selection for increased growth on the low plane resulted in only small changes, whereas indirect selection gave no response.

On the high plane direct selection for reduced growth was much more effective than indirect selection. No difference could be demonstrated between the two selection methods for reducing growth on the low planes of nutrition.

A study of the chemical composition of mice from the H+ and L+ lines reared on the high plane of nutrition showed that the increase in growth was due to different factors in the two lines. The H+ mice were consistently shorter and fatter than the L+ animals whose body tissues in addition had a higher water content. Selection for high body-weight on the two planes of nutrition was therefore operating on slightly different gene systems which led to differences in the nature of growth.

From comparisons between the direct and correlated selection results, Falconer calculated the genetic correlation between the two traits, growth on high and growth on low planes of nutrition. (See Chapter 4, p. 134 concerning the relationship between the genetic correlation and the correlated selection response.) Four different estimates of the genetic correlation could be made. When the calculation was based on the results from the first four generations, a genetic correlation of 0·65 between the two traits was obtained. The different estimates were in

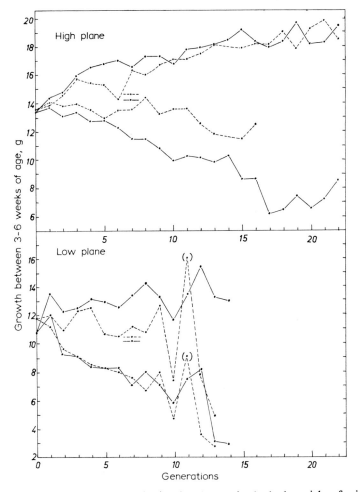

Fig. 16:5. The response to selection for changes in the body weight of mice on high and low planes of nutrition. (Falconer 1960)
——— The response to direct selection; – – – – the result of correlated selection. The horizontal lines in generation 7 are the unselected controls. (See text.)

good agreement with each other. Selection continued for as many as twenty-two generations in some of the lines. When all the generations were included, widely differing estimates of the genetic correlation were obtained in the lines, nor did the response to selection agree with that theoretically expected. Probably the heritabilities and the genetic correlations altered during the course of the experiment.

The experiment referred to above was, in respect of the selection for increased growth, a replication of an earlier experiment by FALCONER and LATYSZEWESKI (1952). Both experiments yielded essentially the same result; which gives them increased importance.

FOWLER and ENSMINGER (1961), U.S.A., carried out a selection experiment with pigs, the design of which was very similar to that of Falconer's mice

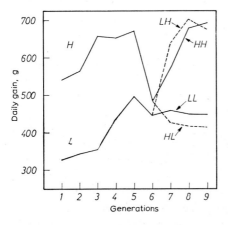

Fig. 16:6. The result of selection for increased daily gain in pigs on high and low planes of nutrition. After the sixth generation the lines were split so that some animals from the line on the high plane were reared on the low plane and vice versa. (See text.) (FOWLER and ENSMINGER 1961)

experiment. Selection was made for increased growth rate on high and low planes of nutrition. After six generations the response to selection of pigs from each line reared on the alternative planes of nutrition was studied. From a population formed by crossing the Danish Landrace and Chester White breeds, two genetically similar groups of about 60 pigs each were formed. One group was allotted to a high plane of nutrition (line H), the other group (line L) was fed at 70 per cent of line H. The three best gilts and the best boar of each litter were used for breeding. In addition to growth rate, attention was paid to the number of pigs alive at weaning. The matings were planned to keep inbreeding at a low level. After six generations some pigs from the H line were reared on the low plane (line HL), and some pigs from the L line formed a new line (LH) now reared on the high plane. It was thereby possible to estimate the correlated response of the selection practised on the one plane by measuring the growth of the pigs from this line when they were reared on the other plane. The experiment continued for nine generations and included a total of 1705 pigs. The results of the experiment are presented in Fig. 16:6.

In the first generation the daily growth rate was 540 g on the high plane and 327 on the low plane. In the ninth generation the corresponding figures were

694 and 449 g. The drop in growth rate in the sixth generation was due to disease, and for reasons unknown the pigs on the high plane were affected most. From a comparison of the total selection response and the accumulated selection differential during the nine generations, the heritability for growth rate was estimated at 0·52 for the high plane and 0·49 for the low plane. The corresponding heritabilities, calculated from the regression of progeny of dams within boars, were 0·41 and 0·54 respectively. The daily growth rate on the high plane could be increased equally well by selection on the high and on the low plane of nutrition. On the other hand, selection for high rate of growth on the high plane gave a poor response once the animals were reared on the low plane of nutrition. The genetic correlation between the growth rate on the two planes of nutrition was estimated at 0·7. Partially different gene systems, therefore, were causing rapid growth rate in the two cases. The selected pigs on a low plane of nutrition had a better feed conversion than those on the H line. Pigs in the HH and LH lines had approximately the same daily gain, but here again the pigs selected on the low plane (LH) required less feed per kg growth.

Like Falconer's mice experiments, this experiment indicates that the best response to selection is obtained when it is carried out under relatively unfavourable conditions, or under conditions which are comparable with those in which the animal may be expected to produce.

In most West European countries, selection in pig breeding is based on the results obtained at special testing stations which, in general, apply standards of feeding and management far above those used in commercial practice As was mentioned previously, the testing stations in Denmark use individual feeding, and this gives the pigs a decidedly more sheltered existence than pigs in commercial practice where there is competition at the feed trough. It is possible, therefore, that the results at the testing stations give the majority of pig breeders a poorer guide to the selection of breeding animals than they would if the environment at the stations was more in keeping with conditions in commercial practice. For a number of reasons, however, it is difficult to recommend a more extensive husbandry at the stations, since they are required also to serve as an example of what can be achieved with the available breeding stock when the feeding and management is properly applied. Nevertheless, care should be taken to see that the environment at the stations does not deviate too far from practical conditions.

Another example of the necessity of relating the environment in which selection is carried out to the environment in which the progeny will live is provided by the experiment referred to in Chapter 8, in which the resistance of the fowl to various diseases was increased by selection. HUTT (1958) and his co-workers succeeded in producing a poultry strain with good production capacity which also had such a high general resistance that in this respect it was superior to all other strains tested in the official random sample test in the State of New York (cf. p. 233). Undoubtedly, one reason for this good result was that the birds in the selected strain were continually exposed to different forms of infection, i.e. they were selected in an environment unfavourable to good health.

SELECTION FOR PRODUCTION TRAITS IN FARM ANIMALS

For hundreds of years man has sought to improve the production traits of farm animals by selection. Certainly, a great deal of progress has been made. Various types of stock have acclimatised themselves to the changes which domestication imposed. In some cases the adaptation has gone so far that the animals could no longer survive without the aid of man. One example is the broodiness of Leghorn hens which is so poor that the breed could hardly survive without incubators and artificial brooders.

In spite of the continuous selection which has taken place there is still a considerable reservoir of genetic variation for the majority of production traits. Reference can be made to Chapters 8-13 for the heritability of these traits. It is only in the past hundred years that reliable information about the production capacity of farm animals has been collected. This information shows that, even in this relatively short period of time, pronounced improvements have taken place over the whole series of production traits.

In Chapter 11 the changes which have been brought about in the production traits of the Danish Landrace pigs during the past sixty years were dealt with (Fig. 11:8 and Table 11:2). Further examples of similar changes will be presented in Chapter 18. It is obvious, however, that these changes cannot be attributed solely to genetic causes. Simultaneously with selection there has been improvement in nutrition; if the environmental factors had not been improved, it is probable that the genetic improvements could not have found expression.

The selection intensity and generation interval

In Chapter 4 (p. 133) it was shown that the selection response, Re depended upon the heritability of the trait and the selection differential, S. It was also stated that for practical reasons it was often more convenient to express the selection differential in standard deviation units (the selection intensity $= i$); $Re = h^2 i \sigma_p$; where $\sigma_p =$ the phenotypic standard deviation of a trait showing normal distribution. If the proportion of animals which are to be used for breeding is known, the selection intensity can easily be calculated from tables of the ordinates and area of the normal curve; $i = z/v$, where z represents the height of the ordinate of the normal curve where the group of animals selected for breeding are separated and v is the selected fraction of the population (cf. Fig. 16:7). This assumes that all animals which exceed a particular production level are used for breeding. Table 16:1 shows the selection intensity when varying fractions of the progeny are selected for breeding.

When 40 per cent of the animals are used for breeding and selection is based on a single trait, the mean value of these animals will exceed the breed or group mean by approximately one standard deviation, i.e. with respect of milk yield, by about 750 kg. If it is only necessary to save 5 per cent of the animals for breeding the selection intensity will be doubled. The selection response is therefore highly dependent on the number (proportion) of animals which are required for breeding.

A low heritability reduces the selection response; it is more difficult to judge

Table 16:1 Changes in the selection intensity when different fractions of the population are used for breeding. Normal distribution is assumed

Fraction (v) used for breeding	Distance in σ units between the mean and the value on the X-axis which separates the group used for breeding	Height of the ordinate (z) where fraction (v) is truncated	Selection in σ units: intensity $i = z/v$
0·90	−1·28	0·1758	0·195
0·80	−0·84	0·2803	0·350
0·70	−0·52	0·3485	0·498
0·60	−0·25	0·3867	0·645
0·50	0·00	0·3989	0·798
0·40	+0·25	0·3867	0·967
0·30	+0·52	0·3485	1·162
0·20	+0·84	0·2803	1·402
0·10	+1·28	0·1758	1·758
0·05	+1·65	0·1023	2·050
0·01	+2·33	0·0264	2·640

the genotype on the basis of the phenotype and more mistakes are made in selecting animals for breeding. From an economic point of view it is not the selection response per generation but the result per year which is important. In Chapter 4 the concept of *generation interval* was briefly mentioned. This is the average time interval between the birth of parents and their progeny. The selection response per year (Re_y) is obtained by dividing the response per generation by the generation interval, thus, $Re_y = \dfrac{h^2 S}{y} = \dfrac{h^2 i \sigma_P}{y}$, where $y =$ the generation interval in years. The length of the generation interval for the larger farm animals is shown in Table 16:2. The figures are based on Swedish data, but the intervals ought to be very much the same in other West European countries. In modern poultry breeding the generation interval is a little over one year.

Table 16:2 Length of the generation interval in years for herdbook registered animals of different species of farm animals in Sweden (JOHANSSON, 1949)

Species	Sire-progeny years	Dam-progeny years	Average years
Horses	9·5	8·9	9·3
Cattle			
Swedish Friesian	4·6	5·5	5·1
Swedish Red and White	4·8	6·1	5·4
Swedish Polled	4·3	6·9	5·6
Sheep			
Landrace	3·2	4·1	3·7
Pigs			
Yorkshire	2·7	3·3	3·0
Landrace	2·2	2·8	2·5

With the exception of horses, the generation interval is consistently higher between dam and progeny than between sire and progeny. In cattle, 50–70 per cent of the female progeny must be saved for breeding in order to keep the numerical size of the breed constant. In the fowl and pigs, only 10–20 per cent of the females need be saved for breeding. The selection intensity among females of horses, cattle and sheep is consequently low. Intensity of selection and generation interval are highly interdependent; changes in one will be accompanied by changes in the other. In cattle, the generation interval between dam and female progeny is approximately six years. Even so, there is not much to be gained by shortening the generation interval. With a constant population size, selection intensity will automatically be lowered and the net effect on selection response per year will therefore be negligible.

It is not possible to give any reasonably accurate figures for the proportion of male animals required for breeding. This depends entirely on whether natural mating or artificial insemination is used; if the latter, it depends how intensively the individual males are used. It is clear, however, that selection intensity can be much higher in males than in females in all the classes of farm animals. Where males are concerned, the selection intensity and generation interval are not highly interdependent. With artificial insemination, only a small percentage of the bull calves need to be reserved for breeding. The generation interval of progeny-tested sire-son can be reduced to a minimum if the A.I. bulls are mated to a number of cows, selected as dams of bull calves, about 2 years before the progeny-tests of the bulls are available. If the progeny-test is good, the next generation of young bulls can then be used for breeding shortly after the progeny test of their sires is available. If the progeny-test of the bulls is poor, their sons can be slaughtered before they are used for breeding. In this way, the generation interval of progeny-tested sire-son can be reduced to 4·5–5 years.

SELECTION METHODS

The breeding merit of an individual animal or group of animals is seldom determined by a single trait, but more often on the basis of several traits simultaneously. Pig-breeding provides good examples of this. Economic production of pigmeat is affected not only by the growth rate, feed conversion and carcass quality but also by the litter size, the mothering ability of the sows and the resistance of the pigs to various diseases. With the aid of the information obtained from the individual itself and/or its relatives (ancestors, collaterals or progeny), it is possible, in the manner described earlier, to estimate the breeding value of each individual trait or the total economic value (cf. Chapter 15, pp. 395 and 419).

Selection aims at improving the animals' total value and not just one trait. It can be carried out according to any one of three main methods (LUSH, 1945):

1. *Tandem selection*, in which selection is made for one trait at a time until this is improved, then for a second trait and later for a third. Selection may thereafter be switched back to the first trait, then to the second, etc., until finally each reaches a desired level.

2. *Independent culling levels*, where a minimum phenotypic value is set up for each trait and individuals which do not achieve this minimum value are culled, whether they reach the value required for other traits or not.

3. *Selection index*, constructed in the manner described in Chapter 15 (p. 420),

where the various traits are weighted so that selection on the basis of the index gives the best possible economic result.

In the tandem method the distribution curve is truncated at a particular value and all animals above this value are saved for breeding (Fig. 16:7). Let us assume that in a breed of pigs the selection is based entirely on back-fat thickness —measured by ultrasonics—at 90 kg liveweight. It is further assumed that the

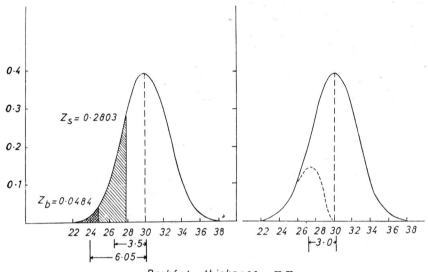

Fig. 16:7. Selection for decrease in backfat thickness of the pig.
Left: Selection by truncation of the normal distribution. No pigs with a backfat thickness above a certain value will be used for breeding; 20% of the sows and 2% of the young boars must be kept. *Right*: selection is made simultaneously for several traits; 20% of the sows will still be saved for breeding, but their distribution with regard to backfat thickness will approach the normal distribution. (See text.)
The figures on the Y-axis (0·1, 0·2 etc.) refer to the proportions of the total population.

pigs have been fed and housed under standardised conditions, that the average back-fat thickness is 30 mm with a standard deviation of 2·5 mm, and that the heritability of the character is 0·5. If 20 per cent of the females are to be saved for breeding the selection intensity will be $-1·402$ (Table 16:1) and the average back-fat thickness of the selected gilts will be less than the population average by $1·402 \times 2·5 = 3·50$ mm. The boars can be selected much more intensively. If it is assumed that 2 per cent are to be used for breeding, then the selection differential will be $-2·42 \times 2·5 = -6·05$ mm. The mean value of the selection differentials will be $-\frac{1}{2}(3·50 + 6·05) = -4·775$ mm. The selection response that can be theoretically expected is thus $-0·5 \times 4·775 = -2·39$ mm, since the heritability for back-fat thickness is 0·5.

Now let us assume that in addition to back-fat thickness attention is to be paid to growth rate and points for type, and that selection is to be based on independent culling levels or a selection index. Many of the animals which previously would have been culled on account of high back-fat thickness will now be retained on the basis of their other two traits. The distribution of the animals selected for breeding, classified on the basis of back-fat thickness, will not be truncated as before, but will approximate the normal curve (Fig. 16:7). On the other hand, the selection differential for back-fat thickness will be less than when selection was confined to this trait alone; how much less, will depend on the demands placed on the other two traits. In our hypothetical case we may assume that the female's differential is reduced from 3·5 to 3·0 mm. It is quite clear that the more traits to be considered, the less will be the selection differential for any one trait.

LUSH, and recently also YOUNG (1961), have discussed in detail the relative response under these three methods. Tandem selection is by far the least efficient. If selection is made for n independent and equally important traits with the same heritability and variance, then the average response per generation for each trait, when the tandem method is used, will be only one-nth of the response obtained when selection is for only one trait. On the above premise, the selection index method is $\sqrt{n}$ as efficient as the tandem method.

When selection is based on independent culling levels, the selection intensity for a single trait is reduced as the number of traits to be considered increases. Let us assume that selection is made for two uncorrelated traits, A and B, and that 10 per cent of the animals are required for breeding. If selected only for A, the best tenth of the A animals will be kept. The same will apply to B. However, only 1 per cent of the individuals can be expected to be among the best 10 per cent for both A and B. If equal attention is paid to both traits, the breeding animals will have to be taken from the 32 per cent which have the highest A values and the 32 per cent with the best B values. The selection intensity against the individual traits will thus be proportional to the factor $\sqrt[n]{v}$, where n is the number of traits and v the fraction which must be saved for breeding. Selection based on independent culling levels is more efficient than the tandem method, but less efficient than the selection index. In the latter case, unusually high merit in one trait is allowed to compensate for slight inferiority in others.

The basic reason for the superiority of the selection index method over the method of independent culling levels is illustrated in Fig. 16:8. It assumes that a fraction v of the animals are saved for breeding and that selection is made for two uncorrelated and equally important traits X and Y. When independent culling levels are applied, the selection intensity for X is as if the fraction $\sqrt{v}$ of the population were retained for breeding (rectangle ECDG) and similarly the rectangle HBCI represents the selection intensity when selection is based on Y. However, only the fraction v, included in the area ECIF, fulfil both the stipulated requirements simultaneously. The total merit of many of the animals saved for breeding and lying close to the point F is lower than that of many of the animals included in the rectangles BEFH and GFID. If a selection index, reflecting the total economic merit of the animals and at the same time observing that a fraction v of the animals are to be saved for breeding, is calculated, all animals above the line KL will be selected. All individuals in the areas KEM and NIL have higher total merit than animals in the triangle FMN. Selection on the basis of an index that considers the total merit of the individual, is thus more efficient than selection based on independent culling levels. If the two traits have different economic importance such that a unit change in X is worth

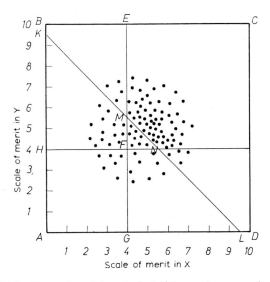

Fig. 16:8. Illustration of the principal difference between selection based on independent culling levels and a selection index. (See text.) (Modified from LUSH 1945)

more than a corresponding change in Y, the line KL will be steeper, i.e. the number in the area NIL will be larger than that in KEM.

YOUNG (1961) extended the comparison of three selection methods to cover cases where the traits had unequal variances, heritabilities and economic values. No details will be given here, but the reader is referred to the original paper. Factors such as the selection intensity, the number of traits under selection and their relative importance (i.e. the product of the economic weight, heritability and phenotypic standard deviation) were found to influence the relative efficiency of the methods. Index selection is never less efficient than independent culling though in some cases it is no more efficient. Independent culling is never less, but in some cases no more, efficient than tandem selection. With an increasing number of traits the superiority of the index method increases, and its superiority is at a maximum when the traits are of equal importance. With increasingly intense selection, the superiority of index selection over independent culling decreases while its efficiency, as compared with that of the tandem method, remains unchanged. The outcome of the three methods is strongly influenced by the phenotypic correlation between the traits under selection. The relative efficiency of index selection is higher when the phenotypic correlation is low or negative.

The greatest difficulty in selecting for two or more traits at the same time is that the possibility that strong negative genetic correlations may occur. DICKERSON (1954) has shown that the heritability for a combination of n negatively correlated traits with the same phenotypic and genetic variance approaches zero as the mean genetic correlation between all possible pairs of characters approaches $-1/(n-1)$. Selection will then become ineffective. However negative genetic correlations between a few traits does not make it impossible for simultaneous progress to be made with them, irrespective of whether selection is based on an index or not. The simultaneous improvement in milk yield and butterfat content with several different breeds of cattle—traits which show negative genetic correlation may be mentioned as an example.

Individual and family selection

As was mentioned in the previous chapter, selection can be based on the individual's own performance or on the performance of various kinds of relatives. The more information that is available, the more accurate will be the assessment. The extent to which the increased accuracy of calculating the mean or the combined information from relatives and individual can assist the progress per year, depends mainly on the following three factors:

1. the heritability of the trait;
2. changes in the variation of selection criteria and thereby in the selection differential;
3. the effect on the generation interval.

If selection is based on the individual's mean production over n years, then the heritability of the mean will be $h_n^2 = nh^2/\{1+(n-1)r\}$, where r is the coefficient of repeatability, on the assumption that the heritability is the same for all the production results included in the mean. The selection response, however, is not proportional to the increased accuracy. The standard deviation of the mean is $\sigma \sqrt{\{1+(n-1)r\}/n}$ instead of σ_p and the progress per generation is therefore proportional to $\sqrt{n/\{1+(n-1)r\}}$ instead of $n/\{1+(n-1)r\}$. Fig. 16:9 shows the effect of calculation of the mean on the heritability and selection response per generation. Calculation of the mean is of most importance at low heritabilities.

Even if the selection of animals for breeding is less accurate on the basis of one production result than on several, there is the advantage of shortening the generation interval and, with individual selection at least, this may outweigh the disadvantages (cf. p. 398). There is, however, one disadvantage with early selection and that is that the persistency of the animal is not considered. Advantages and disadvantages must be weighed and, with regard to the various traits, different principles can be applied to male and females. For example, many heifers may be reared in order to make an early selection on the basis of the first lactation result, but in the selection of bull calves for rearing much weight is placed on the dams having several years of production and on the sires having been progeny-tested.

In selecting pigs and fowls for breeding, it is often possible to consider the individual's own performance and also the mean result of full or half sibs. In properly planned breeding, it is necessary to be aware of those factors which influence the relative response when selection is based on the individual's own and the family group's performance. The selection of males for sex-limited traits, such as milk yield and egg production, must be based on information from different types of female relatives, e.g. the progeny or groups of full or half sibs. It is of less interest to compare the response of this type of selection with the results that could have been obtained by individual selection if the traits had been

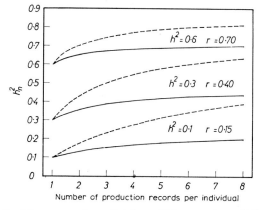

Fig. 16:9. Diagram to show, for three different values of h^2, how the heritability of the mean increases with the number of records from which the mean is calculated (dotted line) and the relative gain in selection response per generation when the reduced variation is taken into consideration (solid line), r = coefficient of repeatability.

directly measurable in the males. We shall therefore confine our comparison to cases where individual and family selection constitute two alternative possibilities and it is desired to determine which is most effective.

Let us assume that we have recorded the daily growth rate of five groups of gilts, each consisting of 4 litter mates and that the growth rate of the individuals and the groups were as shown in Fig. 16:10. If eight pigs are to be saved for breeding, then selection may be based either on the individual's own performance and/or the mean performance of the litter group. If selection were made between litters then all the breeding animals would come from litters 2 and 4, since they have the highest means. If selection on individual performance were practised, then the pigs H, P, O, L, T, G and K with either D or S would be selected. The two latter pigs had the same rate of growth, 665 g per day. The decision as to which of the two is to be selected depends on the importance given to the litter mean. If there is no environmental correlation between the litter mates, then S should be selected, since the higher mean value for litter 5 indicates that pig S has a better genotype than pig D. If the difference between litter means is largely due to litter 1 having had poorer environmental conditions than litter 5, then pig D should be selected instead of S. The reasoning in this case

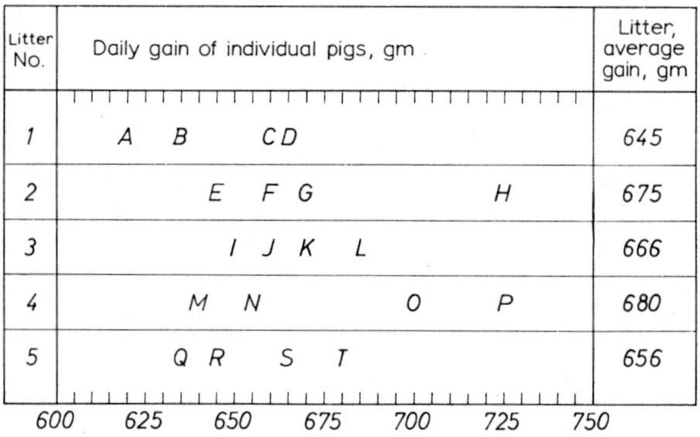

Litter No.
Daily gain of individual pigs, gm
Litter, average gain, gm
1
A B C D
645
2
E F G H
675
3
I J K L
666
4
M N O P
680
5
Q R S T
656
600 625 650 675 700 725 750

Fig. 16:10. The principles of individual selection and selection between full sib groups. (See text.)

is that pig D has performed rather well in spite of the relatively poor environmental conditions. If the environmental correlation is very high between members of the same family, it is obviously meaningless to select on the basis of the family mean. In such a case, selection should be made within each family group separately.

Relative response in relation to individual selection
2·0
1·9
1·8
1·7
1·6
1·5
1·4
1·3
1·2
1·1
1·0
4b
4a
3
2
1
0 0·05 0·10 0·20 0·30 0·40 0·50
h^2

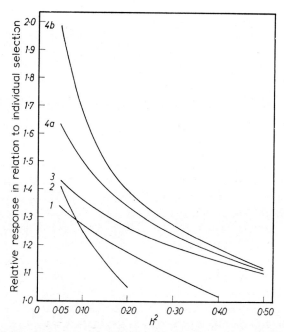

Fig. 16:11. Response with selection on (1) the average of full sib groups, (2) the average of half sib groups consisting of 4 full sib groups each, (3) individual performance and the full sib group mean (the individual is included in the mean of the full sib group), (4) performance and average for full and half sib groups (*a*) with 4 full sib groups per half sib group and (*b*) with 16 full sib groups per half sib group. In all cases the full sib group consists of 6 individuals. (Partly after OSBORNE 1957)

The relative efficiency of the different methods can be directly determined if the number of individuals per family, the relationship r_g between individuals in the same family and the phenotypic correlation t are known. If there is no environmental correlation then $t = r_g h^2$. In some cases it is not possible to ignore completely the environmental correlation between members of a family, e.g. litter mates. In practical breeding the environmental correlation is seldom so high that selection within families is more efficient than individual selection. In the following, therefore, within-family selection will not be considered.

As has been mentioned several times previously, the selection response with individual selection is $ih^2\sigma_p$, where i is the selection intensity and σ_p is the standard deviation of the phenotypic value. The selection response (Re_f) in selection between families, can be expressed in a similar way, $Re_f = ih^2{}_f\sigma_{pf}$. The standard deviation of the family means (σ_{pf}) will of course be less than the standard deviation of the individual values. Consequently, for selection between families to give the same, or a better, response than individual selection, the increase in the heritability must be large enough to outweigh the effect of the smaller standard deviation. Selection between families is most important for traits with low heritability, and when the genetic correlation (relationship) between members of the family is high (cf. Fig. 16:11).

When non-random environmental differences occur between full or half sib groups, the response to family selection is very much reduced. The phenotypic correlation between members of the same family is then no longer $r_g h^2$ but $r_g h^2 + c^2$. Suppose that selection is made for a trait with a heritability of 0·3, for example, the number of eggs laid in the first year by surviving hens. Further suppose that the environmental correlation between full sibs is 0·2, that the population consists of full sib groups of 10 individuals each and that the breeding males are selected from the best tenth of the full sib groups. The progress from between family selection of breeding females will then be only 85 per cent of the expected progress from selection on individual performance, as compared with 114 per cent if c^2 had been 0. It is therefore very important that the environment is randomised, in order to make selection between families efficient.

Selection is most efficient when attention is paid to the individual's own performance and also the average performance of the family to which the individual belongs. It is then necessary to construct an index, in the manner described earlier (Chapter 15, p. 405), in which the individual's own performance and the family mean are weighted so that the correlation between the index and the breeding value is at a maximum.

In poultry breeding, use is made of populations consisting of full sib groups which are in turn part of larger groups of half sibs with the same sire. In this case the mean production of the full and half sib groups can be considered in addition to the individual's own production. OSBORNE (1957) compared the effects of the following four methods of selection with the effect of individual selection:

1. Selection between full sib groups.
2. Selection between paternal half sib groups.

3. Index selection, taking into account the individual's own production and the mean of the full sib group, the individual's result being included in the mean.
4. Index selection, taking into account the individual's result and the mean result of the full and half sib groups.

The selection response with the different methods is illustrated graphically in Fig. 16:11. It is assumed that the number of paternal half sib groups is large. At low heritabilities the selection efficiency can be increased by considering the family mean. Selection between paternal half sib groups, consisting of four groups of full sibs each with six individuals, is more efficient than individual selection at heritabilities less than 0·2. Combined selection based on the individual's own performance and the means of the full and half sib group is always the most efficient; with low heritability and relatively large families, the superiority of this method is considerable. It should perhaps be stressed once more that, with regard to sex-limited traits, family selection on progeny-testing is the only way in which selection pressure can be put on males. For traits which can be measured in both sexes (e.g. growth rate of the pig) a move from individual selection to progeny-testing or family selection will mean either a considerable increase in the rate of inbreeding or a reduction in the selection intensity.

Selection methods for exploiting non-additive genetic variation

It has already been shown that, when continuous selection is applied to closed lines, the selection response gradually declines and finally ceases. In many cases, however, the lines continue to show genetic variation; but this may then be partly, even largely, of a non-additive nature. The breeding of the larger farm animals has seldom been taken so far that the non-additive genetic variation constitutes a decisive component of the variation of the production traits. In the fowl, there are several examples of continuous selection in a closed group causing a reduction in the additive genetic variation for egg production (cf. p. 430), i.e. a plateau value has been reached, where further selection is ineffective. For certain traits, such as viability and fertility, the non-additive genetic variation is probably very important in all species of farm animals. The inbreeding depression for both these traits has been large in all the investigations which have been carried out (cf. Chapter 14).

Several methods of exploiting non-additive genetic variation have been suggested, all of them based on a combination of selection and crossing. These methods have found their greatest application in maize breeding but they are also used in poultry and pigs. A general discussion of the methods was given in Chapter 14.

In crosses between inbred lines originating from the same base population, no gene combination can occur (apart from mutation effects) that could not have been produced in the original population. The line crosses may, however, bring to light some valuable gene combinations which can then be multiplied and mass-produced commercially. There are practical difficulties in finding such lines, and it is necessary to cross many lines with each other in order to discover which ones give rise to superior gene combinations. How similar or variable the progeny of a line cross are, depends on the degree of inbreeding of the lines. If the lines to be crossed with each other are 100 per cent inbred, there is no genetic variation in the F_1 generation. The more inbred the lines, the greater will be the portion of the variation between the line-crosses caused by non-additive genetic variation.

In line breeding and crossing, it is usual to distinguish between the *general* and *specific* combining ability of the lines. A measure of the former is the average of

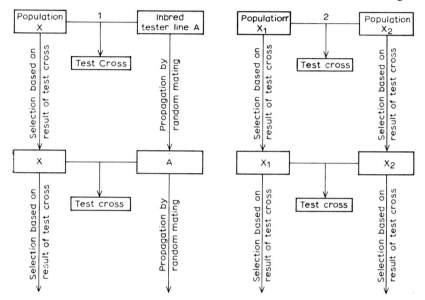

Fig. 16:12. (*1*) Crossing with recurrent selection to an inbred 'tester' line; (*2*) crossing with reciprocal recurrent selection. (See text.)

the progeny when a particular line is crossed with a number of other lines selected at random. The specific combining ability refers to the deviation of a cross from that expected from the general combining abilities of the parent lines. The general combining ability is caused mainly by additive gene effects, whereas the specific combining ability is due entirely to non-additive gene effects. Of the variance between line crosses the part which is caused by the lines' general combining ability is for the most part directly proportional to the degree of inbreeding of the lines. The variance component due to specific combining ability increases as the square, or higher powers, of the inbreeding coefficient. If the specific combining ability is to be utilised in breeding, it is necessary to work with relatively highly inbred lines.

Recurrent selection. According to this method, the value of the new lines is tested by crossing them with a relatively highly inbred 'tester' line, known to have good general combining ability. The 'tester' line is reproduced without selection, whereas in the tested line, individuals which have given the best result in crossing are used. The method has been used in the breeding of poultry (Fig. 16:12).

The difficulty is, however, to produce and maintain highly inbred lines, and attempts have been made to utilise less inbred populations as 'tester lines'. If such lines are allowed to mate at random, there is a risk of the genetic constitution changing from generation to generation. The result of recurrent selection entirely within one line will then be insignificant, and therefore resort has been made to other methods.

Theoretically, the method of *reciprocal recurrent selection* is of special interest. This implies progeny-testing each of two lines by crossing with the other line (Fig. 16:12). Those animals which give the best result in crossing are used for the multiplication of their own lines. The crossbred animals are not used for further breeding. The aim is gradually to increase the general and specific combining ability of both lines. The improvement in the specific combining ability is brought about by the selection altering the gene frequencies on both sides, so that crossing between the lines produces heterozygosity in the gene system, which by overdominance has a positive effect on the particular production trait. Each of the lines, which should not be inbred, should have rather different gene constitutions at the beginning; otherwise no great positive effect can be expected of the method. The selection cannot be expected to lead to any great improvement within the lines for those traits for which selection is conducted, since the selection is designed to make both lines homozygous for individual genes in loci which exhibit overdominance. On the other hand, the danger of a depression in the productivity within the lines should be less than with consequent inbreeding, since this increases homozygosity in all gene systems and not just in those which, by overdominance, influence the production traits.

Methods which make use of crossing with recurrent selection have one big disadvantage. The generation interval is extended since the crossbred animals are not utilised for breeding. It is, of course, possible to mate the males of the one line with the females of the other line and at the same time randomly with females of their own line. This requires tremendous resources of animals and accommodation. Even if the methods discussed here appear attractive from a theoretical point of view, experiments in the laboratory (with *Drosophila* and mice) have not produced convincing evidence that they have any advantage over other, more conventional, breeding methods.

Crosses between inbred lines. The results of the extensive inbreeding and crossing experiments which have been carried out in different parts of the world were reviewed in Chapter 14. Only in so far as they apply to pigs and poultry, will the breeding methods be briefly dealt with here. Usually intensive inbreeding is carried out (generally full sib mating) for a number of generations, until the coefficient of inbreeding is about 0·5. Crosses are then made between the lines, or with a particular line, which has shown good general combining ability. Selection is made all the time both within and between lines. Many lines 'succumb' due to lack of viability.

The production of commercial animals from the inbred lines usually implies crossing four different lines, whereby the negative effects of inbreeding on reproductive capacity are neutralised. Animals from lines A and B are crossed together as are animals from lines C and D. The commercial animals are then produced by the crossing $F_{AB} \times F_{CD}$. It should be mentioned that breeding based on the utilisation of the superior crossing results of a particular line is a gamble.

If a breeder has developed many lines, testing all four-way crosses may be virtually impossible. From n lines, $n(n-1)$ single cross combinations can be produced and the number of possible four-way crosses amounts to $n(n-1)(n-2)(n-3)$. With 10 lines, 90 single crosses and no less than 5040 four-way crosses are possible. It would therefore be an advantage if the production of the four-way crosses could be predicted from the results of the single crosses of the lines making up the four-way crosses. HILL and NORDSKOG (1958) studied this problem in crosses between four White Leghorn lines and four lines of heavy breeds. Viability and age at first egg in the four-way crosses were not highly predictable from the results of the single crosses Body-weight and broodiness were predicted quite accurately, while hen-day egg production was intermediate. Good predictability for the latter three characters in multiple cross performance was obtained with 20–30 birds in each single cross.

Strain breeding. Strain breeding was defined in Chapter 14 as a milder form of inbreeding than that practised in the formation of inbred lines. The method usually aims at producing commercial animals by crosses between the strains. Either a relatively large number of strains can be tested in crosses with each other, selection being made between and within the strains on the basis of the results of crossing, or a few selected strains are maintained, whereby the larger part of the commercial animals in the population are obtained by rotational crosses between the strains. The former method is no different in principle from crossing between inbred lines; the difference is only in the degree of inbreeding.

The efficiency of the other kind of strain breeding is highly dependent on the size of the strains. If the strains are numerically small and closed, there is automatically a certain amount of inbreeding depression, and apart from this the possibilities for selection are much less than with a large population. ROBERTSON and RENDEL (1950) calculated the maximum annual progress in the milk yield of cows in closed units of varying size. The units were assumed to be optimum with regard to the age of the cows and the distribution between progeny-tested bulls and young bulls on test. The annual progress with a population of 120 individuals was calculated to be 1·10 per cent of the herd mean. The corresponding figures for populations of 2000 and 10 000 were 1·69 and 2·05 respectively.

Justification of the latter kind of strain breeding requires that the increase in vitality and productivity, to be expected when the males from the various strains are used in rotation crossing, outweigh the disadvantages that may arise with inbreeding depression and reduced selection possibilities within the strains. A programme for strain breeding of cattle, making use of numerically large strains which are not entirely closed will be discussed in the next chapter.

Adaptation to local conditions. A related question in strain breeding is whether there is anything to be gained by dividing the breeds into strains which, by selection, are adapted to various local conditions within the territory inhabited by the breeds. Cattle and sheep are most sensitive to adaptation since their production is based mainly on pasture and roughage, whereas pigs are less affected in this respect. In recent years the question of local adaptation has arisen in poultry husbandry, since the large poultry breeders now distribute commercial birds from the same lines over a large area with widely differing conditions.

As far as sheep are concerned, breed formation has for a long time been aimed at adaptation to local conditions. In the more fertile agricultural regions of Great Britain there are the lowland breeds, e.g. South Down, whereas the more hardy breeds, e.g. Cheviot, predominate in the hilly and less fertile areas of England and Scotland; and finally, in the mountainous areas of Scotland are even hardier breeds, such as the Scottish Blackface. In cattle too, there has been a tendency for breeds to be adapted to local conditions but not to the same extent as with sheep. Friesian cattle are found in areas with widely differing conditions of pasture and fodder production, and a valid question is how much progress would be achieved by dividing the breed into locally adapted strains. The answer depends upon whether there is a non-linear interaction between genetic and environmental factors for production. In Chapters 4 and 5 (cf. pp. 117 and 149) it was stated that almost certainly such interactions occur when the environmental differences are very large. In the tropics, for example, Zebu cattle may have a higher milk yield than the European breeds, which are unadapted to tropical conditions.

With the environmental variation which occurs within the normal distribution area of a cattle breed, however, the non-linear interaction between genetic and environmental factors seems to be of minor importance, at least with regard to milk production. A study by ROBERTSON, O'CONNOR and EDWARDS (1960) will be mentioned as an example. They divided the progeny records of 76 A.I. bulls into three groups (low, medium and high) according to the yield levels of the herds in which the heifers produced their first lactation. Independent progeny-tests (contemporary comparisons, cf. p. 412) were carried out at each of the three yield levels. The correlation between the true breeding values of the bulls at the different levels was very close to unity. It was concluded that there is no need to provide special strains within breeds to suit particular management levels or to confine progeny-testing to the higher producing herds.

Clear evidence of the interaction between heredity and environment has been obtained in investigations with poultry. ABPLANALP and MENZI (1961) studied the egg production and mortality of pullets from different lines when they were placed in a number of different flocks in Switzerland. The experiment was in progress for four years and included 7 series. In each series a comparison was made between all participating groups in each locality. The interaction variance could therefore be estimated for egg production in the first year of surviving pullets. Interaction variance accounted for about one-third of the variance between lines. The interaction effect for mortality was quite considerable, approximately three times larger than the variance between lines. The environmental factors responsible for the changes in the ranking order or relative production capacity of the lines between localities could not be identified.

Environmental factors show continuous variation, and it is certainly impossible to produce lines or breeds which are superior in all the possible conditions. It is equally impossible to produce closed strains specially adapted for each individual environment in the continuous environmental spectrum. Division into numerically small strains means, as was mentioned before, a reduction of selection possibilities. It therefore remains to aim at producing animals which are suited to a rather wide range of environmental conditions. It may often be better to improve environmental factors, as nutrition and management, than to attempt to alter the genetic constitution of the animals. Another possibility is to utilise existing breeds already adapted to particular agricultural conditions in areas with similar conditions.

THE GENETIC CHANGE IN POPULATIONS OF FARM ANIMALS

During the past decades a considerable increase in production has taken place in most species of farm animals. Since *both* genetic merit and environmental factors are often improved simultaneously, it is difficult to distinguish one cause from the other. The response to selection per year with a constant environment has already been defined as Re_y, i.e. the genetic selection differential, h^2S, divided by the generation interval. In selection programmes with farm animals, Re_y will be confounded with parallel changes in environment. We therefore now discuss some methods by which the true genetic change may be measured.

In experiments with laboratory animals the effect of environmental variation between years can be eliminated by the use of control populations or two-way selection. Control populations have also been used in poultry selection. As was mentioned earlier, Dickerson and co-workers designed a repeat mating system for use in poultry selection, whereby birds of the same average genetic merit could be compared in successive years. The system made it possible to separate the genetic and environmental changes per year. The studies showed that the improvement in genetic merit may be much less than expected from the genetic selection differential.

For obvious reasons, two-way selection and control populations cannot be utilised in the estimation of genetic change in the larger species of farm animals. However, approaches similar to the repeat mating system may be used. In cattle, for instance, cows have several lactations over a consecutive period of years. It can be taken for granted that the genotype of a cow does not change with the years. The variation in production from year to year is therefore due to age effects and environmental changes within the herd. Provided that 'true' age corrections can be made, it should be possible to estimate the environmental trend and to separate the genetic from the environmental changes. A maximum-likelihood method for doing so has been developed by HENDERSON (1958). Several studies using this method have now been performed; but as most of them were limited to single herds, the results quite naturally have been variable, ranging from a negative genetic change of 0·7 per cent to an increase of 1·75 per cent of the average yield per year. ARAVE and co-workers (1964) estimated the genetic change over a period of thirty years in 12 Jersey herds taking part in a dairy herd improvement programme at the University of California. The data included 12 000 lactation records from 3900 cows. The estimates of the average annual genetic change in each herd were made by computing the linear regression of FCM yield on year of records adjusted for the yearly environmental effects. The annual genetic change for all herds was 74 lb FCM, or 0·7 per cent of the average of the herd yields. However, the variation in the genetic change between herds ranged from -51 lb to $+145$ lb. The age correction factors were found to differ considerably from herd to herd, and in some cases they also differed markedly from the standard age corrections used in the U.S.A. A weakness of the maximum-likelihood method is the difficulty of applying 'true' age corrections.

A different method for measuring genetic change has been applied by SMITH (1962) in studies of the genetic change in selection of pigs. The progeny of a particular sire will each have a random half of the sire's genes, while the other half will come from the dams. The change in the average phenotype of the progeny of the sire is therefore a measure of the environmental change and half the genetic change, provided that the dams are all approximately the same age when mated to the sire. The difference between the change in the population average and the change within sires will thus measure half the change in genetic merit in the period studied. The average genetic change over y years can be estimated as follows:

$$2[(\bar{x}_{p_y} - \bar{x}_{p_0}) - (\bar{x}_{s_y} - \bar{x}_{s_0})]/y$$

where $\bar{x}_p$ and $\bar{x}_s$ are the population and repeated sire means at year 0 and year y.

Smith applied this method to a set of pig records collected over nine years by a private breeder. Practically the whole of the phenotypic change in back-fat thickness and carcass length appeared to be genetic, while only a small fraction (if any) of the change in age at slaughter could be ascribed to genetic improvement. The results of another study by Smith, on data from the Danish progeny-testing stations, were less encouraging. In the years from 1952 to 1960 the average back-fat thickness of tested pigs was reduced from 34·2 mm to 28·5 mm, i.e. by 0·63 mm per year. The overall estimate of the genetic change was 0·15 mm annually—about 20 per cent of the observed change. Considering the high heritability of back-fat thickness, this is a surprisingly low figure.

The method of estimating genetic change from changes within sire families requires that all matings be unselected with regard to age and genetic merit. This requirement is probably not always fulfilled, and there is probably a tendency to use older and better females as mates of older sires with good progeny-test records. Such selection invalidates the estimate of genetic progress, and careful examination of the nature of the data is therefore necessary. Recent developments in the deep-freeze technique of storing semen, which enable the semen from one animal to be used over long periods, will probably facilitate investigations into the changes of genetic merit.

Selected References

Abplanalp, H. and Menzi, M. 1961. Genotype-environment interactions in poultry. *Br. Poult. Sci.*, **2**: 71–80.

Arave, C. W., Laben, R. C. and Mead, S. W. 1964. Measurement of genetic change in twelve California dairy herds. *J. Dairy Sci.*, **47**: 278–283.

Dickerson, G. E. 1965. Experimental evaluation of selection theory in poultry genetics today. *Proc. XI Int. Congr. Genetics*, **3**: 747–761.

Falconer, D. S. 1960. *Introduction to Quantitative Genetics* (356 pp.). Oliver and Boyd, Edinburgh.

——, 1960. Selection of mice for growth on high and low planes of nutrition. *Genetic Res.*, **1**: 91–113.

Fowler, S. H. and Ensminger. 1960. Interaction between genotype and plane of nutrition in selection for rate of gain in swine. *J. Anim. Sci.*, **19**: 434–449.

Lerner, I. M. 1958. *The Genetic Basis of Selection* (298 pp.). Wiley and Sons.

Lush, J. L. 1945. *Animal Breeding Plans* 3rd ed. (443 pp.). Iowa State University Press, Ames, Iowa.

Nordskog, A. W. and Giesbrecht, F. G. 1964. Regression in egg production in the domestic fowl when selection is relaxed. *Genetics*, **50**: 407–416.

Osborne, R. 1957. The use of sire and dam family averages in increasing the efficiency of selective breeding under a hierarchical breeding system. *Heredity*, **11**: 96–116.

Robertson, F. W. 1955. Selection response and the properties of genetic variation. *Cold Spring Harb. Symp. quant. Biol.*, **20**: 166–177.

Smith, C. 1962. Estimation of genetic change in farm livestock using field records. *Animal Prod.*, **4**: 239–251.

Young, S. S. Y. 1961. A further examination of the relative efficiency of three methods of selection for genetic gain under less restricted conditions. *Genetic Res.*, **2**: 106–121.

17 Some Special Problems in Breeding and Selection

Some special problems, which have not been covered in the previous chapters, will be dealt with here. These include breeding and selection methods with artificial insemination and the question of individual performance versus progeny-testing with regard to the nature of the product and the species of animal. There will be no further discussion of sheep.

HORSES

The special problems encountered in horse-breeding are associated with the fact that each individual leaves relatively few offspring and the generation interval is long. The most important traits of both draft and saddle horses (strength, speed, endurance and temperament) are manifested in both sexes; and this, combined with the length of the generation interval, means that selection based on the individual's own phenotype may be expected to give quicker results than selecting on the basis of the performance of the progeny.

What is needed first and foremost in horse-breeding is systematic measurement of the animal's utility characteristics. Competitions of training and performance have long been the mode in 'Blood Horse' breeding; this is especially true of the English Thoroughbred and the American Trotter. The speed of these two breeds in flat races, in steeplechases, and on the trotting tracks has been decisive in the selection of breeding animals. Some heritability estimates of the performance of Thoroughbred horses in flat races have been made, and a study by HINRICHSEN and BORMANN (1964) may be mentioned as an example. Data from about 30 000 races, in which 2000 horses, the progeny of 35 stallions, took part, were analysed. The performance was corrected for distance, age and turf conditions and the heritability, as estimated from half sib correlations, was found to be 0·173, 0·087 and 0·166 for 2-, 3- and 4-year olds respectively. The repeatability of a race performance was 0·2.

Various pulling tests for draft horses have been developed, using stationary or movable equipment, but systematic selection of breeding animals on the basis of such tests has been the exception rather than the rule. The breeders of heavy draft horses, e.g. the Belgian, Percheron and Clydesdale breeds, have been content with the information gained from visual appraisal, such as body proportions, movement and soundness of legs. In the 1920s, however, Professor TERHO in Finland developed not only a pulling-test device applied to a motor vehicle, but also a testing scheme, which, with minor modifications, has been used regularly as a prerequisite to the herdbook registration of the Finnish horse—a dual-purpose draft-trotting horse. The Finnish horse is tested for speed also when drawing a sulky and is measured and scored according to a special points system. VARO (1965) analysed the records of about 6000 female horses and, based on the intra-class correlation of paternal half sibs within age groups and districts, arrived at the following estimates of heritability.

	h^2
Pulling-power (the resistance of the load increased by 5% of the body-weight of the horse for each 50 metres), kg	0·26
Pulling-power as a percentage of the body-weight of the horse	0·14
Walking speed	0·41
Trotting speed	0·43
Points for movement	0·41
Points for temperament	0·23
Withers height, cm	0·26
Heart girth, cm	0·32
Hip width, cm	0·34

Since World War II the number of draft horses in north-western Europe and North America has decreased to such an extent that there would appear to be little need for pulling tests in the future. Concerning all types of horses, however, it may be assumed that suitable performance tests provide the best foundation for breeding improvement through selective breeding.

CATTLE

The following discussion deals mainly with dairy cattle under artificial insemination; but the same general principles apply also to beef cattle and with natural breeding practice.

It has been shown by several authors, originally by ROBERTSON and RENDEL (1950), that the possibility for genetic improvement of a population of dairy cattle increases, although at a diminishing rate, with increasing size of the breeding unit, mainly because the progeny-testing and selection of bulls can be organised much more efficiently than in a small unit. Individual herds are usually rather small, seldom exceeding 100–200 cows. SPECHT and MCGILLIARD (1960) conclude that in a herd of less than 100 cows, progeny-testing is less efficient for improvement of the genetic gain than the use of young sires selected on the basis of their dams' production. In such herds, the genetic gain would hardly exceed 1 per cent per year, whereas in a breeding unit of 10 000 cows, with efficient organisation of milk recording, progeny-testing and selection of bulls, it would be possible to reach a yearly gain of 2·5 per cent. SKJERVOLD (1965) has calculated that the possibilities for genetic progress are 50 per cent greater in a breeding unit of 20 000 milk-recorded cows than in a unit of 2000 cows. The gain in rate of genetic progress by a further increase in size of the breeding unit—above 20 000 milk-recorded cows—would be comparatively small. Skjervold estimated the percentage contribution to the total genetic gain that may be achieved by efficient selection of parents for the next generation of cows and bulls within two breeding units of different sizes to be as follows:

Parents chosen	Size of the cow population	
	2000 cows	400 000 cows
Sires of young bulls for test	44%	62%
Dams of bulls	26%	23%
Sires of cows	15%	10%
Dams of cows	15%	5%

The figures show that in an A.I. unit the greater part of the genetic progress comes from the breeding of new young bulls from proven sires and élite cows, and that the relative importance of the bull sires increases with increasing size of the breeding unit. It is assumed here that the owners of the herds select only dams of their own cows and that the rest of the assessment of breeding values and selection is centralised to the headquarters of the association. The following points will be taken up for further discussion:

1. Organisation of progeny-testing and selection among the tested bulls;
2. selection of dams of bulls;
3. selection and rearing of young bulls for testing;
4. selection among cows within herds.

Progeny-testing and selection among the tested bulls

It is very important that the bulls be progeny-tested at the earliest possible age and that an efficient selection be made among the tested bulls. The age of the bull at the completion of the test depends on (*a*) the age at which the young bull is capable of producing semen of satisfactory fertility and (*b*) the time taken to obtain a sufficient number of first lactation records. In an experiment conducted by the Milk Marketing Board (MMB) with four young Friesian bulls, it was found that semen collection could start when the bulls were about 9 months old. When two collections were taken every fourth day the quantity of semen per ejaculate was about half that of mature bulls. The quality of the semen was passable during the first 3–4 weeks of collection, but thereafter it improved so that the 30–60 days non-return rate became about normal. If the resultant daughters are inseminated to calve down at about two years of age, at least a preliminary progeny test for milk yield could possibly be made when the bull has reached $4\frac{1}{2}$ years of age. This may be difficult to accomplish in practice, but it should be quite possible to complete the inseminations for a progeny-test when the bulls are about 15 months old, so that a sufficient number of daughters calve down before $2\frac{1}{2}$ years of age and complete their 305-day first lactation ten months later. This means that the bulls can be progeny-tested when they are about $5\frac{1}{2}$ years old. A preliminary test on the milk yield of the daughters during the first 4 months after calving could be made six months earlier (cf. Chapter 15, p. 413).

Usually the test inseminations are spread more or less at random over the entire cow population. Difficulties are often encountered in securing satisfactory testing facilities, however, especially in regions where only a minor fraction of the cows are milk-recorded. In addition, breeders prefer to raise heifers from progeny-tested bulls, and they select heavily on milk yield before the first lactation is completed. In England, the Milk Marketing Board has therefore introduced a new scheme for progeny-test inseminations based on individual contracts between the Board and members with tested herds. The MMB undertakes to inseminate an agreed number of cows in each herd, free of charge, with semen from young bulls. For each available first lactation by a heifer resulting from the testing inseminations, the Board pays £5 to the owner of the herd. At least 12 cows must be offered from each contract herd. The intention is to breed each test-bull to about 300 recorded cows, so that 50–100 daughters with a first lactation record will be available for the progeny-test. When the required number of daughters are inseminated, the bull is withdrawn from service until an assessment can be made of the yield and conformation of the daughters. Planned matings in the contract herds will ensure that in each herd contemporary daughters of several other bulls are available for comparison.

The policy of handling the young bulls during the interval between the last test matings and the completion of the progeny-test varies between countries, and perhaps, also between regions in the same country. Sometimes the bulls 'in waiting' are used to a limited extent in A.I.; sometimes they are let out for natural service in private herds, and sometimes they are 'laid-off' completely. With the advance in deep-freezing it is now possible to collect semen regularly from these bulls and store it until the test is completed. A decision can then be made whether the stored semen should be used or not. This latter method would, of course, allow the most extensive use of superior bulls and is therefore preferable from a genetic point of view. It has also been suggested that as much semen as possible should be harvested as early as possible from the young bulls. When enough semen for about 50 000 inseminations has been collected and deep-frozen, the bulls could be slaughtered in order to reduce the cost of keeping them until the progeny-test is completed. Under any circumstances, the procedure has to be chosen with due regard to available facilities and operational cost.

In organising the progeny tests three important questions must be considered: (1) the testing capacity, i.e. the total number of milk-recorded cows available for the test matings; (2) the number of young bulls that should be tested each year, m; and (3) the number of daughters with a first lactation record that are required for the progeny-test of each bull (the group size $= n$). The total number of daughters required for the progeny-tests would then be $nm = M$. The problem is to determine the group size that will lead to the highest possible selection differential with regard to the breeding value of the progeny-tested sires. The accuracy of the progeny-test increases with increasing group size but concurrent with this increase the possibility of testing enough bulls for an efficient selection decreases. The question of the optimal balance between the number of tested young bulls and the size of the progeny groups has been analysed by ROBERTSON (1957).

As an example, let us suppose that $M = 1000$, which would correspond to at least 4000 pregnancies in tested herds when 50 per cent of the heifer calves are raised and complete their first lactation. Further, suppose that the four best of the tested bulls are added to the stud each year for use in A.I. according to their capacity ($S = 4$). If m bulls are tested the fraction of selected bulls will be $p = S/m$. The genetic superiority of the selected sires is given by $\Delta_A = i r_{IA} \sigma_A$ (cf. p. 397), where r_{IA} denotes the accuracy of the test, i the selection intensity in standard deviation units, and $\sigma_A =$ the standard deviation of the breeding value of the young bulls in the population. The value of $r_{IA} = \sqrt{\dfrac{0 \cdot 25 n h^2}{1 + (n-1) 0 \cdot 25 h^2}}$ and the value of i can be read from tables of ordinates and areas of the normal curve from the relationship $i = z/p$, where z is the ordinate at the point of truncation and $p =$ the fraction of bulls selected when the progeny-test is completed, i.e. the area to the right of z. On these premises the value of n which maximises Δ_A can be estimated by the following type of calculations where h^2 is assumed to be 0·3:

n =	5	10	20	30	40	50	100
m =	200	100	50	33	25	20	10
p =	0·02	0·04	0·08	0·12	0·16	0·20	0·40
r_{IA} =	0·537	0·669	0·786	0·842	0·873	0·896	0·944
i =	2·42	2·15	1·86	1·67	1·52	1·40	0·97
$i r_{IA}$ =	1·300	1·438	1·462	1·406	1·330	1·254	0·916

The values of $i r_{IA}$ show the genetic selection differential in standard deviation units. The optimum size of the progeny groups would seem to be a little below 20; which means that somewhat more than 50 bulls should be tested per year in order to reach maximum genetic effect of the selection.

Robertson has presented a general solution of the problem where he applies the notion 'testing ratio' $= K = m/S$. The fraction of bulls selected from those tested would then be $p = \dfrac{S}{m/n} = \dfrac{n}{K}$ and $n = Kp$. Further, $r_{IA} = \sqrt{\dfrac{n}{n+a}} = \sqrt{\dfrac{p}{p+a/K}}$ where $a = \dfrac{4-h^2}{h^2}$; $\Delta_A = \dfrac{z}{p} \sqrt{\dfrac{p}{p+a/K}} \sigma_A$. Thus the superiority of the selected sires is a function of p and K/a, and for any value of K/a we are concerned with finding a value of p that maximises Δ_A. At values of K/a greater than 3, it was found that the values of p and n approximate closely to $p = 0 \cdot 28 \sqrt{a/K}$, and $n = 0 \cdot 56 \sqrt{K/h^2}$. If these two formulae are applied to the above example the following values are obtained: $p = 0 \cdot 062$, $n = 16$ and $m = 62$. In this case, about 15 times as many young bulls should be tested as those selected after a completed test. When n varies between 10 and 25, however, the effect of this variation on $i r_{IA}$ is rather small.

The effect of group size on the accuracy of the progeny-test ($b = r_{IA}^2$) and the genetic selection differential ($i r_{IA}$), when the total testing capacity makes it possible to obtain 2000 milk-recorded daughters of young test-bulls per year is shown in Fig. 17:1. In this case, the optimum group size is about 40 daughters (tolerable range 30–60) depending on the fact that m is twice, and S only half, as large as in the example referred to above. The 'testing ratio', $K = m/S$ will therefore be four times as large. With decreasing value of K, the group size must be reduced to prevent the genetic selection differential from decreasing. The optimal group size increases with increasing testing capacity and increasing selection intensity. It decreases with increasing heritability; but moderate changes have only a minor influence. The testing capacity is the most important factor.

From an analysis of factors affecting the optimum structure of A.I. in dairy cattle breeding, SKJERVOLD and LANGHOLZ (1964) conclude that at least 50 per

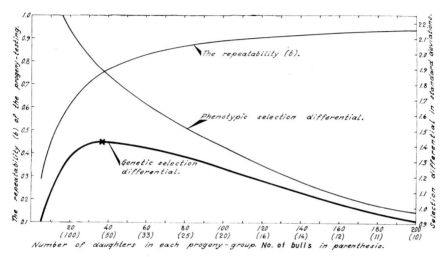

Fig. 17:1. The effect of group size on accuracy (b) of the progeny test and on the phenotypic and genetic selection differentials when the testing capacity is limited to 2000 milk-recorded daughters of the bulls. In this case the maximum genetic differential can be obtained when about 50 bulls are tested, each one on 40 daughters. (After SKJERVOLD 1963)

cent of all inseminations should be carried out with semen from young bulls. If the testing capacity is reasonably large, however, this percentage can vary considerably without any serious effect on the genetic progress. It was further concluded that, irrespective of the number of tested bulls, only the best two or three bulls from each annual group should be used for production of new young bulls for testing. Some of the next-best of the tested bulls could, of course, be used as sires of cows. We return to these points in a following section.

Available data show that about 20 per cent of the young bulls used for testing inseminations never reach the age when their progeny-tests are completed. This has to be taken into consideration in the planning of the testing operations. The average productive life of the tested bulls seems to be about four years.

Selection of dams of bulls

The selection of dams of bulls must be based on the cows' own merits and on the merits of their parents. That a cow has produced several high milk records in succession must be considered as an indication of merit not only with regard to the records as such but also because it gives an indication of the general constitution (stamina) of the animal. It is rather uncertain, however, how the milk records should be evaluated. Investigations in Great Britain, Holland, Germany and Sweden have shown that the correlation between the average milk yield of the dams and the progeny-test of their sons is surprisingly low. POLITIEK (1965) studied data for 156 progeny-tested bulls which all had progeny-tested sires and dams with records for several lactations. He found no correlation between the dams' average milk yield and their sons' progeny-test, whereas the correlation between the average fat content of the dams' milk and the corresponding average for their sons' daughters was 0·44. The correlation between the sires' and their sons' progeny-tests was about 0·35 for milk as well as for fat content. O'CONNOR (1963) obtained similar results in an analysis of MMB records. Data were available from progeny-tests of 736 Friesian bulls, and the 305-day lactation records of their 640 dams. As far as possible, the first five of the dams' records were included and the regression of the sons' progeny-test on dams' milk yield (taken as deviations from the contemporary herd average) was calculated for each lactation of the dams separately, and also for the average of all her records up to five. In the case of fat percentage, the dams' actual records and the first lactation daughter average of their sons were used. The regressions for fat percentage were in good agreement with those expected in theory, but the regressions for milk yield, especially for the 2nd–5th lactations, were considerably lower. The regression for the first lactation was the highest and even higher than that for the average of all available lactations of the dams. There was considerable variation in the dams' records.

The most important reason for the results obtained is probably that the dams had been subjected to better care and feeding than their contemporaries in the same herd, especially when, during the first lactation, they had shown themselves as promising mothers of bulls. This would be a systematic error of considerable importance in estimating the dams' breeding value, and it would influence the first record less than the following records, which agrees with the results actually obtained by O'Connor. Furthermore, the bull-dams in private herds are a highly selected sample which may affect not only the correlations referred to but also the regressions. In cases where it is shown that the correlation between the dams' records and the progeny-test of their sons is much lower than theoretically expected, this must be considered in estimating the dams' breeding value. To estimate the heritability of the dams' average milk yield during n lactations by using the formula $h_n^2 = nh^2/\{1 + (n-1)r\}$, where r is the repeatability of individual records (cf. p. 121), would give erroneous results.

The Angler Breed Society in Schleswig-Holstein has practised station testing of presumptive mothers of bulls. The cows are selected on the basis of their milk records during the first two lactations, as well as on conformation and other characters which are considered important. They are brought to the testing station early in the autumn when they are pregnant for the third time and expected to calve in the interval between 1st October and 31st December. At the station, the feeding is the same for all cows from one month before calving until two weeks after calving. Thereafter the ration is adjusted according to the milk yield. The station test of the cows covers the first eleven months after calving. The heifer calves are returned to the owners at one week of age, but the bull calves are reared at the station under standardised environmental conditions with individual recording of growth rate and feed consumption. At thirteen months of age each of the young bulls approved for breeding is used for 500 inseminations for progeny-test. The difficulty with this scheme is that it is rather expensive to conduct, and consequently the possibilities of testing sufficient young bulls for an effective selection after completion of the progeny-test will be limited.

Another scheme has been put into practice in Great Britain, Sweden, and some other countries. The Milk Marketing Board and also the Swedish A.I. Breeders' Association have organised a selection of 'élite cows' in recorded herds. The preliminary selection of these cows is based on the comparison of their milk records with that of the contemporary herd average. Final selection is made by a special panel or by the A.I. officers, who take due consideration not only of the milk records but also of the environment in which the records are made, the conformation of the animals and other characters which are of interest from a breeding point of view. An agreement is then made with the owners that the élite cows will be inseminated from the best progeny-tested bulls in the A.I. stud and that the A.I. association has the right to buy the bull calves at an agreed price and rear them under its own regime if it wishes to do so. The élite cows can easily be subjected to special veterinary examinations and to tests for such characters as udder proportions, size and shape of teats, milking rate, etc. This system can be applied at reasonable cost, and it will probably increase the accuracy of selection.

Since the cow's own records are very modifiable by feeding and management, however, it is most important to consider the progeny-test of the sire of the cow. Attention should also be paid to the records of the dam when an élite cow is selected.

The selection of young bulls for testing

A young bull should be chosen on the merits of its sire and dam as well as on its own phenotype. The sire should be one of the best available progeny-tested bulls and the dam should be an elite cow. It is not necessary to consider the sire's ancestry, but, as was pointed out above, both grandparents on the dam's side should be considered. The relative weights that should be given to the dam's own records, the progeny-test of the maternal grandsire and the records of the maternal granddam were discussed on p. 401 and need not be repeated here. The phenotypic characters of the young bulls can best be assessed if the bulls are reared in large batches on the same farm under standardised conditions of feeding and management. The growth rate and feed consumption of the individuals can then be recorded and their mating ability and semen quality can be studied. With the modern ultrasonic measuring technique, it is also possible to obtain some information on muscular development which, together with the growth rate, is of primary interest for meat production. On the basis of these data, the young bulls can be selected before they are utilised for inseminations for progeny-testing for milk yield and milk composition. A realisation of this scheme would undoubtedly contribute to an increase in the breeding value of the young bulls selected for progeny-testing.

Selection of cows within herds

The selection of cows within herds may be important, even if the average genetic effect of such selection is rather modest compared with that of bull selection. The trend in modern farming seems to be to get the heifers in calf rather early and then select intensely on the milk yield during the first months of the lactation. The poor milkers are sold for meat without being remated, or at least without any serious attempt to get them pregnant. This selection has a considerable temporary effect on the herd average, and when the culled heifers are well paid for by the slaughter-houses the procedure may improve the economy of milk production. In Sweden and Great Britain it has been shown that for each lactation (first, second, third, etc.) the average yield of the culled cows is considerably lower than the corresponding average yield of the cows that are kept for further lactations. It would seem that the genetic effect of culling based solely on the last record tends to decrease as the number of previous records increases. The culling of cows within a herd should preferably be based on all available records, due consideration being paid to known environmental factors which may have influenced the various records. An illness may have reduced a cow's future milking ability, and in such a case it is justified to cull her on the basis of the last record only.

The organisation of breed improvement within A.I. units

In the early days of artificial insemination the fear was expressed that a drastic reduction in the number of bulls would lead to close inbreeding and deterioration of the stock. On the contrary, several investigations have shown that the rate of inbreeding has decreased rather than increased when a change was made from natural breeding to A.I. With natural breeding in the bull-producing herds, it is quite common to linebreed to a certain famous sire, and when bulls are repeatedly purchased from the same strain for use in commercial herds these herds will also be more or less inbred. The A.I. studs seem to have bought their bulls from more varied sources.

However, there is a tendency to reduce the number of bulls in A.I. As was mentioned before, Skjervold has recommended using only the best two of the progeny-tested bulls in each batch (year) as sires of new bulls for a large breeding unit. If each selected sire is used in A.I. for a period of four years on an average, this would mean that 8 bull-sires are used at the same time. If all the 8 sires are used to the same extent, except for chance fluctuations, and only for the purpose of breeding new bulls for testing, and about 200 other bulls are used in a 4-year period for producing cows, then the increase in homozygosity per generation would be about 0·45 per cent, according to the formula of ROBERTSON (1954):

$F_x = \dfrac{1}{8}\left[\dfrac{1}{4N_B} + \dfrac{3}{4N_C}\right]$ where N_B = number of bull-sires and N_C = number of bulls used for producing cows. On the other hand, if only 4 of the old proven bulls are used for breeding bulls and the other 4 are used for insemination of 50 per cent of the rest of the cow population, the increase in homozygosity would be about 1·4 per cent assuming only a slight change in generation interval (SKJERVOLD and LANGHOLZ, 1964). The figures are not alarming, but they do indicate—at least the last one—that the frequency of close consanguineous matings would be fairly high, unless such matings are carefully avoided by the inseminators.

The Milk Marketing Board in England is practising a rotation of progeny-tested bulls between centres in order to avoid inbreeding. After a bull has served two or three years in one area, it is moved to another according to a certain plan, so that closely related bulls are not used consecutively on the same cow population. Plans have been made to divide the breeding area of each major breed into three or four regions, and to develop a number of broadly unrelated groups of bulls which would be used in rotation between these regions for insemination of commercial cows. By this policy, it should be possible to keep inbreeding among the commercial cows at a very low level and avoid the close matings which might result from the extensive use of a number of sons of an A.I. bull in the same area as their father.

Some of the larger A.I. associations in Sweden have practised a somewhat different method. The bulls at stud are divided into groups according to the principle of minimal relationship between the groups. Generally three groups of bulls are used, which may be denoted A, B and C. The inseminators carry semen from all three groups, and a cow sired by a group A bull will receive semen from a bull in group B; semen from group C is used for a resulting daughter, and the next generation is inseminated from a group A bull, and so on. To keep this scheme working through a series of years the sires of new bulls must be bred within the unit or brought from outside. Usually the latter possibility has been utilised.

However, a long-term programme of rotational breeding could easily be carried through if the A.I. association bred all the sires needed from élite cows. In this case the élite cows should also be divided into three groups according to their sire. Élite cows sired by an A-bull would be mated to an A-bull, cows sired by a B-bull would be mated to a B-bull and cows sired by a C-bull would be mated to a C-bull, preferably avoiding close matings. When semen is taken from the young bulls for progeny-test this semen should be distributed at random to the whole cow population, excluding the élite cows and avoiding close matings. When the élite cows are partly recruited from the main population of cows there would not be any real strain formation; on the contrary the level of heterozygosity of the 'commercial cows' would be increased rather than decreased. The programme is open to the criticism that the selection of bull-sires would be less intense than according to Skjervold's system. If six bull-sires are selected, two for each group instead of two for the whole breeding unit, the genetic selection differential decreases. How much it decreases depends on the size of the breeding unit and the organisation of the progeny-testing. If the testing capacity in the whole breeding unit corresponds to 6000 milk-recorded heifers, the optimal group size would be 56 daughters for each one of 107 bulls tested ($p = 0.0188$) when only 2 bull-sires are selected, and 32 daughters from each one of 187 test-bulls ($p = 0.0321$) when 6 bull-sires are selected. The genetic selection differential in the former case would be 2·21 and in the latter 1·90 This difference would probably be outweighed by the avoidance of inbreeding depression in the latter case

Another possibility would be to breed closed strains for bull production within a large A.I. unit and use these strains in rotational crossbreeding for production of 'commercial cows'. This would be fairly easy to organise if each strain was represented by a number of élite herds where a mild inbreeding would be permissible. It would also be possible by planned inseminations of selected élite cows distributed over many different herds. The young bulls would be tested in matings to the 'commercial cows'. With only two strains within an A.I. unit it would be a modified type of reciprocal recurrent selection.

When a numerically large breed, comprising perhaps half a million cows or more, is distributed over a large area, the breed could be divided into several non-interbreeding units, each with its own selection programme. In one unit perhaps selection would be made almost exclusively for milk and/or fat yield; in another region more attention might be paid to meat qualities. Furthermore, if environmental differences between the regions were fairly pronounced, the adaptation of each population to its particular environment would gradually increase. After a number of generations, crossbreeding between the populations could be organised and a certain amount of heterosis could perhaps be obtained. Even if selection were carried out in the same way in all the populations, they would be expected to deviate genetically a little bit from each other for each generation they were kept apart, although at a slower rate than if the improvement programmes had been different. The random drift could be quite rapid in small populations but would proceed more slowly in large populations (cf. p. 104).

Breed improvement can be organised in many different ways. The most important thing is that efficient selection for economically important characters is carried out, and that inbreeding is avoided as far as possible. In order to utilise the possibilities for breed improvement under artificial insemination, a centralised direction of planning and execution of the work is necessary. This applies to the recording of production traits and processing of the records as well as to progeny-testing, bull selection and mating schemes.

Breeding for beef traits

The most important traits for economic beef production are the growth rate and feed conversion, both of which can be measured with the live animals. A direct test of feed conversion is rather costly since it requires an accurate measurement of the feed consumption of the individuals. For practical purposes, recording of the growth rate is usually considered sufficient because growth rate and feed conversion are highly correlated. Next in importance are body conformation and muscular development. These traits can also be judged or measured on the live animals, although less accurately than on the carcass after slaughter. If measurements of meat quality are desired, for example marbling and tenderness, then the traits concerned must be measured on the carcasses. Performance tests and heritability of beef traits were discussed in Chapter 11.

Progeny-tests for beef traits, which can be measured with satisfactory accuracy on the live animals, are not as important as are such tests for milk yield or milk composition since the performance test can be made on males as well as females. Furthermore, the heritability of growth rate, and probably also body conformation and muscular development, seems to be somewhat higher than that of milk yield. If young dairy bulls are reared in large batches with standardised feeding and management, comparable measurements can be obtained on their growth rate, conformation and muscular development, as discussed before, and an efficient selection can be made on the basis of the data obtained. When, in addition, progeny-tests of the same bulls are desirable, this test can be made on the male progeny, castrated or uncastrated, and it can be completed earlier than the daughter test for milk yield and milk composition. The progeny-tests can be carried out either at special stations or in the field. The same principles apply to the organisation of these tests as in progeny-tests for milk yield. Many young bulls must be tested in order to make an efficient selection possible. If the station method is applied, the whole annual batch of progeny groups should preferably be tested at the same time and at the same station. In a field test, the best arrangement would be to distribute the animals on test between a certain number of large farms with, say, two progeny of each bull on each farm, so that farm differences in feeding and management would be equalised between the groups. If the élite cows, supposed to be dams of breeding bulls, could be judged and selected for 'meat type' this would probably add a little to the progress.

Selection on individual performance for growth rate, body conformation and muscular development would seem to be of primary importance with regard to beef breeds; the progeny-testing of bulls is less needed, although it may aid progress. When beef bulls are used in cross breeding with dairy cattle for meat improvement, a progeny-test based on the crossbred offspring would give valuable information about the sires.

Crossbreeding

In most European countries there is a tendency to reduce the number of cattle breeds by fusion or grading processes. With modern housing and well organised feed supply for the animals throughout the year, there is not the same need as in bygone times to adapt the breeds or strains to local soil and climatic conditions. When the variations in this respect do not exceed certain limits, it is probably more profitable to adjust housing and feed resources to the breed than the reverse. As stated before, the possibility for genetic improvement increases with increasing size of the breeding unit.

When breed fusion is suggested, the actual procedure should be carefully planned before the process is started. In most cases each breed probably has some special virtues, or—expressed in another way—it carries some favourable genes which are not possessed by another breed. The fusion should therefore be preceded by planned crosses and analysis of the results. Too often a breed is eliminated by a grading-up process to another breed which is supposed to be superior in all respects. Experimental crossbreeding is easy to organise, with artificial insemination and transport of frozen semen from one region to another.

A certain amount of improvement of the beef traits in a dairy breed, e.g. Friesians, is apparently possible without a noticeable loss of the milking capacity. How far this improvement can proceed without adverse effect on the dairy traits is not known, but this question deserves thorough investigation. Selection experiments could be arranged with relatively small breeding units. Perhaps the method adopted by the Milk Marketing Board is preferable, viz. to breed the Friesians mainly for milk, and to cross cows that yield less than average with beef bulls for the production of calves which can be reared for beef.

PIGS

In most west European countries selection of pigs for breeding is based largely on the Danish system of progeny-testing at special testing stations (cf. p. 414). Where properly utilised the method has undoubtedly led to considerable improvement in the genetic potential of pigs, both with regard to growth rate and carcass quality. Genetic progress would probably have been still greater, however, if the individual's own performance and that of litter mates had also been considered in selection. It was shown in Chapter 15 and 16 (cf. Figs. 15:4 and 16:11 on pp. 404 and 444) that the relative weight to be accorded to the individual itself increases as the heritability of the trait increases. Several of the traits which can be measured on the live animal with satisfactory accuracy have relatively high heritabilities, e.g. back-fat thickness ($h^2 = 0.5$) and growth rate (with group feeding, $h^2 = 0.3$).

For a number of reasons the progeny-testing of boars is inaccurate. In Chapter 15 (Fig. 15:7) it was shown how the accuracy of assessing the breeding value of a boar increased with the number of litters and the number of progeny in the test litter when he was mated at random to the sows. It is seldom that the theoretically expected accuracy is reached in practice, because the dams of the litters are often related and are always selected to a greater or lesser degree. Consequently, a large part of the assumed difference between boars is probably due instead to differences between the sows they are mated with. The error is much more serious in the progeny-testing of boars than of bulls, because the number of dams is necessarily small when the boars are assessed on the basis of four pigs per litter, due to the limited accommodation at testing stations.

An advantage of selection based on individual performance is that the generation interval is reduced. In progeny-testing for growth rate the boars are at least 18 months old before selection can be made, whereas with individual performance testing, the selection can be almost completed by the time the boars are ready for breeding at 7–8 months of age.

Since several carcass quality traits can be assessed only after slaughter, the evaluation of boars on their individual performance is incomplete in these respects. It is possible, however, to obtain some idea of a boar's breeding value for these traits from the carcasses of his litter mates. We shall now discuss the selection possibilities when based on the combined information of the individual's own performance, its litter mates' performance and progeny in a way which maximises progress per year.

Selection based on sib and own performance

The accuracy of assessing the breeding value of an individual on the basis of the mean of litter mates increases with the number of sibs tested. The correlation between the individual's breeding value and the mean of litter mates increases with the number of sibs according to the formula

$$\sqrt{\frac{0 \cdot 25nh^2}{1+(n-1)(0 \cdot 5h^2+c^2)}}, \text{(cf. Table 15:2)}.$$

It is seldom that the correlation (c^2), caused by the particular litter environment, can be ignored in the testing of litter mates. The larger c^2, the less there is to be gained in accuracy by increasing the test litter size.

The selection response is proportional to selection intensity and the accuracy with which the breeding values can be estimated. Tests for carcass traits will automatically reduce the possible selection intensity among the remaining pigs and this reduction will increase as more pigs are used for testing. In practice, it is necessary to compromise between the demand for accuracy of litter-mate assessment and the demand for selection possibilities among the remaining full sibs which are tested only for traits that do not require slaughter. FREDEEN (1954) was probably the first to study the effect of size and sex composition of test litters when selection is based both on the individual's own performance and on the growth rate and carcass characters of litter mates. He estimated the possible genetic progress in a hypothetical breeding unit of 20 sows and 2 boars where 16 gilts and 2 boars were needed for replacements each year. The results are

Table 17:1 Possible rates of genetic improvement when varying numbers of each sex are slaughtered and selection is based on own performance or on sib performance (FREEDEN, 1954)

Selection based on	Heritability	Composition of test litter (male/female)									
		2/2	1/0	0/1	2/0	1/1	0/2	3/0	2/1	1/2	0/3
Own performance	Any value	100	125·0	123·1	115·7	119·1	113·2	104·8	109·9	109·3	92·1
Sib performance	0·5	100	82·8	76·7	100·8	93·3	87·1	110·0	101·9	95·1	73·0
	0·4	100	79·8	73·8	98·9	91·6	85·5	109·1	101·0	94·4	72·5
	0·3	100	76·5	70·8	96·8	89·6	83·7	108·2	100·2	93·5	71·8
	0·1	100	69·7	64·5	91·9	85·0	79·4	105·8	98·0	91·4	70·2

summarised in Table 17:1. The basis for comparison of progress is the traditional test litter of 2 castrated males and 2 females. It is apparent from the table that the traditional test litter is by no means the most efficient. More gilts than boars are required for herd replacements, and it is therefore more advantageous to use male pigs for carcass equality evaluation. If it is assumed that slaughter is necessary for the evaluation of carcass traits, then the maximum progress seems to be made when the test litters for slaughter consist of 3 males, or 2 males and 1 female. In the past decade, however, there have been considerable improvements in the technique of measuring carcass quality on live pigs by means of ultrasonic measurements. This means that more reliance can be given to performance testing, and Fredeen's results indicate that test litters might well consist of only 2 male pigs.

These new principles for the selection of breeding pigs have begun to find application in several countries. In Norway, for example, SKJERVOLD and ÖDEGARD have developed a selection scheme which is now being tried experimentally. Selection is based on a combination of individual performance, sib testing and pedigree information. Young boars, from good parents in breeding herds, are performance-tested for feed conversion, growth rate and back-fat thickness; they are also scored for bacon type. Information about the boar's breeding value for other carcass traits is obtained from full and half sibs, which are tested at the progeny-testing stations. These stations are still operating with the traditional two males and two females in each test litter. Finally, information about the fertility of the dams, and possibly the maternal and paternal grandparents, is available from sow-recording in the breeding herds.

This information about the boar's production capacity is then condensed into a selection index in which, as mentioned in Chapter 15, consideration is given to the relative economic value of the traits and the weight to be given to each piece of information, according to whether it is derived from the individual itself and/or the various types of relatives. The selection index is constructed in such a way that an individual which corresponds to the population mean in all respects scores 100. Traits which can be measured on the live animal account for between 60 and 70 per cent of the index.

Methods of obtaining maximum genetic progress in pig breeding were recently considered by a study group of the Pig Industry Development Authority (PIDA) in Britain. The study group contained a number of pig specialists and animal geneticists. The suggestions they made are of considerable interest, and merit a brief summary here. PIDA has accepted the suggestions and a national scheme based on them is now in operation.

A progeny-testing scheme along the lines of the Danish system had been operating in the U.K. since 1957. The testing capacity was about 300 boars annually, and it was open to approximately 1250 'élite' herds; which means that each breeder could, on the average, get one boar tested every fourth year. In very few herds has much use been made of the information obtained from the tests, and the genetic progress due to progeny-testing has therefore been negligible.

According to the new system recently introduced by PIDA use is made of both performance and progeny testing. The unit of testing is a group of four litter mates consisting of one castrate, one gilt and two boars. The castrate and gilt are penned and fed together and after slaughter at 200 lb the carcasses are examined in detail for carcass quality. The two boars are penned together but fed separately. At 200 lb they are assessed for rate of growth and feed conversion. In addition, their back-fat thickness is measured by ultrasonics to supplement the carcass information of their littler mates.

The intention is to increase the number of litter groups for a complete progeny test of boars from four to six. The available testing capacity allows the performance testing of about 6500 individual boars and a simultaneous progeny test of 500 sires.

A group of approximately 70 élite herds have been selected, and these make extensive use of the available testing capacity. In addition, there is a group of about 200 'accredited' herds which use boars from the élite herds and thereby multiply the improved stock to provide a sufficient number of good boars to the commercial pig producer. The accredited herds also make some use of the testing facilities.

The boars which complete the performance test are classified into three groups: (1) those judged good enough for use in élite or accredited herds, (2) those which will be sold for use in other herds at the discretion of the breeder and (3) those which will have to be slaughtered. This classification is not based on fixed standards but rather on competition between contemporary boars on test.

In an efficient breeding programme the objectives should be simple and clearly defined. In the PIDA system selection is based on two characters; carcass quality and economy of performance. Lean percentage as estimated by progeny testing or ultrasonic measurements, is the principle method of assessing carcass quality. Other carcass characters will be recorded, so that it will be possible to detect any deterioration. Daily gain and feed conversion will be recorded separately to be later combined into a single figure representing economy of performance.

Extra care is taken to avoid the spreading of contagious diseases by boars which, after selection go back as breeding boars to élite or accredited herds. The use of pigs of both sexes for carcass traits eliminates the risk of selection bringing about undesirable sex differences in the carcass quality.

As most of the progeny for the sib and performance-testing will come from the restricted number of élite herds, the breeding boars in these herds will automatically be progeny-tested. This will serve as a check as to whether the initial selection was right or wrong.

The advantages of this selection system over selection based mainly on progeny-testing are obvious and can be summarised as follows:

(a) Performance-testing is quick, cheap and convenient.

(b) The generation interval can be reduced to a minimum.

(c) The accuracy of assessing the majority of traits is as great as with progeny-testing. The system has a 'built-in' progeny-test, which will serve as a check as to the correctness of the performance-test.

(d) The selection between boars can be intensified. With the Danish system very little is known about the boars before they are progeny-tested, and the possibilities for selection after progeny-testing are very small, since, for economic reasons, only a very few potential stock boars can be retained in the herd.

Organised crossbreeding

Artificial insemination is making progress in pig-breeding, even if the rate is much slower than in cattle-breeding. Several of the advantages of this method of reproduction, which were discussed in connection with cattle-breeding, would appear to be equally applicable to pig-breeding. A special advantage in pig-breeding is that artificial insemination should make possible systematic cross-breeding (rotational crossing) with first class boars on a large scale. The experimental results discussed in Chapters 11 and 14 clearly demonstrated the advantages of crossbreeding.

The A.I. boars could be selected in the manner described above on the basis of their own performance and their sibs' carcass quality and growth rate. Following this, the boars could be progeny-tested on the basis of crossbred litters, selected at random from gilts in commercial herds. In this way it would be possible to avoid the present bias in most progeny-testing schemes, which is due to relationship between dams and use of dams selected on the basis of previous test results. Replacement A.I. boars would be obtained from purebred élite herds associated with the A.I. station.

POULTRY

As was mentioned earlier, the modern trend in poultry breeding is specialisation on egg *or* meat production. The dual-purpose breeds have markedly decreased in popularity.

Selection for egg production

Flocks of laying hens consist mainly of full and half sib groups of about the same age. The first selection of males can therefore be made on the production traits of their sib groups, and the selection of females can be based on their individual records as well as on those of their sib groups. The problem is to combine the various criteria of the individuals breeding value so that the selection effect is maximised. Progeny-testing of the males means that a certain fraction of the two-year-old males will be used for breeding. Thus the question arises as to what is the optimum combination of progeny and sib group records, and the optimum age structure of the flocks.

It has been shown that the importance of family selection increases with decreasing heritability of the traits (cf. p. 444), and many production traits in poultry have low heritability (p. 355). For example, selection for higher viability must be based on averages for large sib groups or progeny groups. When Hutt selected for increased resistance to leucosis, he progeny-tested each prospective breeding male on 40–50 daughters (p. 232). The increased length of the generation interval, however, is always a disadvantage in the progeny-testing schemes.

The hatching season is concentrated mainly in the first five months of the year, when the pullets have not yet completed their first laying year. *Dempster* and *Lerner* theoretically analysed the relation between age structure of breeding flocks and the progress by selection. Pullets were selected for breeding on the basis of their own egg production to the 1st of January and on the corresponding part-time record of their full sibs; the young males were selected on their full sibs' record. When two-year-old females were used for breeding, they were selected on individual records and on the records of their full sibs during the whole laying year. Three-year-old females were selected on the records of their progeny. Two-year-old males were selected on the progeny-test of their daughters' records to the 1st of January and also on the records of their full sibs in their completed laying year. Each breeding group consisted of 85 females and 11 males. The greatest yearly progress (5·9 eggs per female) would be expected when about 90 per cent of the females used for breeding and 82 per cent of the males were only one year old. The greatest possible selection differential was applied to select the few two- and three-year-old breeding animals. When only one-year-old animals were used for breeding, the yearly progress in number of eggs per hen fell to 5·4; and when only two-year-old or older hens were used, the corresponding progress was reduced to 4·2 eggs in spite of the fact that the male population had optimal age structure. In modern poultry breeding the major part of the eggs used for maintaining the flocks are produced by hens in their first laying year. By reducing the individual trap-nesting as far as possible and basing the selection mainly on part-time records (to the 1st of January), a considerable amount of labour can be saved—a necessary saving during the hatching season. The loss in selection efficiency by not trap-nesting during the whole of the first laying year is considered to be comparatively small.

The egg production of the young hens during the hatching season is about 10 per cent higher than that of the two-year-old hens, and the fertility and hatchability of their eggs is higher; which is another advantage in using the one-year-old hens for breeding. Lerner has estimated that, whereas the young hens in the month of March produce 18 chicks, the two- and three-year-olds produce only 14 and 12 respectively. Therefore it is possible to apply a higher selection differential to the younger than to the older hens.

One argument against the intensive use of young hens for breeding has been that there is little information on their viability or persistency of production. However, heritability of the viability of layers is rather low (0·05–0·15). Furthermore, in the first selection of breeding birds the viability of their full and half sib groups may be considered, and later on progeny from parents or families with low viability may be excluded from the breeding flocks.

Strains of poultry which have been subject to selection for high egg production during many generations, according to conventional methods, may be plateauing.

thus indicating that the remaining genetic variance is mainly of a non-additive nature (DICKERSON, 1965). In such cases, recurrent or reciprocal recurrent selection for high combining ability between lines or strains may be resorted to for further progress. It may then be expected that progeny-test selection for heterosis effects on overall performance under varied environments will be more effective than selection under a single location environment.

Selection for meat

As was previously mentioned, the most important traits in broiler production are growth rate and feed utilisation. In poultry, as well as in farm mammals, there is a high genetic correlation between these two traits, and therefore it is not necessary for selection purposes to keep records on both. Efficient recording of feed consumed is expensive and difficult to accomplish, even for families or other groups of animals, whereas weight recording is comparatively easy and much less expensive. Recording may therefore be concentrated on the growth rate of the animals. It has been shown that the heritability for this trait is fairly high and, furthermore, is expressed in both sexes.

Another important trait is carcass quality. The muscular development on the individuals can be judged on living birds, but some other components of meat quality can be appraised only on the carcasses. Such traits are, for example, the ratio of edible meat to bone and the taste of the meat. To obtain a more accurate record of the complex trait, meat quality, it may be necessary to determine several quality components on broiler carcasses. In commercial broiler selection programmes, the trait mainly selected for is growth rate, and in some cases also breast width and depth. Individual records may be made more efficient by consideration of full or half sib groups. Some strains of broiler chickens have been subjected to such high selection pressure that the heritability for growth rate is below 10 per cent. In such cases it may be necessary to use progeny-testing or to attempt to create new variation by amalgamating several nearly plateaued populations into a single composite population and again use individual selection on the composite.

Random sample tests

The trade in living birds (chickens and pullets) plays a much greater role in poultry production than in other lines of animal production. It is therefore important to apply accurate tests of the quality of marketed birds from various firms or hatcheries. The buyers are interested in the average quality rather than in the quality of highly selected samples. Methods have been developed for 'random sample tests' in various countries, chiefly in the U.S.A. These tests differ from progeny or other family tests, since the object is to show the average production value of certain strains or commercial 'hybrid chicks' rather than the breeding value of individuals.

The method used for such tests in the U.S.A. and Britain is as follows. A random sample of fertilised eggs is picked by an impartial person from one flock of several nominated by the breeder (or firm) who wishes to enter the test. The eggs are sent to a central testing station, where they are hatched. If the test applies to egg production, the male chicks are killed and up to 500 female chicks are raised. Just before sexual maturity, a certain number of pullets (between 25 and 400, depending on the test) are transferred to the laying house and either trap-nested or recorded in battery cages or flock-recorded in litter pens for a certain length of time. In order to avoid environmental differences between groups, several groups are kept together in the laying house.

Records are kept of age at sexual maturity, number, size quality of the eggs produced, the hatchability of the eggs and mortality of the chicks, as well as the viability of the pullets and, in some cases, feed consumption. Usually, the net income per hen in the tested groups is also calculated.

The random sample test was a great improvement over previous methods in testing groups of pullets from the competing flocks, but it does not usually take into consideration the effect of genotype-environment interactions. Some newer tests are carried out by keeping birds on litter and in cages; and, by comparing results from several tests, it is possible to reduce the bias caused by interactions. The samples in some tests (25 pullets for testing) are rather small, and samples of identical breeding may therefore give quite different results in successive years or at different testing stations. Single location test results could be very misleading.

HILL and NORDSKOG have carried out random sample tests for egg production and viability in the first laying year for 10 different crosses between inbred lines. The sample from each line was randomly divided on four poultry establishments; 18 pullets from each cross were tested per year in each location, and the test was repeated in three successive years in order to study the efficiency of the method. The coefficient of repeatability of egg yield per surviving hen in a group was only 0.12 when the crosses were tested at different locations and in different years; it was 0.17 when the tests were made at the same location but in different years; and 0.25 when the crosses were tested at different locations but in the same year. The repeatabilities for mortality were somewhat higher, but their rank order was the same in the different comparisons as for egg production.

In order to establish significant differences between two crosses ($P < 0.05$), the difference in egg production per surviving hen had to be at least 61 eggs, when the comparison was made at one location and in one year; when the comparison was repeated in four successive years at one location a difference of 31 eggs was required; and when comparison was made in four successive years and at four locations, significance was reached for a difference of 21 eggs. The corresponding differences in mortality were 33, 20 and 13 percentage units respectively. Hill and Nordskog point out that the same efficiency of the test was reached when the crosses were tested in one location in three successive years as at two locations under two successive years, or in 13 locations in only one year. Repetition in time was apparently more important than testing at different locations.

Random sample tests are used also in broiler breeding. In the U.S.A. a fairly large number of chicks are used in tests (about 250), and this covers the time from hatching to 8 weeks of age (in earlier years 10–12 weeks). Experience has shown that in ideal environmental conditions quite a number of broilers reach slaughter weight at 8 weeks of age.

It may be questioned whether random sample tests have now outlived their usefulness, particularly in the United States. Even in the larger tests, the differences in performance in the production both of eggs and of broiler strains are so small that they are not statistically of any great significance. Even when the results are summarised across several tests, as in the report of the egg production tests in the United States by the U.S. Department of Agriculture, it is extremely difficult to decide which breeder's chickens are superior.

SELECTED REFERENCES

DICKERSON, G. E. 1962. Random sample performance testing of poultry in the U.S.A. *Anim. Br. Abstr.*, **30**: 1–8.

FREDEEN, H. T. 1954. Rate of genetic improvement in swine as influenced by the size and sex composition of the test litters. *Can. J. agric. Sci.*, **34**: 121–130.

HILL, J. E. and NORDSKOG, A. W. 1956. Efficiency of performance testing in poultry. *Poult. Sci.*, **35**: 256–265.

LERNER, I. M. 1958. *The Genetic Basis of Selection* (298 pp.). John Wiley and Sons, New York.

ROBERTSON, A. 1957. Optimum group size in progeny testing and family selection. *Biometrics*, **13**: 442–450.

ROBERTSON, A. and RENDEL, J. M. 1950. The use of progeny testing with artificial insemination in dairy cattle. *J. Genet.*, **50**: 21–31.

SKJERVOLD, H. and LANGHOLZ, H. J. 1964. Factors affecting the optimum structure of A.I. breeding in dairy cattle. *Z. Tierzücht ZüchtBiol.*, **80**: 25–40.

MILK MARKETING BOARD. 1965. *In 25 years 25 million cattle have been bred by A.I. in Great Britain* (20 pp.).

Study group on herd improvement and testing, 1965 (26 pp.). The Pig Industry Development Authority, PIDA House, London.

18 Retrospects and Prospects

The first chapter in this book reviewed a number of older breeding theories and the development of practical breeding before the advent of the modern science of genetics. Among other things, mention was made of the introduction, towards the end of the nineteenth century, of the measuring of body development and performance in the various species of farm animals to supplement the judging of external appearance which had predominated earlier. It is now apparent that breeding work was most successful in those species of farm animals in which subjective judging was mainly replaced by selection on performance measurements. A few examples may be cited of the changes in body type and production traits which have taken place in the last half century.

SOME RESULTS OF SELECTION COMBINED WITH IMPROVED ENVIRONMENT

For a long time the Friesian cattle in Holland and north-west Germany were regarded as the world's foremost dairy breed. During the past two decades, however, the breed has undergone radical changes in body type; the aim has been to combine better meat characteristics with high milk production; and selection has therefore been made for a low-set, wider and more compact animal than previously. There has also been intensive selection for higher butterfat content, which at the turn of the century was very low—a little over 3 per cent. In the Dutch province of Friesland, a number of body measurements have been taken in connection with herdbook registration ever since the 1890s. These measurements, together with milk yield and butterfat, have been recorded in the herdbooks. It is therefore possible to demonstrate by figures the changes that the breed has undergone. The present author (JOHANSSON, 1967) selected at random 100 cows and about the same number of bulls from the Friesian herdbooks for each of the years 1913, 1928, 1943 and 1958. The bulls are first recorded in the herdbooks when they are about 17 months and cows at an average of $3\frac{3}{4}$ years old. Table 18:1 shows the averages of the more important measurements for cows and bulls at the various ages.

During the past thirty years, the height at the withers has decreased quite considerably. Body length has also decreased, especially in the cows. All breadth measurements on the bulls have increased appreciably, whereas in the cows the changes in this respect are insignificant. Rump length decreased, whilst thurl width has increased, especially in the bulls, so that the proportion of thurl width to rump length has been greatly altered. The animals have become more compact and 'blocky'. That this is more evident in the bulls is probably due to

Table 18:1 *Average body measurements of cows and bulls registered in the Friesian herdbook in the years 1913, 1928, 1943 and 1958*

Herdbook year	Body measurements in cm							Ratio (%) Thurl width: length of pelvis
	Height at withers	Body length	Heart girth	Depth of chest	Width of chest	Hip width	Thurl width	
Cows (average age 45 months)								
1913	135·9	153·9	183·7	70·0	42·1	54·1	50·1	96·9
1928	136·6	154·9	186·6	70·9	42·5	54·1	50·6	98·1
1943	133·2	152·3	182·7	70·0	41·4	54·2	50·3	99·1
1958	129·1	150·2	183·0	69·6	42·7	54·7	51·4	101·7
Bulls (average age 17 months)								
1913	134·1	150·8	179·6	67·0	43·5	47·5	48·3	96·9
1928	133·6	151·6	185·4	68·5	45·4	49·7	50·8	99·1
1943	130·1	149·7	183·1	67·7	45·3	50·7	51·5	100·9
1958	127·6	149·3	187·7	68·3	47·5	51·9	52·5	102·6

the fact that they are more highly selected than the cows for herdbook registration.

To some extent, the type alterations are probably due to improved nutrition. It can be seen that heart girth and the breadth measurements were negatively affected during the wartime (1943) shortage of feedstuffs. Selection is nevertheless an important reason for the alterations in conformation; in other words, the genotype of the animals has been changed.

A similar study has been carried out on the Swedish Red and White cattle. A study of the measurements, which are published in the herdbooks, shows that the withers height, hip and pin bone width, as well as heart girth, have increased considerably since the beginning of the century. The changes in body type are much less than in the Friesian cattle, but there has been some increase in the width of the rump in relation to the withers height.

The milk records are also worth studying. The following figures are taken from the Friesian herdbooks and apply to the first lactation of the same cows as those whose body measurements were presented in Table 18:1. The average age at first calving was about 26 months.

Herdbook Year	Average yield in first lactation	
	Milk, kg	Butterfat, %
1913	3118	3·34
1928	3319	3·64
1943	3110	3·79
1958	3687	4·05

The way in which the alteration of type may have influenced the milk yield cannot be ascertained from these figures, since milk yield is much more modifiable by feeding and management than body measurements. The improvement of the environment of the cows from 1913 to 1958 has certainly been considerable.

Fig. 18:1 shows how the average milk yield and butterfat percentage for the total number of recorded cows in Sweden increased between 1911 and 1965. It is worth noting that the milk yield showed greater sensitivity to feed shortage during the two world wars than the butterfat content, which was hardly affected at all. This is in agreement with other observations that milk yield is much more dependent on the environmental conditions than is the butterfat content. It would appear quite probable that the greater part of the increase in butterfat content between 1911 and 1965 was due to genetic changes in the cow population, brought about by selection. The increase in milk yield may be caused mainly by environmental improvements. Only some 7 per cent of the total cow population was recorded in 1911, as compared with 34 per cent in 1965. Present-day cows, however, are probably better able to utilise good feeding and management than the cows of fifty years ago.

In Chapter 11 mention was made of how the growth rate, body length and back-fat thickness of pigs was changed in the Danish Landrace during the period between 1910 and 1960. The changes in the Swedish Landrace show a similar picture. The following averages refer to groups of Danish Landrace pigs tested at Danish stations between 1926 and 1960 (JONSSON, 1961).

The change in type of the pigs (at 90 kg liveweight) is particularly notable in

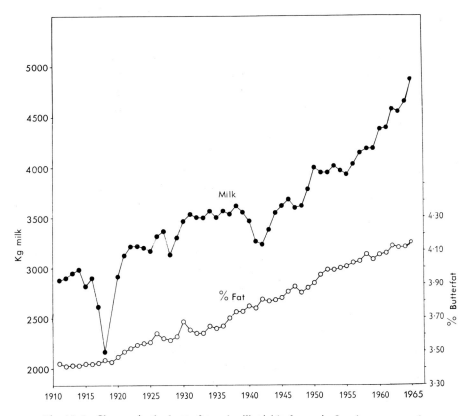

Fig. 18:1. Changes in the butterfat and milk yield of cows in Sweden per recording year from 1911 to 1965. The figures refer to the annual average of all recorded cows irrespective of breed.

	Av. daily gain	Body length	Back-fat thickness	Belly thickness
	g	cm	cm	cm
1926/27	623	88·9	4·05	3·06
1936/37	628	92·8	3·49	3·26
1949/50	672	93·6	3·40	3·28
1959/60	685	95·6	3·40	3·28

the body length and back-fat thickness. SMITH (1963) made an analysis of Danish testing data from 1952 to 1960 and estimated that about 20 per cent of the change in back-fat thickness during that period was due to genetic changes. This is a surprisingly low figure considering the high heritability of the character.

It is in poultry breeding that the greatest progress has been made, and this has been achieved in a relatively short period of time. The initial experiments and the basic research have been carried out mainly at Federal and State research stations in the U.S.A. The results have been exploited by commercial firms with large capital resources, and these now conduct their own research under scientific guidance. Several of them employ a complete staff of scientists: geneticists, nutritionists and veterinarians. A corresponding development has taken place in Great Britain since World War II.

Between 1925 and 1959 the annual egg production per hen in the U.S.A. increased from 120 to 209, but even more striking results have been achieved in poultry meat production. Twenty years ago it would have been regarded as impossible to produce broilers with the growth rate and muscle development that are now achieved. Improved breeding methods have played a very big part in this development.

PROSPECTS FOR FURTHER PROGRESS

Chapters 5 to 13 reviewed the effect of inheritance on the more important qualitative and quantitative traits of farm animals, and in Chapters 14 to 17 breeding and selection methods were discussed together with the theoretical foundations and their experimental and practical applications.

Numerous experiments with farm and laboratory animals have shown how it is possible to bring about a considerable shift in the average value of one or several quantitative traits by planned selection for a limited number of generations. In some cases the shift has been beyond the limit of the variation at the commencement of the planned breeding. The effects of inbreeding have been explained, at least in broad outline, and also the advantages which can be expected from crosses between inbred lines, strains and breeds. By applying this knowledge to practical breeding it should be possible to increase the efficiency of animal production quite appreciably.

It is general knowledge that plant breeding has been, and still is, of great importance for crop production and agricultural development. What is not generally known is that animal breeding offers the same possibilities for animal production though it may take longer to achieve the objectives, especially in cattle which have a long generation interval. As with crop production, breeding work and environmental improvement must work together in order that the end result may be the best possible.

A great deal has been accomplished in farm animal breeding with the knowledge and technical aids now available. Let us once again recall the intensification of breeding work made possible by artificial insemination, which has led to a central organisation of genetic improvement within large breeding units. Modern data machines play an important part, since individuals and groups in large populations can be followed continuously and information, which is needed as a basis for selection, e.g. the fertility of bulls and the yields of their daughters, can be rapidly extracted.

From time to time, the results of research and new knowledge are made available. As far as research in population genetics is concerned, perhaps the most important tasks at the moment are to increase our knowledge of the genetic and environmentally determined correlations between different quantitative traits, and to test in practice the efficiency of selection indexes which summarise the economic importance and correlations of a number of traits. The possibilities of taking advantage not only of the additive genetic variation within populations, but also of the non-additive by suitable breeding and selection methods, require further investigation. Studies are at present in progress both in the U.S.A. and in Great Britain.

Biochemical genetics is another line of research that can be expected to yield valuable information for animal breeding. Apart from blood-grouping research, great opportunities present themselves for the study of, for example, the composition and properties of animal proteins and fat. In Chapters 7 and 10 mention was made of the discovery that certain serum globulins as well as lacto-albumins and globulins are genetically determined and show simple Mendelian inheritance. The refinement of biochemical techniques will continue to increase the possibilities of elucidating the physiological basis of metabolic disturbances and disease resistance. This in turn will increase the possibilities for genetic analyses of these important phenomena.

Since the end of World War II consumers have shown a growing interest in the quality of animal products, such as milk, meat and eggs. This interest will undoubtedly continue and increase. Newer methods of measuring the quality of products are being devised, and these increase the possibilities of analysing the importance of heredity in the variation of these characteristics. Though in many cases it is the feeding of the animals that has the greatest effect on the quality of the product, it is nevertheless the inheritance which determines the 'norm of reaction'. In some cases a quality trait, e.g. yellow depot fat in Icelandic sheep, can be determined by a single gene, but more often one is concerned with polygenic inheritance such as the butterfat content of milk or the protein content. In fact, it would be difficult to find a single quality trait which, on closer examination, exhibited no degree of genetic variation and consequently was subject to no change by selection. Crossbreeding can be expected to increase in importance in connection with product quality, perhaps primarily because the F_1 generation generally is less variable than the parental breeds. This applies particularly when highly improved breeds are crossed with each other. Examples can be drawn not only from broiler production but also from the rearing of pigs, lambs and beef cattle for slaughter.

SELECTED REFERENCE

LERNER, I. M. and DONALD, H. P. 1966. *Modern Developments in Animal Breeding* (294 pp.). Academic Press, London and New York.

Index

Accessory glands, 18
Acclimatisation, 260
Acquired characters, 6
Achondroplasia, 218, 388
Adaptation to local conditions, 449
Additive effect of gene, 112
Adrenal virilism, 253, 254
Adrenocorticotrophic hormone, 25
Agglutination, 186, 194
A. I. units, 462
Albinism, 86, 166
Albino, 69, 160, 167, 177, 179
Albumen quality, 353
  heritability, 354
Aleutian mink, 46, 177
  disease, 76, 177
Alkaptonuria, 86
Allantochorion, 22
Allantois, 22
Alleles, 42
Alopex lagopus, 178
Amnion, 22
Ancestors, performance of, 399
Ancestral inheritance, law of, 8
Ancon sheep, 67
Androgen, 28
Aneuploidy, 72
Angora, 182
Animal Breeding Research Organization (ABRO), 151, 373
Antibody, 186
Anti-serum, 186
Antigen, 186
Artificial insemination, A.I., 11
  cattle breeding with, 455
  pig breeding with, 470
Astrakan pelt, 338
Atresia ani, 227, 389
Atrophic rhinitis, 234, 405
Autosexing, 61, 63
Autosomes, 37

Back-cross, 43, 366
Back-fat thickness, pigs, 319, 416
Bakewell, Robert, 2
Bateson, William, 9
Belly thickness, pigs, 319
Binomial, 90
Biochemical genetics, 84, 479
Birth weight, 300
  heritability, 313

Blastocyst, 22
Blending inheritance, 9
Blind teats, 268
Blood antigens, 185
Blood groups, 185
  and production, 206
Blue fox, 178
Body measurements, cattle, 313
  pig, 313
  sheep, 313
Body proportion, 307
Body size, correlations between various measurements of, 311
  effect of inbreeding on, 378
  heritability, 313
  poultry, 349, 421
Blood spots in eggs, 354
Breast angle in broilers, 358
Breast blisters in broilers, 358
Breast width in broilers, 358
Breed, 362
  improvement within A.I. units, 462
Breeding, for beef traits, 464
  cattle, 455
  horse, 454
  for increased heterozygosity, 365
  method, 361
  poultry, 470
  season, 20
  tests, 388
  unit; size of, 455
  value, 116
    accuracy of estimation, 393, 402
    assessment of the individual's, 393
    concept, definition, 392
    correlation between, 107
    of dams, 401, 460
    estimation of, 387
    general, 116, 332, 392
    of parents, 401
    special breeding, 116, 332, 392
Breitschwanz, 338
Broad-tail lamb, 338
Broiler chickens, 418
  production, 356, 472
  strains, 359
Broodiness, 345
Bull, rating of the, 411
  -testing station, 409
Butterfat percentage; time trend in recorded cows in Sweden, 476

481

Butterfat tests, 280
    yield, heritability, 287, 288

Calving interval, 256, 281
    season, 283
Carcass conformation, 314
    quality, 358, 414
        cattle, 322
        effect of sex on, 322
        heritability of, 323
        pig, 319
        technique of measuring, 314, 468
Carotene, 318
Casein types, 294
Castration, effect of, 304
Cattle, birth weight, 300
    blood groups, 188
    breeding for beef traits, 464
    carcass quality, 314, 322
    chromosome number, 39
    climatic sensitivity, 230
    colour, 166
    crossbreeding, 380, 465
    disease resistance, 231
    foetal mortality, 256
    gametic sterility, 245
    generation interval, 437
    genetic progress in milk production, 451
    gestation length, 24
    gonad hypoplasia, 239
    growth rate, 304
    horn, 180
    improvement within A.I. units, 462
    inbreeding depression and heterosis, 378
    long term changes in body size, 476
        in milk production, 477
    malformations, 218, 220, 228
    milking characteristics, 264
    milk production, 279
    multiple birth, 136
    progeny test of bulls, 406, 456
    reduced fertility, 248
    reproductive cycle, 20
    selection index, 420
Cell division (mitosis), 32
Centimorgan, 57
Centromere, 32
Centrosome, 30
Chiasmata, 34
Chinchilla, 69
Chorion, 17
Chromatids, 32
Chromosomal aberrations, 247
    change, 68
Chromosome, 30
    acentric, 70
    diploid number of, 39
    homologous, 57
    map, 61
Chromosome number, change in, 74

Chromomeres, 30
Climatic sensitivity, 230
Clitoris, 16
Cloaca, 17
Coates Herd Book, 4
Cockerel, progeny test of, 417
Co-dominant, 295
Collateral relatives, 107, 403
Colling, Charles, 3
Colour inheritance, 158
    cattle, 166
    foxes, 178
    horses, 164
    mink, 176
    pigs, 173
    poultry, 179
    rodents, 160
    sheep, 171
Combining ability, 392
    general, 447
    specific, 447
Conception results, repeatability of, 250
Congenital malformations, 64, 214
Contemporaneity, effect on heritability, 153, 288
    effect on likeness between twins, 153
Contemporary comparison, 412
    lactation, 154
Control population, 364, 451
Copulatory inability, 252
Corpus luteum, 20
Correlated selection response, 134, 297, 426
Correlation between breeding values, 107
    between breeding value and phenotype, 395
    between characters, 131
    coefficient, 94
    between full sibs, 415
Cotyledons, 17
Covariation, 94
Cows, culling of, 462
Cow-testing associations, 409
Creeper fowl, 388
Crick, Francis, H. C., 80
Crimp, 332
Criss-crossing, 366
Criss-cross inheritance, 61
Cross-breeding, 363, 374, 446, 465, 470
    double, 365
    four-way, 365
    reciprocal, 65
    rotational, 366, 377, 464
    three-way, 365, 430
    two-way, 365, 430
Crossing, dairy breeds, 380
    inbred lines, 371
    -over, 34, 55, 82
    for the production of a new breed, 367
Crosswise inheritance, 62

INDEX

Cryptorchid, 16
Cryptorchidism (cryptorchism), 227, 241
Culling levels, independent, 438
Cystic ovaries, 30, 139, 253, 262
Cytogenetics, 31
Cytology, 31
Cytoplasm, 30
Cytoplasmic influence, 67

Daily gain, cattle, 304
    pigs, 302, 416
Dam-daughter comparison, 409
    correlation, 122, 124
Danish bull testing stations, 289
Darwin, Charles, 6, 383
Daughter-dam regression, 122, 155
    -level, milk production, 407, 412
Defect, acquired, 214
    congenital, 214
    morphological, 214
    physiological, 214
Deficiency, 70
Deletion, 69, 70
Deoxyribonucleic acid (DNA), 78, 81
Dermatosis vegetans, 225
Determination, coefficient of, 94
    of parentage, 202
Dexter cattle, 218
Diallel mating, 418
Dicentric, 70
Diethylstilbestrol, 28
Dihybrid segregation, 45
Disease, resistance to, 214, 231, 347
DNA, 78, 81
Dominance deviations, 112
    theory, 382
Dominant, 42
Double hybrids, 372
Dry period, length of, 282
Duck, chromosome number, 39
Duplication, 70

Economic weight factors in selection indices, 479
Effective population size, 106
    daughters, 412
Egg-production, 340, 342
    correlations between components of, 349
    heritability, 355
    repeatability of, 473
    -record, part time, 417
    -shape, 350
    -shell, characteristics, 352
    specific gravity of, 352
    transplantation, 13
    quality, characters, 350
    weight, 346, 421, 428
        heritability of, 347
Ejaculate, volume of, 249
Electrophoresis, 196

Elite cow, selection of, 461
Embryonic mortality, 257, 259
Endocrine imbalance, 253
Environment, 41
Epididymis, 17, 214
Epistasis, 49, 64
Erotomania, 253
Erythrocyte antigens, 185
    mosaicism, 191
Eumelanin, 159
Evacuation of udder, 276
Expressivity, 50, 244
Eye muscle, area of, 315

Fallopian tube, 17
Family, 362, 363
Fat, colour of, 318
    content of carcass, 315
Feather characters, 179, 183, 358
Feathering, fast or slow, 359
Feed conversion, cattle, 313, 323
    pigs, 303
    poultry, 358
Female fertility, measures of, 255
    variation in, 253
    genitalia, underdevelopment of, 241
    sex hormones (oestrogen), 28
Fertilisation, 20, 36
    rate, 257
Fertility, 341
    of male animals, 249
    reduced, 238, 248, 259
Fleece colour, 338
    weight, heritability, 334
Fisher, Ronald A, 10
Fitness, reproductive, 428
Flushing, 141
Foetal mortality, 256
Follicle stimulating hormone (FSH), 25, 26, 253
Formalism, in animal breeding, 5
Fowl, blood groups, 195
    chromosome number, 39
    colour, 179
    crossbreeding, 446
    disease resistance, 229, 231
    egg-production, 340
    generation interval, 437
    inbreeding depression and heterosis, 369
    malformations, 225, 228
    meat production, 356
    random sample tests, 472
    reproductive fitness, 428
    selection for egg production, 428
        for increased shank length, 428
        index, 421
        for meat, 472
        plateau, 429
Fox, colour, 178
    chromosome number, 39

Fox, gestation length, 24
Freemartins, early diagnosis of, 142, 205
Friesian cattle, changes in body measurements, 475
Front-to-rear index of udder, 266
Fruit-fly, selection experiments, 424

Galactosaemia, 86
Galton, Francis, 7
Gametes, 17
Gametic sterility, 245, 259
Gene, 9
  concept, 77
  frequency, 96, 97, 361
  function, 77
  interaction, 64, 115
  mutation, 68
  additive effect of, 112, 392
  non-additive effect of, 115
  semi-lethal, 216, 388
General Stud Book, The, 4
Generation interval, 133, 398, 436, 437
Genetic change, average per year, 452
  code, 81
  correlation between two characters, 132
  defects, 347
  equilibrium, 96
  polymorphism, 206
Genetics, 9
Genital malformations, 239
Genom, 40
Genotype, 9, 77
  -environment interaction, 117, 149, 393, 450, 451
Germ cell chimerism, 142
Gestation, 21
  length of, 24, 262
Goat, chromosome number, 39
  gestation length, 24
  reproductive cycle, 21
Gonad hypoplasia, 239
gonadotrophic hormones, 25
Graafian follicle, 20
Grading-up, 367
Group feeding, 302, 416
Growth gradient, 307
  hormone, 25, 27
  rate, 302, 356, 414
    heritability, 313
    post-weaning, 302
Guard hair, 328
  medullated, 336

Haemoglobin variation, 87, 197
Haemolysis, 186
Haemophilia, 64, 73
Hair characteristics, 182
  follicle, 328

Hairlessness, 72, 220
Haploid chromosome number, 40
Hardy-Weinberg law, 10, 96
Hatchability, 341, 370, 428
Haugh units, 354
Heat, intensity of, 255
  period, 21
Hedlund white, 76
Hemizygous, 61
Hen day production, 343
  housed production, 343
Herd difference, 284
  heritability of, 408
Hereditary defect, 214, 387
Heritability, body characteristics in the horse, 455
  body size in cattle, pigs, sheep, 313
  butterfat yield, 287
  carcass quality, cattle and pigs, 323
    poultry, 358
  egg albumen traits, 354
  egg production, 355
  egg shell quality, 352
  egg weight, 347
  feed conversion, 313, 358
  growth rate, cattle, 313
    sheep, 313
    pigs, 313, 416
    poultry, 356
  lactation yield, 284
  meat quality, 323
  methods of estimating, 121
  milk composition, 294
  milk flow, 278
  milk yield, 152, 153, 280, 288
  muscle area in pigs, 320
  pulling-power in horses, 455
  trotting speed, 455
  weaning weight, cattle and sheep, 302
  wool quality, 337
  wool yield, 337
Hernia, 374
Hermaphroditism, 16, 60. 242
Heteroalleles, 83
Heterogametic, 38, 59
Heterosis, 368
Heterozygous, 42
Homoalleles, 82
Homogametic, 38, 59
Homozygosity per generation, increase in, 106, 462
Homozygous, 42
Hormone balance, 248
Horns, 180
Horse, blood groups, 194
  -breeding: special problems, 454
  chromosome number, 39
  colour, 164
  generation interval, 437
  gestation length, 24

INDEX

Horse, heritability in, 455
  reproductive cycle, 20
Hybrid boars, 377
Hybrid chicks, 371
Hybrid vigour, 383
Hypophysis, 25
Hypoplasia, ovarian, 239
Hypospadia, 241
Hypostatic, 49
Hypothalamus, 25

Immunisation technique, 185
Immunogenetics, 185
Impotence, 251, 252
Impotentia coeundi, 252
Inbred line, 363, 446
  lines, crosses between, 446
Inbreeding, 106, 364, 368
  close, 364
  coefficient of, 106
  depression, 368, 374, 382, 383
  effect on fertility, 259
  experiments, 371
    cattle, 378
    goats, 377
    pigs, 373
    poultry, 369
    sheep, 377
  systematic, 110
Incest, 364
Incomplete penetrance, 50, 99, 239
Incubation time, 25
Independent assortment, law of, 42
Individual feeding of test pigs, 416
Individual performance, 398
Infertility, due to overfeeding, 261
Inheritance, intermediate, 44
Inseminations per conception, 255
Interaction between genotype and environment, 117, 149, 151, 393, 450
Intersexuality, 16, 59, 60, 239, 242
Intra-muscular fat, 318
Intra-pair contemporaneity, 154
Inversion, 70, 247
  -heterozygotes, 247
Ionising irradiation, 74
Iso-enzyme, 199
Iso-genic line, 121
Ivanov, Elia, 11

Johannsen, Wilhelm, 8, 77
Judging animals, 5

Karakul, 171, 330, 337
Karyogram, 38
Kemp, 171, 328
Kinky tail, 374
Klinefelter's syndrome, 60
Knobbed sperm, 245

α-lactalbumin, 294

Lactation curve, shape of 283
  yield, heritability, 284
β-lactoglobulin, 294
Lactose, 290
Lamarck, 6
Laying intensity, 345
  persistency, 345
  rate, 421
Lean, colour of, 315
Let-down reflex, 28, 275
Lethal genes, automatic test for, 391
  economic importance of, 391
Leydig cell, 17
Libido, 251, 252
Life cycle, 14
Line, 362, 363
  breeding, 364
Linkage, 55
Litter size and growth rate, 302
Lush, Jay L, 11
Luteinising hormone (LH), 25, 26, 253
Luteotrophic hormone, 27

Macrochromosomes, 38
Male sex hormone, androgen, 29
Malformation, anatomical, 214-228
  of genitalia, 239
Marbling, of meat, 318
Mass selection, 388, 398, 424
Maternal impression, 1
  influence, 65, 151
Mating behaviour, 250
  polyallel, 418
  technique, 251
Mature weight, heritability, 313
Meat, organoleptic characteristics, 318
  production, 300, 356
  quality, 314
    heritability of, 323, 358
  spots in eggs, 354
  tenderness of, 318
  texture of, 318
Melanins, 159
Mendel, Gregor, 8, 41
Mendel's law, 41
Mendelian ratio, modified, 49
Microchromosomes, 38
Migration, 101
Milk composition, 290, 413
  constituents, correlation between, 297
    heritability, 294
  equivalent, 420
  fat-corrected (FCM), 279
  fat yield, 279
  flow, heritability of, 278,
  protein composition, 296
  yield, heritability of, 152, 154, 280, 287, 288
    non-genetic causes of variation, 281
    persistency of, 283

Milk yield, recorded cows in Sweden, 476
  trend in, 287
  and daily gain, correlation between, 326
  and fat content, correlation between, 298
Milk Marketing Board, 382
Milking interval, length of, 282
  rate, 272
  trait, 413
Milkograph, 272
Mink, chromosome number, 39
  colour, 176
  gestation length, 24
Modification, 41
Modifier, 50
Monohybrid segregation, 42
Monozygosity, diagnosis of, 146, 204
Monozygous twins in research, 145
Monotocous animals, 20, 136
Morphological defects, 214
Mouse, selection experiments, 426, 432
Mullerian ducts, 16
Multiple birth, 136, 138
  disadvantage of, 141
  frequency of, 138
Multiple correlation, 95, 397, 419
Muscle area, heritability of, 320
  degeneration, pigs, 315
Mutation, 67, 84, 101
  artificially induced, 72
  frequency, 72
Mutagenic agent, 74
Mutant, viability of, 75
Mutational site, 82
Muton, 82

Natural selection, 6, 260
Net merit, 419
Neurospora, 78
Nicking, 392
Nilsson-Ehle, H., 9
non-additive inheritance, 393
Non-disjunction, 60
Non-linear interaction, 450
Non-return, 250
Normal curve, 91
Nucleolus, 30
Nucleotide, 80
Nucleus, 36
Nymphomania, 253

Oestrogen, 28
Oestrous, 27
  cycle, 20
    hormonal control of, 29
    length of, 21
    regularity of, 255
Oligogenes, 50, 74
Oogenesis, 35

Outbreeding, 365
Ovarian hypoplasia, 239
Ovary, 17
Overdominance, 64, 383
Ovum, viability of the unfertilised, 22
Oxytocin, 26

Panmixia, 361
Paresis, puerperal, 261
Part-lactation record, 280, 413
Parthenogenesis, 40
Partial correlation, 94
Partial dominance, 44
Pastel, mink, 177
Path coefficient, 95, 395, 399
Pea comb, 47
Pearson, Karl, 10
Pedigree, 399
Pelger's nuclear anomaly, 217
Pelt characteristics, 337
Penetrance, 53, 240
Perinatal mortality, 256
Perosis, 229
Persian pelt, 338
Phaeomelanin, 159
Phenocopy, 64, 216
Phenogroup, 189
Phenotype, 9, 41, 77
Phenotypic resemblance, 122, 365
  value, 112, 391
Phenylketonuria, 86
Phosphatase, 199
Physiological genetics, 84
Pig Industry Development Authority (PIDA), 468
Pigs, birth weight, 301
  blood group, 195
  carcass quality, 315, 319
  chromosome number, 39
  colour, 173
  crossbreeding, 470
  disease resistance, 233
  foetal mortality, 258
  generation interval, 437
  genetic progress in carcass traits, 452
  gestation length, 24
  growth rate, 302
  inbreeding depression and heterosis, 373
  intersexuality, 242
  litter size, 143
  long term changes in production traits, 303, 478
  malformations, 224, 227
  progeny test of boars, 414
  reproductive cycle, 21
  selection on different planes of nutrition, 434
  selection methods, 466
  selection for less back-fat thickness, 439
Pituitary hormone, 26

Placenta, 17
Placentom, 17
Plateaued population, 356
Platinum fox, 50, 72
  mink, 46, 176
Pleiotropy, 76, 207, 341
Plumage, 183
Point mutation, 69
Polydactylism, 73
Polygenes, 50, 74
Polygenic inheritance, 53, 255
Polyhybrid segregation, 45
Polyovulation, 140
Polypeptide, 82
Polyploidy, 71, 74
Polytocous animals, 17, 136, 258
Porphyria, 86, 388
Position effect, 70
Poultry, see fowl
Pre-weaning growth, 301
Primary gametocytes, 33
Principle of constancy, 6
Probability, 89
Production test, 4
Progeny-test, 406, 456, 470
  accuracy of, 457
  boars, 414, 466
  bull, 409
  milk composition, 413
  milking trait, 413
  milk yield, 409
  optimum group size, 458
  organisation of, 456
  poultry, 470
  repeatability of, 415
Progesterone, 20, 27, 29
Prolactin, 27
Prolonged gestation, 263
Protein content of milk, 290
Przewalski horse, 164
Puberty, age, 21
Pulling test, draft horse, 454
Pullorum disease, 231
Pure breed, 362
Pure line, 8, 383

Qualitative trait, 41, 89
Quantitative trait, 41, 53, 89, 110
  variation of, 116

Rabbit, chromosome number, 39
  colour, 160
  fur characteristics, 182
  gestation length, 24
Random drift, 101, 464
  mating, 361, 364
  sample test, 472
Recessive, 42
  defect, 220
  lethal, 220

genes, test for, 390
Reciprocal recurrent selection, 367, 448
Recurrent selection, 447
Reduction division, meiosis, 33
Regional Swine Breeding Laboratory, 373
Regression, 7
  coefficient, 94
Relationship, 106
  coefficient of, 107, 364
Repeatability, coefficient of, 120
Reproductive disorders, 238
  disturbance; stress, 260, 428
  organs, 16
Residual milk, 291
Resistance, biological basis, 235
  to infectious disease, 231
  to nutritional deficiency, 229
Retained placenta, 254
Return breeders, 257
Rex, rabbit, 183
Ribonucleic acid, RNA, 79
Ribosome, 81
Rotational breeding, 463
  crossing, 366, 377
Rose comb, 47
Royal pastel, 48

Sapphire mink, 46
Scrotal hernia, 241
Segregation, law of, 42
Selection, 101, 133
  artificial, 102, 361
  asymmetric, response to, 425
  cattle, 455, 462
  combined with improved environment, 475
  differential, 133, 397, 425, 440
    effective, 428
    expected, 428
    genetic, 451, 458
  egg production, 470
  equilibrium between natural and artificial, 429
  experiments, fowl, 426
    fruit flies, 424
    mice, 426
    swine, 434
  family, 442, 470
  fowl, 232, 470
  index, 419, 438, 446
  individual performance, 398, 424, 441, 443, 465
Selection, intensity, 134, 397, 436, 467
  limit, 424, 431
  for, meat, 472
  method, 438
  natural, 6, 102, 260, 361
  pause, 424
  pigs, 434, 466
  plateau, 424, 429

Selection, production trait, 436
　reciprocal recurrent, 367, 448
　recurrent, 447
　response, 397, 425, 431, 467
　　correlated, 426
　　on different planes of nutrition, 432
　　per year, 437
　reverse, 424
　sib and own performance, 467
　suspended, 424, 430
　tandem, 438, 440
Semen, 18
　quality of, 249
　volume of, 19
Seminiferous tubule, 17
Service period, 254
Sex chromatin, 60
　chromosomes, 37
　determination, 58
　dimorphism, 64
　hormones, 28
　limited manifestation, 64
　linked gene, 62
　　inheritance, 58, 61, 63
　reversal, 61
Sexing, 62
Sexual drive, 251
　maturity, age at, 345
Sheep, birth weight, 300
　blood groups, 194
　chromosome number, 39
　colour, 171
　generation interval, 437
　gestation length, 24
　growth rate, 304
　horn, 180
　inbreeding depression and heterosis, 377
　malformations, 224, 228
　multiple birth, 140
　pelt characteristics, 337
　reproductive cycle, 21
　wool production, 328
Sickle-cell haemoglobin, HbS, 87
Silver fox, 50, 72, 178
Similarity index, twins, 46
Single comb, 47
Skeletal deformities, 214-228, 374
Skin character, poultry, 179, 358
Solids-not-fat, 290
Socklot, mink, 177
Spallanzani, Lazzaro, 11
Species, 362
Spinning counts, 330
Sperm abnormality, 245
　viability of, 22
Spermatocytes, primary, 35
　secondary, 35
Spermatogenesis, 35
Spermiostasis, 241
Standard deviation, 91
　error, 93
　partial regression coefficient, 399
　progeny record, 403
Staple length, 330
Stimulation theory, 382
Strain, 362, 363
　breeding, 363, 449
Sterility, 248
　gametic, 245, 259
Superfoetation, 136
Superovulation, 140
Survival rate, chicks, 358
Survivor production, poultry, 343
Synapsis, 34

Teat, 264
　anomalies, 268
　number, 272
　shape of, 270
　size of canal, 275
Telegony, 1
Telocentric, 32
Teratogenic agent, 215
Test, for recessive genes, 390
　insemination, 456
Testicle, 17
Testicular hypoplasia, 239
Testing for recessive genes, 389
　capacity, 457
　interval, effect on error of milk yield, 280
　ratio, 458
　station, 410
Testosterone, 29
Thoroughbred, horses, 4, 454
Thyrotrophic hormone, 25, 27
Three-way crosses, 430
Topcrossing, 367
Transduction, 79
Transferrin, 197, 211
Transgressive segregation, 53
Translocation, 70, 247
　-heterozygote, 247
Trap-nesting, 342, 418, 471
　reliability of, 344
Triplet birth, 139
Trisomy, 72
Turkey, 359
　chromosome number, 359
　production characteristics, 359, 429
Turner's syndrome, 60
Twins, 136, 191, 204
　conjoined, 138
　dizygotic (DZ), 136
　identical, 136
　monozygous (MZ), 136, 284
　one-egg, 136
　Siamese, 138
　two-egg, 136
　efficiency value, 148
Tyrosinosis, 86

Udder, 264
    anomalies of, 268
    evacuation curves, 272
    index, 266
    proportions, 267
    types, 265
    infection, susceptibility to, 233, 278
Ultrasonic measuring, 317
Unfavourable genes, testing for, 387
Uniformity trial, 148
Uterus, 16

Vagina, 16
    artificial, 251
Variance, 93
Vas deferens, 17
Vascular anastomosis, 239
Vertebrae, number in pigs, 319
Viability, 215-237, 347, 358
    heritability of, 127, 349
Vitamin A, 318
de Vries, Hugo, 67
Vulpes vulpes, 178

Warner-Bratzler shear, 319

Watson-Crick's model, 80
Weaning weight, heritability, 302
Weight factor in indices, 400
Weinberg, Hardy-Weinberg law, 10
Weismann, August, 6
White heifer disease, 241
Wiad Research Station, 148, 304, 376
Winter pause in egg production, 345
Wolffian ducts, 16, 241
Wool covering, 335
    fibre, 328
    quality, 330, 335
        heritability, 336
    traits, genetic correlation, 337
    types of, 328
    yield, 332
        heritability, 336
Wright, Sewall, 10

Xanthophyll, 159

Yolk character, 355

Zygote, 39
    frequency, 96, 361